Yeasts in Natural Ecosystems: Diversity

Pietro Buzzini • Marc-André Lachance •
Andrey Yurkov
Editors

Yeasts in Natural Ecosystems: Diversity

Springer

Editors
Pietro Buzzini
Department of Agricultural,
Environmental and Food Sciences
University of Perugia
Perugia, Italy

Marc-André Lachance
Department of Biology
University of Western Ontario
London, Ontario
Canada

Andrey Yurkov
German Collection of Microorganisms
and Cell Cultures
Leibniz Institute DSMZ
Braunschweig, Germany

ISBN 978-3-319-62682-6 ISBN 978-3-319-62683-3 (eBook)
DOI 10.1007/978-3-319-62683-3

Library of Congress Control Number: 2017952595

Printed on acid-free paper

This Springer imprint is published by Springer Nature
The registered company is Springer International Publishing AG
The registered company address is: Gewerbestrasse 11, 6330 Cham, Switzerland

Preface

Since the pioneer investigations of Pasteur in the 1860s, the study of the yeast world has seen important advances, especially in terms of their phenotypic, taxonomic, and genetic characterization and commercial exploitation. Research on the application of yeasts in traditional and advanced biotechnologies or their use as model organisms in science overshadows taxonomic studies and assessments of biodiversity. As a result, yeasts are often equated with the species *Saccharomyces cerevisiae* (baker's yeast), and the primary role of yeasts is thus reduced to the production of fermented foods and beverages. The truth is that the domesticated species represents only a tiny fragment of the vast diversity of the yeast world, especially those genera and species inhabiting natural ecosystems. Progress in the discovery of new species and the application of modern techniques for the rapid detection and identification of yeasts are only beginning to change the situation.

Yeasts are widely distributed throughout all biomes and most of the world's ecosystems, where they occur together in communities or guilds of species that share similar ecosystem functions, properties, and interactions. Yeasts participate in the degradation of complex organic substances but also synthesize, accumulate, and release organic molecules into the environment and act as primary and secondary decomposers in ecosystems. This volume reviews the diversity and distribution of yeasts across different habitats, including but not limited to nutrient-rich sources such as fruits, flower nectar, and decaying cactus tissues. Aquatic habitats are extremely diverse in size and properties and may well harbor the largest pool of yeast cells as suggested by recent studies. Soil has been among the first substrates sampled for yeasts. Although yeasts often occur in soils in low numbers, the global species diversity in this habitat is likely to be large in view of the wide range of soil properties and vegetation. Numerous species-rich communities associated with living plants undergo successional changes as senescence and decomposition proceed. As a result, the number of yeast cells in the phylloplane and decaying plant material is possibly as large as in water. Yeasts are also common inhabitants of extreme environments such as hypersaline habitats and cold polar and nonpolar regions. A large proportion of yeasts are associated with invertebrate animals and birds that may reflect the co-evolution and physiological adaptations of yeasts and their vectors.

The book presents a comprehensive overview of yeast diversity in natural ecosystems and constitutes the second volume of a whole monograph on *Yeasts in Natural Ecosystems*, of which the first volume (assembled by the same editors) is dedicated to yeast ecology. Several books and book series such as *The Life of Yeasts* (Phaff, Miller, Mrak), *The Yeasts* (Rose and Harrison), *Yeasts in Natural and Artificial Habitats* (Spencer and Spencer), and *Biodiversity and Ecophysiology of Yeasts* (Rosa and Péter) published in the last four decades covered major advances in yeast biodiversity. With this book, we attempt to give an update to topics covered in previous books, introduce new subjects, and provide novel views on selected aspects of yeast diversity in natural ecosystems. We hope that the book will introduce readers to the history, tools, and most recent developments in this field.

In view of rapid decline of many natural habitats due to anthropogenic activities (agriculture, deforestation, urbanization) and climate change, the need to study biodiversity is pressing. Rising temperatures threaten species inhabiting cold and aquatic environments, and species in terrestrial ecosystems are endangered by habitat fragmentation or loss. Soil erosion (i.e., topsoil degradation or loss) is an important concern to the global community since it affects both below- and aboveground communities, including soil-related habitats, as well as invertebrate and plant associates. Soil erosion leads to increased pollution and eutrophication of water bodies, thereby changing local communities and sometimes favoring unwanted or pathogenic species. The protection of genetic resources and their natural habitats is a priority of current conservation activities. However, unlike animals and plants, microscopic organisms and fungi are rarely, if ever, considered in conservation programs. The most common practice today is to conserve microbial species ex situ, namely, to preserve pure cultures or consortia in culture collections. Most of our knowledge of intrinsic properties (autoecology) of yeasts reported throughout this book is derived from laboratory experiments with pure cultures. Accordingly, the importance of culture collections for ecological studies is highlighted by presenting an overview of worldwide available yeast strains and their origins.

The chapters of this book review the knowledge accumulated during more than 60 years of yeast biodiversity research. Yeast species cited in these works were often identified by different techniques and criteria that may not be as accurate as the current sequence-based approaches. Many species names cited in the early literature are not current. Accordingly, all original taxonomic designations reported in the cited references were checked and, if necessary, updated following the latest taxonomic guidelines published in Kurtzman et al. (2011), Liu et al. (2015), Wang et al. (2015a, b), or more recent literature. A unified list of abbreviations was prepared to assist readers in following species names throughout the book.

The selection of topics and invitation of potential contributors were made by the three editors. Chapters were edited and managed by P. Buzzini and A. Yurkov. The editors thank all the authors for their excellent contributions. We also acknowledge researchers for granting access to public repositories of publications and sharing unpublished results.

P. Buzzini is grateful to Ann Vaughan-Martini and dedicates this book to the memory of his teacher (and friend) Alessandro Martini.

A. Yurkov is grateful to his teachers, soil microbiologists, and yeast ecologists Inna Babjeva and Ivan Chernov. A few sections of the book review their work and are dedicated to the memory of Ivan Chernov, who studied distribution of yeasts across many terrestrial biomes.

A. Yurkov acknowledges the research network of yeast scientists promoted by the van Uden International Advanced Course on Molecular Ecology, Taxonomy and Identification of Yeasts. Many of the authors of this book were participants and later lecturers in this course in various years.

M. A. Lachance is grateful to P. Buzzini and A. Yurkov for their invitation to join the editorial team in a mostly advisory capacity.

Finally, the editors would like to thank the Springer team, especially Isabel Ullmann and Dr. Andrea Schlitzberger, for their valuable and continuous support during the preparation of this book.

Perugia, Italy — Pietro Buzzini
London, Western Ontario, Canada — Marc-André Lachance
Braunschweig, Germany — Andrey Yurkov
May 2017

References

Kurtzman CP, Fell JW, Boekhout T (2011) The yeasts. A taxonomic study, 5th edn. Elsevier B.V., Amsterdam

Liu XZ, Wang QM, Göker M, Groenewald M, Kachalkin AV, Lumbsch HT, Millanes AM, Wedin M, Yurkov AM, Boekhout T, Bai FY (2015) Towards an integrated phylogenetic classification of the Tremellomycetes. Stud Mycol 81:85–147

Phaff HJ, Miller MW, Mrak EM (1978) The life of yeasts, 2nd edn. Harvard University Press, Cambridge

Rosa CA, Péter G (2006) Biodiversity and ecophysiology of yeasts. Springer, Berlin

Rose AH, Harrison JS (1987) The yeasts, 2nd edn. Academic Press, London

Spencer JFT, Spencer DM (1997) Yeasts in natural and artificial habitats. Springer, Berlin

Wang QM, Begerow D, Groenewald M, Liu XZ, Theelen B, Bai FY, Boekhout T (2015a) Multigene phylogeny and taxonomic revision of yeasts and related fungi in the Ustilaginomycotina. Stud Mycol 81:55–83

Wang QM, Yurkov AM, Göker M, Lumbsch HT, Leavitt SD, Groenewald M, Theelen B, Liu XZ, Boekhout T, Bai FY (2015b) Phylogenetic classification of yeasts and related taxa within Pucciniomycotina. Stud Mycol 81:149–189

Contents

Abbreviations

The following abbreviations are used for names of microbial genera (including synonyms) cited in the text.

Apiotrichum	*Ap.*
Aureobasidium	*A.*
Barnettozyma	*Barn.*
Bullera	*B.*
Candida	*C.*
Citeromyces	*Cit.*
Clavispora	*Cl.*
Colletotrichum	*Coll.*
Cryptococcus	*Cr.*
Cutaneotrichosporon	*Cut.*
Cyberlindnera	*Cyb.*
Cystobasidium	*Cyst.*
Cystofilobasidium	*Cystofil.*
Debaryomyces	*Deb.*
Dipodascus	*Dip.*
Diutina	*Diut.*
Escherichia	*E.*
Exophiala	*Ex.*
Filobasidium	*F.*
Galactomyces	*Gal.*
Ganoderma	*Gan.*
Goffeauzyma	*Goff.*
Guehomyces	*Gu.*
Hannaella	*Hann.*
Hanseniaspora	*H'spora*
Hortaea	*Ho.*

Kazachstania	*Kaz.*
Kluyveromyces	*K.*
Kodamaea	*Kod.*
Lachancea	*Lach.*
Leucosporidium	*Leuc.*
Lipomyces	*L.*
Magnusiomyces	*Magn.*
Metschnikowia	*M.*
Meyerozyma	*Mey.*
Middelhovenomyces	*Midd.*
Millerozyma	*Mill.*
Moesziomyces	*Moesz.*
Mrakia	*Mr.*
Myxozyma	*Myx.*
Naganishia	*Na.*
Ogataea	*O.*
Ophiocordyceps	*Oph.*
Papiliotrema	*Pa.*
Phaeotheca	*Phaeoth.*
Phaffia	*Ph.*
Phaffomyces	*Phaff.*
Phenoliferia	*Phen.*
Pichia	*P.*
Rhinocladiella	*Rhin.*
Rhodosporidiobolus	*Rhod.*
Rhodosporidium	*Rhodosp.*
Rhodotorula	*Rh.*
Rhynchogastrema	*Rhy.*
Saccharomyces	*S.*
Saitozyma	*Sa.*
Scheffersomyces	*Scheff.*
Schwanniomyces	*Schw.*
Solicoccozyma	*Sol.*
Spathaspora	*Spath.*
Spencermartinsiella	*Spenc.*
Sporidiobolus	*Sporid.*
Sporobolomyces	*Sp.*
Sporopachydermia	*Sporop.*
Starmera	*Starm.*
Sterigmatomyces	*Sterig.*
Sugiyamaella	*Sug.*
Suhomyces	*Su.*
Symbiotaphrina	*Symb.*
Tausonia	*Ta.*
Tortispora	*Tort.*
Torulaspora	*T'spora*

Trichosporon	*Tr.*
Trimmatostroma	*Trimm.*
Vanrija	*Va.*
Vishniacozyma	*Vishn.*
Wickerhamomyces	*W.*
Xylona	*Xyl.*
Yamadazyma	*Yam.*
Yarrowia	*Y.*
Zygosaccharomyces	*Zygosacch.*

Chapter 1
Yeasts in Continental and Seawater

Diego Libkind, Pietro Buzzini, Benedetta Turchetti, and Carlos A. Rosa

Abstract Even though yeasts are normal inhabitants of almost any type of aquatic environment, in comparison to other type of substrates, relatively little research has been carried out on the factors affecting their biodiversity and distribution patterns. The distinction of a yeast species as transient or resident element of an aquatic habitat has long been challenging and has been one of the main difficulties in the study of yeast diversity in, for example, continental lakes and rivers. The present chapter will provide an overview of our current knowledge on yeast diversity and ecology in continental freshwater and marine environments; in particular habitats like tropical and temperate rivers and lakes, seawater, and glacial melting water bodies will be reviewed. Water temperature and trophic state are major factors determining the yeast community composition in water bodies, and as they get more extreme due to the increase of stress factors such as cold temperatures, UV radiation, and scarce nutrient availability, the prevalence of basidiomycetous yeast gets more notorious. As a result of the evolutionary adaptation to extreme conditions, certain biotechnologically relevant traits became evident in extremophilic aquatic yeasts such as the production of carotenoid pigments, UV sunscreens, extracellular cold-adapted enzymes, etc.

Keywords Aquatic environments • Biodiversity • Ecology • Taxonomy

D. Libkind (✉)
Laboratorio de Microbiología Aplicada, Biotecnología y Bioinformática, Instituto Andino-Patagónico de Tecnologías Biológicas y Geoambientales (IPATEC), CONICET – UNComahue, Quintral 1250, 8400 Bariloche, Río Negro, Argentina
e-mail: libkindfd@comahue-conicet.gob.ar

P. Buzzini • B. Turchetti
Dipartimento di Scienze Agrarie, Alimentari e Ambientali, Università degli Studi di Perugia, Borgo XX Giugno, 74, Perugia, Italy

C.A. Rosa
Departamento de Microbiologia, ICB, Universidade Federal de Minas Gerais, Belo Horizonte, MG, Brazil

P. Buzzini et al. (eds.), *Yeasts in Natural Ecosystems: Diversity*,
DOI 10.1007/978-3-319-62683-3_1

Contents

1.1 Introduction

Yeasts, single-celled and free-floating, occur in water habitats, attached to substrates or within animal hosts (Jones and Slooff 1966; Pore and Sorenson 1990; Libkind et al. 2003; Kurtzman and Fell 2006; Yamaguchi et al. 2009). Yeasts are common inhabitants of aquatic environments and their population density, and species diversity depends on the water type and purity (Hagler and Mendoça-Hagler 1981). The biodiversity and distribution of yeasts have been largely overlooked, partly because of the prevailing ideas among microbiologists about the ubiquity of microorganisms (Reche et al. 2005; Yurkov 2017). As a result, yeasts are poorly documented in most reviews regarding freshwater fungi (e.g., Goh and Hyde 1996; Shearer et al. 2007; Wurzbacher et al. 2010). But it is now recognized that yeasts are among the taxa that deserve urgent research on aspects such as species richness and distribution patterns in most ecosystems (Lachance and Starmer 1998). Freshwater lakes and streams, brackish water, sewage-contaminated water, glacier meltwater, and wastewater have been the source of many basidiomycetous yeasts (Cooke 1976; Pore and Sorenson 1990; de García et al. 2010a, b, 2015; Jones et al. 2014) and made them the prevailing taxonomic group of yeasts in surveys of many aquatic ecosystems (de García et al. 2007; Brandão et al. 2011a, b).

Studies on the geographic distribution of aquatic yeast species are scarce, and most of them are focused on polluted water (Nagahama 2006). Yeast species exclusively associated with aquatic habitats are not numerous. For example, the most common ascomycetous yeast isolated from marine waters is *Debaryomyces hansenii*, a yeast species considered ubiquitous and isolated from many different environments and regions. Other ubiquitous species frequently observed also in aquatic habitats are *Aureobasidium pullulans*, *Rhodotorula mucilaginosa*, and

Vishniacozyma victoriae (former *Cryptococcus victoriae*) (Vaz et al. 2011; Brandão et al. 2011a, b, 2017; Buzzini et al. 2012; de García et al. 2012). Some species are endemic to determined regions: for example, *Metschnikowia australis* is associated with algae, marine invertebrates, and seawater in Antarctica (Lachance 2011b; Furbino et al. 2014). These observations suggest that geographical patterns and local conditions could influence the ecological distribution of aquatic yeast communities. An overview of geographic distribution patterns of yeasts has been recently published (Yurkov 2017).

Although many of the yeast species collected from water bodies may be truly aquatic species (i.e., autochthonous species), it is also conceivable that some species reached the aquatic environment through runoff from the surrounding watershed (i.e., allochthonous species). For yeasts recovered from aquatic environments, such distinction is always challenging, mainly because of our lack of understanding of factors limiting the distribution of most taxa.

The present chapter will provide an overview of yeast diversity and ecology in continental freshwater (lakes and rivers) and in seawater environments.

1.2 Yeasts in Freshwater of Tropical Rivers and Lakes

1.2.1 General Aspects of Tropical Lakes

There is a high diversity in different aquatic environments in tropical ecosystems. These water bodies can be surrounded by forests with huge species richness, or be situated in altitudinal regions with low plant diversity. In addition, these aquatic environments can be located near cities, resulting in a high anthropogenic influence in their water bodies. Lakes and rivers located in protected areas harbor yeast communities where the species composition reflects inputs from terrestrial sources such as soil and plant debris (Medeiros et al. 2008). These yeasts contribute to the recycling of the plant litter in these environments, through the action of extracellular enzymes, obtaining low weight organic carbon and making them available also to other organisms (Medeiros et al. 2012). According to Hagler and Ahearn (1987), oligotrophic clean lakes usually contain less than 100 yeasts l^{-1}, and mesotrophic lakes and rivers have total yeast counts in the range of 100–500 yeasts l^{-1}, whereas eutrophic aquatic environments usually have more than 10^3 yeast cells l^{-1}. However, Medeiros et al. (2008) reported counts around 3.5×10^3 yeasts l^{-1} in a pristine lake of Southeastern Brazil. This result suggests that the yeast populations in clean lake waters could be influenced by the influx of allochthonous organic matter in these environments. The majority of species isolated from tropical lakes and rivers are polytrophic generalists (i.e., species which assimilate a wide range of carbon sources). Nutritionally versatile yeasts are likely to colonize aquatic environments with low nutrient concentrations more efficiently (Rosa et al. 1995). In eutrophic water bodies, yeasts with narrow nutritional profiles may occur in high densities due to utilization of simple carbon sources available in these environments. Species as *Pichia kudriavzevii* (former *Candida krusei* = *Issatchenkia orientalis*), *Pichia membranifaciens*, *Candida glabrata* (*Nakaseomyces* clade), and *Hanseniaspora guilliermondii* may be prevalent in eutrophic aquatic environments (Medeiros et al. 2008; Brandão et al. 2010). These species assimilate a low number of carbon sources and probably survive and grow in eutrophic aquatic environments using simple sugars as glucose, fructose, and sucrose.

In eutrophic tropical lakes, the total yeast counts have been suggested as an indicator of anthropogenic eutrophication, i.e., organic matter concentration (Hagler et al. 1986; Brandão et al. 2010; Carneiro et al. 2015). The densities of total culturable yeasts also correlate positively with the abundance of *Escherichia coli* and of total coliform bacteria in several tropical water bodies, suggesting that total culturable yeast counts could be considered indicator of the abundance of these bacteria and, as consequence, of fecal pollution in freshwater habitats (Hagler et al. 1986; Brandão et al. 2010; Medeiros et al. 2012; Stone et al. 2012; Carneiro et al. 2015).

The most limiting nutrients in tropical lakes are nitrogen and phosphorus (Lewis 2000). The availability of these compounds can influence the colonization of these environments by yeasts. Rosa et al. (1990, 1995) reported significant positive correlation between total yeast counts and total dissolved phosphate in a paleo-karstic (i.e., karstified rock or area that has been buried by later sediments) tropical lake in Brazil. The species *A. pullulans*, *Deb. hansenii*, *Naganishia albida* (*Cryptococcus albidus*), *Papiliotrema laurentii*, *Rhodotorula glutinis*, *Rh. mucilaginosa*, and *Cutaneotrichosporon cutaneum* occurred in the highest frequencies, and a higher yeast diversity occurred during August and February, corresponding to the end of the isothermal and the beginning of the stratification periods in this tropical lake (Rosa et al. 1995). Morais et al. (1996) reported that the yeast species diversity decreased in the other paleo-karstic tropical lake with depth probably due to the absence of fermentative metabolism of most of the predominant species that could limit their distribution to the highly oxygenated surface waters. A probable correlation could exist between yeast counts and zones of nutrient accumulation determined by the thermal stratification of the water column of the lake. Yeasts are heterotrophic microorganisms that tend to be more prominent in habitats where nutrients are available. According to Morais et al. (1996), the predominance of oxidative polytrophic yeasts and pigmented species suggested that these microorganisms were probably carried from soils and foliar surfaces.

The yeast species richness in lakes and rivers is in general higher in tropical than in temperate and cold environments. This highest species richness could probably be related to the occurrence of dense and diverse surrounding plant communities (Fig. 1.1).

Fig. 1.1 Pictures of an Atlantic rain forest (**a**) and a glacially originated (**b**) lakes in Brazil (Dom Helvécio lake) and Argentina (Steffen lake), respectively. The latter photograph courtesy of Andrea Trochine

1.2.2 Yeast Communities Associated with Tropical Lakes and Rivers

Several works determined the yeast communities associated with tropical freshwater environments using morphological and physiological tests to identify these microorganisms. Most of these studies were done before the year 2000 (Hagler and Mendonça-Hagler 1981; Hagler et al. 1986; Rosa et al. 1995; Morais et al. 1996), and probably several species were misidentified (or not distinguished), becoming difficult to estimate the species richness in these environments. After the sequencing methodologies utilizing regions of the rRNA gene for identification of yeast species become routine, some works determining yeast species richness and diversity in tropical lakes and rivers were published (Medeiros et al. 2008, 2012; Brandão et al. 2010, 2017; Silva-Bedoya et al. 2014). These papers showed that the yeast communities from tropical lakes are dominated by basidiomycetous yeasts, mainly from the genera *Apiotrichum*, *Hannaella*, *Moesziomyces*, *Naganishia*, and *Papiliotrema*, as well as to the former polyphyletic genera *Rhodotorula* and *Sporobolomyces* (Medeiros et al. 2008; Brandão et al. 2011b, 2017). Ascomycetous yeasts occur in minor densities and are mainly represented by the genera *Aureobasidium*, *Debaryomyces*, *Meyerozyma*, and *Pichia* (mainly *P. kudriavzevii*) and species of the clade *Candida albicans/Lodderomyces*. Most common yeast species isolated from tropical freshwaters are *A. pullulans*, *Deb. hansenii*, *P. kudriavzevii*, *Pa. laurentii*, *Rh. mucilaginosa*, and *Sporobolomyces japonicus* (Medeiros et al. 2008; Silva-Bedoya et al. 2014; Brandão et al. 2017). These species correspond to around 50% of the total yeast isolates of the freshwater environments studied until now. Other species occur in minor frequencies but are consistently isolated from freshwater bodies, i.e., *Hannaella luteola*, *Hannaella pagnoccae*, *Meyerozyma guilliermondii*, *Moesziomyces aphidis*, *Na. albida*, and *Rhynchogastrema complexa* (Medeiros et al. 2008; Brandão et al. 2010, 2017; Silva-Bedoya et al. 2014). Most of the prevalent yeast species isolated from tropical freshwater are considered ubiquitous because they are not restricted to water but found in different environments and sampling sites. These yeasts are also frequently isolated from phyllosphere of terrestrial plants and soils, and their occurrence in the adjacent lakes and rivers might be the result of runoff from surrounding plants and soil particles. Higher occurrence of basidiomycetous (over ascomycetous) yeasts in these environments could be explained by the metabolic versatility of these yeasts, assimilating on average a broader range of carbon sources and being more tolerant to the variation of the physicochemical properties of these water bodies (Brandão et al. 2011b, 2017).

The presence or absence of some yeast species in tropical water bodies has been related to the anthropogenic impact on these environments. The species *Kluyveromyces aestuarii* is a marine organism and was observed in very high numbers in mangroves in Rio de Janeiro, so that could be considered an indicator species typical for mangroves; its presence in a specific environmental sample suggests the influence of the mangrove habitat, while its absence in mangroves could be related to a non-fecal pollution or other habitat alterations that modify the natural environment (Araújo and Hagler 2011). For example, *C. albicans* has been isolated on CHROMagar *Candida* from tropical lakes and rivers subjected to fecal pollution in Southeastern Brazil (Rosa, unpublished data). This species occurred in counts around 5.0 CFU ml^{-1} in rivers subjected to fecal pollution, and it was absent in lakes and rivers located in protected ecological reserves. Brandão et al. (2010) studied the yeast occurrence in three lakes of Southeastern Brazil and reported that the yeast densities, determined by the multiple-tube fermentation technique, were significantly correlated only with the density of fecal coliforms. Clinically relevant yeasts, *P. kudriavzevii*, *Mey. guilliermondii*, and *Candida tropicalis*, were the most frequently isolated species in this work and are associated with fecal contamination of water by warm-blooded animals.

Medeiros et al. (2008) and Brandão et al. (2010) tested the yeasts isolated at 37 °C in relation to their susceptibility to commonly used antifungal drugs: ketoconazole, fluconazole, itraconazol, and amphotericin B. These authors verified that several species were resistant to all antifungals tested, and approximately 20% of the isolates were resistant to amphotericin B. Several species isolated from lakes and rivers in Southeastern Brazil showed the virulence factors such as production of proteinases and phospholipases and were able to adhere to human buccal epithelial cells (Rosa, unpublished data). The presence of yeast strains resistant to commonly used antifungal drugs and isolation of strains producing virulence factors suggest that these environments, when affected by fecal pollution, can pose potential health risks for people utilizing these waters for recreation.

Some new species were recently described based on isolates obtained from tropical aquatic environments. For example, *Saturnispora serradocipensis* was isolated as a minor component of a yeast community from leaf detritus immersed in a tropical stream in the National Park of the Serra do Cipó, Minas Gerais State, Brazil (Canelhas et al. 2011). *Rhynchogastrema* (originally *Bandoniozyma*) *aquatica* was isolated from a freshwater lake surrounded by Atlantic Rain Forest, whereas *Rhy.* (*Bandoniozyma*) *complexa* was isolated in two lakes of the

Amazonian region and other substrates, including bromeliad leaves (Valente et al. 2012). *Yarrowia porcina* was obtained from sediment of a tropical freshwater river in Southeastern Brazil (Nagy et al. 2014). This river was exposed to different types of human influence, namely, cattle farming, mining, and domestic and industrial effluents (Medeiros et al. 2008; Nagy et al. 2014).

1.3 Yeasts in Freshwater of Temperate Lakes and Rivers

1.3.1 Yeast Diversity in Temperate Lakes and Lagoons

Due to a combination of solar heating and wind mixing of surface water layers, most lakes and lagoons in temperate regions are characterized by a fairly predictable seasonal pattern, with alternate cycles of layering and complete mixing. A number of studies investigated the yeast diversity in temperate lakes and lagoons using both culture-dependent and culture-independent approaches. The occurrence of yeasts in lake and lagoon ecosystems is common and is frequently influenced by allochthonous species external sources, including living and decaying plants and animals (Kurtzman and Fell 2004). The complete list of yeast species found in habitats associated with temperate lakes and lagoons is reported in Table 1.1. None of the papers reported the dominance of either Ascomycota or Basidiomycota phyla.

European lakes and lagoons were studied since the early 2000s. Bogusławska-Was and Dabrowski (2001) investigated the yeast diversity in strongly eutrophic waters and bottom sediments sampled in the Szczecin Lagoon, Poland. Yeast abundance in sediments and waters reached maximum in May and July, respectively. A total of 21 species of fungi including yeasts and yeastlike dimorphic fungi were found: most species were shared between both environments, while *A. pullulans* was only observed in water samples, and *Candida inconspicua*, *Candida utilis* (now *Cyberlindnera jadinii*), and *Pichia carsonii* (*Priceomyces carsonii*) were characteristic of bottom sediments only. *Candida famata* (*Deb. hansenii*) and *Rh. mucilaginosa* were the dominant species.

Culture-independent approaches were also used to study yeasts in aquatic samples. A 18S rRNA PCR survey was applied to study the eukaryotic community of the Lake Pavin, France. Of the 16 environmental fungal sequences, two were putatively identified as belonging to the basidiomycetous species *Rhodosporidium diobovatum* (currently *Rhodotorula diobovata*) and *Filobasidium globisporum* and one close to the ascomycetous yeast *Taphrina letifera* (Lefèvre et al. 2007). More recently, Monchy et al. (2011) studied fungal diversity in lake waters collected along transects from the shore to the center of Lake Pavin and Lake Aydat, France, using a twofold approach, including both cloning/sequencing of the 18S, ITS1, 5.8S, ITS2, and partial 28S region and the pyrosequencing of 18S rRNA hypervariable V2, V3, and V5 regions. Fungi represented about 50% of the total operational

Table 1.1 Diversity of yeasts and yeastlike dimorphic fungi in freshwater of temperate lakes, lagoons, and ponds

Species	Original taxonomic designation	Source	Locality	References
Ascomycetous yeasts				
Candida pseudolambica		Freshwater marshes	Florida Everglades—USA	Fell et al. (2011)
Barnettozyma californica	*Hansenula californica*, *Williopsis californica*	Lake and pond water	Iberian Pyrite Belt, Sao Domingos, Portugal; St. Lawrence, Quebec, Canada	Simard and Blackwood (1971a, b), Gadanho et al. (2006)
Candida sp.		Pond water	Lake St. Clair, Canada; Iberian Pyrite Belt, Sao Domingos, Portugal	Kwasniewska (1988), Gadanho et al. (2006)
Candida amphicis	*Candida amphixiae*	Freshwater marshes	Florida Everglades, USA	Fell et al. (2011)
Candida carpophila		Freshwater marshes	Florida Everglades, USA	Fell et al. (2011)
Candida dendronema		Freshwater marshes	Florida Everglades, USA	Fell et al. (2011)
Candida fluviatilis		pond water	Iberian Pyrite Belt, Sao Domingos, Portugal	Gadanho et al. (2006)
Candida fructus	*Candida musae*	Lake water	Lowland Zahorie, Bratislava, Slovakia	Slávikovà et al. (1992)
Candida glabrata		Lagoon water	Szczecin Lagoon, Poland	Bogusławska-Was and Dabrowski (2001)
Candida inconspicua		Lagoon water	Szczecin Lagoon, Poland	Bogusławska-Was and Dabrowski (2001)
Candida maris		Freshwater marshes	Florida Everglades, USA	Fell et al. (2011)
Candida melibiosica		Freshwater marshes	Florida Everglades, USA	Fell et al. (2011)
Candida norvegica	*Torulopsis norwegica*, *Torulopsis vanzylii*	Lake water	St. Lawrence, Quebec, Canada	Simard and Blackwood (1971a, b)
Candida parapsilosis		Lake and lagoon water	Douglas Lake, Cheboygan County, Michigan, USA; St. Lawrence, Quebec, Canada; Szczecin Lagoon, Poland	van Uden and Ahearn (1963), Simard and Blackwood (1971a, b), Bogusławska-Was and Dabrowski (2001)

(continued)

Table 1.1 (continued)

Species	Original taxonomic designation	Source	Locality	References
Candida pini	*Torulopsis pinus*	Lake water	St. Lawrence, Quebec, Canada	Simard and Blackwood (1971a, b)
Candida rancensis		Pond water	Iberian Pyrite Belt, Sao Domingos, Portugal	Gadanho et al. (2006)
Candida sake	*Torulopsis sake*	Lake and lagoon water	St. Lawrence, Quebec, Canada; Szczecin Lagoon, Poland	Simard and Blackwood (1971a, b), Bogusławska-Was and Dabrowski (2001)
Candida sharkiensis		Freshwater marshes	Florida Everglades—USA	Fell et al. (2011)
Candida silvanorum		Freshwater marshes	Florida Everglades—USA	Fell et al. (2011)
Candida tenuis		Lake water	St. Lawrence, Quebec, Canada	Simard and Blackwood (1971a, b)
Candida tropicalis		Lake water	Lowland Zahorie, Bratislava, Slovakia	Sláviková et al. (1992)
Candida versatilis	*Torulopsis anomala*	Lake water	St. Lawrence, Quebec, Canada	Simard and Blackwood (1971a, b)
Candida zeylanoides		Lagoon water, freshwater marshes	Szczecin Lagoon, Poland; Florida Everglades—USA	Bogusławska-Was and Dabrowski (2001), Fell et al. (2011)
Cyberlindnera saturnus	*Hansenula saturnus*, *Williopsis saturnus*	lagoon water	Szczecin Lagoon, Poland	Bogusławska-Was and Dabrowski (2001)
Debaryomyces sp.		Lake water	Douglas Lake, Cheboygan County, Michigan, USA; Lake St. Clair, Canada	van Uden and Ahearn (1963), Kwasniewska (1988)
Debaryomyces hansenii[b]	*Candida famata*, *Torulopsis famata*	Lake, lagoon and pond water	St. Lawrence, Quebec, Canada; Lowland Zahorie, Bratislava, Slovakia; Szczecin Lagoon, Poland; Iberian Pyrite Belt, Sao Domingos, Portugal; Lake Biwa, Japan	Simard and Blackwood (1971a, b), Sláviková et al. (1992), Bogusławska-Was and Dabrowski (2001), Gadanho et al. (2006), Ishida et al. (2015)

(continued)

Table 1.1 (continued)

Species	Original taxonomic designation	Source	Locality	References
Debaryomyces maramus	*Debaryomyces marama*	Lake water	St. Lawrence, Quebec, Canada	Simard and Blackwood (1971a, b)
Galactomyces candidus	*Geotrichum candidum*	Lake water	Lowland Zahorie, Bratislava, Slovakia	Sláviková et al. (1992)
Hanseniaspora guilliermondii	*Kloeckera apis*	Lake water	Lowland Zahorie, Bratislava, Slovakia	Sláviková et al. (1992)
Hanseniaspora uvarum	*Kloeckera apiculata*	Lake water, freshwater marshes	Lowland Zahorie, Bratislava, Slovakia; Florida Everglades, USA	Sláviková et al. (1992), Fell et al. (2011)
Hyphopichia burtonii		Lake water	Lowland Zahorie, Bratislava, Slovakia	Sláviková et al. (1992)
Kluyveromyces aestuarii		Freshwater marshes	Florida Everglades, USA	Fell et al. (2011)
Kluyveromyces lactis var. *drosophilarum*	*Kluyveromyces marxianus* var. *drosophilarum*	Lake water	Lowland Zahorie, Bratislava, Slovakia	Sláviková et al. (1992)
Kregervanrija fluxuum		Freshwater marshes	Florida Everglades, USA	Fell et al. (2011)
Lachancea kluyveri	*Saccharomyces kluyveri*	Lake and lagoon water	Lowland Zahorie, Bratislava, Slovakia; Szczecin Lagoon, Poland	Sláviková et al. (1992), Bogusławska-Was and Dabrowski (2001)
Lachancea meyersii		Freshwater marshes	Florida Everglades, USA	Fell et al. (2011)
Lindnera jadinii	*Candida utilis* (anamorph)	Lagoon water	Szczecin Lagoon, Poland	Bogusławska-Was and Dabrowski (2001)
Magnusiomyces capitatus	*Trichosporon capitatum*	Lake water	St. Lawrence, Quebec, Canada	Simard and Blackwood (1971a, b)
Martiniozyma asiatica	*Candida asiatica*	Freshwater marshes	Florida Everglades, USA	Fell et al. (2011)
Metschnikowia pulcherrima	*Candida pulcherrima*	Lake water	Douglas Lake, Cheboygan County, Michigan, USA; Lowland Zahorie, Bratislava, Slovakia	van Uden and Ahearn (1963), Sláviková et al. (1992)
Meyerozyma guilliermondii	*Candida guilliermondii*, *Pichia guilliermondii*	Lake water	Lowland Zahorie, Bratislava, Slovakia	Sláviková et al. (1992)
Nakazawaea ernobii	*Torulopsis ernobii*	Lake water	St. Lawrence, Quebec, Canada	Simard and Blackwood (1971a, b)

(continued)

Table 1.1 (continued)

Species	Original taxonomic designation	Source	Locality	References
Nakazawaea holstii	*Candida silvicola*, *Hansenula holstii*	Lake water	St. Lawrence, Quebec, Canada	Simard and Blackwood (1971a, b)
Pichia fermentans	*Candida lambica*	Lake water	Lowland Zahorie, Bratislava, Slovakia	Slávikova et al. (1992)
Pichia kluyveri		Freshwater marshes	Florida Everglades, USA	Fell et al. (2011)
Pichia kudriavzevii	*Candida krusei*, *Issatchenkia orientalis*	Lake and pond water, freshwater marshes	Florida Everglades, USA; Lowland Zahorie, Bratislava, Slovakia	Slávikova et al. (1992), Fell et al. (2011)
Priceomyces carsonii	*Pichia carsonii*	Lagoon water	Szczecin Lagoon, Poland	Bogusławska-Was and Dabrowski (2001)
Saccharomyces cerevisiae	*Saccharomyces italicus*	Lake and lagoon water	St. Lawrence, Quebec, Canada; Szczecin Lagoon, Poland	Simard and Blackwood (1971a, b), Bogusławska-Was and Dabrowski (2001)
Saccharomycopsis fibuligera		Lake water	Lowland Zahorie, Bratislava, Slovakia	Slávikova et al. (1992)
Saturnispora silvae	*Candida silvae*	Freshwater marshes	Florida Everglades, USA	Fell et al. (2011)
Schwanniomyces vanrijiae		Freshwater marshes	Florida Everglades, USA	Fell et al. (2011)
Taphrina letifera[a]		Lake water	Lake Pavin, Massif Central, France	Lefèvre et al. (2007)
Torulaspora delbrueckii	*Candida colliculosa*	Lagoon water	Szczecin Lagoon, Poland	Bogusławska-Was and Dabrowski (2001)
Trichomonascus ciferrii	*Candida ciferrii*	Lake water	Lowland Zahorie, Bratislava, Slovakia	Slávikova et al. (1992)
Wickerhamiella domercqiae	*Torulopsis saccharum*	Lake water	St. Lawrence, Quebec, Canada	Simard and Blackwood (1971a, b)
Wickerhamomyces anomalus	*Hansenula anomala*, *Pichia anomala*	Lake water	Lowland Zahorie, Bratislava, Slovakia	Slávikova et al. (1992)
Wickerhamomyces bovis	*Pichia bovis*	Lake water	St. Lawrence, Quebec, Canada	Simard and Blackwood (1971a, b)
Wickerhamomyces hampshirensis		Freshwater marshes	Florida Everglades, USA	Fell et al. (2011)

(continued)

Table 1.1 (continued)

Species	Original taxonomic designation	Source	Locality	References
Wickerhamomyces sydowiorum		Freshwater marshes	Florida Everglades, USA	Fell et al. (2011)
Yarrowia lipolytica	*Candida lipolytica*	Lagoon water	Szczecin Lagoon, Poland	Bogusławska-Was and Dabrowski (2001)
Basidiomycetous yeasts				
Anthracocystis flocculosa	*Pseudozyma flocculosa*	Freshwater marshes	Florida Everglades, USA	Fell et al. (2011)
Buckleyzyma aurantiaca	*Rhodotorula aurantiaca*	Lake water	St. Lawrence, Quebec, Canada	Simard and Blackwood (1971a, b)
Bullera alba		Lake water	Lowland Zahorie, Bratislava, Slovakia	Slávikovà et al. (1992)
Bullera unica		Pond water	Iberian Pyrite Belt, Sao Domingos, Portugal	Gadanho et al. (2006)
Cryptococcus sp.		Pond water	Iberian Pyrite Belt, Sao Domingos, Portugal	Gadanho et al. (2006)
Cryptococcus neoformans		Lagoon water	Szczecin Lagoon, Poland	Bogusławska-Was and Dabrowski (2001)
Cutaneotrichosporon curvatus	*Candida curvata*	Lake water	St. Lawrence, Quebec, Canada	Simard and Blackwood (1971a, b)
Cutaneotrichosporon cutaneum	*Trichosporon cutaneum*	Lake water	Szczecin Lagoon, Poland	Bogusławska-Was and Dabrowski (2001)
Cystobasidium sp.	*Rhodotorula cassiicola*	Freshwater marshes	Florida Everglades, USA	Fell et al. (2011)
Cystobasidium laryngis	*Rhodotorula laryngis*	Lake water	St. Lawrence, Quebec, Canada	Simard and Blackwood (1971a, b)
Cystobasidium minutum	*Rhodotorula minuta*	Lake water, deep igneous rock aquifers, freshwater marshes	Lowland Zahorie, Bratislava, Slovakia; Lake St. Clair, Canada; Aspo HRL, Sweden; Florida Everglades, USA	Kwasniewska (1988), Slávikovà et al. (1992), Ekendahl et al. (2003), Fell et al. (2011)
Cystobasidium slooffiae	*Rhodotorula slooffiae*	Freshwater marshes	Florida Everglades, USA	Fell et al. (2011)
Cystofilobasidium bisporidii		Freshwater marshes	Florida Everglades, USA	Fell et al. (2011)
Cystofilobasidium macerans	*Rhodotorula macerans*	Lake water	St. Lawrence, Quebec, Canada	Simard and Blackwood (1971a, b)

(continued)

Table 1.1 (continued)

Species	Original taxonomic designation	Source	Locality	References
Dioszegia zsoltii		Freshwater marshes	Florida Everglades, USA	Fell et al. (2011)
Filobasidium floriforme		Freshwater marshes	Florida Everglades, USA	Fell et al. (2011)
Filobasidium globisporum[a]		Lake water	Lake Pavin, Massif Central, France	Lefèvre et al. (2007)
Filobasidium magnum	*Cryptococcus ater*, *Cryptococcus magnus*	Lake water, freshwater marshes	Lowland Zahorie, Bratislava, Slovakia; Florida Everglades, USA	Sláviková et al. (1992), Fell et al. (2011)
Hannaella luteola	*Cryptococcus luteolus*	Lake water	St. Lawrence, Quebec, Canada	Simard and Blackwood (1971a, b)
Hannaella sinensis	*Bullera sinensis*	Freshwater marshes	Florida Everglades, USA	Fell et al. (2011)
Hasegawazyma lactosa	*Rhodotorula lactosa*	Lake water	St. Lawrence, Quebec, Canada	Simard and Blackwood (1971a, b)
Leucosporidium muscorum	*Candida muscorum*	Lake water	St. Lawrence, Quebec, Canada	Simard and Blackwood (1971a, b)
Leucosporidium scottii	*Candida scottii*	Lake water	St. Lawrence, Quebec, Canada; Lowland Zahorie, Bratislava, Slovakia; Iberian Pyrite Belt, Rio Tinto, Spain	Simard and Blackwood (1971a, b), Sláviková et al. (1992), Gadanho et al. (2006)
Moesziomyces aphidis	*Pseudozyma aphidis*	Freshwater marshes	Florida Everglades, USA	Fell et al. (2011)
Moesziomyces parantarcticus	*Pseudozyma parantarctica*	Freshwater marshes	Florida Everglades, USA	Fell et al. (2011)
Mrakia aquatica	*Candida aquatica*	Lake water scums	Malham Tarn, Yorkshire, UK	Jones and Slooff (1966)
Naganishia albida	*Cryptococcus albidus*, *Cryptococcus albidus* var. *albidus*	Lake and lagoon water, pond water	Florida Everglades, USA; Lake Okeechobee, Florida, USA; St. Lawrence, Quebec, Canada; Lowland Zahorie, Bratislava, Slovakia; Szczecin Lagoon, Poland	van Uden and Ahearn (1963), Simard and Blackwood (1971a, b), Sláviková et al. (1992), Bogusławska-Was and Dabrowski (2001)

(continued)

Table 1.1 (continued)

Species	Original taxonomic designation	Source	Locality	References
Naganishia albidosimilis	*Cryptococcus albidosimilis*	Pond water	Iberian Pyrite Belt, Sao Domingos, Portugal	Gadanho et al. (2006)
Naganishia diffluens	*Cryptococcus diffluens*	Lake water, freshwater marshes	Douglas Lake, Cheboygan County, Michigan, USA; Florida Everglades, USA	van Uden and Ahearn (1963), Fell et al. (2011)
Naganishia gastrica	*Cryptococcus gastricus*	Lake water	Douglas Lake, Cheboygan County, Michigan, USA	van Uden and Ahearn (1963)
Naganishia globosa	*Hansenula amylofaciens*	Lake water	St. Lawrence, Quebec, Canada	Simard and Blackwood (1971a, b)
Naganishia liquefaciens	*Cryptococcus liquefaciens*	Freshwater marshes	Florida Everglades, USA	Fell et al. (2011)
Papiliotrema flavescens	*Cryptococcus flavescens*	Freshwater marshes	Florida Everglades, USA	Fell et al. (2011)
Papiliotrema laurentii	*Cryptococcus laurentii*	Lake and lagoon water, pond water, fresh-water marshes	Douglas Lake, Cheboygan County, Michigan, USA; Florida Everglades, USA; St. Lawrence, Quebec, Canada; Lowland Zahorie, Bratislava, Slova-kia; Szczecin Lagoon, Poland; Florida Everglades, USA	van Uden and Ahearn (1963), Simard and Blackwood (1971a, b), Sláviková et al. (1992), Bogusławska-Was and Dabrowski (2001), Bogusławska-Was and Dabrowski (2001), Fell et al. (2011)
Papiliotrema pseudoalba	*Bullera pseudoalba*	Freshwater marshes	Florida Everglades, USA	Fell et al. (2011)
Papiliotrema taeanensis	*Cryptococcus taeanensis*	Freshwater marshes	Florida Everglades, USA	Fell et al. (2011)
Piskurozyma capsuligena	*Filobasidium capsuligenum*, *Candida japonica*	Lake water	St. Lawrence, Quebec, Canada	Simard and Blackwood (1971a, b)
Pseudohyphoz-yma bogoriensis	*Rhodotorula bogoriensis*	Pond water	Iberian Pyrite Belt, Sao Domingos, Portugal	Gadanho et al. (2006)
Rhodosporidi-obolus ruineniae	*Sporidiobolus ruineniae*	Freshwater marshes	Florida Everglades, USA	Fell et al. (2011)

(continued)

Table 1.1 (continued)

Species	Original taxonomic designation	Source	Locality	References
Rhodotorula sp.		Pond water	Iberian Pyrite Belt, Sao Domingos, Portugal	Gadanho et al. (2006)
Rhodotorula babjevae	*Rhodosporidium babjevae*	Freshwater marshes	Florida Everglades, USA	Fell et al. (2011)
Rhodotorula diobovata[b]	*Rhodosporidium diobovatum*	Lake water, freshwater marshes	Lake Pavin, Massif Central, France; Florida Everglades, USA	Lefèvre et al. (2007), Fell et al. (2011)
Rhodotorula evergladiensis		Freshwater marshes	Florida Everglades, USA	Fell et al. (2011)
Rhodotorula glutinis		Lake and lagoon water, pond water	Douglas Lake, Cheboygan County, Michigan, USA; Florida Everglades, USA; St. Lawrence, Quebec, Canada; Lake St. Clair, Canada; Lowland Zahorie, Bratislava, Slovakia; Szczecin Lagoon, Poland	van Uden and Ahearn (1963), Ahearn et al. (1968), Simard and Blackwood (1971a, b), Kwasniewska (1988), Slávikovà et al. (1992), Bogusławska-Was and Dabrowski (2001)
Rhodotorula graminis		Lake water	St. Lawrence, Quebec, Canada	Simard and Blackwood (1971a, b)
Rhodotorula mucilaginosa	*Rhodotorula pilimanae*, *Rhodotorula rubra*	Lake and lagoon water, pond water, fresh-water marshes	Douglas Lake, Cheboygan County, Michigan, USA; Florida Everglades, USA; St. Lawrence, Quebec, Canada; Lake St. Clair, Canada; Lowland Zahorie, Bratislava, Slovakia; Szczecin Lagoon, Poland; Iberian Pyrite Belt, Sao Domingos, Portugal	van Uden and Ahearn (1963), Simard and Blackwood (1971a, b), Kwasniewska (1988), Slávikovà et al. (1992), Bogusławska-Was and Dabrowski (2001), Gadanho et al. (2006), Fell et al. (2011)
Rhodotorula paludigena	*Rhodosporidium paludigenum*	Freshwater marshes	Florida Everglades, USA	Fell et al. (2011)
Rhodotorula toruloides		Pond water	Iberian Pyrite Belt, Sao Domingos, Portugal	Gadanho et al. (2006)

(continued)

Table 1.1 (continued)

Species	Original taxonomic designation	Source	Locality	References
Sakaguchia cladiensis	*Rhodotorula cladiensis*	Freshwater marshes	Florida Everglades, USA	Fell et al. (2011)
Sampaiozyma ingeniosa	*Torulopsis ingeniosa*	Lake water	St. Lawrence, Quebec, Canada	Simard and Blackwood (1971a, b)
Sampaiozyma vanillica	*Rhodotorula vanillica*	Freshwater marshes	Florida Everglades, USA	Fell et al. (2011)
Solicoccozyma aeria	*Cryptococcus albidus* var. *aerius*, *Cryptococcus aerius*	Lake water	Lowland Zahorie, Bratislava, Slovakia	Slávikovà et al. (1992)
Solicoccozyma terreus	*Cryptococcus terreus*	Freshwater marshes	Florida Everglades, USA	Fell et al. (2011)
Sporidiobolus pararoseus		Freshwater marshes	Florida Everglades, USA	Fell et al. (2011)
Sporobolomyces sp.		Lake water	Lake St. Clair, Canada	Kwasniewska (1988)
Sporobolomyces beijingensis		Freshwater marshes	Florida Everglades, USA	Fell et al. (2011)
Sporobolomyces blumeae		Freshwater marshes	Florida Everglades, USA	Fell et al. (2011)
Sporobolomyces carnicolor		Freshwater marshes	Florida Everglades, USA	Fell et al. (2011)
Sporobolomyces japonicus		Freshwater marshes	Florida Everglades, USA	Fell et al. (2011)
Sporobolomyces roseus		Freshwater marshes	Florida Everglades, USA	Fell et al. (2011)
Sporobolomyces ruberrimus		Freshwater marshes	Florida Everglades, USA	Fell et al. (2011)
Sporobolomyces salmonicolor	*Sporidiobolus salmonicolor*	Lake water	Lowland Zahorie, Bratislava, Slovakia	Slávikovà et al. (1992)
Symmetrospora marina	*Rhodotorula marina*	Lake water; freshwater marshes	St. Lawrence, Quebec, Canada; Florida Everglades, USA	Simard and Blackwood (1971a, b), Fell et al. (2011)
Udeniomyces pyricola	*Bullera pyricola*	Freshwater marshes	Florida Everglades, USA	Fell et al. (2011)
Vanrija humicola	*Candida humicola*	Lake water	St. Lawrence, Quebec, Canada	Simard and Blackwood (1971a, b)
Vishniacozyma tephrensis	*Cryptococcus tephrensis*	Freshwater marshes	Florida Everglades, USA	Fell et al. (2011)
Yeastlike dimorphic fungi				

(continued)

Table 1.1 (continued)

Species	Original taxonomic designation	Source	Locality	References
Aureobasidium pullulans	*Pullularia pullulans*	Lake and lagoon water, fresh-water marshes	St. Lawrence, Quebec, Canada; Lowland Zahorie, Bratislava, Slovakia; Szczecin Lagoon, Poland; Florida Everglades, USA	Simard and Blackwood (1971a, b), Kwasniewska (1988), Slávíková et al. (1992), Bogusławska-Was and Dabrowski (2001), Fell et al. (2011)

[a]Only non-culturable
[b]Both culturable and non-culturable

taxonomic units (OTUs) identified in Lake Pavin and 30% in Lake Aydat and were dominated by sequences of Chytridiomycota. Pyrosequencing yielded Saccharomycetales as the sole Ascomycota in both lakes, while Basidiomycota sequences were mainly retrieved in the riparian areas of Lake Aydat.

Yeasts and filamentous fungi were collected from a number of stations throughout Lake St. Clair, Canada. The predominant basidiomycetous yeast isolates were identified as *Rh. mucilaginosa*, *Cystobasidium minutum* (*Rhodotorula minuta*), *Rh. glutinis*, and *Sporobolomyces* sp. (Kwasniewska 1988). Besides, Ishida et al. (2015) studied eukaryotic diversity of mesotrophic Lake Biwa and eutrophic Lake Inba, Japan. Results revealed that aquatic fungi included not only members of Chytridiomycota but also members of Aphelida, Cryptomycota, and yeasts, in particular the species *Deb. hansenii*, which was found on the surface of diatom cells from Lake Biwa (Ishida et al. 2015).

1.3.2 Yeast Diversity in Temperate Rivers and Creeks

Although a number of studies have found yeasts in temperate rivers and creeks, the identification of autochthonous populations was sometimes problematic because several yeast species seem to be very versatile, and, therefore, their isolation is not considered sufficient to infer any type of ecological association with aquatic sources. Yeasts represent a regular component of eukaryotic populations in freshwaters of temperate rivers and creeks (Kurtzman and Fell 2004). The complete list of yeast species found in temperate rivers and creeks (and associated habitats) is reported in Table 1.2.

Table 1.2 Diversity of yeasts and yeastlike dimorphic fungi in temperate rivers and creeks

Species	Original taxonomic designation	Source	Locality	References
Ascomycetous yeasts				
Barnettozyma californica	*Hansenula californica*, *Williopsis californica*	River and creek water	Danube, Bratislava, Slovakia; Iberian Pyrite Belt, Sao Domingos, Portugal; Iberian Pyrite Belt, Rio Tinto, Spain	Sláviková and Vadkertiová (1997), Gadanho et al. (2006)
Candida sp.		Creek water	Iberian Pyrite Belt, Sao Domingos, Portugal; Tagus river, Portugal	Gadanho et al. (2006), Coelho et al. (2010)
Candida bertae		Leaves submerged in a stream	River Olo, Alvao Natural Park, Portugal	Sampaio et al. (2007)
Candida boidinii		River water	Danube, Bratislava, Slovakia; Tagus river, Portugal	Sláviková and Vadkertiová (1997), Coelho et al. (2010)
Candida etchellsii		Leaves submerged in a stream	River Olo, Alvao Natural Park, Portugal	Sampaio et al. (2007)
Candida fluviatilis		River and creek water	Iberian Pyrite Belt, Sao Domingos, Portugal; Iberian Pyrite Belt, Rio Tinto, Spain	Gadanho et al. (2006)
Candida glabrata		River water	Tagus river, Portugal	Coelho et al. (2010)
Candida haemulonis	*Candida haemulonii*	Leaves submerged in a stream	River Olo, Alvao Natural Park, Portugal	Sampaio et al. (2007)
Candida inconspicua		River water	Danube, Bratislava, Slovakia	Sláviková and Vadkertiová (1997)
Candida intermedia		River water	Danube, Bratislava, Slovakia; Tagus river, Portugal	Sláviková and Vadkertiová (1997), de Almeida (2005)
Candida maltosa		River water, leaves submerged in a stream	Danube, Bratislava, Slovakia; River Olo, Alvao Natural Park, Portugal	Sláviková and Vadkertiová (1997), Sampaio et al. (2007)

(continued)

Table 1.2 (continued)

Species	Original taxonomic designation	Source	Locality	References
Candida oleophila		River water	Iberian Pyrite Belt, Rio Tinto, Spain; Tagus river, Portugal	Gadanho et al. (2006), Coelho et al. (2010)
Candida palmioleophila		River water	Tagus river, Portugal	Coelho et al. (2010)
Candida parapsilosis		River water	Danube, Bratislava, Slovakia; Tagus river, Portugal	Slávikova and Vadkertiová (1997), de Almeida (2005), Coelho et al. (2010)
Candida pseudoglaebosa		River water	Iberian Pyrite Belt, Rio Tinto, Spain	Gadanho et al. (2006)
Candida pseudointermedia		River water	Tagus river, Portugal	Coelho et al. (2010)
Candida pseudolambica		River water	Tagus river, Portugal	Coelho et al. (2010)
Candida rancensis		Creek water	Iberian Pyrite Belt, Sao Domingos, Portugal	Gadanho et al. (2006)
Candida saitoana		River water	Tagus river, Portugal	Gadanho and Sampaio (2004)
Candida sake	*Torulopsis sake*	Leaves submerged in a stream	River Olo, Alvao Natural Park, Portugal	Sampaio et al. (2007)
Candida tropicalis		River water	Danube, Bratislava, Slovakia	Slávikova and Vadkertiová (1997)
Candida vartiovaarae		Leaves submerged in a stream	River Olo, Alvao Natural Park, Portugal	Sampaio et al. (2007)
Candida versatilis		Leaves submerged in a stream	River Olo, Alvao Natural Park, Portugal	Sampaio et al. (2007)
Candida zeylanoides		River water	Tagus river, Portugal	Coelho et al. (2010)
Citeromyces matritensis		Leaves submerged in a stream	River Olo, Alvao Natural Park, Portugal	Sampaio et al. (2007)
Clavispora lusitaniae		River water	Tagus river, Portugal	de Almeida (2005), Coelho et al. (2010)

(continued)

Table 1.2 (continued)

Species	Original taxonomic designation	Source	Locality	References
Cyberlindnera saturnus	*Hansenula saturnus*, *Williopsis saturnus*	Leaves submerged in a stream	River Olo, Alvao Natural Park, Portugal	Sampaio et al. (2007)
Debaryomyces hansenii	*Candida famata*, *Torulopsis famata*	River and creek water, leaves submerged in a stream	Danube, Bratislava, Slovakia; Tagus river, Portugal; Iberian Pyrite Belt, Sao Domingos, Portugal; River Olo, Alvao Natural Park, Portugal; Tagus river, Portugal	Sláviková and Vadkertiová (1997), Gadanho and Sampaio (2004), Gadanho et al. (2006), Sampaio et al. (2007), Coelho et al. (2010)
Diutina catenulata	*Candida catenulata*	River water	Tagus river, Portugal	de Almeida (2005)
Galactomyces candidum	*Geotrichum candidum*	River water	Danube, Bratislava, Slovakia	Sláviková and Vadkertiová (1997)
Galactomyces geotrichum		River water	Tagus river, Portugal	Gadanho and Sampaio (2004)
Geotrichum klebahnii		River water	Danube, Bratislava, Slovakia	Sláviková and Vadkertiová (1997)
Hanseniaspora guilliermondii		River water	Tagus river, Portugal	Gadanho and Sampaio (2004)
Hanseniaspora uvarum	*Kloeckera apiculata*	River water	Danube, Bratislava, Slovakia; Tagus river, Portugal	Sláviková and Vadkertiová (1997), Gadanho and Sampaio (2004)
Hyphopichia burtonii		River water	Danube, Bratislava, Slovakia	Sláviková and Vadkertiová (1997)
Kluyveromyces lactis		River water	Tagus river, Portugal	Coelho et al. (2010)
Kluyveromyces marxianus		Leaves submerged in a stream	River Olo, Alvao Natural Park, Portugal	Sampaio et al. (2007)
Lipomyces tetrasporus		Leaves submerged in a stream	River Olo, Alvao Natural Park, Portugal	Sampaio et al. (2007)

(continued)

Table 1.2 (continued)

Species	Original taxonomic designation	Source	Locality	References
Metschnikowia bicuspidata		Leaves submerged in a stream	River Olo, Alvao Natural Park, Portugal	Sampaio et al. (2007)
Metschnikowia pulcherrima	*Candida pulcherrima*	River water	Danube, Bratislava, Slovakia	Sláviková and Vadkertiová (1997)
Meyerozyma caribbica	*Candida fermentati*	River water	Tagus river, Portugal	Coelho et al. (2010)
Meyerozyma guilliermondii	*Candida guilliermondii*, *Pichia guilliermondii*	River water	Tagus river, Portugal	Gadanho and Sampaio (2004), de Almeida (2005), Coelho et al. (2010)
Nakazawaea ishiwadae	*Candida ishiwadae*	Leaves submerged in a stream	River Olo, Alvao Natural Park, Portugal	Sampaio et al. (2007)
Pichia sp.		River water	Tagus river, Portugal	Coelho et al. (2010)
Pichia fermentans	*Candida lambica*	River water	Danube, Bratislava, Slovakia	Sláviková and Vadkertiová (1997)
Pichia kudriavzevii	*Candida krusei*, *Issatchenkia orientalis*	River water	Danube, Bratislava, Slovakia; Tagus river, Portugal	Sláviková and Vadkertiová (1997), Coelho et al. (2010)
Saccharomyces cerevisiae	*Saccharomyces italicus*	River water	Danube, Bratislava, Slovakia; Tagus river, Portugal	Sláviková and Vadkertiová (1997), Coelho et al. (2010)
Schwanniomyces occidentalis	*Debaryomyces occidentalis*	Leaves submerged in a stream	River Olo, Alvao Natural Park, Portugal	Sampaio et al. (2007)
Sugiyamaella castrensis	*Candida castrensis*	Leaves submerged in a stream	River Olo, Alvao Natural Park, Portugal	Sampaio et al. (2007)
Sugiyamaella valdiviana	*Candida valdiviana*	Leaves submerged in a stream	River Olo, Alvao Natural Park, Portugal	Sampaio et al. (2007)
Wickerhamiella domercqiae	*Torulopsis saccharum*	Leaves submerged in a stream	River Olo, Alvao Natural Park, Portugal	Sampaio et al. (2007)
Wickerhamomyces anomalus	*Hansenula anomala*, *Pichia anomala*	River water, leaves submerged in a stream	Danube, Bratislava, Slovakia; River Olo, Alvao Natural Park, Portugal; Tagus river, Portugal	Sláviková and Vadkertiová (1997), Sampaio et al. (2007), Coelho et al. (2010)

(continued)

Table 1.2 (continued)

Species	Original taxonomic designation	Source	Locality	References
Yarrowia lipolytica	*Candida lipolytica*	River water	Tagus river, Portugal	Coelho et al. (2010)
Basidiomycetous yeasts				
Buckleyzyma aurantiaca	*Rhodotorula aurantiaca*	Leaves submerged in a stream	River Olo, Alvao Natural Park, Portugal	Sampaio et al. (2007)
Bullera alba		Leaves submerged in a stream	River Olo, Alvao Natural Park, Portugal	Sampaio et al. (2007)
Bullera penniseticola		Leaves submerged in a stream	River Olo, Alvao Natural Park, Portugal	Sampaio et al. (2007)
Bullera unica		Creek water	Iberian Pyrite Belt, Sao Domingos, Portugal	Gadanho et al. (2006)
Colacogloea foliorum	*Rhodotorula foliorum*	Leaves submerged in a stream	River Olo, Alvao Natural Park, Portugal	Sampaio et al. (2007)
Cryptococcus sp.		River and creek water	Iberian Pyrite Belt, Rio Tinto, Spain; Iberian Pyrite Belt, Sao Domingos, Portugal; Tagus river, Portugal	Gadanho et al. (2006), Coelho et al. (2010)
Cryptococcus amylolentus		Leaves submerged in a stream	River Olo, Alvao Natural Park, Portugal	Sampaio et al. (2007)
Cryptococcus neoformans	*Filobasidiella neoformans*	Leaves submerged in a stream	River Olo, Alvao Natural Park, Portugal	Sampaio et al. (2007)
Curvibasidium cygneicollum	*Rhodotorula fujisanensis*	Leaves submerged in a stream	River Olo, Alvao Natural Park, Portugal	Sampaio et al. (2007)
Cutaneotrichosporon cutaneum	*Trichosporon cutaneum*	River water	Danube, Bratislava, Slovakia	Sláviková and Vadkertiová (1997)
Cutaneotrichosporon cutaneum	*Trichosporon cutaneum*	Leaves submerged in a stream	River Olo, Alvao Natural Park, Portugal	Sampaio et al. (2007)
Cutaneotrichosporon moniliiforme	*Trichosporon moniliiforme*	Leaves submerged in a stream	River Olo, Alvao Natural Park, Portugal	Sampaio et al. (2007)
Cutaneotrichosporon mucoides	*Trichosporon mucoides*	Leaves submerged in a stream	River Olo, Alvao Natural Park, Portugal	Sampaio et al. (2007)

(continued)

Table 1.2 (continued)

Species	Original taxonomic designation	Source	Locality	References
Cystobasidium minutum	*Rhodotorula minuta*	River water, leaves submerged in a stream	Danube, Bratislava, Slovakia; River Olo, Alvao Natural Park, Portugal	Sláviková and Vadkertiová (1997), Sampaio et al. (2007)
Cystofilobasidium sp.		Leaves submerged in a stream	River Olo, Alvao Natural Park, Portugal	Sampaio et al. (2007)
Cystofilobasidium bisporidii		Leaves submerged in a stream	River Olo, Alvao Natural Park, Portugal	Sampaio et al. (2007)
Cystofilobasidium capitatum		River water, leaves submerged in a stream	Danube, Bratislava, Slovakia; River Olo, Alvao Natural Park, Portugal	Sláviková and Vadkertiová (1997), Sampaio et al. (2007)
Cystofilobasidium ferigula	*Cryptococcus ferigula*	Leaves submerged in a stream	River Olo, Alvao Natural Park, Portugal	Sampaio et al. (2007)
Cystofilobasidium infirmominiatum	*Rhodotorula infirmo-miniata*	Leaves submerged in a stream	River Olo, Alvao Natural Park, Portugal	Sampaio et al. (2007)
Cystofilobasidium macerans	*Cryptococcus macerans*	Leaves submerged in a stream	River Olo, Alvao Natural Park, Portugal	Sampaio et al. (2007)
Derxomyces mrakii	*Bullera mrakii*	Leaves submerged in a stream	River Olo, Alvao Natural Park, Portugal	Sampaio et al. (2007)
Fellomyces polyborus		Leaves submerged in a stream	River Olo, Alvao Natural Park, Portugal	Sampaio et al. (2007)
Fibulobasidium inconspicuum		Leaves submerged in a stream	River Olo, Alvao Natural Park, Portugal	Sampaio et al. (2007)
Filobasidium floriforme		Leaves submerged in a stream	River Olo, Alvao Natural Park, Portugal	Sampaio et al. (2007)
Filobasidium oeirense	*Cryptococcus oeirensis*	River water	Tagus river, Portugal	Coelho et al. (2010)
Filobasidium uniguttulatum		River water	Tagus river, Portugal	Coelho et al. (2010)
Hannaella luteola	*Cryptococcus luteolus*	Leaves submerged in a stream	River Olo, Alvao Natural Park, Portugal	Sampaio et al. (2007)
Leucosporidium creatinivorum	*Rhodotorula creatinovora*	Leaves submerged in a stream	River Olo, Alvao Natural Park, Portugal	Sampaio et al. (2007)

(continued)

Table 1.2 (continued)

Species	Original taxonomic designation	Source	Locality	References
Leucosporidium fragarium	*Rhodotorula fragaria*	Leaves submerged in a stream	River Olo, Alvao Natural Park, Portugal	Sampaio et al. (2007)
Leucosporidium scottii	*Candida scottii*	Leaves submerged in a stream	River Olo, Alvao Natural Park, Portugal	Sampaio et al. (2007)
Leucosporidium yakuticum	*Rhodotorula yakutica*	Leaves submerged in a stream	River Olo, Alvao Natural Park, Portugal	Sampaio et al. (2007)
Microstroma phylloplanum	*Rhodotorula hinnulea*	Leaves submerged in a stream	River Olo, Alvao Natural Park, Portugal	Sampaio et al. (2007)
Naganishia albida	*Cryptococcus albidus*	River water, leaves submerged in a stream	Danube, Bratislava, Slovakia; River Olo, Alvao Natural Park, Portugal; Tagus river, Portugal	Slávíková and Vadkertiová (1997), Sampaio et al. (2007), Coelho et al. (2010)
Naganishia albidosimilis	*Cryptococcus albidosimilis*	River and creek water, leaves submerged in a stream	Iberian Pyrite Belt, Sao Domingos, Portugal; River Olo, Alvao Natural Park, Portugal	Gadanho et al. (2006), Sampaio et al. (2007)
Papiliotrema laurentii	*Cryptococcus laurentii*	River water, leaves submerged in a stream	Danube, Bratislava, Slovakia; River Olo, Alvao Natural Park, Portugal; Tagus river, Portugal	Slávíková and Vadkertiová (1997), Sampaio et al. (2007), Coelho et al. (2010)
Phaeotremella pseudofoliacea	*Tremella foliacea*	Leaves submerged in a stream	River Olo, Alvao Natural Park, Portugal	Sampaio et al. (2007)
Phaeotremella skinneri	*Cryptococcus skinneri*	Leaves submerged in a stream	River Olo, Alvao Natural Park, Portugal	Sampaio et al. (2007)
Pseudohyphozyma bogoriensis	*Rhodotorula bogoriensis*	Creek water	Iberian Pyrite Belt, Sao Domingos, Portugal	Gadanho et al. (2006)
Rhodosporidiobolus colostri	*Rhodotorula colostri*	Leaves submerged in a stream	River Olo, Alvao Natural Park, Portugal	Sampaio et al. (2007)

(continued)

Table 1.2 (continued)

Species	Original taxonomic designation	Source	Locality	References
Rhodotorula sp.		River and creek water	Iberian Pyrite Belt, Sao Domingos, Portugal; Iberian Pyrite Belt, Rio Tinto, Spain; River Olo, Alvao Natural Park, Portugal; Tagus river, Portugal	Gadanho et al. (2006), Coelho et al. (2010)
Rhodotorula babjevae	*Rhodosporidium babjevae*	River water, leaves submerged in a stream	River Olo, Alvao Natural Park, Portugal; Tagus river, Portugal	Sampaio et al. (2007), Coelho et al. (2010)
Rhodotorula diobovata	*Rhodosporidium diobovatum*	River water	Tagus river, Portugal	de Almeida (2005), Coelho et al. (2010)
Rhodotorula glutinis		River water, leaves submerged in a stream	Danube, Bratislava, Slovakia; River Olo, Alvao Natural Park, Portugal; Tagus river, Portugal	Slávikovà and Vadkertiová (1997), Sampaio et al. (2007), Coelho et al. (2010)
Rhodotorula graminis		Leaves submerged in a stream	River Olo, Alvao Natural Park, Portugal	Sampaio et al. (2007)
Rhodotorula kratochvilovae	*Rhodosporidium kratochvilovae*	River water	Tagus river, Portugal	Coelho et al. (2010)
Rhodotorula mucilaginosa	*Rhodotorula rubra*	River and creek water, leaves submerged in a stream	Danube, Bratislava, Slovakia; Tagus river, Portugal; Tagus river, Portugal; River Olo, Alvao Natural Park, Portugal; Iberian Pyrite Belt, Sao Domingos, Portugal; Iberian Pyrite Belt, Rio Tinto, Spain	Slávikovà and Vadkertiová (1997), Gadanho and Sampaio (2004), de Almeida (2005), Gadanho et al. (2006), Sampaio et al. (2007), Coelho et al. (2010)
Rhodotorula sphaerocarpa	*Rhodosporidium sphaerocarpum*	Leaves submerged in a stream	River Olo, Alvao Natural Park, Portugal	Sampaio et al. (2007)

(continued)

Table 1.2 (continued)

Species	Original taxonomic designation	Source	Locality	References
Rhodotorula toruloides	*Rhodosporidium toruloides*	Creek water, leaves submerged in a stream	Iberian Pyrite Belt, Sao Domingos, Portugal; River Olo, Alvao Natural Park, Portugal	Gadanho et al. (2006), Sampaio et al. (2007)
Saitozyma podzolica	*Cryptococcus podzolicus*	Leaves submerged in a stream	River Olo, Alvao Natural Park, Portugal	Sampaio et al. (2007)
Solicoccozyma aeria	*Cryptococcus albidus* var. *aerius*, *Cryptococcus aerius*	Leaves submerged in a stream	River Olo, Alvao Natural Park, Portugal	Sampaio et al. (2007)
Solicoccozyma fuscescens	*Cryptococcus fuscescens*	Leaves submerged in a stream	River Olo, Alvao Natural Park, Portugal	Sampaio et al. (2007)
Solicoccozyma terreus	*Cryptococcus terreus*	Leaves submerged in a stream	River Olo, Alvao Natural Park, Portugal	Sampaio et al. (2007)
Spencerozyma crocea	*Rhodotorula crocea*	Leaves submerged in a stream	River Olo, Alvao Natural Park, Portugal	Sampaio et al. (2007)
Sporobolomyces sp.		Leaves submerged in a stream	River Olo, Alvao Natural Park, Portugal; Tagus river, Portugal	Sampaio et al. (2007), Coelho et al. (2010)
Sporobolomyces roseus		River water, leaves submerged in a stream	River Olo, Alvao Natural Park, Portugal; Tagus river, Portugal	Sampaio et al. (2007), Coelho et al. (2010)
Sporobolomyces salmonicolor	*Sporidiobolus salmonicolor*	River water, leaves submerged in a stream	Danube, Bratislava, Slovakia; River Olo, Alvao Natural Park, Portugal	Sláviková and Vadkertiová (1997), Sampaio et al. (2007)
Sporobolomyces shibatanus		River water	Danube, Bratislava, Slovakia	Sláviková and Vadkertiová (1997)
Sterigmatosporidium polymorphum		Leaves submerged in a stream	River Olo, Alvao Natural Park, Portugal	Sampaio et al. (2007)
Trigonosporomyces hylophilus	*Rhodotorula hylophila*	Leaves submerged in a stream	River Olo, Alvao Natural Park, Portugal	Sampaio et al. (2007)
Udeniomyces megalosporus	*Bullera megalosporus*	Leaves submerged in a stream	River Olo, Alvao Natural Park, Portugal	Sampaio et al. (2007)

(continued)

Table 1.2 (continued)

Species	Original taxonomic designation	Source	Locality	References
Udeniomyces pyricola	*Bullera pyricola*	Leaves submerged in a stream	River Olo, Alvao Natural Park, Portugal	Sampaio et al. (2007)
Vanrija humicola	*Cryptococcus humicolus*	Leaves submerged in a stream	River Olo, Alvao Natural Park, Portugal	Sampaio et al. (2007)
Vanrija longa	*Cryptococcus longus*	River water	Tagus river, Portugal	Gadanho and Sampaio (2004)
Xanthophyllomyces dendrorhous		Leaves submerged in a stream	River Olo, Alvao Natural Park, Portugal	Sampaio et al. (2007)
Yeastlike dimorphic fungi				
Aureobasidium pullulans		River water, leaves submerged in a stream	Danube, Bratislava, Slovakia; River Olo, Alvao Natural Park, Portugal	Slávikovå and Vadkertiová (1997), Sampaio et al. (2007)

[a]Both culturable and non-culturable

The first studies on yeast diversity in temperate rivers and creeks are dated back to the early 1970s. Simard and Blackwood (1971a, b) found that culturable yeast abundance in water samples collected at St. Lawrence River, Canada, was considerably higher in July and somewhat higher in September. A wide range of ascomycetous and basidiomycetous yeasts was identified with no phylum prevalence. In particular species belonging to the polyphyletic genera *Candida* and *Rhodotorula* were the most abundant, and the species *Rh. glutinis* was the most frequently found (about 58% of the total isolates). However, the taxonomical identity of these isolates should be analyzed with care given that the correct identification of those species is only possible using molecular techniques.

The Danube is the second longest European river and is located in Central and Eastern Europe. Slávikova and Vadkertiová (1997) studied the culturable yeast diversity in water samples of Danube collected in the area of Bratislava. Yeast abundance ranged from 10^2 to 10^3 CFU l^{-1}. The species *A. pullulans*, *Candida maltosa*, *C. krusei* (*P. kudriavzevii*), *Geotrichum candidum* (*Galactomyces candidum*), and *Saccharomyces cerevisiae* among Ascomycota and *Cystofilobasidium capitatum* and *Rh. glutinis* among Basidiomycota were the most frequently found. Among them, *S. cerevisiae* was the dominant species (isolated from 50% of samples and about 25% of isolates). However, it should be kept in mind that the eutrophy of the Danube passing through different cities along the course could have influenced the natural composition of yeast communities.

Some rivers and creeks of the Iberian Peninsula have received particular attention since the 2000s. The Tagus is the longest river on the Iberian Peninsula. It is 1038 km long and spans 716 km in Spain, 47 km along the border between Portugal and Spain, and 275 km in Portugal, where it flows into the Atlantic Ocean. Temperature gradient gel electrophoresis (TGGE), including nested PCR to obtain fungal amplicons containing the D2 domain of the 26S rRNA gene, TGGE band extraction, re-amplification, and sequencing, was used for assessing yeast diversity in the estuary of the Tagus. Fourteen fungal taxa were detected and all except one were yeasts. Most yeast sequences corresponded to members of the Ascomycota and only three belonged to the Basidiomycota. The dominant species were *Deb. hansenii*, *Rh. mucilaginosa*, and *Cryptococcus longus* (currently *Vanrija longa*) (Gadanho and Sampaio 2004). More recently, de Almeida (2005) found that yeast occurrence in waters sampled at the Tagus estuary did not seem to depend upon tides, but river discharge had a dramatic impact on both the density and diversity of the community. Seasonal changes and yeast abundance were the highest around February/March as the result of the winter peak of the river flow. The main species found were *Candida catenulata* (currently *Diutina catenulata*), *Candida intermedia*, *Candida parapsilosis*, *Clavispora lusitaniae*, *Deb. hansenii*, and *Pichia guilliermondii* (currently *Mey. guilliermondii*) among Ascomycota and *Rh. mucilaginosa* and *Rhodosp. diobovatum* (currently *Rh. diobovata*) among Basidiomycota. Interestingly, Coelho et al. (2010) found a correlation between the occurrence of both *C. parapsilosis* and *P. guilliermondii* and the incidence of the bacterial fecal species *E. coli* in waters sampled in Tagus estuary.

Other Iberian rivers have been also taken into consideration. Sampaio et al. (2007) studied the culturable yeast populations occurring on submerged leaves (alder, eucalyptus, and oak) in a natural mountain stream of River Olo, Portugal. Seventy-two yeast taxa were found, and basidiomycetous species predominated over ascomycetous ones. In all leaf litter types, three ubiquitous yeast species, such as *Cr. albidus* (*Na. albida*), *Deb. hansenii*, and *Rh. glutinis*, were found.

The occurrence and diversity of culturable yeasts in water sampled in River Tinto, which is located in the Iberian Pyrite Belt, a vast geographical area with particular geological features that stretches along much of the south of the Iberian Peninsula, from Portugal to Spain, was studied. In the Iberian Pyrite Belt, acid rock drainage gives rise to aquatic habitats with low pH and high concentrations of heavy metals and causes important environmental peculiarities. Twenty-seven yeast species were detected, 48% of which represented undescribed taxa. A few species, namely, *Candida fluviatilis* and *Rhodosporidium toruloides* (currently *Rhodotorula toruloides*), occurred in situations of intermediate environmental stress (Gadanho et al. 2006). Comparable results were found in a similar environment but in this case of volcanic origin, the River Agrio as described by Russo et al. (2008). Several species/lineages were common to both acidic rivers and showed interesting physiological adaptations to such extreme environments (Gadanho et al. 2006; Russo et al. 2008, 2010, 2016).

1.4 Yeast Diversity in Seawater

1.4.1 General Aspects

The presence of yeasts has been recognized in all worldwide oceans, from coastal regions to offshore environments (including oceanic surface, deep waters, and sediments). Marine fungi including yeasts are known to inhabit a wide variety of habitats within the marine environments (Mouton et al. 2012). The first observation of yeasts in marine water was reported by Fischer and Brebeck (1894). Marine yeasts are believed to have physiological adaptations to overcome the adverse effect of salinity and high hydrostatic pressure and to grow and interact with other microbial marine communities (Lorenz and Molitoris 1997; Kandasamy et al. 2012; Rédou et al. 2015). However, the ecological significance of the presence of viable yeast cells in these habitats as the result of a mere survival under the harsh environmental conditions occurring in seawater environment or, on the contrary, the result of their ability to play an active role in this niche (or even a combination of both) is still an open question (Mouton et al. 2012). Researches on ecology and diversity of marine yeasts have been reviewed in recent years (Kutty and Philip 2008; Nagano et al. 2010; Fell 2012; Kandasamy et al. 2012). The genera and species found in worldwide marine habitats are reported in Table 1.3.

1.4.2 Yeasts in Coastal Marine Environments

Coastal marine waters are subjected to terrestrial influxes due to natural drainage and human activities, and most fungi, including yeasts, can be considered as facultative marine fungi with a presumably terrestrial origin. Terrestrial and human-associated allochthonous yeasts are introduced into these waters by rains and rivers, and their persistence is related to the ability to sustain the saline conditions occurring in seawater. The cell abundance in some cases can be as high as thousands of cells l^{-1} (Fell 2012). A number of ascomycetous and basidiomycetous yeasts have been found as inhabitant of both terrestrial and nearshore marine environments. The most frequently isolated species were *C. parapsilosis*, *Deb. hansenii*, and a few members of the genus *Malassezia* (Gao et al. 2008; Fell 2012) (Table 1.3). Many of these species were most likely introduced from the surrounding terrestrial vegetation, including mangrove habitats (Statzell-Tallman et al. 2008). The yeast diversity in aquatic ecotone habitats (estuaries, mangroves, salt marshes, bogs, and phytotelmata) is also reported in Chap. 2 of this book.

Some species have been regularly found in coastal marine habitats, namely, *Cryptococcus marinus* (currently *Bandonia marina*) and *K. aestuarii* (van Uden and ZoBell 1962; Fonseca et al. 2011), while a few *Metschnikowia* species, namely, *M. australis*, *Metschnikowia bicuspidata*, and *Metschnikowia krissii*, are either endemic or restricted to marine nearshore habitats, sometimes associated with

Table 1.3 Diversity of yeasts and yeastlike dimorphic fungi in marine environments

Species	Original taxonomic designation	Isolation source	Locality	References
Ascomycetous yeasts				
Barnettozyma californica	*Williopsis californica*	Seawater	Faro, Portugal	Gadanho et al. (2003)
Blastobotrys adeninivorans	*Trichosporon adeninivorans*	Beach sand	Pernambuco, Brazil	Loureiro et al. (2005)
Brettanomyces anomalus		Seawater	Pernambuco, Brazil	Loureiro et al. (2005)
Brettanomyces bruxellensis		Seawater; beach sand	Pernambuco, Brazil	Loureiro et al. (2005)
Brettanomyces custersianus		Beach sand	Pernambuco, Brazil	Loureiro et al. (2005)
Candida sp.[b]		Sea sediment; shrimp; hydrothermal vent; mussel	Sepetiba Bay, Rio de Janeiro, Brazil; Southwest cost of India; Mid-Atlantic Ridge; South Pacific Basins; East Pacific Rise; Sagami Bay, Japan; Pacific Ocean	Pagnocca et al. (1989), Prabhakaran and Gupta (1991), Bass et al. (2007), Takishita et al. (2007), Burgaud et al. (2010), Xu et al. (2014, 2016)
Candida albicans		Sea sediment; seawater; beach sand	Indian Ocean; Southwest cost of India; Greece; Pernambuco, Brazil; Florida, USA	Fell (1967), Prabhakaran and Gupta (1991), Papadakis et al. (1997), Loureiro et al. (2005), Vogel et al. (2007)
Candida aquaetextoris		Seawater	Arabian Sea	Babu et al. (2013)
Candida atlantica	*Trichosporon atlanticum*	Shrimp; mussel; seawater, hydrothermal vent	Azores Archipelago; North Atlantic Ocean; Portugal; Mid-Atlantic Ridge; South Pacific Basins; East Pacific Rise	Siepmann and Hohnk (1962), Gadanho et al. (2003), Gadanho and Sampaio (2005), Burgaud et al. (2010)
Candida atmosphaerica		Seawater; seafish	Indian Ocean and Island of Mauritius; Azores Archipelago	Fell (1967), Gadanho and Sampaio (2005)
Candida blankii		Seawater	Pernambuco, Brazil	Loureiro et al. (2005)
Candida boidinii		Seawater	Faro, Portugal	Gadanho et al. (2003)

Candida diddensiae		Seawater	Pacific Ocean (equator); Pernambuco, Brazil	Yamasato et al. (1974), Loureiro et al. (2005)
Candida etchellsii[a]		Sediment	East Indian Ocean	Zhang et al. (2014)
Candida fennica		Seawater; beach sand	Pernambuco, Brazil	Loureiro et al. (2005)
Candida geochares		Seawater	Pernambuco, Brazil	Loureiro et al. (2005)
Candida glabrata	*Torulopsis glabrata*	Seawater; sea sediment	Sepetiba Bay, Rio de Janeiro, Brazil; Greece; Taiwan	Pagnocca et al. (1989), Papadakis et al. (1997), Chen et al. (2009)
Candida inconspicua[a]		Sediment	East Indian Ocean	Zhang et al. (2014)
Candida intermedia		Seawater; sediment; shrimp	Sepetiba Bay, Rio de Janeiro, Brazil; Pernambuco, Brazil	Pagnocca et al. (1989), Loureiro et al. (2005)
Candida maltosa		Beach sand	Pernambuco, Brazil	Loureiro et al. (2005)
Candida maris	*Torulopsis maris*	Seawater	Torres Strait, Australia	van Uden and ZoBell (1962)
Candida maritima		Seawater	Greece	Papadakis et al. (1997)
Candida melibiosica		Seawater; beach sand	Pernambuco, Brazil	Loureiro et al. (2005)
Candida membranifaciens subsp. *flavinogenie*		Seawater	China Eastern Sea	Wang et al. (2008)
Candida metapsilosis[a]		Sediment	Pacific Ocean	Xu et al. (2014)
Candida naeodendra		Beach sand	Florida, USA	Vogel et al. (2007)
Candida oceani		Hydrothermal vent	Mid-Atlantic Ridge	Burgaud et al. (2011)
Candida orthopsilosis[a]		Sediment	Central Indian Basin	Singh et al. (2012b)
Candida palmioleophila		Seawater	Pernambuco, Brazil	Loureiro et al. (2005)
Candida parapsilosis[b]		Sea sediment; sea-water; beach sand; sea fish; seaweeds; sediment shrimp	Biscayne Bay, Florida, USA; Indian Ocean; Sepetiba Bay, Rio de Janeiro, Brazil; Camp Varnum, Rhode Island, USA; Southwest cost of India; Greece; Faro, Portugal; Azores Archipelago; Pernambuco, Brazil; Pacific Ocean, close to Japan	Roth et al. (1962), Fell (1967), Seshadri and Sieburth (1971), Pagnocca et al. (1989), Prabhakaran and Gupta (1991), Papadakis et al. (1997), Gadanho et al. (2003), Gadanho and Sampaio (2005),

(continued)

Table 1.3 (continued)

Species	Original taxonomic designation	Isolation source	Locality	References
				Loureiro et al. (2005), Nagano et al. (2010)
Candida pseudointermedia		Seawater; sediment	Sepetiba Bay, Rio de Janeiro, Brazil	Pagnocca et al. (1989)
Candida rhagii		Seawater; beach sand	Pernambuco, Brazil	Loureiro et al. (2005)
Candida rugopelliculosa		Beach sand	Pernambuco, Brazil	Loureiro et al. (2005)
Candida saitoana	*Torulopsis candida*	Seawater	Pernambuco, Brazil	Loureiro et al. (2005)
Candida sake[b]	*Candida salmonicola*	Seawater; beach sand; sediment	Greece; Pernambuco, Brazil; East Indian Ocean	Papadakis et al. (1997), Loureiro et al. (2005), Zhang et al. (2014)
Candida sorboxylosa		Seawater	Faro, Portugal	Gadanho et al. (2003)
Candida tenuis		Seawater and sea fish	Indian Ocean	Fell (1967)
Candida torresii	*Torulopsis torresii*	Seawater	Torres Strait, Australia	van Uden and ZoBell (1962)
Candida tropicalis[b]		Seawater; sea fish; beach sand; sea sediment; shrimp; marine sponge	Biscayne Bay, Florida, USA; Indian Ocean; Southwest cost of India; Sepetiba Bay, Rio de Janeiro, Brazil; Greece; Florida, USA; Island of Oahu, Hawaii; Taiwan	Roth et al. (1962), Fell (1967), Pagnocca et al. (1989), Prabhakaran and Gupta (1991), Papadakis et al. (1997), Vogel et al. (2007), Gao et al. (2008), Chen et al. (2009)
Candida vaccinii		Beach sand	Pernambuco, Brazil	Loureiro et al. (2005)
Candida viswanathii	*Candida lodderae*	Sea fish; seawater; shrimp; mussel; hydrothermal vent	Indian Ocean; Azores Archipelago; Mid-Atlantic Ridge; South Pacific Basins; East Pacific Rise	Fell (1967), Gadanho and Sampaio (2005), Burgaud et al. (2010)
Candida xylopsoci[a]		Sediment	East Indian Ocean	Zhang et al. (2014)
Candida zeylanoides	*Candida krissii*	Seaweeds; seawater	Camp Varnum, Rhode Island, USA; Pacific Ocean (equator); Greece	Seshadri and Sieburth (1971), Yamasato et al. (1974), Papadakis et al. (1997)

Clavispora lusitaniae		Beach sand	Florida, USA	Vogel et al. (2007)
Cutaneotrichosporon curvatus[a]	*Cryptococcus curvatus*	Sediment	East Indian Ocean	Zhang et al. (2014)
Cutaneotrichosporon moniliiforme[a]	*Trichosporon moniliiforme*	Sediment	East Indian Ocean	Zhang et al. (2014)
Debaryomyces hansenii[b]	*Candida famata*; *Debaryomyces nicotianae*; *Debaryomyces kloeckerii*; *Torulopsis famata*	Seawater; sea fish; sea sediment; shrimp; mussel; hydrothermal vent	Biscayne Bay, Florida, USA; Indian Ocean and Island of Mauritius; Sepetiba Bay, Rio de Janeiro, Brazil; Southwest cost of India; Faro, Portugal; China South Sea; Mid-Atlantic Ridge; South Pacific Basins; East Pacific Rise; Central Indian Basin	Roth et al. (1962), Fell (1967), Pagnocca et al. (1989), Prabhakaran and Gupta (1991), Gadanho et al. (2003), Bass et al. (2007), Gao et al. (2007), Burgaud et al. (2010), Singh et al. (2012a)
Dipodascus australiensis[a]		Sediment	East Indian Ocean	Zhang et al. (2014)
Diutina catenulata	*Candida catenulata*	Seawater; beach sand	Greece; Pernambuco, Brazil; Florida, USA	Papadakis et al. (1997), Loureiro et al. (2005), Zhang et al. (2014), Vogel et al. (2007)
Diutina rugosa	*Candida rugosa*	Seawater; beach sand	Indian Ocean; Greece	Fell (1967), Papadakis et al. (1997)
Galactomyces sp.		beach sand	Florida, USA	Vogel et al. (2007)
Geotrichum sp.		Seawater; sediment; shrimp	Sepetiba Bay, Rio de Janeiro, Brazil	Pagnocca et al. (1989)
Geotrichum candidum[b]		Sea sediment	Southwest cost of India; East Indian Ocean	Prabhakaran and Gupta (1991); Zhang et al. (2014)
Hanseniaspora uvarum	*Kloeckera apiculata*	Sea fish; seawater; sediment; shrimp	Biscayne Bay, Florida, USA, USA; Indian Ocean; Sepetiba Bay, Rio de Janeiro, Brazil; Taiwan	Roth et al. (1962), Fell (1967), Pagnocca et al. (1989), Chen et al. (2009)
Kazachstania sp.[a]	*Kazachstania zonata* like	Sediment	Worldwide Seas	Bass et al. (2007)
Kazachstania humilis	*Candida milleri*	Seawater	Pernambuco, Brazil	Loureiro et al. (2005)

(continued)

Table 1.3 (continued)

Species	Original taxonomic designation	Isolation source	Locality	References
Kazachstania jiainica	*Kazachstania jiainicus*	Seawater	Taiwan	Chen et al. (2009)
Kazachstania yakushimaensis	*Saccharomyces yakushimaensis*	Seawater	Taiwan	Chen et al. (2009)
Kluyveromyces aestuarii	*Saccharomyces aestuarii*	Marine mud; seawater	Biscayne Bay, Florida, USA; Torres Strait, Portugal; Shroud Cay, Bahamas	Fell (1961), Lachance (2011a)
Kluyveromyces nonfermentans		Sediment	Suruga or Sagami Bay, Japan	Nagahama et al. (1999)
Kodamaea sp.[a]		Sediment	Arabian Sea	Jebaraj et al. (2010)
Kodamaea ohmeri		Beach sand; seafish; seawater	Florida, USA; Pacific Ocean; Taiwan	Vogel et al. (2007), Li et al. (2008), Chen et al. (2009), Dong et al. (2015)
Kregervanrija fluxuum	*Candida mycoderma*	Seawater	Indian Ocean	Fell (1967)
Metschnikowia sp.[a]		Sediment	Pacific Ocean, close to Japan	Nagano et al. (2010)
Metschnikowia bicuspidata		Beach sand; seawater	Florida, USA; Point Reyes, California	Vogel et al. (2007), Lachance (2011b)
Metschnikowia krissii		Seawater	La Jolla, California, USA	Lachance (2011b)
Metschnikowia reukaufii		Sediment	South China Sea	Li et al. (2010a)
Metschnikowia zobellii		Seawater	La Jolla, California, USA; Clyde estuary, Scotland	van Uden and Castelo-Branco (1961), Miller and Phaff (1998)
Meyerozyma sp.[a]		Sediment	Pacific Ocean	Xu et al. (2016)
Meyerozyma guilliermondii[b]	*Candida guilliermondii*; *Pichia guilliermondii*	Seawater; sea fish; sediment; shrimp; beach sand; marine algae; mussel; hydrothermal vent	Biscayne Bay, Florida, USA; Indian Ocean and Island of Mauritius; Sepetiba Bay, Rio de Janeiro, Brazil; Southwest cost of India; Greece; Azores Archipelago; Qingdao, China; Mid-Atlantic Ridge; South Pacific Basins; East Pacific Rise; Pacific Ocean; Canterbury Basin, New Zealand	Roth et al. (1962), Fell (1967), Pagnocca et al. (1989), Prabhakaran and Gupta (1991), Papadakis et al. (1997), Gadanho and Sampaio (2005), Gao et al. (2007), Burgaud et al. (2010), Xu et al. (2014), Rédou et al. (2015)

Nakazawaea ishiwadae	*Candida ishiwadae*	Beach sand	Florida, USA	Vogel et al. (2007)
Pichia sp.[b]		Seawater; sediment	Sepetiba Bay, Rio de Janeiro, Brazil	Pagnocca et al. (1989), Bass et al. (2007)
Pichia fermentans		Seawater	Indian Ocean	Fell (1967)
Pichia kudriavzevii	*Candida krusei*; *Issatchenkia orientalis*	Seawater; sediment; shrimp; beach sand	Sepetiba Bay, Rio de Janeiro, Brazil; Greece; Florida, USA; Taiwan	Pagnocca et al. (1989), Papadakis et al. (1997), Vogel et al. (2007), Chen et al. (2009)
Pichia membranifaciens	*Candida valida*	Sea sediment	Sepetiba Bay, Rio de Janeiro, Brazil	Pagnocca et al. (1989)
Pichia occidentalis	*Candida sorbosa*	Sea sediment; shrimp	Sepetiba Bay, Rio de Janeiro, Brazil	Pagnocca et al. (1989)
Saccharomyces sp.[b]		Sea sediment	Southwest cost of India; Central Indian Basin	Prabhakaran and Gupta (1991), Singh et al. (2012a)
Saccharomyces cerevisiae	*Saccharomyces fructuum*; *Saccharomyces steineri*	Seawater	Indian Ocean	Fell (1967)
Scheffersomyces spartinae	*Pichia spartinae*	Oyster grass (*Spartina alterniflora*) marshes	Barataria Bay, Louisiana, USA	Meyers et al. (1975)
Starmerella bombicola	*Candida bombicola*	Seawater	Pernambuco, Brazil	Loureiro et al. (2005)
Torulaspora delbrueckii	*Saccharomyces rosei; Torulopsis colliculosa*	Seawater; beach sand	Greece; Florida, USA; Taiwan	Papadakis et al. (1997), Vogel et al. (2007), Chen et al. (2009)
Wickerhamomyces anomalus	*Hansenula anomala*; *Pichia anomala*	Sea fish; seawater; sediment; shrimp; beach sand; gut of Sea squirts	Biscayne Bay, Florida, USA; Island of Mauritius; Sepetiba Bay, Rio de Janeiro, Brazil; Florida, USA; Taiwan; Coast of Yantai, China	Roth et al. (1962), Fell (1967), Pagnocca et al. (1989), Vogel et al. (2007), Chen et al. (2009), Guo et al. (2013)
Wickerhamomyces onychis	*Pichia onychis*	Beach sand	Florida, USA	Vogel et al. (2007)

(continued)

Table 1.3 (continued)

Species	Original taxonomic designation	Isolation source	Locality	References
Wickerhamomyces pijperi	*Pichia pijperi*	Seawater	Faro, Portugal	Gadanho et al. (2003)
Yamadazyma barbieri		Seawater	Ipanema beach, Rio de Janeiro, Brazil; mid-Atlantic ridge	Burgaud et al. (2016)
Yamadazyma triangularis	*Candida polymorpha*	Seawater	Indian Ocean; Island of Mauritius	Fell (1967)
Yarrowia divulgata		Seawater and ocean fish		Nagy et al. (2013)
Yarrowia keelungensis		Seawater	Keelung City off the northern coast of Taiwan	Chang et al. (2013)
Yarrowia lipolytica		Beach sand; marine algae; seawater; sediment	Florida, USA; Qingdao, China; Persian Gulf, Iran; St. Helena Bay, South Africa	Vogel et al. (2007), Hassanshahian et al. (2012), Mouton et al. (2012)
Basidiomycetous yeasts				
Apiotrichum dulcitum	*Trichosporon dulcitum*	Beach sand	Pernambuco, Brazil	Loureiro et al. (2005)
Bandonia marina	*Candida marina*	Seawater	Torres Strait, Australia	van Uden and ZoBell (1962)
Bensingtonia sp.		Seawater	Faro, Portugal	Gadanho et al. (2003)
Buckleyzyma aurantiaca	*Rhodotorula aurantiaca*	Sediment	Pacific Ocean	Nagahama et al. (2001a)
Bullera alba		Seawater	Faro, Portugal	Gadanho et al. (2003)
Bullera unica		Sediment	Canterbury Basin, New Zealand	Rédou et al. (2015)
Cryptococcus sp.[b]		Seawater; sediment	Faro, Portugal; Pacific Ocean	Gadanho et al. (2003), Xu et al. (2014, 2016)
Cutaneotrichosporon curvatus[b]	*Candida curvata; Cryptococcus curvatus*	Seawater; sediment; cold methane seep	Greece; Kuroshima Knoll, Japan; Sagami Bay, Japan	Papadakis et al. (1997), Takishita et al. (2006, 2007)

Cutaneotrichosporon cutaneum	*Trichosporon cutaneum*	Sea fish	Biscayne Bay, Florida, USA	Roth et al. (1962)
Cutaneotrichosporon dermatis[b]	*Trichosporon dermatis*	Seawater; sediment	Azores Archipelago; Pacific Ocean	Gadanho and Sampaio (2005), Xu et al. (2014)
Cutaneotrichosporon mucoides[a]	*Trichosporon mucoides*	Sediment	Pacific Ocean, close to Japan	Nagano et al. (2010)
Cystobasidium sp.[a]		Sediment	Arabian Sea; Pacific Ocean, close to Japan	Jebaraj et al. (2010), Nagano et al. (2010)
Cystobasidium benthicum	***Rhodotorula benthica***	Tubeworm	Sagami Bay and Iheya Ridge, Japan	Nagahama et al. (2003a)
Cystobasidium calyptogenae[b]	*Rhodotorula calyptogenae*	Clam; sediment	Sagami Bay, Japan; Central Indian Basin	Nagahama et al. (2003a), Singh et al. (2012b)
Cystobasidium minutum	*Rhodotorula minuta*	Sea fish; sediment; seawater; tubeworm; beach sand	Biscayne Bay, Florida, USA; Southwest cost of India; Pacific Ocean; Faro, Portugal; Sagami Bay and Iheya Ridge, Japan; Pernambuco, Brazil	Roth et al. (1962), Prabhakaran and Gupta (1991), Nagahama et al. (2001a, 2003a), Gadanho et al. (2003), Loureiro et al. (2005)
Cystobasidium pallidum	*Rhodotorula pallida*	Seawater	Indian Ocean	Fell (1967)
Cystobasidium slooffiae[b]	*Rhodotorula slooffiae*	Clam; sediment	Sagami Bay, Japan; Central Indian Basin; Pacific Ocean	Nagahama et al. (2003a), Singh et al. (2012a), Xu et al. (2014)
Cystofilobasidium sp.[a]	*Cystofilobasidium infirmominiatum like*	Sediment	Worldwide Seas	Bass et al. (2007)
Dioszegia antarctica		Rock surface	Vailulu'u active submarine volcano, Samoan volcanic chain	Jebaraj et al. (2010)
Filobasidium sp.[a]		Sediment	Worldwide Seas	Bass et al. (2007)
Filobasidium magnum	*Cryptococcus magnus*	Seawater	Faro, Portugal	Gadanho et al. (2003)
Filobasidium uniguttulatum	*Cryptococcus uniguttulatus*	Seawater	Faro, Portugal	Gadanho et al. (2003)
Goffeauzyma sp.[a]		Marine sponge	Yongxing Island, South China Sea	Jin et al. (2014)

(continued)

Table 1.3 (continued)

Species	Original taxonomic designation	Isolation source	Locality	References
Hannaella surugaensis	*Cryptococcus surugaensis*	Sediment	Suruga Bay, Japan	Nagahama et al. (2003b)
Hasegawazyma lactosa	*Rhodotorula lactosa*	Seaweeds	Camp Varnum, Rhode Island, USA	Seshadri and Sieburth (1971)
Kondoa malvinella	*Rhodosporidium malvinellum*	Seawater	Southern Pacific; Indian Ocean	Sampaio (2011)
Leucosporidium scottii		Hydrothermal vent	Mid-Atlantic Ridge; South Pacific Basins; East Pacific Rise	Burgaud et al. (2010)
Malassezia sp.[a]		Sediment; marine sponge	Island of Oahu, Hawaii; Arabian Sea; Pacific Ocean	Bass et al. (2007), Gao et al. (2008), Jebaraj et al. (2010), Xu et al. (2014, 2016)
Malassezia furfur	*Pityrosporum orbiculare*	Seawater; beach sand	Greece	Papadakis et al. (1997)
Malassezia restricta[a]		Sediment	Central Indian Basin	Singh et al. (2012a)
Malassezia slooffiae[a]		Sediment	Central Indian Basin	Singh et al. (2012a)
Moesziomyces aphidis	*Pseudozyma aphidis*	Seawater	Faro, Portugal	Gadanho et al. (2003)
Naganishia sp.[a]	*Cryptococcus vishniacii* like	Sediment	Worldwide Seas	Bass et al. (2007)
Naganishia albida	*Cryptococcus albidus*	Sediment; seawater	Indian Ocean; Pacific Ocean; Sepetiba Bay, Rio de Janeiro, Brazil; Greece; Japanb Trench; Faro, Portugal	Fell (1967), Yamasato et al. (1974), Pagnocca et al. (1989), Papadakis et al. (1997), Abe et al. (2001), Gadanho et al. (2003)
Naganishia globosa	*Cryptococcus saitoi*	Rock surface	Vailulu'u active submarine volcano, Samoan volcanic chain	Jebaraj et al. (2010)
Naganishia liquefaciens	*Cryptococcus liquefaciens*	Sediment	Japan Trench; Pacific Ocean	Abe et al. (2006), Xu et al. (2014)
Naganishia uzbekistanensis	*Cryptococcus uzbekistanensis*	Hydrothermal vent	Mid-Atlantic Ridge; South Pacific Basins; East Pacific Rise	Burgaud et al. (2010)

Papiliotrema aurea	*Cryptococcus aureus*	Sediment	China South Sea	Gao et al. (2007)
Papiliotrema laurentii	*Cryptococcus laurentii*	Seawater	Island of Mauritius; Sepetiba Bay, Rio de Janeiro, Brazil; Faro, Portugal; Pacific Ocean, close to Japan	Fell (1967), Pagnocca et al. (1989); Gadanho et al. (2003), Nagano et al. (2010)
Phaeotremella sp.[a]		Sediment	Pacific Ocean, close to Japan	Nagano et al. (2010)
Rhodosporidium sp.[a]		Sediment	Pacific Ocean	Xu et al. (2016)
Rhodotorula sp.[b]	*Rhodosporidium diobovatum* like; *Rhodotorula mucilaginosa* like	Sediment; seawater	Zhanjiang, China; Central Indian Basin; Pacific Ocean	Yang et al. (2011), Singh et al. (2012a, b), Xu et al. (2016)
Rhodotorula babjevae	*Rhodosporidium babjevae*	Seawater	Faro, Portugal	Gadanho et al. (2003)
Rhodotorula diobovata	*Rhodosporidium diobovatum*	Seawater; shrimp; mussel; hydrothermal vent	Faro, Portugal; Azores Archipelago; Mid-Atlantic Ridge; South Pacific Basins; East Pacific Rise; Zhanjiang, China	Gadanho et al. (2003); Gadanho and Sampaio (2005), Burgaud et al. (2010), Yang et al. (2011)
Rhodotorula glutinis		Sea fish; seawater; sediment; beach sand; shrimp	Biscayne Bay, Florida, USA; Indian Ocean; Pacific Ocean; Sepetiba Bay, Rio de Janeiro, Brazil; Southwest cost of India; Azores, Portugal; Florida, USA; Pernambuco, Brazil; Zhanjiang, China	Roth et al. (1962), Fell (1967), Yamasato et al. (1974), Pagnocca et al. (1989), Prabhakaran and Gupta (1991), Nagahama et al. (2001a), Gadanho and Sampaio (2002), Loureiro et al. (2005), Yang et al. (2011)
Rhodotorula graminis		Seawater; sediment	Sepetiba Bay, Rio de Janeiro, Brazil; Southwest cost of India; Atlantic Ocean	Pagnocca et al. (1989), Prabhakaran and Gupta (1991), Gadanho and Sampaio (2002)
Rhodotorula kratochvilovae	*Rhodosporidium kratochvilovae*	Seawater	Arrabida, Portugal; Faro, Portugal	Gadanho and Sampaio (2002), Gadanho et al. (2003)

(continued)

Table 1.3 (continued)

Species	Original taxonomic designation	Isolation source	Locality	References
Rhodotorula mucilaginosa[b]	*Rhodotorula pilimanae*; *Rhodotorula rubra*	Seawater; sediment; seaweeds; shrimp; mussel; hydrothermal vent; beach sand; sea fish;	Biscayne Bay, Florida, USA; Indian Ocean; Camp Varnum, Rhode Island, USA; Sepetiba Bay, Rio de Janeiro, Brazil; Southwest cost of India; Pacific Ocean; Faro, Portugal; Azores Archipelago; Pernambuco, Brazil; Queenscliff region, Victoria, Australia; Mid-Atlantic Ridge; South Pacific Basins; East Pacific Rise; Bohai Sea, China; St. Helena Bay, South Africa; Zhanjiang, China; Canterbury Basin, New Zealand; Central Indian Basin	Roth et al. (1962), Fell (1967), Seshadri and Sieburth (1971), Yamasato et al. (1974), Pagnocca et al. (1989), Prabhakaran and Gupta (1991), Nagahama et al. (2001a), Gadanho et al. (2003), Gadanho and Sampaio (2005), Loureiro et al. (2005), Vogel et al. (2007), Burgaud et al. (2010), Li et al. (2010b), Yang et al. (2011), Gupta et al. (2012), Mouton et al. (2012), Xu et al. (2014), Rédou et al. (2015), Singh et al. (2012a)
Rhodotorula nothofagi		Seawater	Faro, Portugal	Gadanho et al. (2003)
Rhodotorula pacifica		Sediment	Pacific Ocean	Nagahama et al. (2006)
Rhodotorula paludigena	*Rhodosporidium paludigenum*	Beach sand; seawater and shrimp	Florida, USA; Zhanjiang, China	Vogel et al. (2007), Yang et al. (2011)
Rhodotorula sphaerocarpa	*Rhodosporidium sphaerocarpum*	Seawater; shrimp	Faro, Portugal; Azores Archipelago; Florida, USA; Bahamas, USA; Zhanjiang, China	Gadanho et al. (2003), Gadanho and Sampaio (2005), Sampaio (2011), Yang et al. (2011)
Rhodotorula toruloides	*Rhodosporidium toruloides*	Seawater; shrimp	Azores Archipelago; Southern Ocean; Zhanjiang, China	Gadanho and Sampaio (2005), Sampaio (2011), Yang et al. (2011)
Saitozyma podzolica[a]	*Cryptococcus podzolicus*	Sediment	East Indian Ocean	Zhang et al. (2014)
Sakaguchia dacryoidea		Seawater	Faro, Portugal	Gadanho et al. (2003)
Sakaguchia lamellibrachiae	*Rhodotorula lamellibrachiae*	Sediment; deep-sea floor; tubeworm	Sagami Bay, Japan	Nagahama et al. (2001b, 2003a)
Sampaiozyma ingeniosa	*Rhodotorula ingeniosa*	Seawater	Pernambuco, Brazil	Loureiro et al. (2005)

Spencerozyma crocea	*Rhodotorula crocea*	Seawater	Indian Ocean	Fell (1967)
Sporidiobolus pararoseus		Shrimp	Zhanjiang, China	Yang et al. (2011)
Sporobolomyces lactosus[a]		Sediment	East Indian Ocean	Zhang et al. (2014)
Sporobolomyces roseus		Shrimp; mussel	Faro, Portugal	Gadanho et al. (2003)
Sporobolomyces salmonicolor	*Sporobolomyces odorus*; *Sporobolomyces hispanicus*	Seawater; sediment	Indian Ocean; Pacific Ocean	Fell (1967), Nagahama et al. (2001a)
Sporobolomyces shibatanus		Sediment	Pacific Ocean	Nagahama et al. (2001a)
Sterigmatomyces halophilus[b]		Seawater	Indian Ocean; Pacific Ocean	Fell (1967, 1970), Xu et al. (2014)
Sterigmatomyces sp.[a]		Sediment	Pacific Ocean	Xu et al. (2016)
Symmetrospora marina	*Rhodotorula marina*	Seawater	Pacific Ocean (equator)	Yamasato et al. (1974)
Tausonia sp.[a]	*Trichosporon pullulans* like	Sediment	Worldwide Seas	Bass et al. (2007)
Tausonia pullulans[a]	*Trichosporon pullulans*	Sediment	Pacific Ocean	Xu et al. (2014)
Trichosporon sp.[b]		Seawater; shrimp; sediment	Sepetiba Bay, Rio de Janeiro, Brazil; Pacific Ocean	Pagnocca et al. (1989), Xu et al. (2016)
Trichosporon aquatile		Beach sand	Pernambuco, Brazil	Loureiro et al. (2005)
Trichosporon asahii[b]		Beach sand; sediment	Florida, USA; Central Indian Basin	Vogel et al. (2007), Singh et al. (2012a, b)
Trichosporon beigelii		Beach sand	Pernambuco, Brazil	Loureiro et al. (2005)
Trichosporon coremiiforme		Beach sand	Florida, USA	Vogel et al. (2007)

(continued)

Table 1.3 (continued)

Species	Original taxonomic designation	Isolation source	Locality	References
Vanrija fragicola[a]	*Cryptococcus fragicola*	Sediment	East Indian Ocean	Zhang et al. (2014)
Vishniacozyma sp.[a]	*Cryptococcus carnescens* like	Sediment	Pacific Ocean	Bass et al. (2007), Xu et al. (2014)
Vishniacozyma victoriae	*Cryptococcus victoriae*	Seawater	Faro, Portugal	Gadanho et al. (2003)
Yeastlike microorganisms				
Aureobasidium sp.[a]		Sediment; marine sponge	Island of Oahu, Hawaii; Yongxing Island, South China Sea; Pacific Ocean	Bass et al. (2007), Gao et al. (2008), Jin et al. (2014), Xu et al. (2016)
Aureobasidium pullulans[b]		Sediment	Pacific Ocean	Xu et al. (2014)
Exophiala sp.[b]		Sediment	Pacific Ocean; Canterbury Basin, New Zealand	Xu et al. (2014, 2016), Rédou et al. (2015)
Exophiala dermatitidis		Seawater; sediment	Azores Archipelago	Gadanho and Sampaio (2005), Xu et al. (2014)
Sympodiomycopsis sp.[a]		Sediment	Pacific Ocean	Xu et al. (2016)

[a]Only non-culturable
[b]Both culturable and non-culturable

marine invertebrates and fishes (Fell and Hunter 1968; Seki and Fulton 1969; Donachie and Zdanowski 1998; Ebert et al. 2000a, b; Moore and Strom 2003; Wang et al. 2008; Lachance 2011b).

Coastal marine sediments represent peculiar environmental niches. Mouton et al. (2012) studied several fungal isolates obtained from marine sediments collected close to St. Helena Bay Western Cape, South Africa. Among them, *Rh. mucilaginosa* and *Yarrowia lipolytica* were found. More recently, Hassanshahian et al. (2012) isolated some *Y. lipolytica* strains from oil-polluted sediment and seawater samples collected in the Persian Gulf (Iran). The strains showed a high ability to degrade aliphatic hydrocarbons.

Several studies also reported the direct pollution of marine nearshore habitats with clinically important yeasts: *C. albicans*, *C. tropicalis*, and some *Trichosporon* spp. strains including *Trichosporon asahii* were isolated from worldwide bathing beaches (Velegraki-Abel et al. 1987; Papadakis et al. 1997; Vogel et al. 2007; Chen et al. 2009; Sabino et al. 2011) and polluted sediments and shrimps (Hagler et al. 1986; Pagnocca et al. 1989; Soares et al. 1997).

1.4.3 Yeasts in Offshore and Deep-Sea Marine Environments

Due to the high costs involved in offshore sampling, oceanographic studies are limited in number. Although the frequency of occurrence is depending on the source and geographical origin of samples, Kandasamy et al. (2012) reported that the yeast abundance increases with increasing distance from the coastline and increasing depth of coastal sea. Conversely, Fell (2012) reported that in offshore regions there is a reduction in the diversity of species and population densities. Bass et al. (2007) postulated that yeasts are the prevalent form of fungi in the deep sea, as revealed with a culture-independent assay. This hypothesis is consistent with that reported by Fell (2012) who suggested that the unicellular lifestyle is apparently better adapted to the aqueous environment than fungal hyphae. Numerically, yeast cells in open ocean waters ranged from 0 to 10 cells l^{-1}, although regions of high organic activity can result in intensive yeast proliferation (Fell 2012). The yeast genera and species found in worldwide offshore marine water are reported in Table 1.3.

Generally, basidiomycetous yeasts account for the majority of the total yeast population in oligotrophic oceanic water, while the ascomycetous yeasts constitute the majority of the total yeast population in the offshore marine sediments (Kandasamy et al. 2012). For example, a number of basidiomycetous yeasts were isolated in seawater collected in the Atlantic Ocean (southern Portugal). Specifically, *Rhodosporidium babjevae*, *Rhodosp. diobovatum* (currently *Rhodotorula babjevae* and *Rh. diobovata*, respectively), and *Pseudozyma aphidis* (currently *Moesz. aphidis*) were the most frequently found species (Gadanho et al. 2003). Among the ascomycetes, some *Candida* species and *Deb. hansenii* were found as inhabitant of offshore and deep-sea oceanic habitats (Kandasamy et al. 2012), in agreement with the early study of Hagler and Ahearn (1987), who postulated that *Deb. hansenii* can be considered the most common ascomycete in marine waters.

Although they have been occasionally isolated from nonmarine sources, some species were found to inhabit specific water masses or oceanographic regions. Among them, *Blastobotrys parvus* (formerly *Sympodiomyces parvus*) was found in warmed Antarctic and sub-Antarctic waters (Fell and Statzell-Tallman 1971), *Candida natalensis* was isolated in the Indo-Pacific ocean (Fell 2012), and *Candida norvegica* was found in a narrow geographical area southward from the polar front (Fell 2012). Strains of the species *Leucosporidium antarcticum* (currently *Glaciozyma antarctica*) have been isolated in Antarctic waters adjacent to the ice pack (Fell 2012). The yeast diversity in cold Polar, sub-Polar, and non-Polar habitats is also reviewed in Chaps. 11 and 12 of this book.

Nagano et al. (2010) investigated deep-sea sediments (1200–10,000 m) collected off the coast of Japan by a culture-independent approach. A number of unknown species together with OTUs (operational taxonomic units) close to *C. parapsilosis*, *Cryptococcus skinneri* (currently *Phaeotremella skinneri*), *Metschnikowia colocasiae*, *Metschnikowia continentalis*, and *Metschnikowia kamakouana* were found. Singh et al. (2012a) investigated fungal diversity in two deep-sea (5000 m) sediment cores collected in the Central Indian Basin by both culture-dependent and culture-independent approaches. A total of 19 culturable fungi including yeasts and 46 OTUs were found. The most represented yeast species were *Deb. hansenii*, *Rhodotorula slooffiae* (currently *Cystobasidium slooffiae*), *Rh. mucilaginosa*, and the black yeast *Hortaea* sp. The detection of amplicon sequences belonging to Exobasidiomycetes and Cystobasidiomycetes from deep-sea samples was also reported (Singh et al. 2012b). Zhang et al. (2014) studied yeast diversity of deep-sea sediment samples collected in the East Indian Ocean. OTUs were related to some yeast species, namely, *Candida etchellsii*, *C. inconspicua*, *Candida sake*, *Candida xylopsoci*, *Dipodascus australiensis*, and *Gal. candidum* among Ascomycota and *Cryptococcus curvatus*, *Cryptococcus fragicola*, *Cryptococcus podzolicus* (currently *Cutaneotrichosporon curvatus*, *Vanrija fragicola*, and *Saitozyma podzolica*, respectively), *Guehomyces pullulans* (*Tausonia pullulans*), *Cyst. slooffiae*, *Sporobolomyces lactosus*, *Sterigmatomyces halophilus*, and *Trichosporon moniliiforme* (currently *Cutaneotrichosporon moniliiforme*) among Basidiomycota and the black yeast *Hortaea werneckii*. A recent study on fungi and yeast diversity in the deep-sea biosphere (water depth ranging from about 5000 to 7000 m) of the Pacific Ocean also reported that the 38.5% of environmental sequences were closely related to phylogenetic lineages comprised by yeasts, namely, to the genera *Candida*, *Erythrobasidium*, *Meyerozyma*, *Tilletiopsis*, *Rhodotorula*, and unspecified members of the former polyphyletic genera *Cryptococcus* and *Rhodotorula*. Several OTUs were also classified as species of the genera *Malassezia*, *Trichosporon* (*Cutaneotrichosporon*), and *Sterigmatomyces*, which are well-known pathogens or parasites of marine animals (Zhang et al. 2014; Xu et al. 2016). Yeasts as parasites of animals, plants, and other fungi have been recently reviewed (Begerow et al. 2017).

Hydrothermal vents are fissures in a planet's surface from which geothermally heated water issues. Their presence strongly affects chemistry and biology of the adjacent habitats. The yeast diversity found in marine hydrothermal vents is reported in Table 1.3. Gadanho and Sampaio (2005) explored the culturable yeast diversity in thermal systems of the Mid-Atlantic Ridge. Species identifications reported both marine- and terrestrially associated species. Unknown species represented 33% of the total yeast taxa. The authors reported some species frequently found in marine waters, namely, *Candida atlantica*, *Candida atmosphaerica*, *C. parapsilosis*, *Rh. diobovata*, and *Rhodosporidium sphaerocarpum* (currently *Rhodotorula sphaerocarpa*). Similar results were obtained by Burgaud et al. (2010) from a series of studies at vents in the Mid-Atlantic Ridge, South Pacific Basins, and East Pacific Rise (Table 1.3). On the contrary, Le Calvez et al. (2009) observed the predominant presence of unknown fungal groups, including species of the former polyphyletic genera *Cryptococcus* and *Filobasidium* in vents of the Pacific and Atlantic Oceans.

Jin et al. (2014) studied fungal diversity in two South China Sea sponges (*Theonella swinhoei* and *Xestospongia testudinaria*) which resulted in 26 OTUs that were assigned to Ascomycota, Basidiomycota, and Blastocladiomycota, including a taxon closely related to the yeast species *Cryptococcus gastricus* (currently *Goffeauzyma gastrica*). The fungal composition of both sponges was significantly different from that of seawater.

A few new species inhabiting different offshore marine habitats were recently described. For example, *Yarrowia keelungensis* was isolated from the sea-surface micro-layer off the northern coast of Taiwan (Chang et al. 2013), while *Yamadazyma barbieri* was isolated from Mid-Atlantic Ridge ocean water samples located in the direct vicinity of black smokers near the Rainbow deep-sea hydrothermal vent and one from Brazilian marine water samples off the Ipanema beach (Burgaud et al. 2016). Also, *Yam. barbieri* is phylogenetically related to a few species of the *Yamadazyma* clade, which were isolated from marine habitats including deep-sea hydrothermal vents, i.e., *C. atmosphaerica*, *C. atlantica*, *Candida oceani*, *Candida spencermartinsiae*, and *Candida taylorii*. These species are widespread in marine habitats and represent an ecologically and phylogenetically defined cluster of marine species in the *Yamadazyma* clade (Burgaud et al. 2016).

1.5 Glacially Originated Water Bodies

1.5.1 General Aspects

Glacial ice is the largest reservoir of freshwater on Earth and many glaciers, because seasonal climates store water as ice during the colder seasons and release it later in the form of meltwater when warmer summer temperatures occur. As glaciers retreats, many water bodies (lakes, ponds, etc.) are left behind and retain direct connection to the glacier through rivers or streams. Even though most glacial ice is located in the polar regions, continental glaciers may be also found in mountain ranges between 35°N and 35°S like the Himalayas, Andes, Rocky Mountains, and in a few other high mountains, being Greenland and Patagonia the biggest expanses of continental glaciers. When a glacier erodes the land, and then melts, it fills the hole or space that it has created originating glacial lakes. In tropical areas, glacial meltwater is an important water resource for major cities (Cook et al. 2016). As glaciers retreat, which is a global tendency, ice-marginal lakes (or proglacial lakes) become distant water bodies which in most cases remain connected by rivers and streams. Most glacial lakes are in elevated locations and this increases in lower latitudes. Normally they are in remote or even protected areas, with little human influences, such as local pollution or land use change, and can be considerate open systems with surface in- and outflow. Most of these aquatic environments considered oligotrophic are characterized by the presence of very low nutrient levels and low rates of external supply.

1.5.2 Yeast Diversity and Ecology in Patagonian Glacial Water Bodies

An interesting example of oligotrophic water bodies is lakes and rivers of Andean Patagonia in South America. Patagonia, the southernmost region of the United States, extends approximately from 40°S—where the width of the continent is about 1000 km—and gradually narrows southward until it disappears in Cape Horn at 56°S (Coronato 2016). These lakes are mostly of glacial origin, reaching in the case of lakes, depths frequently greater than 100 m. They are extremely transparent due to their oligotrophic or even ultra-oligotrophic condition and have an extended euphotic zone (i.e., layer of seawater that receives enough sunlight for photosynthesis to occur) of about 50 m (Modenutti et al. 1998). Quirós and Drago (1985) classified Andean lakes as warm monomictic with a period of summer stratification. Their characteristic low nutrient concentrations is a consequence of the low chemical weathering of the dominant igneous bedrock, as well as low rates of atmospheric deposition. It has been reported that the presence of marine cyclic cations, dissolved inorganic nitrogen, is much lower than the world average (Pedrozo et al. 1993). Microorganisms thriving in such glacial-originated water bodies are typically exposed to multiple stress factors, namely, low temperatures, lack of nutrients, and increased UV exposure as a result of high elevation and transparency.

An additional description of South American cold habitats and a comprehensive list of psychrophilic and psychrotolerant yeast species recovered from such environments are given in Chap. 12 of this book. In this paragraph, we will focus on specific interesting cases illustrating diversity and ecology of yeasts in the northern Andean Patagonia (Argentina) where a great variety of glacially formed water bodies exist. They are characterized by ultra-oligotrophic to mesotrophic conditions and range from small to large lakes, including small high elevation lakes, sometimes surrounded by dense forest (Quirós and Drago 1985; Díaz et al. 2000) (Fig. 1.1). These habitats are normally exposed to extended daylight (at latitude 41–42°S) and consequently increased UV radiation, due also to ozone layer depletion and a clean atmosphere. As already mentioned, these aquatic environments are highly transparent due to their ultra-oligotrophic character and, thus, substantially affected by UV radiation (Villafañe et al. 2001). It is expected then that yeasts autochthonous to these habitats should display cold tolerance and nutritional plasticity and possess efficient strategies for photoprotection. Thus, although yeasts are ubiquitous components of these aquatic environments, they are often present at much lower concentrations than can be expected in terrestrial habitats, such as soils (10^3–10^4 cells g^{-1}) (Spencer and Spencer 1997; Slávikova and Vadkertiova 2000) and phylloplane [10^2–10^4 cells $(cm^2)^{-1}$] (Last and Price 1969; Inácio et al. 2002). Yeasts inhabiting soils and phylloplane are reviewed in Chaps. 3, 4, 6, and 7 of this book.

Even though a portion of the yeast species present in a water body might be native to the aquatic system, it is known that a significant proportion of the yeasts present in a water sample had reached the aquatic environment through runoff from the surrounding watershed (Hagler and Ahearn 1987). If this is entirely true, and assuming the absence of microbial growth in the water column, a few hypotheses can be drawn for the yeast cell dynamics in the water column of ultra-oligotrophic and pristine lakes like those discussed here. First, a higher number of yeast cells might be expected in coastal areas than in open waters (pelagic sites) due to dilution effect and higher availability of the organic matter. Second, allochthonous species can be expected to be more abundant in small lakes having a low lake-area-watershed-area ratio, as well as in the littoral (as opposite to pelagic) areas of larger lakes. One can speculate that the proportion of these allochthonous yeasts should be higher in coastal waters, as well as in small lakes having a low lake-area-watershed-area ratio. In contrast, the proportion of autochthonous species must be higher in the pelagic zone of large lakes given they are better adapted to the harsh conditions and can prevail. Finally, if photoprotective compounds (i.e., carotenoids and mycosporines, MYCs) provide adaptation advantages in the pelagic zone of a highly UV-exposed region, then differences in the proportion of species producing either carotenoids or MYCs between these two types of habitats can also be expected.

During studies of several water bodies in Northwestern Patagonia, including lakes, rivers, and ponds, yeasts were found in almost all samples (Libkind et al. 2003, 2009a; Brandão et al. 2011a, 2017), and they normally presented average abundances around 220 $\pm$ 389 CFU l^{-1}. Average yeast abundances ranged from 2 to 250 CFU l^{-1}, rarely exceeding 200 CFU l^{-1}, and these numbers are consistent with those reported for other clear lakes (Hagler and Ahearn 1987). A few exception were found in the high-altitude lake Laguna Negra (890 CFU l^{-1}, 93% of red yeasts) and in a few samples from an anthropogenically impacted coast (1668 CFU l^{-1}). The former case cannot be attributed to a high anthropogenic influence because of the secluded location of this lake, but given 87.5% of the pigmented yeasts were *Rh. mucilaginosa*, it is conceivable that an occasional surge of organic matter caused a temporary increase of this red yeast population (Libkind et al. 2009a) and the carotenoid pigments probably provided this yeast photoprotection (Moliné et al. 2010). In fact, a strain of *Rh. mucilaginosa* from a high-altitude lake was found as a good carotenoid producer for applied purposes (Libkind et al. 2004a). The latter case (coastal water) showed yeast counts characteristic of eutrophic waters (Simard and Blackwood 1971a, b; Meyers et al. 1970; Hagler and Ahearn 1987) and might indicate an increased availability of organic matter due to the coastal condition (higher availability of vegetal and animal residues). Though in a minor scale, a higher level of viable yeasts was found in anthropogenically affected coastal sites located at San Carlos Bariloche City, coast in the Nahuel Huapi Lake. These sites had higher yeast values (97–141 CFU l^{-1}) than other ones far off the influence of the city (22–73 CFU l^{-1}). Even though further studies should be performed in order to confirm this observation, for oligotrophic lakes such as those from Andean Patagonia, it seems that the total number of yeasts in the water sample serve as a good indicator of anthropogenic impact, given that normally unpolluted waters have very low numbers of yeasts. All other Andean lakes surveyed showed yeast values typical for open waters of non-polluted lakes (Hagler and Ahearn 1987; Nagahama 2006). In a yet unpublished work, 20 water bodies were pooled into two groups depending on whether they were coastal (8) or pelagic sites (12) (Libkind et al. unpublished data). The former group showed higher yeast numbers (442 ± 558 CFU l^{-1}) than the latter group (71 ± 62 CFU l^{-1}). Other studies reported also that open waters of clear lakes yield generally yeast counts below 100 CFU l^{-1} (van Uden and Ahearn 1963; Meyers et al. 1970; Hagler and Ahearn 1987) and that this value increases with the proximity to the coast (Hagler and Ahearn 1987).

Basidiomycetous yeasts are the predominant group in these type of aquatic cold environments (Libkind et al. 2003, 2005b, 2009b, 2010; de García et al. 2007, 2012; Brandão et al. 2011a). This is similar to other cold habitats and an exhaustive overview of the biodiversity of cold-adapted yeasts (see Chap. 12 of this book). A few authors have suggested that this could be due to a higher nutritional versatility and a higher tolerance to extreme environmental conditions of the basidiomycetous compared to ascomycetous yeasts (Sampaio 2004; Frisvad 2008). As already stated above, UV radiation is a major environmental factor in clear lakes and even more pronounced in Patagonian lakes. Thus, we compared the UV susceptibility of a large representative set of ascomycetous and basidiomycetous yeasts using different culture media (Moliné 2004). Even though results were greatly influenced by the nutritional level of the media, in general ascomycetous yeasts were significantly less tolerant to any of the radiation treatments used in the experiment (PAR, PAR + UVA, and PAR + UVB). This study also suggested that the carotenoid-accumulating yeasts (often referred to as red yeasts) were generally more tolerant than nonpigmented species. Later, using naturally occurring albino strains of *Sporobolomyces ruberrimus* and *Cystofil. capitatum* (two species typically found in glacially originated water bodies in Patagonia), Moliné et al. (2009)

demonstrated that carotenoid pigments provide protection against UVB. Further experimental evidence of the utility of carotenoids as photoprotectants arose from studies performed with one of the prevailing yeasts in aquatic environments of glacial origin: the ubiquitous red yeast *Rh. mucilaginosa* (Moliné et al. 2010). In this work, the accumulation of carotenoids, in particular torularhodin, was demonstrated to contribute substantially to enhance UVB tolerance in yeasts. The protective mechanism was shown to be indirect, probably by quenching reactive oxygen species (ROS) as a result of the antioxidant properties of carotenoid pigments. The ability of certain yeast species to synthesize mycosporines, compounds able to directly protect cells from UV radiation (natural sunscreens), was discovered for the first time studying glacially formed aquatic environments (Libkind et al. 2004b, 2006). Mycosporines are water-soluble compounds composed of a cyclohexenone attached to an amino acid (or amino alcohol). Fungal mycosporines absorb light in the UV spectrum with a maximum at 310 nm wavelength (UVB). The primary role assigned to mycosporines was to act as photoprotective UV filters (Shick and Dunlap 2002; Torres et al. 2004), although other roles were also attributed including antioxidant activity, osmoregulation, resistance to thermal stress, and to serve as intracellular nitrogen storage (Oren and Gunde-Cimerman 2007). Most basidiomycetous yeasts isolated from Patagonian lakes were found to synthesize a UV-absorbing compound when grown under photosynthetically active radiation (Libkind et al. 2004b), and this occurred more frequently in species of the subphylum Pucciniomycotina (Libkind et al. 2011b) and Agaricomycotina (Libkind et al. 2005a, 2011c). The main mycosporine found in yeast so far is mycosporine-glutaminol-glucoside (MGG) (Sommaruga et al. 2004) for which the UVB photoprotective role was recently experimentally demonstrated (Moliné 2010; Moliné et al. 2011). Biochemical characterization of yeast MGG further revealed that possesses high photostability and antioxidant properties (Moliné et al. 2011) and thus showing its value as natural UV protectants and as biotechnological relevant compound (Colabella et al. 2014; Libkind et al. 2016).

The occurrence of MGG-positive yeast in glacially originated water bodies goes through a wide range, from 14% to near 90% of total cultivable yeast community (Libkind et al. 2006, 2009a; Brandão et al. 2011a). MGG synthesis was more frequent in yeasts that were not able to accumulate carotenoid pigments such as *Cr. albidus* (currently *Na. albida*), *Cryptococcus antarcticus* (*Naganishia antarctica*), *Cryptococcus saitoi* (*Naganishia globosa*), *Cryptococcus festucosus* (*Holtermanniella festucosa*), *Cryptococcus adeliensis* (*Naganishia adeliensis*), *Cryptococcus magnus* (*Filobasidium magnum*), *Gu. pullulans* (*Ta. pullulans*), and *A. pullulans* (Libkind et al. 2009a; Brandão et al. 2011a), and red yeasts capable to produce MGG were *Cyst. minutum* (*Rh. minuta*), *Cystobasidium laryngis* (*Rhodotorula laryngis*), and *Dioszegia* spp. Like for red yeasts, the proportion of mycosporine-positive species was higher for lakes with higher transparency or in pelagic zones (Brandão et al. 2011a). In glaciers meltwater and ice, yeasts, which are able to synthesize MGG, are less frequent and include species like *Dioszegia crocea* and *Dioszegia fristingensis* as the most important ones (de García et al. 2012). *Dioszegia* species were proven to be considerably higher resistant to UVB damage than most other aquatic yeasts (Moliné 2004; Libkind et al. 2009a), probably as a result of its particular carotenoid composition (plectaniaxanthin) (Madhour et al. 2005) and the high levels of MGG they accumulate (Libkind et al. 2005a, 2009a), both explaining their ability to thrive and prevail in extreme environments like those described in this paragraph.

A few studies reported yeasts biodiversity indices (e.g., Shannon-Weaver index; see also Yurkov and Pozo (2017)) of aquatic biotopes in Patagonia and showed that the yeast diversity in these environments (expressed as Shannon diversity index values) seems comparable or even higher than the ones of some in Patagonian forest soils or from tropical lakes (Brandão et al. 2011a, 2017; Mestre et al. 2011; de García et al. 2012). Yeast biodiversity in Patagonian aquatic environments comprises a large number of species. Among carotenoid-producing yeasts, the ubiquitous species *Rh. mucilaginosa* was the most frequent and commonly found species (Libkind et al. 2003, 2008). Species of the genera *Cystofilobasidium*, *Dioszegia*, *Rhodosporidiobolus*, *Rhodotorula*, *Sporidiobolus*, and *Sporobolomyces* were also isolated, though less frequently. Interestingly, ballistospore-producing yeasts (a characteristic typical for yeasts living on leaf surfaces) were more frequent in aquatic environments under relatively low human impact and in lakes surrounded with dense forest (Libkind et al. 2003). Nonpigmented yeasts were studied in a less extent, but species belonging to the genera *Candida* and *Torulaspora*, as well as to the former polyphyletic genus *Cryptococcus*, were found (Brizzio and van Broock 1998). Other genera like *Cutaneotrichosporon*, *Debaryomyces*, *Hanseniaspora*, *Leucosporidium*, and *Pichia* may also occur in waters (de García et al. 2007; Libkind et al. 2009a; Brandão et al. 2011a, 2017). The majority of these species have been already reported to be present in other aquatic environments; however, several novel species of the genera *Sporobolomyces* and *Cystofilobasidium* have been described (for recent review, see Buzzini et al. 2012; de García et al. 2014). Recently, *Cystobasidium psychroaquaticum* was described from psychrophilic aquatic habitats, including oligotrophic lakes and swamps, although a few strains were also found in terrestrial habitats (Yurkov et al. 2015). To be specially mentioned is the finding in coastal areas of glacially formed water bodies of a few isolates of biotechnologically relevant yeasts such as *Saccharomyces eubayanus* (brewing) and *Phaffia rhodozyma* (astaxanthin) (Libkind et al. 2007; Brandão et al. 2011a), although these yeasts relevant to the beer and aquaculture industries, respectively, are clearly related to forest substrates rather than water and thus represent an allochthonous species in aquatic environments (Libkind et al. 2011a, d).

1.6 Concluding Remarks

Most yeasts recovered from water samples are actually associated with plants and soils and arrive to the aquatic ecosystems through runoff phenomena, rather than being true aquatic yeasts. This is why quantitative analysis of aquatic yeast communities is complicated due to large variability of data and a large fraction of presumably transient microbiota. The proportion of allochthonous yeasts is influenced by the trophic state of the lake, the surrounding vegetation, water body size, and the number and type of effluents, among other factors. Despite these complications, various interesting niches are present in aquatic environments, especially in marine environments, that deserve special attention due to its fundamental relevance or to the lack of available knowledge. For example, yeasts associated to deep sea (i.e., hydrothermal vents) have been mostly overlooked, and the few studies performed revealed promising results (Gadanho and Sampaio 2005). Macroalgae-associated yeasts are also an interesting topic of research given that species like *M. australis* show a strong association with this substrate in Antarctica (Godinho et al. 2013). The potential ecoclade (Gadanho and Sampaio 2009) of the genus *Yamadazyma* is also interesting given these yeast species are mostly found in seawater (Burgaud et al. 2016), and there is evidence that they might be metabolically active, participating in the carbon cycle in these ecosystems.

The application of culture-independent strategies (metagenomics) will allow to have a more clear picture of yeast diversity in aquatic environments. The low number of yeast cells found in many aquatic habitats (10–100 CFU l^{-1}), mostly oligotrophic or ultra-oligotrophic, complicates the use of metagenomic approaches and the achievement of a comprehensive picture of the yeast community. However, these might still be useful for the detection of novel taxa and the relative quantification of the main species of yeasts.

Yeasts inhabiting aquatic environments can be very often subjected to multiple stress factors such as cold temperatures, UV radiation, ultra-oligotrophicity, salinity, etc. Species adapted to one or several of these factors have naturally developed mechanisms to reduce the negative effects of the harsh environmental conditions. Metabolic and physiological characterization of isolated yeasts represents a valuable tool for the identification of potentially autochthonous species and the detection of biotechnologically relevant traits such as the production of carotenoid pigments, UV sunscreens, extracellular cold-adapted enzymes, etc. In the present chapter, we have shown that aquatic environments can act as reservoir of many yeast species bearing physiological adaptations interesting from both fundamental and applied perspectives and that many additional yeast diversity studies are needed in order to increase our incipient knowledge on the factors affecting yeast distribution and composition in water habitats.

References

Abe F, Miura T, Nagahama T, Inoue A, Usami R, Horikoshi K (2001) Isolation of a highly copper-tolerant yeast, *Cryptococcus* sp., from the Japan Trench and the induction of superoxide dismutase activity by Cu^{2+}. Biotechnol Lett 23:2027–2034

Abe F, Minegishi H, Miura T, Nagahama T, Usami R, Horikoshi K (2006) Characterization of cold- and high-pressure-active polygalacturonases from a deep-sea yeast, *Cryptococcus liquefaciens* strain N6. Biosci Biotechnol Biochem 70:296–299

Ahearn DG, Roth FJ, Meyers SP (1968) Ecology and characterization of yeasts from aquatic regions of South Florida. Mar Biol 1(4):291–308

Araújo FV, Hagler AN (2011) *Kluyveromyces aestuarii*, a potential environmental quality indicator yeast for mangroves in the state of Rio de Janeiro, Brazil. Braz J Microbiol 42:954–958

Babu DT, Antony SP, Joseph SP, Bright AR, Philip R (2013) Marine yeast *Candida aquaetextoris* S527 as a potential immunostimulant in black tiger shrimp *Penaeus monodon*. J Invertebr Pathol 112:243–252

Bass D, Howe A, Brown N, Barton H, Demidova M, Michelle H, Li L, Sanders H, Watkinson SC, Willcock S, Richards TA (2007) Yeast forms dominate fungal diversity in the deep oceans. Proc Biol Sci 274:3069–3077

Begerow D, Kemler M, Feige A, Yurkov AM (2017) Parasitism in Yeasts. In: Buzzini P, Lachance MA, Yurkov AM (eds) Yeasts in natural ecosystems: ecology. Springer, Heidelberg, pp 179–210

Bogusławska-Was E, Dabrowski W (2001) The seasonal variability of yeasts and yeast-like organisms in water and bottom sediment of the Szczecin Lagoon. Int J Hyg Environ Health 203:451–458

Brandão LR, Medeiros AO, Duarte MC, Barbosa AC, Rosa CA (2010) Diversity and antifungal susceptibility of yeasts isolated by multiple-tube fermentation from three freshwater lakes in Brazil. J Water Health 8:279–289

Brandão LR, Libkind D, Vaz ABM, Espírito Santo L, Moliné M, de García V, van Broock M, Rosa CA (2011a) Yeasts from an oligotrophic lake in Patagonia (Argentina): diversity, distribution and synthesis of photo-protective compounds and extracellular enzymes. FEMS Microbiol Ecol 76:1–13

Brandão LR, Rosa CA, Medeiros AO (2011b) Yeast diversity in freshwater ecosystems. In: Daniels JA (ed) Advances in environmental research. Science Publisher, Hauppauge, pp 207–222

Brandão LR, Vaz AB, Espírito Santo LC, Pimenta RS, Morais PB, Libkind D, Rosa LH, Rosa CA (2017) Diversity and biogeography patterns of yeast communities from lakes of Antarctica, Argentinean Patagonia, and tropical Brazil. Fungal Ecol 28:33–43

Brizzio S, van Broock M (1998) Characterization of wild yeast killer from Nahuel Huapi National Park (Patagonia, Argentina). Food Technol Biotechnol 4:273–278

Burgaud G, Arzur D, Durand L, Cambon-Bonavita MA, Barbier G (2010) Marine culturable yeasts in deep-sea hydrothermal vents: species richness and association with fauna. FEMS Microbiol Ecol 73:121–133

Burgaud G, Arzur D, Sampaio JP, Barbier G (2011) Candida oceani sp. nov., a novel yeast isolated from a Mid-Atlantic Ridge hydrothermal vent (-2300 meters). Antonie van Leeuwenhoek 100:75–82

Burgaud G, Coton M, Jacques N, Debaets S, Maciel NO, Rosa CA, Gadanho M, Sampaio JP, Casaregola S (2016) *Yamadazyma barbieri* f.a. sp. nov., an ascomycetous anamorphic yeast isolated from a Mid-Atlantic Ridge hydrothermal site (−2300 m) and marine coastal waters. Int J Syst Evol Microbiol 66:3600–3606

Buzzini P, Branda E, Goretti M, Turchetti B (2012) Psychrophilic yeasts from worldwide glacial habitats: diversity, adaptation strategies and biotechnological potential. FEMS Microbiol Ecol 82:217–241

Canelhas MR, Barbosa AC, Medeiros AO, Lee CF, Huang LY, Lachance MA, Rosa CA (2011) *Saturnispora serradocipensis* sp. nov. and *Saturnispora gosingensis* sp. nov., tow ascomycetous yeasts from ephemeral habitats. A van Leeuwenhoek 99:241–247

Carneiro MT, Silva DM, Chagas TGP, Zahner V, Asensi MD, Hagler AN (2015) Complementary bioindicators to colorimetric assays in the water quality analysis: the potential of yeasts in Juturnaíba lake/Brazil. Sistem Gestão 10:542–552

Chang CF, Chen CC, Lee CF, Liu SM (2013) Identifying and characterizing *Yarrowia keelungensis* sp. nov., an oil-degrading yeast isolated from the sea surface microlayer. A van Leeuwenhoek 104:1117–1123

Chen Y-S, Yanagida F, Chen LY (2009) Isolation of marine yeasts from coastal waters of northeastern Taiwan. Aquat Biol 8:55–60

Coelho MA, Almeida JM, Martins IM, da Silva AJ, Sampaio JP (2010) The dynamics of the yeast community of the Tagus river estuary: testing the hypothesis of the multiple origins of estuarine yeasts. A van Leeuwenhoek 98:331–342

Colabella F, Moliné M, Libkind D (2014) UV Sunscreens of microbial origin: mycosporines and mycosporine-like aminoacids. Recent Pat Biotechnol 8:189–193

Cook SJ, Kougkoulos I, Edwards LA, Dortch J, Hoffmann D (2016) Glacier change and glacial lake outburst flood risk in the Bolivian Andes. Cryosphere 10:2399–2413

Cooke WEB (1976) Fungi in sewages. In: Jones EBG (ed) Recent advances in aquatic mycology. Wiley, New York, pp 389–434

Coronato F (2016) Introducing Patagonia: an approach from geo-history. In: Olivera N, Libkind D, Donati R (eds) Biology and biotechnology of patagonian microorganisms. Springer, pp 1–14

de Almeida JM (2005) Yeast community survey in the Tagus estuary. FEMS Microbiol Ecol 53:295–303

de García V, Brizzio S, Libkind D, Buzzini P, van Broock M (2007) Biodiversity of cold-adapted yeasts from glacial meltwater rivers in Patagonia, Argentina. FEMS Microbiol Ecol 59:331–341

de García V, Brizzio S, Libkind D, Rosa CA, van Broock M (2010a) *Wickerhamomyces patagonicus* sp. nov., an ascomycetous yeast species from Patagonia, Argentina. Int J Syst Evol Microbiol 60:1693–1696

de García V, Brizzio S, Russo G, Rosa CA, Boekhout T, Theelen B, Libkind D, van Broock M (2010b) *Cryptococcus spencermartinsiae* sp. nov., a basidiomycetous yeast isolated from glacial waters and apple fruits. Int J Syst Evol Microbiol 60:707–711

de García V, Brizzio S, van Broock MR (2012) Yeasts from glacial ice of Patagonian Andes, Argentina. FEMS Microbiol Ecol 82:540–550

de García V, Libkind D, Moliné M, Rosa CA, Giraudo MR (2014) Cold-adapted yeasts in Patagonian habitats. In: Buzzini P, Margesin R (eds) Cold adapted yeasts. Springer, Berlin, Heidelberg, pp 123–148

de García V, Coelho MA, Maia TM, Rosa LH, Vaz AM, Rosa CA, Sampaio JP, Goncalves P, van Broock M, Libkind D (2015) Sex in the cold: taxonomic reorganization of psychrotolerant yeasts in the order Leucosporidiales. FEMS Yeast Res 15:1–11

Díaz M, Pedrozo F, Baccala N (2000) Summer classification of Southern Hemisphere temperate lakes (Patagonia, Argentina). Lakes Reservoirs 5:213–229

Donachie SP, Zdanowski MK (1998) Potential digestive function of bacteria in krill *Euphausia superba* stomach. Aquat Microb Ecol 14:129–136

Dong X, Jiang W, Li C, Ma N, Xu Y, Meng X (2015) Patulin biodegradation by marine yeast *Kodameae ohmeri*. Food Addit Contam Part A 32:352–360

Ebert D, Lipsitch M, Mangin KL (2000a) The effect of parasites on host population density and extinction: experimental epidemiology with *Daphnia* and six micro-parasites. Am Nat 156:459–477

Ebert D, Zschokke-Rohringer CD, Carius HJ (2000b) Dose effects and density-dependent regulation of two microparasites of *Daphnia magna*. Oecologia 122:200–209

Ekendahl S, O'Neill AH, Thomsson E, Pedersen K (2003) Characterisation of yeasts isolated from deep igneous rock aquifers of the Fennoscandian Shield. Microb Ecol 46:416–428

Fell JW (1961) A new species of *Saccharomyces* isolated from a subtropical estuary. A van Leeuwenhoek 27:27–30

Fell JW (1967) Distribution of yeasts in the Indian Ocean. Bull Mar Sci 17:454–470

Fell JW (1970) The genus *Sterigmatomyces*. In: Lodder J (ed) The yeasts, a taxonomic study. North Holland, Amsterdam, pp 1229–1234

Fell JW (2012) Yeast in marine environments. In: Jones EBG, Pang K-L (eds) Marine fungi and fungal-like microorganisms marine and freshwater botany. de Gruyter, Berlin, Boston, pp 91–102

Fell JW, Hunter IL (1968) Isolation of heterothallic yeast strains of *Metschnikowia* Kamienski and their mating reactions with *Chlamydozyma* Wickerham spp. A van Leeuwenhoek 34:365–376
Fell JW, Statzell-Tallman A (1971) *Sympodiomyces* gen. n., a yeast-like organism from southern marine waters. A van Leeuwenhoek 37:359–367
Fell JW, Statzell-Tallman A, Scorzetti G, Gutiérrez MH (2011) Five new species of yeasts from fresh water and marine habitats in the Florida Everglades. A van Leeuwenhoek 99:533–549
Fischer B, Brebeck C (1894) Zur Morphologie, Biologie, und Systematik der Kahmpilze, der *Monilia candida* Hansen und des Soorerregers. C. Fischer, Jena
Fonseca A, Boekhout T, Fell JW (2011) *Cryptococcus* Vuillemin (1901). In: Kurtzman CP, Fell JW, Boekhout T (eds) The yeasts, a taxonomic study, 5th edn. Elsevier, Amsterdam, pp 1661–1738
Frisvad JC (2008) Fungi in cold ecosystems. In: Margesin R, Schinner F, Marx JC, Gerday C (eds) Psychrophiles: from biodiversity to biotechnology. Springer, Berlin, Heidelberg, pp 137–156
Furbino LE, Godinho VM, Santiago IF, Pellizzari FM, Alves TM, Zani CL, Jinion PA, Romanha AJ, Carvalho AG, Gil LH, Rosa CA, Minnis AM, Rosa LH (2014) Diversity patterns, ecology and biological activities of fungal communities associated with the endemic macroalgae across the Antarctic peninsula. Microb Ecol 67:775–787
Gadanho M, Sampaio JP (2002) Polyphasic taxonomy of the basidiomycetous yeast genus *Rhodotorula*: *Rh. glutinis sensu stricto* and *Rh. dairenensis* comb. nov. FEMS Yeast Res 2:47–58
Gadanho M, Sampaio JP (2004) Application of temperature gradient gel electrophoresis to the study of yeast diversity in the estuary of the Tagus river, Portugal. FEMS Yeast Res 5:253–261
Gadanho M, Sampaio JP (2005) Occurrence and diversity of yeasts in the Mid-Atlantic Ridge hydrothermal fields near the Azores archipelago. Microb Ecol 50:408–417
Gadanho M, Sampaio JP (2009) *Cryptococcus ibericus* sp. nov., *Cryptococcus aciditolerans* sp. nov. and *Cryptococcus metallitolerans* sp. nov., a new ecoclade of anamorphic basidiomycetous yeast species from an extreme environment associated with acid rock drainage in Sao Domingos pyrite mine, Portugal. Int J Syst Evol Microbiol 59:2375–2379
Gadanho M, Almeida JMF, Sampaio JP (2003) Assessment of yeast diversity in a marine environment in the south of Portugal by microsatellite-primed PCR. A van Leeuwenhoek 84:217–227
Gadanho M, Libkind D, Sampaio JP (2006) Yeast diversity in the extreme acidic environments of the Iberian Pyrite Belt. Microb Ecol 52:552–563
Gao L, Chi Z, Sheng J, Wang L, Li J, Gong F (2007) Inulinase-producing marine yeasts: evaluation of their diversity and inulin hydrolysis by their crude enzymes. Microb Ecol 54:722–729
Gao Z, Li B, Zheng C, Wang G (2008) Molecular detection of fungal communities in the Hawaiian marine sponges *Suberites zeteki* and *Mycale armata*. Appl Environ Microbiol 74:6091–6101
Godinho VM, Furbino LE, Santiago IF, Pellizzari FM, Yokoya NS, Pupo D, Alves TM, Junior PA, Romanha AJ, Zani CL, Cantrell CL, Rosa CA, Rosa LH (2013) Diversity and bioprospecting of fungal communities associated with endemic and cold-adapted macroalgae in Antarctica. ISME J 7:1434–1451
Goh TK, Hyde KD (1996) Biodiversity of freshwater fungi. J Ind Microbiol 17:328–345
Guo FJ, Ma Y, Xu HM, Wang XH, Chi ZM (2013) A novel killer toxin produced by the marine-derived yeast *Wickerhamomyces anomalus* YF07b. A van Leeuwenhoek 103:737–746
Gupta A, Vongsvivut J, Barrow CJ, Puri M (2012) Molecular identification of marine yeast and its spectroscopic analysis establishes unsaturated fatty acid accumulation. J Biosci Bioeng 114:411–417
Hagler AN, Ahearn DG (1987) Ecology of aquatic yeasts. In: Rose AH, Harrison JS (eds) The yeasts, Yeasts and the environment, vol 2. Academic Press, London, pp 181–205
Hagler AN, Mendonça-Hagler LC (1981) Yeasts from marine and estuarine waters with different levels of pollution in the state of Rio de Janeiro, Brazil. Appl Environ Microbiol 41:173–178

Hagler AN, Mendonça-Hagler LC, Santos EA, Farage S, Silva Filho JB, Schrank A, de Oliveira RB (1986) Microbial pollution indicators in Brazilian tropical and subtropical marine surface waters. Sci Total Environ 15:151–160

Hassanshahian M, Tebyanian H, Cappello S (2012) Isolation and characterization of two crude oil-degrading yeast strains, *Yarrowia lipolytica* PG-20 and PG-32, from the Persian Gulf. Mar Pollut Bull 64:1386–1391

Inácio J, Pereira P, de Carvalho M, Fonseca Á, Amaral-Collaço MT, Spencer-Martins I (2002) Estimation and diversity of phylloplane mycobiota on selected plants in a Mediterranean-type ecosystem in Portugal. Microb Ecol 44:344–353

Ishida S, Nozaki D, Grossart HP, Kagami M (2015) Novel basal, fungal lineages from freshwater phytoplankton and lake samples. Environ Microbiol Rep 7:435–441

Jebaraj CS, Raghukumar C, Behnke A, Stoeck T (2010) Fungal diversity in oxygen-depleted regions of the Arabian Sea revealed by targeted environmental sequencing combined with cultivation. FEMS Microbiol Ecol 71:399–412

Jin L, Liu F, Sun W, Zhang F, Karuppiah V, Li Z (2014) Pezizomycotina dominates the fungal communities of South China Sea sponges *Theonella swinhoei* and *Xestospongia testudinaria.* FEMS Microbiol Ecol 90:935–945

Jones EBG, Slooff WC (1966) *Candida aquatica* sp. nov. isolated from water scums. A van Leeuvenhoek 32:223–228

Jones G, Southworth EB, Libkind D, Marvanová L (2014) Freshwater basidiomycota. In: Jones G, Pang K-L, Hyde K (eds) Freshwater fungi and fungus-like organisms. De Gruyter, Berlin, pp 73–108

Kandasamy K, Alikunhi NM, Subramanian M (2012) Yeasts in marine and estuarine environments. J Yeast Fungal Res 3:74–82

Kurtzman CP, Fell JW (2004) Yeast. In: Muller GM, Bills GF, Foster MS (eds) Biodiversity of fungi. Elsevier Academic Press, London, pp 337–342

Kurtzman CP, Fell JW (2006) Yeast systematics and phylogeny-implications of molecular identification methods for studies in ecology. In: Rosa CA, Péter G (eds) Biodiversity and ecophysiology of yeasts. Springer, Berlin, pp 11–30

Kutty SN, Philip R (2008) Marine yeasts – a review. Yeast 25:465–483

Kwasniewska K (1988) Horizontal distribution and density of yeasts and filamentous fungi in lake St. Clair water. J Great Lakes Res 14:438–443

Lachance MA (2011a) *Kluyveromyces* van der Walt (1971). In: Kurtzman CP, Fell JW (eds) The yeasts, a taxonomic study. Elsevier, Amsterdam, pp 471–482

Lachance MA (2011b) *Metschnikowia* Kamienski (1899). In: Kurtzman CP, Fell JW (eds) The yeasts, a taxonomic study. Elsevier, Amsterdam, pp 575–621

Lachance MA, Starmer WT (1998) Ecology and yeasts. In: Kurtzman CP, Fell JW (eds) The yeasts, a taxonomic study. Elsevier, Amsterdam, pp 21–30

Last FT, Price D (1969) Yeasts associated with living plants and their environs. In: Rose AH, Harrison JS (eds) The yeasts. Academic Press, New York, pp 183–218

Le Calvez T, Burgaud G, Mahé S, Barbier G, Vandenkoornhuyse P (2009) Fungal diversity in deep-sea hydrothermal ecosystems. Appl Environ Microbiol 75:6415–6421

Lefèvre E, Bardot C, Noël C, Carrias JF, Viscogliosi E, Amblard C, Sime-Ngando T (2007) Unveiling fungal zooflagellates as members of freshwater picoeukaryotes: evidence from a molecular diversity study in a deep meromictic lake. Environ Microbiol 9:61–71

Lewis WM (2000) Basis for the protection and management of tropical lakes. Lakes Reserv Res Manag 5(1):35–48

Li X, Chi Z, Liu Z, Yan K, Li H (2008) Phytase production by a marine yeast *Kodamaea ohmeri* BG3. Appl Biochem Biotechnol 149:183–193

Li J, Peng Y, Wang X, Chi Z (2010a) Optimum production and characterization of an acid protease from marine yeast *Metschnikowia reukaufii* W6b. J Ocean Univ China 9:359–364

Li M, Liu G-L, Chi Z, Chi Z-M (2010b) Single cell oil production from hydrolysate of cassava starch by marine-derived yeast *Rhodotorula mucilaginosa* TJY15a. Biomass Bioenergy 34:101–107

Libkind D, Brizzio S, Ruffini A, Gadanho M, van Broock MR, Sampaio JP (2003) Molecular characterization of carotenogenic yeasts from aquatic environments in Patagonia, Argentina. A van Leeuwenhoek 84:313–322

Libkind D, Brizzio S, van Broock M (2004a) *Rhodotorula mucilaginosa*, a carotenoid producing yeast strain from a Patagonian high-altitude lake. Folia Microbiol 49:19–25

Libkind D, Perez P, Sommaruga R, Dieguez MC, Ferraro M, Brizzio S, Zagarese H, van Broock M (2004b) Constitutive and UV-inducible synthesis of photoprotective compounds (carotenoids and mycosporines) by freshwater yeasts. Photochem Photobiol Sci 3:281–286

Libkind D, Sommaruga R, Zagarese H, van Broock M (2005a) Mycosporines in carotenogenic yeasts. Syst Appl Microbiol 28:749–754

Libkind D, Gadanho M, van Broock M, Sampaio JP (2005b) *Sporidiobolus longiusculus* sp. nov. and *Sporobolomyces patagonicus* sp. nov., novel yeasts of the Sporidiobolales isolated from aquatic environments in Patagonia, Argentina. Int J Syst Evol Microbiol 55:503–509

Libkind D, Dieguez MC, Moliné M, Perez P, Zagarese HE, van Broock M (2006) Occurrence of photoprotective compounds in yeasts from freshwater ecosystems of northwestern Patagonia (Argentina). Photochem Photobiol 82:972–980

Libkind D, Ruffini A, van Broock M, Alves L, Sampaio JP (2007) Biogeography, host-specificity, and molecular phylogeny of *Phaffia rhodozyma* and its sexual form, *Xanthophyllomyces dendrorhous*. Appl Environ Microbiol 73:1120–1125

Libkind D, Gadanho M, van Broock M, Sampaio JP (2008) Studies on the heterogeneity of the carotenogenic yeast from Patagonia, Argentina. J Basic Microbiol 48(2):93–98

Libkind D, Moliné M, Sampaio JP, van Broock M (2009a) Yeasts from high-altitude lakes: influence of UV radiation. FEMS Microbiol Ecol 69:353–362

Libkind D, Gadanho M, van Broock M, Sampaio JP (2009b) *Cystofilobasidium lacus-mascardii* sp. nov., a basidiomycetous yeast species isolated from aquatic environments of the Patagonian Andes, and *Cystofilobasidium macerans* sp. nov., the sexual stage of *Cryptococcus macerans*. Int J Syst Evol Microbiol 59:622–630

Libkind D, Sampaio JP, van Broock M (2010) Cystobasidiomycetes yeasts from Patagonia (Argentina): description of *Rhodotorula meli* sp. nov. from glacial meltwater. Int J Syst Evol Microbiol 60:2251–2256

Libkind D, Hittinger CT, Valerio E, Goncalves C, Dover J, Johnston M, Goncalves P, Sampaio JP (2011a) Microbe domestication and the identification of the wild genetic stock of lager-brewing yeast. Proc Natl Acad Sci U S A 108:14539–14544

Libkind D, Moliné M, Sommaruga R, Sampaio JS, van Broock M (2011b) Phylogenetic distribution of fungal mycosporines within Pucciniomycotina (Basidiomycota). Yeast 28:619–627

Libkind D, Moliné M, van Broock M (2011c) Production of the UVB absorbing compound Mycosporine-glutaminol-glucoside by *Xanthophyllomyces dendrorhous* (*Phaffia rhodozyma*). FEMS Yeast Res 11:52–59

Libkind D, Tognetti C, Ruffini A, Sampaio JP, van Broock M (2011d) *Xanthophyllomyces dendrorhous* (*Phaffia rhodozyma*) on stromata of *Cyttaria hariotii* in Patagonian *Nothofagus* forests. Rev Argent Microbiol 43:226–232

Libkind D, Moliné M, Bellora N, Trochine A, de García V (2016) Patagonian yeasts of biotechnological relevance. In: Olivera N, Libkind D, Donati R (eds) Biology and biotechnology of patagonian microorganisms. Springer, pp 325–351

Lorenz R, Molitoris HP (1997) Cultivation of fungi under simulated deep sea conditions. Mycol Res 101:1355–1365

Loureiro STA, de Queiroz-Cavalcanti MA, Neves RP, de Oliveira Passavante JZ (2005) Yeasts isolated from sand and sea water in beaches of Olinda, Pernambuco state, Brazil. Braz J Microbiol 36:1–8

Madhour A, Anke H, Mucci A, Davoli P, Weber RW (2005) Biosynthesis of the xanthophyll plectaniaxanthin as a stress response in the red yeast *Dioszegia* (Tremellales, Heterobasidiomycetes, Fungi). Phytochemistry 66:2617–2626
Medeiros AO, Kohler LM, Hamdan JS, Missagia BS, Barbosa FA, Rosa CA (2008) Diversity and antifungal susceptibility of yeasts from tropical freshwater environments in Southeastern Brazil. Water Res 42:3921–3929
Medeiros AO, Missagia BS, Brandão LR, Callisto M, Barbosa FA, Rosa CA (2012) Water quality and diversity of yeasts from tropical lakes and rivers from the Rio Doce basin in Southeastern Brazil. Braz J Microbiol 43:1582–1594
Mestre MC, Rosa CA, Safar SV, Libkind D, Fontenla SB (2011) Yeast communities associated with the bulk-soil, rhizosphere and ectomycorrhizosphere of a *Nothofagus pumilio* forest in northwestern Patagonia, Argentina. FEMS Microbiol Ecol 78:531–541
Meyers SP, Ahearn DG, Cooke WL (1970) Mycological studies of lake Champlain. Mycologia 62:504–515
Meyers SP, Ahearn DG, Alexander SK, Cook WL (1975) *Pichia spartinae*, dominant yeast of the Spartina salt marsh. Dev Ind Microbiol 16:262–267
Miller MW, Phaff HJ (1998) Metschnikowia kamienski. In: Kurtzman CP, Fell JW (eds) The yeasts, a taxonomic study. Elsevier, Amsterdam, pp 256–267
Modenutti BE, Balseiro EG, Queimaliños CP, Añon Suarez D, Diéguez MC, Albariño RJ (1998) Structure and dynamics of food web Andean lakes. Lakes Reserv Res Manag 3:179–186
Moliné M (2004) Carotenogénesis: efecto de la radiación ultravioleta sobre levaduras pigmentadas. Undergraduate Thesis. Universidad Nacional del Comahue, Bariloche, Argentina
Moliné M (2010) Producción de compuestos fotoprotectores (carotenoides y micosporinas) por levaduras. PhD thesis, Universidad Nacional de Tucumán, Tucumán, Argentina
Moliné M, Libkind D, Dieguez M d C, van Broock M (2009) Photoprotective role of carotenoids in yeasts: response to UV-B of pigmented and naturally-occurring albino strains. J Photochem Photobiol B 95:156–161
Moliné M, Flores MR, Libkind D, Dieguez M d C, Farias ME, van Broock M (2010) Photoprotection by carotenoid pigments in the yeast *Rhodotorula mucilaginosa*: the role of torularhodin. Photochem Photobiol Sci 9:1145–1151
Moliné M, Arbeloa EM, Flores MR, Libkind D, Farías ME, Bertolotti SG, Churio MS, van Broock MR (2011) UVB photoprotective role of mycosporines in yeast: photostability and antioxidant activity of mycosporine-glutaminol-glucoside. Radiat Res 175:44–50
Monchy S, Sanciu G, Jobard M, Rasconi S, Gerphagnon M, Chabé M, Cian A, Meloni D, Niquil N, Christaki U, Viscogliosi E, Sime-Ngando T (2011) Exploring and quantifying fungal diversity in freshwater lake ecosystems using rDNA cloning/sequencing and SSU tag pyrosequencing. Environ Microbiol 13:1433–1453
Moore MM, Strom MS (2003) Infection and mortality by the yeast *Metschnikowia bicuspidate* var. *bicuspidate* in Chinook salmon fed live adult brine shrimp (*Artemia franciscana*). Aquaculture 220:43–57
Morais P, Resende MA, Rosa CA, Barbosa FAR (1996) Occurrence and diel distribution of yeast in a paleo-karstic lake of Southeastern Brazil. Rev Microbiol 27:182–188
Mouton M, Postma F, Wilsenach J, Botha A (2012) Diversity and characterization of culturable fungi from marine sediment collected from St. Helena Bay, South Africa. Microb Ecol 64:311–319
Nagahama T (2006) Yeast biodiversity in freshwater, marine and deep-sea environments. In: Rosa CA, Péter G (eds) Biodiversity and ecophysiology of yeasts. Springer, Berlin, pp 241–262
Nagahama T, Hamamoto M, Nakase T, Horikoshi K (1999) *Kluyveromyces nonfermentans* sp. nov., a new yeast species isolated from the deep sea. Int J Syst Bacteriol 49:1899–1905
Nagahama T, Hamamoto M, Nakase T, Takami H, Horikoshi K (2001a) Distribution and identification of red yeasts in deep-sea environments around the northwest Pacific Ocean. A van Leeuwenhoek 80:101–110

Nagahama T, Hamamoto M, Nakase T, Horikoshi K (2001b) *Rhodotorula lamellibrachii* sp. nov., a new yeast species from a tubeworm collected at the deep-sea floor in Sagami bay and its phylogenetic analysis. A van Leeuwenhoek 80:317–323

Nagahama T, Hamamoto M, Nakase T, Horikoshi K (2003a) *Rhodotorula benthica* sp. nov. and *Rhodotorula calyptogenae* sp. nov., novel yeast species from animals collected from the deep-sea floor, and *Rhodotorula lysiniphila* sp. nov., which is related phylogenetically. Int J Syst Evol Microbiol 53:897–903

Nagahama T, Hamamoto M, Nakase T, Takaki Y, Horikoshi K (2003b) *Cryptococcus surugaensis* sp. nov., a novel yeast species from sediment collected on the deep-sea floor of Suruga Bay. Int J Syst Evol Microbiol 53:2095–2098

Nagahama T, Hamamoto M, Horikoshi K (2006) *Rhodotorula pacifica* sp. nov., a novel yeast species from sediment collected on the deep-sea floor of the north-west Pacific Ocean. Int J Syst Evol Microbiol 56:295–299

Nagano Y, Nagahama T, Hatada Y, Nonoura T, Takami H, Miyazaki J, Takai K, Horikoshi K (2010) Fungal diversity in deep-sea sediments – the presence of novel fungal groups. Fungal Ecol 3:316–325

Nagy E, Niss M, Dlauchy D, Arneborg N, Nielsen DS, Péter G (2013) *Yarrowia divulgata* f.a., sp. nov., a yeast species from animal-related and marine sources. Int J Syst Evol Microbiol 63:4818–4823

Nagy E, Dlauchy D, Medeiros AO, Péter G, Rosa CA (2014) *Yarrowia porcina* sp. nov. and *Yarrowia bubula* f.a., sp. nov., two yeast species form meat and river sediment. A van Leeuwenhoek 105:697–707

Oren A, Gunde-Cimerman N (2007) Mycosporines and mycosporine-like amino acids: UV protectants or multipurpose secondary metabolites? FEMS Microbiol Lett 269:1–10

Pagnocca FC, Mendonca-Hagler LC, Hagler AN (1989) Yeasts associated with the white shrimp *Penaeus schmitti*, sediment and water of Sepetiba Bay, Rio de Janeiro, Brazil. Yeast 5:5479–5483

Papadakis JS, Mavridou A, Richardson SC, Lamprini M, Marcelou U (1997) Bather-related microbial and yeast populations in sand and seawater. Water Res 31:799–804

Pedrozo F, Chillrud S, Temporetti P, Diaz M (1993) Chemical composition and nutrient limitation in rivers and lakes of northern Patagonian Andes (39.5° – 42° S; 71° W) (Rep. Argentina). Verh Internat Verein Limnol 25:207–214

Pore RS, Sorenson WG (1990) *Reniforma strues*, a new yeast from wastewater. Mycologia 82:549–553

Prabhakaran N, Gupta R (1991) Yeasts from the sediment samples of the EEZ along the southwest coast of India. J Mar Biol Assoc India 33:455

Quirós R, Drago E (1985) Relaciones entre variables físicas, morfométricas y climáticas en lagos patagónicos. Rev Asoc Cienc Nat Litoral 16:181–199

Reche I, Pulido-Villena E, Morales-Baquero R, Casamayor EO (2005) Does ecosystem size determine aquatic bacterial richness? Ecology 86:1715–1722

Rédou V, Navarri M, Meslet-Cladière L, Barbier G, Burgaud G (2015) Species richness and adaptation of marine fungi from deep-subseafloor sediments. Appl Environ Microbiol 81:3571–3583

Rosa CA, Resende MA, Franzot SP, Morais PB, Barbosa FAR (1990) Yeasts and coliform distribution in a paleokarstic lake of Lagoa-Santa Plateau-MG, Brazil. Rev Microbiol 21:19–24

Rosa CA, Resende MA, Barbosa FAR, Morais PB, Franzot SP (1995) Yeast diversity in a mesotrophic lake on the karstic plateau of Lagoa Santa, MG-Brazil. Hydrobiology 308:103–108

Roth FJ, Ahearn DG, Fell JW, Meyew SP, Meyer SA (1962) Ecology and taxonomy of yeasts isolated from various marine substrates. Limnol Oceanogr 2:178–185

Russo G, Libkind D, Sampaio JP, van Broock MR (2008) Yeast diversity at the Volcanic acidic environment of the Lake Caviahue and Rio Agrio (Patagonia, Argentina). FEMS Microbiol Ecol 65:415–424

Russo G, Libkind D, Ulloa RJ, de García V, Sampaio JP, van Broock MR (2010) *Cryptococcus agrionensis* sp. nov., a basidiomycetous yeast of the acidic rock drainage ecoclade, isolated from acidic aquatic environments of volcanic origin (River Agrio, Argentina). Int J Syst Evol Microbiol 60:996–1000

Russo G, Libkind D, Giraudo MR, Delgado O (2016) Metal capture by autochthonous yeasts from a volcanic influenced environment of Patagonia. J Basic Microbiol 56:1203–1211

Sabino R, Verissimo C, Cunha MA, Wergikoski B, Ferreira FC, Rodrigues R, Prada H, Falcao L, Rosado L, Pinheiro C, Paixao E, Brandão J (2011) Pathogenic fungi: an unacknowledged risk at coastal resorts? New insights on microbiological sand quality in Portugal. Mar Pollut Bull 62:1506–1511

Sampaio JP (2004) Diversity, phylogeny and classification of basidiomycetous yeasts. In: Agerer R, Piepenbring M, Blanz P (eds) Frontiers in Basidiomycota mycology. IHW Verlag, Eching, Germany, pp 49–80

Sampaio JP (2011) *Rhodosporidium* Banno (1967). In: Kurtzman CP, Fell JW (eds) The yeasts, a taxonomic study. Elsevier, Amsterdam, pp 1523–1539

Sampaio A, Sampaio JP, Leão C (2007) Dynamics of yeast populations recovered from decaying leaves in a nonpolluted stream: a 2-year study on the effects of leaf litter type and decomposition time. FEMS Yeast Res 7:595–603

Seki H, Fulton J (1969) Infection of marine copepods by *Metschnikowia* sp. Mycopathol Mycol Appl 38:61–70

Seshadri R, Sieburth J (1971) Cultural estimation of yeasts on seaweeds. Appl Microbiol 22:507–512

Shearer CA, Descals E, Kohlmeyer B, Kohlmeyer J, Marvanová L, Padgett D, Porter D, Raja HA, Schmit JP, Thorton HA, Voglymayr H (2007) Fungal biodiversity in aquatic habitats. Biodivers Conserv 16:49–67

Shick JM, Dunlap WC (2002) Mycosporine-like amino acids and related gadusols: biosynthesis, accumulation, and UV-protective functions in aquatic organisms. Annu Rev Physiol 64:223–262

Siepmann R, Hohnk W (1962) Uber Hefen und einige Pilze (Fungi imp., Hyphales) aus dem Nordatlantik. Veroeff Inst Meeresforsch Bremerhaven 8:79–97

Silva-Bedoya LM, Ramírez-Castrillón M, Osório-Cadavid E (2014) Yeast diversity associated to sediments and water from two Colombian artificial lakes. Braz J Microbiol 45:135–142

Simard RE, Blackwood AC (1971a) Ecological studies on yeasts in the St Lawrence River. Can J Microbiol 17:353–357

Simard RE, Blackwood AC (1971b) Yeasts from the St. Lawrence River. Can J Microbiol 17:197–203

Singh P, Raghukumar C, Meena RM, Verma P, Shouche Y (2012a) Fungal diversity in deep-sea sediments revealed by culture-dependent and culture-independent approaches. Fungal Ecol 5:543–553

Singh P, Raghukumar C, Verma P, Shouche Y (2012b) Assessment of fungal diversity in deep-sea sediments by multiple primer approach. World J Microbiol Biotechnol 28:659–667

Sláviková E, Vadkertiová R (1997) Seasonal occurrence of yeasts and yeast-like organisms in the river Danube. A van Leeuwenhoek 72:77–80

Slávikova E, Vadkertiova R (2000) The occurrence of yeasts in the forest soils. J Basic Microbiol 40:207–212

Sláviková E, Vadkertiová R, Kocková-Kratochvílová A (1992) Yeasts isolated from artificial lake waters. Can J Microbiol 38:1206–1209

Soares CAG, Maury M, Pagnocca FC, Araujo FV, Mendonca-Hagler LC, Hagler AN (1997) Ascomycetous yeasts from tropical intertidal dark mud of southeast Brazilian estuaries. J Gen Appl Microbiol 43:265–272

Sommaruga R, Libkind D, van Broock M, Whitehead K (2004) Mycosporine-glutaminolglucoside, a UV-absorbing compound of two *Rhodotorula* yeast species. Yeast 21:1077–1081

Spencer JFT, Spencer DM (1997) Yeasts in natural and artificial habitat. Springer, Berlin

Statzell-Tallman A, Belloch C, Fell JW (2008) *Kwoniella mangroviensis* gen. nov., sp. nov. a tremellaceous yeast from mangrove habitats in the Florida Everglades and Bahamas. FEMS Yeast Res 8:103–113

Stone W, Jones BL, Wilsenach J, Botha A (2012) External ecological niche for *Candida albicans* within reducing, oxygen-limited zones of wetlands. Appl Environ Microbiol 78:2443–2445

Takishita K, Tsuchiya M, Reimer JD, Maruyama T (2006) Molecular evidence demonstrating the basidiomycetous fungus *Cryptococcus curvatus* is the dominant microbial eukaryote in sediment at the Kuroshima Knoll methane seep. Extremophiles 10:165–169

Takishita K, Yubuki N, Kakizoe N, Inagaki Y, Maruyama T (2007) Diversity of microbial eukaryotes in sediment at a deep-sea methane cold seep: surveys of ribosomal DNA libraries from raw sediment samples and two enrichment cultures. Extremophiles 11:563–576

Torres A, Hochberg M, Pergament I, Smoum R, Niddam V, Dembitsky VM, Temina M, Dor I, Lev O, Srebnik M, Enk CD (2004) A new UV-B absorbing mycosporine with photo protective activity from the lichenized ascomycete *Collema cristatum*. Eur J Biochem 271:780–784

Valente P, Boekhout T, Landell MF, Crestani J, Pagnocca FC, Sette LD, Passarini MR, Rosa CA, Brandão LR, Pimenta RS, Ribeiro JR, García KM, Lee CF, Suh SO, Péter G, Dlauchy D, Fell JW, Scorzetti G, Theelen B, Vainstein MH (2012) *Bandoniozyma* gen. nov., a genus of fermentative and non-fermentative tremellaceous yeast species. PLoS One 7:e46060

van Uden N, Ahearn DC (1963) Occurrence and population densities of yeast species in a freshwater lake. A van Leeuwenhoek 29:308–312

van Uden N, Castelo-Branco R (1961) *Metschnikowiella zobellii* sp. nov. and *M. krissii* sp. nov. two yeasts from the Pacific Ocean pathogenic for *Daphnia magna*. J Gen Microbiol 26:141–148

van Uden N, ZoBell CE (1962) *Candida marina* nov. spec., *Torulopsis torresii* nov. spec. and *T. maris* nov. spec., three yeasts from the Torres Strait. A van Leeuwenhoek 28:275–283

Vaz ABM, Rosa LH, Vieira MLA, de Garcia V, Brandão LR, Teixeira LCRS, Moliné M, Libkind D, van Broock M, Rosa CA (2011) The diversity, extracellular enzymatic activities and photoprotective compounds of yeasts isolated in Antarctica. Braz J Microbiol 42:937–947

Velegraki-Abel V, Marselou-Kinti U, Richardson D (1987) Incidence of yeasts in coastal sea water of the Attica Peninsula, Greece. Water Res 21:1363–1369

Villafañe VE, Helbling WE, Zagarese HE (2001) Solar ultraviolet radiation and its impact on aquatic systems of Patagonia, South America. Ambio 30:112–117

Vogel C, Rogerson A, Schatz S, Laubach H, Tallman A, Fell J (2007) Prevalence of yeasts in beach sand at three bathing beaches in South Florida. Water Res 41:1915–1920

Wang L, Chi Z, Wang X, Ju L, Chi Z, Guo N (2008) Isolation and characterization of *Candida membranifaciens* subsp. *flavinogenie* W14-3, a novel riboflavin-producing marine yeast. Microbiol Res 163:255–266

Wurzbacher C, Bärlocher F, Grossart H-P (2010) Fungi in lake ecosystems. Aquat Microb Ecol 59:125–149

Xu W, Pang KL, Luo ZH (2014) High fungal diversity and abundance recovered in the deep-sea sediments of the Pacific Ocean. Microb Ecol 68:688–698

Xu W, Luo Z-H, Guo S, Pang K-L (2016) Fungal community analysis in the deep-sea sediments of the Pacific Ocean assessed by comparison of ITS,18S and 28S ribosomal DNA regions. Deep-Sea Res 109:51–60

Yamaguchi K, Degawa Y, Nakagiri A (2009) An aero-aquatic fungus, *Peyronelina glomerulata*, is shown to have teleomorphic affinities with cyphelloid basidiomycetes. Mycoscience 50:156–164

Yamasato K, Goto S, Ohwada K, Okuno D, Araki H, Iizuka H (1974) Yeasts from the Pacific Ocean. J Gen Appl Microbiol 20:289–307

Yang S-P, Wu Z-H, Jian J-C (2011) Distribution of marine red yeasts in shrimps and the environments of shrimp culture. Curr Microbiol 62:1638–1642

Yurkov AM (2017) Temporal and geographic patterns in yeast distribution. In: Buzzini P, Lachance MA, Yurkov AM (eds) Yeasts in natural ecosystems: ecology. Springer, Heidelberg, pp 101–130

Yurkov AM, Pozo MI (2017) Yeast community composition and structure. In: Buzzini P, Lachance MA, Yurkov AM (eds) Yeasts in natural ecosystems: ecology. Springer, Heidelberg, pp 73–100

Yurkov AM, Kachalkin AV, Daniel HM, Groenewald M, Libkind D, de García V, Zalar P, Gouliamova DE, Boekhout T, Begerow D (2015) Two yeast species *Cystobasidium psychroaquaticum* f.a. sp. nov. and *Cystobasidium rietchieii* f.a. sp. nov. isolated from natural environments, and the transfer of *Rhodotorula minuta* clade members to the genus *Cystobasidium*. A van Leeuwenhoek 107:173–185

Zhang XY, Tang GL, Xu XY, Nong XH, Qi SH (2014) Insights into deep-sea sediment fungal communities from the East Indian Ocean using targeted environmental sequencing combined with traditional cultivation. PLoS One 9:e109118

Chapter 2
Yeasts in Aquatic Ecotone Habitats

Allen N. Hagler, Leda C. Mendonça-Hagler, and Fernando C. Pagnocca

Abstract Aquatic ecotone habitats, like wetlands and phytotelmata, contain higher nutrient levels than are found in open waters resulting from degradation of organic materials like leaf litter. This allows much larger autochthonous yeast populations to develop than in more traditionally studied open water habitats. The single-celled morphology of yeasts makes them naturally better adapted than filamentous fungi to fluid habitats. Heavy influence exists from the extensive phylloplane yeast populations, which have also received little study until recently, and of animals attracted to these resources. *Debaryomyces hansenii*, *Pichia membranifaciens*, *Candida* spp., *Papiliotrema laurentii*, *Naganishia albida*, and *Rhodotorula mucilaginosa* were the most common yeasts detected in these habitats. Some species have a strong association with specific types of aquatic ecotone habitat like *Kluyveromyces aestuarii* in mangroves, *Scheffersomyces spartinae* in salt marshes, and *Kazachstania bromeliacearum* in bromeliad phytotelmata. Yeast diversity in aquatic ecotones is very rich in species occurring at low frequency making these habitats good targets for bioprospecting. The studies of estuaries, mangroves, salt marshes, bogs, and phytotelmata resulted in a list of over 270 identified yeasts and many additional unidentified cultures such as those reported as *Candida* spp. and as the former polyphyletic genera *Cryptococcus* spp. and *Rhodotorula* spp., many of which have since been described as new taxa.

Keywords Yeasts • Wetlands • Mangroves • Phytotelmata • Bromeliads • Bogs • Fens • Estuaries

Contents

A.N. Hagler (✉) • L.C. Mendonça-Hagler
Univ. Fed. do Rio de Janeiro (UFRJ) IMPPG, PBV, Ilha do Fundão, CCS, bl I, Rio de Janeiro 21941-970, Rio de Janeiro, Brazil
e-mail: allenhagler@gmail.com

F.C. Pagnocca
São Paulo State University (UNESP), Institute of Biosciences, Rio Claro 13506-900, São Paulo, Brazil

P. Buzzini et al. (eds.), *Yeasts in Natural Ecosystems: Diversity*,
DOI 10.1007/978-3-319-62683-3_2

2.1 Introduction

Yeasts are found in all kinds of aquatic habitats, but in low numbers of less than 10 l^{-1} in open ocean and lake waters, in larger populations up to 500 l^{-1} near terrestrial influence of the shore and often over 1000 l^{-1} in eutrophic waters such as in aquatic ecotones. Open water habitats of oceans, bays, lakes, ponds, rivers, and streams have been the subject of many studies of yeast communities or guilds (Hagler and Ahearn 1987; Nagahama 2006; Kutty and Philip 2008). The yeast diversity in freshwater of worldwide lakes and rivers and in nearshore and offshore seawater is summarized in Chap. 1 of this book. However, natural aquatic ecotones with much higher nutrient levels in their waters have received little attention in studies of yeast ecology. Yeasts survive well in salt- and freshwater in pure culture but compete in mixed cultures (Ahearn et al. 1968; Hagler and Mendonça-Hagler 1979; Starmer and Lachance 2011). They survive so well that suspensions of pure cultures in distilled water are used as a method to maintain fungi, including yeasts, for prolonged periods (McGinnis et al. 1974). The tendency to be unicellular should favor growth of yeasts over filamentous fungi in aquatic habitats (Lachance and Starmer 1998). The prevalent species in some types of aquatic ecotones are compared in Table 2.1. Estuaries are a classic example of ecotone habitats as interphases between marine and freshwaters carrying nutrients and microbial populations from terrestrial to marine ecosystems (Odum 1993).

Human-associated yeasts are frequent in most estuaries and can serve as pollution indicators (Hagler et al. 1986; Hagler 2006; Starmer and Lachance 2011). Artificial ponds for sewage treatment, recreational activities, and cultivation of aquatic life are dominated by species entering with sewage or fertilizers making them more the subject of industrial and public health microbiology and not included in this chapter. Wetland areas have relatively less volume of water and contain high amounts of degrading leaf litter and other organic materials compared with open waters (Fig. 2.1). This concentration of nutrients makes wetlands prime locations for feeding and breeding sites of many animals which can vector yeasts into the habitat. The shallow waters in these habitats are closely associated with organic and inorganic sediments supporting development of large fungal populations, including yeasts, in complex communities associated with the specialized flora and fauna of such regions (EPA 2016). Although these conditions favor diverse natural yeast communities, such habitats have not received much study. Sampling is complicated by uneven distribution in diverse microhabitats. Difficult access, unpleasant odors, mud, and insects are frequent factors. If close to human population centers, wetlands are often drained or contaminated by sewage. Smaller and somewhat ephemeral isolated volumes of water that are not wetlands, but have similar function in

Table 2.1 Prevalent yeasts in aquatic ecotones

	Estuaries	Mangroves	Bogs and fens	Phytotelmata
Yeasts	$N = 167$	$N = 234$	$N = 47$	$N = 238$
Aureobasidium pullulans	2	2	1	2
Candida aff *famata*[a]	3	17	4	8
Candida boidinii	4	8		
Candida glabrata	2	4		
Candida intermedia	4	5		10
Candida parapsilosis	3	10		3
Candida sake		5	1	3
Candida spp.	3	14	4	5
Candida tropicalis	2	15		3
Clavispora lusitaniae	2	4		1
Cyberlindnera saturnus	2	5	1	3
Cystobasidium minutum	2	1	1	2
Debaryomyces hansenii	3	6	4	5
Diutina aff *rugosa*		7		
Diutina rugosa		6		1
Hanseniaspora uvarum	3	6	2	3
Kazachstania bromeliacearum		2		7
Kazachstania exigua	1	4		1
Kluyveromyces aestuarii	1	13		3
Kodamaea ohmeri	2	5		3
Meyerozyma guilliermondii	4	12	1	4
Naganishia albida	3	1	3	8
Papiliotrema laurentii	5	1	4	7
Pichia kudriavzevii	5	11		
Pichia membranifaciens	3	14	3	7
Pichia occidentalis	1	10		3
Pichia spp.	1	3	1	2
Rhodotorula glutinis	4	2	3	4
Rhodotorula mucilaginosa	6	9	6	9
Rhodotorula spp.	5		2	1
Saccharomyces cerevisiae	3	6		5
Saturnispora silvae	1	4		2
Schwanniomyces vanrijiae		6	2	2
Sporobolomyces roseus	1		4	1
Torulaspora delbrueckii		4		4
Torulaspora sp. anamorph		7		4
Wickerhamomyces anomalus	2	6		1
Yarrowia lipolytica	2	6	1	2
Zygoascus hellenicus	1	4		1

$N =$ total number of samples in each type of ecotone in the 30 studies noted in Tables 11.2, 11.3, 11.4, and 11.5

Nomenclature was updated using Kurtzman et al. (2011a), Species Fungorum (2016) and MycoBank Database (2016)

[a]Includes anamorphic cultures of several species phenotypically similar to *Debaryomyces hansenii*

Fig. 2.1 The Sammamish River in Redmond Washington showing typical wetland vegetation with peat bog in the willows behind them

many terrestrial habitats, are phytotelmata (plant ponds). They are aquatic ecotones involving the plants and water but not soil. Although small and ephemeral relative to humans, they are large and permanent enough to support significant microbial populations. Examples are holes in tree trunks and cups formed by leaves and flowers. Bromeliad tanks are a good example and very important in diverse neotropical habitats especially in the forest canopy. These wetland habitats and phytotelmata have in common the degradation of large amounts of lignocellulose and other plant materials in shallow waters or water-saturated sediments. The plant species in such habitats are limited in diversity by exclusion of oxygen from the roots, but wetlands can support large and diverse microbial and animal populations. Hydrothermal vents in the oceans are another notable form of aquatic ecotone but fed by chemoautotrophic bacteria rather than degrading organic matter and not included in this chapter (Gadanho and Sampaio 2005; Nagahama 2006; Le Calvez et al. 2009). The aquatic surface films, sediment-water interface, and associated benthic organisms in all aquatic habitats are ecotones and have more concentrated nutrients than the water (Hagler and Ahearn 1987; Kachalkin 2014) and are part of more conventional marine and freshwater habitats. Diverse yeast populations are consistently present in waters, sediments, and biota of aquatic ecotones.

2.2 Estuaries

Few studies have been made of pristine estuaries. The prevalent species from some examples with different levels of urban influence increasing from left to right are presented in Table 2.2. The yeast species most typical of estuaries are *Candida* spp.

Table 2.2 Estuarine yeasts present in more than five samples

	Florida Suwannee	Florida Suwannee	Florida Biscayne Bay	Portugal Tagus	Portugal Sado	Portugal Tagus	Brazil UFRJ
	Sediment	Water	Sediment	Water	Water	Water	Water
	1	1	2	3	4	4	5
Yeasts	$N = 9$	$N = 9$	$N = 45$	$N = 84$	$N = 4$	$N = 4$	$N = 12$
Aureobasidium pullulans			2				5
Candida boidinii	1		1	1			2
Candida famata	3	1				3	
Candida intermedia			1		9	3	4
Candida parapsilosis			10	9			3
Candida sorboxylosa							7
Candida spp.	3		1				10
Candida tropicalis			19				7
Candida zeylanoides				9			1
Clavispora lusitaniae				5			1
Cutaneotrichosporon cutaneum	1	1	3			1	4
Cyberlindnera saturnus	4	1					
Cystobasidium minutum		1	9				
Debaryomyces hansenii			7	11			9
Debaryomyces spp.					20	1	
Dipodascus klebahnii	1						9
Hanseniaspora uvarum					23	2	9
Kazachstania exigua							8
Meyerozyma guilliermondii		1	2	8			4
Papiliotrema laurentii	4	1	4	2	3		
Pichia fermentans	4					2	

(continued)

Table 2.2 (continued)

	Florida Suwannee	Florida Suwannee	Florida Biscayne Bay	Portugal Tagus	Portugal Sado	Portugal Tagus	Brazil UFRJ
	Sediment	Water	Sediment	Water	Water	Water	Water
	1	1	2	3	4	4	5
Yeasts	$N = 9$	$N = 9$	$N = 45$	$N = 84$	$N = 4$	$N = 4$	$N = 12$
Pichia kudriavzevii	3	1	1	1			12
Pichia membranifaciens	4	1					4
Pichia occidentalis							11
Pichia terricola						1	7
Rhodotorula glutinis	1		9	1		3	
Rhodotorula mucilaginosa	3	1	14	25		3	1
Rhodotorula spp.	2	2	1		9	2	
Wickerhamomyces anomalus							6
Yarrowia lipolytica				1			4

N = number of samples. References: 1 = Lazarus and Koburger (1974); 2 = Fell et al. (1960); 3 = Coelho et al. (2010); 4 = Taysi and van Uden (1964); 5 = Hagler and Mendonça-Hagler (1981). Nomenclature was updated using Kurtzman et al. (2011a), Species Fungorum (2016), and MycoBank Database (2016)

(especially *Candida intermedia*, *Candida parapsilosis*, and *Candida tropicalis*), *Debaryomyces hansenii*, *Pichia kudriavzevii* (*Issatchenkia orientalis* and *Candida krusei*), *Meyerozyma* (*Candida*) *guilliermondii*, *Cutaneotrichosporon* (*Trichosporon*) *cutaneum*, *Papiliotrema* (*Cryptococcus*) *laurentii*, and *Rhodotorula* spp. (especially *Rhodotorula glutinis* and *Rhodotorula mucilaginosa*). Fell et al. (1960) started a sequence of studies of yeasts in estuaries, most of them near major urban centers. Various studies have been made of the Tagus river estuary in Portugal since the 1960s (Taysi and van Uden 1964). Coelho et al. (2010) working in the same estuary, but with various cultivation methods and the identifications of cultures by rDNA gene sequences, estimated the influence from marine, terrestrial runoff, urban effluents, and resident populations from samples taken in a transect downstream from Lisbon. Some species were associated with different sources entering the estuary. *Mey. guilliermondii*, *C. parapsilosis*, and *Clavispora lusitaniae* made up most of the isolates at 37 °C, and the counts at this temperature had a high correlation with counts of the fecal indicator *Escherichia coli*. Hagler et al. (1986) also found a high correlation of 40 °C yeast counts, largely of *P. kudriavzevii* and *C. tropicalis*, with coliform counts in Rio de Janeiro. High-temperature yeast counts can serve as indicators of domestic sewage (Hagler 2006). *Rh. mucilaginosa* was found as the prevalent species by Coelho et al. (2010) and considered to be mostly from terrestrial runoff although this species is also common in seawater and lakes (Hagler and Ahearn 1987). *Deb. hansenii* was considered a marker of marine origin although it is a species associated also with diverse nonmarine habitats. Pollution from urban areas is an important factor in most estuaries, and human-associated yeasts were very evident in the heavily polluted estuary site in Guanabara Bay, Rio de Janeiro (Hagler and Mendonça-Hagler 1981), but much less present in the more pristine Suwannee river estuary (Lazarus and Koburger 1974) and Everglades (Ahearn et al. 1968). An interesting strategy was used to detect yeasts of the Tagus river estuary, Portugal, by Gadanho and Sampaio (2004). The PCR-DGGE method of Muyzer (1999) was applied to analyze two water samples directly from environmental DNA and DNA extracted after cultivation on solid and liquid enrichment media. This combined procedure increased the species richness assessed in the water samples, allowing the detection of species present in low abundance or less competitive in culture. Ascomycetous species were not detected by PCR-DGGE using environmental DNA, but were after cultivation of the samples on YM agar or broth. The prevalent species detected in this study were *Deb. hansenii*, *Rh. mucilaginosa*, and *Vanrija longa* (*Cryptococcus longus*). Sites with brackish water from ice melts and tidal action similar to an estuary were studied in the north of Russia where salinity in the Kandalaksha Gulf does not exceed 20 ppt. Kachalkin (2014) found in water, sand, silt, and a sponge in the intertidal zone some species typical of estuaries in warmer regions, but also various obligate psychrophilic species including a new species *Glaciozyma litoralis*. The largest yeast populations in these two sites were associated with healthy algae and over 10,000 CFU g^{-1} dominated by *Metschnikowia zobellii*, whereas yeast counts in water were about 100 l^{-1} and much lower than in other substrates of the intertidal zone.

2.3 Mangroves

Mangroves are a type of tropical swamp with influence of tidal action and marine water, and like salt marshes are typically found in estuaries where they show a high level of microbial diversity (Ghizelini et al. 2012; Pires et al. 2012; Fig. 2.2). The yeast counts in water entering the mangrove vegetation of the Coroa Grande mangrove in Sepetiba Bay, Rio de Janeiro, had a geometric mean of 340 l^{-1}. Yeast counts were about 10 times higher in sediments than in water and 100 to 1000 times higher in the intestines of invertebrates living in this mangrove. Counts of filamentous fungi were typically one to two orders of magnitude higher than yeast counts in the same mangrove area (Araujo et al. 1995). The more frequent yeasts found in mangroves are presented in Table 2.3. Ahearn et al. (1968) made an extensive study of yeast communities in South Florida saline and freshwater habitats including the Everglades. More than 50 species were identified with the less advanced taxonomic methods of that time, and representative species were tested for survival as pure or mixed cultures in fresh- and seawater. Among them in all estuarine mangrove locations was *Kluyveromyces aestuarii* that had been described by Fell (1961) from a Miami Florida estuary. It has since been found in mangroves of Rio de Janeiro, Brazil (Araujo and Hagler 2011), in China (Chi et al. 2012), and in Thailand where it occurred with another very similar species, *Kluyveromyces siamensis*, described by Am-In et al. (2008). The consistent presence of *K. aestuarii* in this habitat and absence from others suggested its use as an indicator organism (Araujo and Hagler 2011). Urban beaches of mangrove sediments, but cleared of mangrove vegetation, had more human-associated yeasts including *Candida glabrata*, *C. tropicalis*, *C. parapsilosis*, *P. kudriavzevii*, and

Fig. 2.2 Mangrove at Mosqueiro, Aracaju, Sergipe, Brazil, on the margin of the estuary of Rio Vaza-Barris showing typical tree seedlings and aerial roots, but no smaller plants

Table 2.3 Yeasts found in more than five samples from mangroves

Substrate	Water	Water	Various	Sediment	Crabs	Crabs	Mollusks	Mollusks	Mollusks
Country	USA (a)	Thailand (b)	China (c)	Brazil RJ (d)	Brazil RJ (e)	Brazil RJ (f)	Brazil RJ (g)	Brazil RJ (h)	Brazil RJ (i)
References	1	5	4	3, 6	2	2	2	2	3
Yeast	$N = 44$	$N = 7$	$N = 6$	$N = 45$	$N = 16$	$N = 21$	$N = 18$	$N = 14$	$N = 26$
Candida aaseri			11		1				
Candida aff *famata*[a]				2	3	2	6	13	3
Candida boidinii			1	1		1	6	4	3
Candida intermedia			11	1			1	1	
Candida parapsilosis		2	6		1			1	2
Candida spp.				6	1	5	3	1	1
Candida tropicalis		1	74	4		1	1	2	3
Cystofilobasidium infirmominiatum	16								
Debaryomyces hansenii	25		12						
Kluyveromyces aestuarii	14		12	34	14		13	1	
Kluyveromyces siamensis			10						
Kodamaea ohmeri			15		1		1	2	
Meyerozyma guilliermondii		2	3		1	2	2	4	1
Naganishia albida	18								
Papiliotrema laurentii	59								
Pichia kudriavzevii	14		8				7	3	16
Pichia membranifaciens				10	3	3	14	3	8
Pichia occidentalis			1	1		1	3	3	6
Rhodotorula glutinis	21								
Rhodotorula mucilaginosa	18	2	2	3					2

(continued)

Table 2.3 (continued)

Substrate	Water	Water	Various	Sediment	Crabs	Crabs	Mollusks	Mollusks	Mollusks
Country	USA (a)	Thailand (b)	China (c)	Brazil RJ (d)	Brazil RJ (e)	Brazil RJ (f)	Brazil RJ (g)	Brazil RJ (h)	Brazil RJ (i)
References	1	5	4	3, 6	2	2	2	2	3
Yeast	$N = 44$	$N = 7$	$N = 6$	$N = 45$	$N = 16$	$N = 21$	$N = 18$	$N = 14$	$N = 26$
Torulaspora sp. anamorph				4		1	2	3	2
Wickerhamomyces anomalus			9				2	2	2
Yamadazyma triangularis	11								
Yarrowia lipolytica			6				1	3	2

N = number of samples. 1 = Ahearn et al. (1968); 2 = Araujo et al. (1995); 3 = Araujo (1999); 4 = Chi et al. (2012); 5 = Limtong et al. (2008a); 6 = Soares et al. (1997)

Origin of strains: (a) Florida, Everglades; (b) Khao Lumpee-Haad Thaimueang and Mu Ko Ra-Ko Prathong; (c) Fujian, Guangdong, and Hainan provinces; (d) sediment under mangrove vegetation; (e) detritivores crabs *Sesarma rectum* and *Uca* spp.; (f) herbivorous or omnivorous crabs *Goniopsis cruentata* and *Aratus pisonii*; (g) Rio de Janeiro *Anomalocardia brasiliana* and *Tagelus plebeius*; (h) mussel *Mytella guyanensis*; (i) shipworm *Neoteredo reynei*. RJ = Rio de Janeiro

[a]Anamorphs similar to *Debaryomyces hansenii*. Nomenclature was updated using Kurtzman et al. (2011a), Species Fungorum (2016), and MycoBank Database (2016)

Mey. guilliermondii and absence of *K. aestuarii* (Hagler et al. 1982; Soares et al. 1997). The intestines of detritus feeding crabs and a filter-feeding shipworm (mollusk) in the Coroa Grande mangrove in Rio de Janeiro were all found to contain *K. aestuarii*. However, it was not found in predatory and leaf-feeding crabs or a mussel attached to tree roots under the sediment, and it was nearly absent from clams in the mud flat of the same location (Araujo et al. 1995). *K. aestuarii* appears to be endemic to mangroves and associated with the detritus, presumably from degrading mangrove leaves at the sediment surface. The yeasts common in open waters are present in mangroves. But, there are also a striking number of diverse species isolated at low frequency, making a total of over 130 species in the work included in this review, many of which were probable new species. Many new species have been described from mangroves. These include *Candida sharkiensis*, *Candida rhizophoriensis*, *Sakaguchia* (*Rhodotorula*) *cladiensis*, *Rhodotorula evergladiensis*, and *Papiliotrema mangalensis* (*Cryptococcus mangaliensis*) (Fell et al. 2011); *Candida spencermartinsiae*, *Candida taylorii*, and *Ustilago* (*Pseudozyma*) *abaconensis* (Statzell-Tallman et al. 2011); *Candida chanthaburiensis*, *Candida kungkrabaensis*, and *Candida suratensis* (Limtong and Yongmanitchai 2010); *Candida thaimueangensis* (Limtong et al. 2007); *Geotrichum siamensis* and *Geotrichum phurueaensis* (Kaewwichian et al. 2010); *K. siamensis* (Am-In et al. 2008); *Kwoniella mangroviensis* (Statzell-Tallman et al. 2008); *Lachancea meyersii* (Fell et al. 2004); *Martiniozyma* (*Candida*) *asiatica* (Limtong et al. 2010a); *Rhodotorula paludigena* (*Rhodosporidium paludigenum*—Fell and Tallman 1980); *Saturnispora* (*Candida*) *siamensis* (Boonmak et al. 2009); *Candida phangngensis* (Limtong et al. 2008a); *Saturnispora* (*Candida*) *sanitii* and *Saturnispora* (*Candida*) *suwanaritii* (Limtong et al. 2010b); *Tetrapisispora arboricola* (Ueda-Nishimura and Mikata 1999); and *Torulaspora maleeae* (Limtong et al. 2008b).

Recent studies on fungal diversity in mangroves of New Caledonia used 454-pyrosequencing method, in which DNA was extracted directly from the environmental samples. Sequences from four regions of rDNA (ITS1, ITS2, SSU V5, and SSU V7) were obtained for fungi from submerged and aerial parts of trees. Species richness values were dependent on the gene marker used, ranging from 271 to 1001 OTUs (operational taxonomic units) and with the larger values for ITS sequences. Ascomycetes were dominant with 82% of the sequence reads, whereas Basidiomycetes represented 3%, and 15% could not be assigned to known taxa (Arfi et al. 2012a). The fungal diversity associated with anoxic-sulfidic sediments in the same mangrove was assessed by 454-pyrosequencing using the ITS1 and ITS2 regions (Arfi et al. 2012b). Over a hundred distinct OTUs were detected mostly of filamentous fungi but included the yeasts *Dipodascus australiensis*, *Galactomyces geotrichum*, and a few reads of *Malassezia* sp. and *Deb. hansenii*.

Freshwater swamps and ponds near urban centers are mostly polluted showing wide fluctuations in yeast counts compared with pond and swamp waters of unpopulated regions. There were notable studies done by Ahearn et al. (1968) in the Everglades and by van Uden and Ahearn (1963) in a small unpolluted lake in Michigan that give an idea of what yeasts are present in shallow freshwaters of uninhabited and less populated areas. The human-associated yeasts present in many studies from near urban areas were not common in these waters where yeasts belonging to the former genera *Cryptococcus* and *Rhodotorula* (now classified in diverse basidiomycetous genera) were most common. Yeast counts in Everglades freshwater sites were largely in the 150–500 range and with some up to 1200 l^{-1} (Ahearn et al. 1968). Yeast counts in the sites with 0–9 ppt salinity were mostly in the 100–1000 l^{-1} range, whereas with salinity of 25 ppt, most counts were less than 100. The most common species were strictly oxidative with *Pa. laurentii* the most frequent. The principal difference from estuarine mangrove regions in the Everglades was the lack of *K. aestuarii*, lower frequency of *Cystofilobasidium infirmominiatum*, and much higher frequency of *Sporobolomyces*. The only ascomycetes frequent in the freshwater swamps were *Deb. hansenii*, *P. kudriavzevii* (*C. krusei*), *Metschnikowia reukaufii*, and *Yamadazyma triangularis*.

2.4 Salt Marshes

Salt marshes are found frequently in regions protected from the action of the surf in temperate waters of bays and estuaries. Yeasts were studied in a salt marsh in Louisiana in southern USA where a new species was prevalent, *Scheffersomyces* (*Pichia*) *spartinae*, and with concentrations as great as 9×10^7 cells g^{-1} associated with the plant culm of *Spartina alterniflora*, oyster grass, the dominant plant of the habitat. The prevalent yeasts in the sediment rhizosphere were species of the then polyphyletic genera *Trichosporon*, *Rhodotorula*, and *Rhodosporidium* and *Kluyveromyces lactis*, a species similar to *K. aestuarii* that is prevalent in mangroves (Ahearn et al. 1970; Meyers et al. 1975; Hagler and Ahearn 1987). The population of *K. lactis* was followed in 30 samples each of water and sediment using 2% galactose YNB agar with pH adjusted to 4.0 with lactic acid on which it formed deep rose- to maroon-colored colonies and was compared with the overall yeast population growing on YM agar. It was found to be consistently present as a significant portion of the total yeast population which was 10 to 100 times higher in sediments than in water (Meyers et al. 1971). A more recent study using cultivation-independent methods was unable to confirm these species in a salt marsh in Georgia, but such methods have not shown good detection of yeasts (Buchan et al. 2002). Dini-Andreote et al. (2016) reported a comprehensive study using a high-throughput sequencing method to access fungal community dynamics related to marine-terrestrial transition at a pristine salt marsh (Schiermonnikoog Island, The Netherlands). The natural sedimentation process on this island resulted in a chronosequence developed over a hundred years of terrestrial ecosystem succession. The majority of OTUs based on ITS region sequences were assigned to Ascomycota (66.8%), followed by Basidiomycota (4.3%) with Tremellomycetes yeasts mainly represented by species previously assigned to the former polyphyletic genus *Cryptococcus* found especially in the early succession stages of transition from marine to terrestrial habitats. Yeasts have important populations as part of the normal biota of salt marshes, but have not received much study in this habitat.

2.5 Bogs and Fens

Bogs and fens are the dominant wetlands of our planet and important as carbon sinks. Yeast species make up about 10% of all peatland fungi and probably use simple polymers leached from plant materials in the initial phases of decomposition (Thormann et al. 2007). Peat bogs and fens do contain yeasts, and basidiomycetous species tend to be more prevalent and increasingly so in colder climates. The yeast species found more than once in bogs and fens are noted in Table 2.4. *Candida* spp., *Deb. hansenii*, *Rh. mucilaginosa*, and *Sporobolomyces roseus* were most common, and *Goffeauzyma* (*Cryptococcus*) *gilvescens* dominated in the coldest regions. Kachalkin (2010) isolated psychrophilic yeasts *Sterigmatosporidium polymorphum* and *Phenoliferia* (*Rhodotorula*) *psychrophenolica*, and *Aureobasidium pullulans* var. *subglaciale* from the *Sphagnum* mosses and paludal vascular plants in a swamp region near Moscow. Broad assimilation spectrum fungal species, capable of utilization of organic acids and aromatic compounds, were prevalent in the moss-turf (Kurakov et al. 2008). Yeast populations were noted by Babjeva and Chernov (1995) to be lower in the litter complex at about 10^3 g^{-1} compared to the 10^5–10^6 g^{-1} found in tundra epiphyte complex. A study in Canada and Siberia yielded 12 identified and 8 unidentified probable new species from 34 isolates. *Nadsonia starkeyi-henricii* was included among them and is probably a peatland specialist (Thormann et al. 2007).

2.6 Phytotelmata

Phytotelmata are formed from rainwater collected and preserved in structures of some plants including many bromeliad species. Thousands of species of bromeliads are native to diverse tropical habitats of the Americas (Fig. 2.3). These are dynamic and complex microenvironments inhabited by communities of different organisms including endemic species (Benzing 1990; Whittman 2000; Lopez et al. 2009). The phytotelmata in bromeliad leaf rosettes are a major source of nutrients for these organisms and communities associated with them (Richardson et al. 2000). Animals including insects and small mammals, known to carry yeasts, can have mutualistic relationships with tank bromeliads (Abranches et al. 1997; Pagnocca et al. 2008; Duarte et al. 2016; Leroy et al. 2016). Leaves from the bromeliad itself are in contact with the tank water, but the tanks also collect the leaf litter falling into them from surrounding vegetation. Each plant has many partitions formed by the leaf rosette making it like a circular rack of enrichment cultures around the central tank, each with different conditions for yeast growth. The leaf litter and visiting

Table 2.4 Yeasts found in bogs and fens

	Russia	Russia	Alaska	Canada-Russia	Thailand	Russia	Russia
References	1[a]	2	2	3	4[a]	5[a]	6[a]
Yeast	ND	$N = 69$	$N = 9$	$N = 9$	$N = 15$	$N = 3$	$N = 20$
Candida spp.		5		2		3	2
Cryptococcus spp.[b]		2				3	
Cystobasidium minutum	2					1	
Cystofilobasidium capitatum						1	2
Debaryomyces hansenii		4		2	2	2	
Filobasidium magnum				1			3
Goffeauzyma gilvescens	3		3				
Guehomyces pullulans	2	1		1			
Metschnikowia pulcherrima		1				2	
Mrakia frigida	3	1					
Naganishia albida		5	1	4			
Naganishia diffluens	2						1
Papiliotrema laurentii	2	2		1	1		
Pichia membranifaciens				1		1	2
Rhodotorula glutinis		2		1		2	
Rhodotorula mucilaginosa	2	5	2		3	2	3
Rhodotorula spp.		3		1			
Sporobolomyces roseus	1	5				2	2
Sugiyamaella paludigena		1		2			
Trichomonascus ciferrii					3		
Dothidella spp.							3

N = number of samples, ND = number of samples unknown

[a]Relative abundance data expressed as scale of three levels rather than frequency of occurrence 1 = minimal level detected; 2 = intermediate level; 3 = prevalent. References: 1 = Bab'eva and Chernov (1995); 2 = Polyakova et al. (2001); 3 = Thormann et al. (2007); 4 = Jaiboon et al. (2016); 5 = Kachalkin et al. (2008); 6 = Kachalkin and Yurkov (2012)

[b]Can include other Tremellomycetes formerly in the polyphyletic genus *Cryptococcus*. Nomenclature was updated using Kurtzman et al. (2011a), Species Fungorum (2016), and MycoBank Database (2016)

Fig. 2.3 Bromeliads. **A**, rupestrian bromeliads; **B**, bromeliads and other epiphytes on a tree; **C**, bromeliad growing on soil showing various phytotelmata surrounding central tank and leaf litter in the center; **D**, unshaded bromeliad with photoautotrophic growth in its tank

animals vector diverse species to form metapopulations in an extensive matrix of natural aquatic microcosms. Heavy rains can flood the tanks and wash out the existing nutrients and microbial populations, and during prolonged dry periods, the tanks can dry out (Araujo et al. 1998; Araujo 1999; Garcia 2007). More than 112 yeast species have been found in phytotelmata, and the prevalent yeasts in them are presented in Table 2.5. Phytotelmata in direct sunlight have strong algal growth, rather than degradation of organic materials, as a principal source of organic nutrients for microbial growth. Their yeast community is dominated by basidiomycetous species, whereas shaded plants are also rich in ascomycetous species. The species *Kazachstania bromeliacearum* appears to be endemic to bromeliad phytotelmata, and *Kazachstania rupicola* has been found in rupestrian bromeliads. *C. intermedia* was frequent in the phytotelmata and was the dominant

Table 2.5 Yeasts found in more than five samples of phytotelmata

	Coroa G	Bracui	P. Antas	Marica	MG	MG	SP
	Qq	Vp and Na	Np and Qq	Nc	Vm dry	Vm rain	Cs
References	1, 2	1	1	1, 3	4	4	5
Yeast	$N = 50$	$N = 38$	$N = 43$	$N = 22$	$N = 30$	$N = 30$	$N = 11$
Anomalomyces panici					3	6	
Aureobasidium pullulans					9	4	
Candida aff *famata*			3	5			
Candida intermedia	12	10	9	9		1	2
Candida spp.	2	2		1			2
Candida tropicalis	4	2		1			
Cryptococcus spp.[a]				1	5	8	
Cyberlindnera saturnus	3		3				
Debaryomyces hansenii	6	3	17	1			3
Kazachstania bromeliacearum	4	10	15	4			
Metschnikowia spp.	1			1	1		5
Meyerozyma guilliermondii	4			6		4	
Myriangiales spp.					12	22	
Naganishia albida	10	3	2	9			
Occultifur brasiliensis					13	10	
Papiliotrema laurentii	6		6	9	5		
Rhodotorula glutinis	4	1		1			
Rhodotorula mucilaginosa	8	1	4	3	1	1	
Saccharomyces cerevisiae		6	4	1			
Saitozyma podzolica					9	7	
Saturnispora silvae						12	
Schwanniomyces occidentalis	1		3	5			
Schwanniomyces polymorphus	1		6				
Schwanniomyces vanrijiae			9				
Torulaspora delbrueckii	7	2	2				

N = number of samples. Collection sites in Brazil: Coroa G = sand dune in mangrove, Coroa Grande, Itaguaí, RJ; Bracuí = mangrove epiphytes, Ilha do Jorge, Bracui, RJ; P. Antas = swamp, Poço das Antas Biological Reserve, RJ; Maricá = Restinga da Barra de Maricá, Rio de Janeiro, Brazil; MG = Serra da Piedade, Caeté, Minas Gerais; SP = Picinguaba área, an Atlantic rain forest site at the "Serra do Mar" State Park in São Paulo, Brazil

Plant species: An = *Aechmeae nudicaulis*; Cs = *Canistropsis seidelii*; Nc = *Neoregelia cruenta*; Np = *Nidularium procerum*; Qq = *Quesnelia quesneliana*; Vm = *Vriesea minarum*; Vp = *Vriesea procera*.

References: 1 = Araujo (1999); 2 = Hagler et al. (1993); 3 = Garcia (2007); 4 = Gomes et al. (2015); 5 = Ruivo (2005)

[a]Can include other Tremellomycetes formerly in the polyphyletic genus *Cryptococcus*. Nomenclature was updated using Kurtzman et al. (2011a), Species Fungorum (2016), and MycoBank Database (2016)

yeast in fruits of the bromelias *Quesnelia quesneliana*, *Vriesea procera*, and *Aechmea nudicaulis*. *Deb. hansenii*, its anamorph *Candida famata*, and similar species including *Schwanniomyces occidentalis*, *Schwanniomyces polymorphus*, *Schwanniomyces vanrijiae* and *Mey. guilliermondii* were prevalent especially in the shaded plants. An example of phytotelma other than bromeliad tanks is in the more ephemeral flower structures of the wild banana-like plant *Heliconia velloziana*. The 15 ascomycetous yeasts isolated from 14 phytotelmata of *Hel. velloziana* with more than one isolate were four cultures of *Candida heliconiae*, three each of *Candida picinguabensis* and *Metschnikowia* spp. and two each of *Candida apis*, *Candida pseudointermedia*, *Candida restingae*, *Candida saopaulonensis*, and *Debaryomyces* sp. (Ruivo 2005). The more common basidiomycetous species of the phylloplane were also among the prevalent phytotelmata yeasts (Fonseca and Inácio 2006). Phytotelmata are "in situ" enrichment cultures for yeasts making them a good natural source to tap the species richness of the phylloplane. Gomes et al. (2016) screened enzymes produced by yeasts from *Vriesea minarum* phytotelmata. These enzymes would allow them to participate in the degradation of plant and animal materials falling into the tanks. Phytotelmata yeasts have been the source of various new species: *Carlosrosaea* (*Bullera*) *vrieseae* (Landell et al. 2015); *Candida aechmeae* and *Candida vrieseae* (Landell et al. 2010); *Candida bromeliacearum* and *Candida ubatubensis* (Ruivo et al. 2005); *Candida heliconiae*, *C. picinguabensis*, and *C. saopaulonensis* (Ruivo et al. 2006); *Hagleromyces aurorensis* (Sousa et al. 2014); *Hannaella pagnoccae* (Landell et al. 2014); *Kaz. bromeliacearum* (Araujo et al. 2012); *Kaz. rupicola* (Safar et al. 2013); *Kockovaella libkindii* (Gomes et al. 2016); and *Occultifur brasiliensis* (Gomes et al. 2015).

2.7 Concluding Remarks

Yeasts in aquatic ecotones are rich in species diversity and with high population levels compared with open waters. More than 270 yeast species and many other unidentified yeasts, often representing new taxa, were reported from aquatic ecotones covered in this chapter. Cultivation of yeasts from these waters is complicated by large populations of filamentous fungi and competition between different yeasts while growing on culture media. Using various enrichment cultures and solid media with and without different antifungal antibiotics to inhibit parts of the fungal populations, including some yeasts, should improve isolation of yeasts from aquatic ecotones. No medium, even if all yeasts can grow on it in pure culture, will allow cultivation of all yeasts in the mixed populations of an environmental sample. The methods of cultivation, isolation, and enumeration of yeasts have been reviewed by Boundy-Mills (2006) and Kurtzman et al. (2011b). A single nutrient-rich medium favors fast-growing species that can later inhibit development of others on the isolation medium. Various media with inhibitors and nutrients favoring different species should be used, and the frequency of presence in various samples, rather than counts, used to indicate the relative importance of different species. Indicator dyes, such as bromocresol green at pH 4, added to media can inhibit some species and assist in selection of colonies for further study. Culture-independent methods with DNA analysis can detect the species present and their relative population (Xu 2006; Abarenkov et al. 2010), although this methodology does not yield cultures for further studies or applications. Culture-independent methods have detected yeasts in the presence of large populations of filamentous fungi, but yeasts known to be abundant by cultivation methods are often not detected as prevalent by these methods. The ITS region has been proposed as the universal bar code for next-generation sequencing methodology applied in fungal diversity studies and has shown detection of yeasts (Arfi et al. 2012a, b; Schoch et al. 2011). The detection of yeast taxa could be improved when sequences of large subunit rDNA (D1/D2 region) are used, probably due to the large database available on this region for yeasts (Bokulich et al. 2014). A polyphasic approach of cultivation and culture-independent methods should provide better information on yeast communities in aquatic ecotones. The exceptional species richness found, especially for mangroves and phytotelmata, yielded many new species descriptions. This should encourage further studies of yeast ecology and bioprospecting for yeasts in aquatic ecotone habitats.

References

Abarenkov K, Nilsson RH, Larsson KH, Alexander IJ, Eberhardt V, Erland S, HØiland K, KjØller R, Larsson E, Pemmansen T, Sen R, Taylor AFS, Tedersoo L, Ursing M, Vralstad T, Liimatainen K, Peintner U, Kõljalg U (2010) The UNITE database for molecular identification of fungi – recent updates and future perspectives. New Phytol 186:281–285

Abranches J, Valente P, Nóbrega HN, Fernandez FAS, Mendonça-Hagler LC, Hagler AN (1997) Yeast diversity and killer activity dispersed by fecal pellets from marsupials and rodents in a southeast Brazilian tropical habitat mosaic. FEMS Microbiol Ecol 26:27–33

Ahearn DG, Roth FJ, Meyers SP (1968) Ecology and characterization of yeasts from aquatic regions of south Florida. Mar Biol 1:291–308

Ahearn DG, Yarrow D, Meyers SP (1970) *Pichia spartinae* sp. nov. from Louisiana marshland habitats. Antonie Van Leeuwenhoek 36:503–508

Am-In S, Yongmanitchai W, Limtong S (2008) *Kluyveromyces siamensis* sp. nov., an ascomycetous yeast isolated from water in a mangrove forest in Ranong Province, Thailand. FEMS Yeast Res 8:823–828

Araujo, FV (1999) Comunidades de leveduras associadas a sedimentos de manguezais e a bromélias em ecossistemas costeiros do Rio de Janeiro, Brasil. D.Sc Thesis, Inst. Microbiol. PPG, UFRJ. pp 183

Araujo FV, Hagler AN (2011) *Kluyveromyces aestuarii*, a potential environmental quality indicator yeast for mangroves in the State of Rio de Janeiro, Brazil. Braz J Microbiol 42:954–958

Araujo FV, Soares CAG, Hagler AN, Mendonça-Hagler LC (1995) Ascomycetous yeast communities of marine invertebrates in a southeast Brazilian mangrove ecosystem. Antonie Van Leeuwenhoek 68:91–99

Araujo FV, Medeiros RJ, Mendonça-Hagler LC, Hagler AN (1998) A preliminary note on yeast communities of bromeliad-tank waters of Rio de Janeiro, Brazil. Rev Microbiol 29:118–121

Araújo FV, Rosa CA, Freitas LFD, Lachance M-A, Vaughan-Martini A, Mendonça-Hagler LC, Hagler AN (2012) *Kazachstania bromeliacearum* sp. nov., a yeast species from water tanks of bromeliads. Int J Syst Evol Microbiol 62:1002–1006

Arfi Y, Buée M, Marchand C, Devasseur A, Record E (2012a) Multiple markers pyrosequencing reveals highly diverse and host specific fungal communities on the mangrove trees *Avicenia marina* and *Rhizophora stylosa*. FEMS Microbiol Ecol 79:433–444

Arfi Y, Marchand C, Wartel M, Record E (2012b) Fungal diversity in anoxic-sulfidic sediments in a mangrove soil. Fungal Ecol 5:282–285

Babjeva IP, Chernov IY (1995) Geographic aspects of yeast ecology. Physiol Gen Biol Rev 9:1–54

Benzing DH (1990) Vascular epiphytes. General biology and related biota. Cambridge University Press, Cambridge

Bokulich NA, Thomgate JH, Richardson PM, Mills DA (2014) Microbial biography of wine grapes is conditioned by cultivar, vintage, and climate. Proc Natl Acad Sci U S A 111:E139–E148

Boonmak C, Jindamorakot S, Kawasaki H, Yongmanitchai W, Suwanarit P, Nakase T, Limtong S (2009) *Candida siamensis* sp. nov., an anamorphic yeast species in the *Saturnispora* clade isolated in Thailand. FEMS Yeast Res 9:668–672

Boundy-Mills K (2006) Methods for investigating yeast biodiversity. In: Rosa CA, Gabor P (eds) Biodiversity and ecophysiology of yeasts. Springer, Berlin, pp 67–100

Buchan A, Newell SY, Moreta JI, Moran MA (2002) Analysis of internal transcribed spacer (ITS) regions of rRNA genes in fungal communities in a southeastern U.S. salt marsh. Microb Ecol 43:329–340

Chi Z-M, Liu T-T, Chi Z, Liu G-L, Wang Z-P (2012) Occurrence and diversity of yeasts in the mangrove ecosystems in Fujian, Guangdong and Hainan provinces of China. Indian J Microbiol 52:346–353

Coelho MA, Almeida JMF, Martins IM, da Silva AJ, Sampaio JP (2010) The dynamics of the yeast community of the Tagus river estuary: testing the hypothesis of the multiple origins of estuarine yeasts. Antonie Van Leeuwenhoek 98:331–342

Dini-Andreote F, Pylro VS, Baldrian P, van Elsas JD, Salles JF (2016) Ecological succession reveals potential signatures on marine-terrestrial transition in salt marsh fungal communities. ISMEJ 10:1–14

Duarte APM, Ferro M, Rodrigues A, Bacci M Jr, Nagamoto NS, Forti LC, Pagnocca FC (2016) Prevalence of the genus *Cladosporium* on the integument of leaf-cutting ants characterized by 454 pyrosequencing. Antonie Van Leeuwenhoek 109:1235–1243

EPA (2016) Wetlands classification and types in wetlands protection and restoration. https://www.epa.gov/wetlands. Accessed on Nov. 18, 2016

Fell JW (1961) A new species of *Saccharomyces* isolated from a subtropical estuary. Antonie Van Leeuwenhoek 27:27–30

Fell JW, Tallman AS (1980) *Rhodosporidium paludigenum* sp. nov., a basidiomycetous yeast from intertidal waters of south Florida. Int J Syst Evol Microbiol 30:658–659

Fell JW, Ahearn DG, Meyers SP, Roth FJ (1960) Isolation of yeasts from Biscayne Bay, Florida, and adjacent benthic areas. Limnol Oceanogr 5:366–371

Fell JW, Statzell-Tallman A, Kurtzman CP (2004) *Lachancea meyersii* sp. nov., an ascosporogenous yeast from mangrove regions in the Bahama Islands. Stud Mycol 50:359–363

Fell JW, Statzell-Tallman A, Scorzetti G, Gutierrez MH (2011) Five new species of yeasts from fresh water and marine habitats in the Florida Everglades. Antonie Van Leeuwenhoek 99:533–549

Fonseca A, Inácio J (2006) Phylloplane yeasts. In: Rosa CA, Gabor P (eds) Biodiversity and ecophysiology of yeasts. Springer, Berlin, pp 263–301

Gadanho M, Sampaio JP (2004) Application of temperature gradient gel electrophoresis to the study of yeast diversity in the estuary of the Tagus River, Portugal. FEMS Yeast Res 5:253–261

Gadanho M, Sampaio JP (2005) Occurrence and diversity of yeasts in the Mid-Atlantic Ridge hydrothermal fields near the Azores archipelago. Microb Ecol 50:408–417

Garcia, KM (2007) Meios diferenciais para isolamento e triagem de leveduras endofíticas da bromélia *Neoregelia cruenta.* Dissertação, Programa de Pós Graduação em Biotecnologia Vegetal-UFRJ pp 100
Ghizelini AM, Mendonça-Hagler LCS, Macrae A (2012) Microbial diversity in Brazilian mangrove sediments – a mini review. Braz J Microbiol 43:1242–1254
Gomes FCO, Safar SVB, Marques AR, Medeiros AO, Santos ARO, Carvalho C, Lachance MA, Sampaio JP, Rosa CA (2015) The diversity and extracellular enzymatic activities of yeasts isolated from water tanks of *Vriesea minarum,* an endangered bromeliad species in Brazil, and the description of *Occultifur brasiliensis* f.a., sp. nov. Antonie Van Leeuwenhoek 107:597–611
Gomes FCO, Safar SVB, Santos ARO, Lachance M-A, Rosa CA (2016) *Kockovaella libkindii* sp. nov., a yeast species isolated from water tanks of bromeliad. Int J Syst Evol Microbiol 66:5066–5069
Hagler AN (2006) Yeasts as indicators of environmental quality. In: Rosa CA, Gabor P (eds) Biodiversity and ecophysiology of yeasts. Springer, Berlin, pp 519–536
Hagler AN, Ahearn DG (1987) Ecology of aquatic yeasts. In: Rose AH, Harrison JS (eds) The yeasts, vol 1, 2nd edn. Academic Press, London, pp 181–206
Hagler AN, Mendonça-Hagler LC (1979) Sodium chloride tolerance of yeasts from a polluted estuary in Rio de Janeiro. Rev Microbiol 10:30–33
Hagler AN, Mendonça-Hagler LC (1981) Yeasts from marine and estuarine waters with different levels of pollution in the state of Rio de Janeiro, Brazil. Appl Environ Microbiol 41:173–178
Hagler AN, de Oliveira RB, Mendonça-Hagler LC (1982) Yeasts in the intertidal sediments of a polluted estuary in Rio de Janeiro, Brazil. Antonie Van Leeuwenhoek 48:53–56
Hagler AN, Mendonça-Hagler LC, Silva JB, Santos EA, Farage S, Shrank A, de Oliveira RB (1986) Evaluation of microbial pollution indicators in Brazilian tropical and subtropical marine surface waters. Sci Total Environ 58:151–160
Hagler AN, Rosa CA, Morais PB, Mendonça-Hagler LC, Franco GMO, Araujo FV, Soares CAG (1993) Yeasts and coliform bacteria of water accumulated in bromeliads of mangrove and sand dune ecosystems in the southeast Brazil. Can J Microbiol 39:973–977
Jaiboon K, Lertwattanasakul N, Limtong P, Limtong S (2016) Yeasts from peat in a tropical peat swamp forest in Thailand and their ability to produce ethanol, indole-3-acetic acid and extracellular enzymes. Mycol Prog 15:755–770
Kachalkin AV (2010) New data on the distribution of certain psychrophilic yeasts in Moscow oblast. Microbiology 79:840–844
Kachalkin AV (2014) Yeasts of the White Sea intertidal zone and description of *Glaciozyma litorale* sp. nov. Antonie Van Leeuwenhoek 105:1073–1083
Kachalkin AV, Yurkov AM (2012) Yeast communities in *Sphagnum* phyllosphere along the temperature-moisture ecocline in the boreal forest-swamp ecosystem and description of *Candida sphagnicola* sp. nov. Antonie Van Leeuwenhoek 102:29–43
Kachalkin AV, Glushakova AM, Yurkov AM, Chernov IY (2008) Characterization of yeast groupings in the phyllosphere of *Sphagnum* mosses. Mikrobiologiya 77:533–541
Kaewwichian R, Yongmanitchai W, Srisuk N, Fujiyama K, Limtong S (2010) *Geotrichum siamensis* sp. nov. and *Geotrichum phurueaensis* sp. nov., two asexual arthroconidial yeast species isolated in Thailand. FEMS Yeast Res 10:214–220
Kurakov AV, Lavrent'ev RB, Nechitailo TY, Golyshin PN, Zvyagintsev DG (2008) Diversity of facultatively anaerobic microscopic mycelial fungi in soils. Microbiology 77:90–98
Kurtzman CP, Fell JW, Boekhout T (2011a) The yeasts, a taxonomic study, 5th edn. Elsevier, Amsterdam
Kurtzman CP, Fell JW, Boekhout T, Robert V (2011b) Methods for isolation, phenotypic characterization and maintenance of yeasts. In: Kurtzman CP, Fell JW, Boekhout T (eds) The yeasts, a taxonomic study, 5th edn. Elsevier, Amsterdam, pp 97–107
Kutty SN, Philip R (2008) Marine yeasts – a review. Yeast 25:465–483
Lachance MA, Starmer WT (1998) Ecology and yeasts. In: Kurtzman CP, Fell JW (eds) The yeasts, a taxonomic study, 5th edn. Elsevier, Amsterdam, pp 21–30

Landell MF, Billodre R, Ramos JP, Leoncini O, Vainstein MH, Valente P (2010) *Candida aechmeae* sp. nov. and *Candida vrieseae* sp. nov., novel yeast species isolated from the phylloplane of bromeliads in southern Brazil. Int J Syst Evol Microbiol 60:244–248

Landell MF, Brandão LR, Barbosa AC, Ramos JP, Safar SV, Gomes FCO, Sousa FM, Morais PB, Broetto L, Leoncini O, Ribeiro JR, Fungsin B, Takashima M, Nakase T, Lee CF, Vainstein MH, Fell J, Scorzetti G, Vishniac H, Rosa CA, Valente P (2014) *Hannaella pagnoccae* sp. nov., a tremellaceous yeast species isolated from plants and soil. Int J Syst Evol Microbiol 64:1970–1977

Landell MF, Brandão LR, Safar SVB, Gomes FCO, Félix CR, Santos ARO, Pagani DM, Ramos JP, Broetto L, Mott T, Vainstein MH, Valente P, Rosa CA (2015) *Bullera vrieseae* sp. nov., a tremellaceous yeast species isolated from bromeliads. Int J Syst Evol Microbiol 65:2466–2471

Lazarus CR, Koburger JA (1974) Identification of yeasts from the Suwannee River Florida estuary. Appl Microbiol 27:1108–1111

Le Calvez T, Borgaud G, Mahé S, Barbier G, Vandenkoornhuyse P (2009) Fungal diversity in deep-sea hydrothermal ecosystems. Appl Environ Microbiol 75:6415–6421

Leroy C, Petitclerc F, Orivel J, Corbara B, Carrias JF, Dejean A, Cereghino R (2016) The influence of light, substrate and seed origin on the germination and establishment of an ant-garden bromeliad. Plant Biol 19:70–78

Limtong S, Yongmanitchai W (2010) *Candida chanthaburiensis* sp. nov., *Candida kungkrabaensis* sp. nov. and *Candida suratensis* sp. nov., three novel yeast species from decaying plant materials submerged in water of mangrove forests. Antonie Van Leeuwenhoek 98:379–388

Limtong S, Yongmanitchai W, Kawasaki H, Seki T (2007) *Candida thaimueangensis* sp. nov., an anamorphic yeast species from estuarine water in a mangrove forest in Thailand. Int J Syst Evol Microbiol 57:650–653

Limtong S, Yongmanitchai W, Kawasaki H, Seki T (2008a) *Candida phangngensis* sp. nov., an anamorphic yeast species in the *Yarrowia* clade, isolated from water in mangrove forest in Phang-Nga Province, Thailand. Int J Syst Evol Microbiol 58:515–519

Limtong S, Imanishi Y, Jindamorakot S, Ninomiya S, Yongmanitchai W, Nakase T (2008b) *Torulaspora maleeae* sp. nov., a novel ascomycetous yeast species from Japan and Thailand. FEMS Yeast Res 8:337–343

Limtong S, Kaewwichian R, Am-In S, Nakase T, Lee C, Yongmanitchai W (2010a) *Candida asiatica* sp. nov., an anamorphic ascomycetous yeast species isolated from natural samples from Thailand, Taiwan, and Japan. Antonie Van Leeuwenhoek 98:475–481

Limtong S, Kaewwichian R, Am-In S, Boonmak C, Jindamorakot S, Yongmanitchai W, Srisuk N, Kawasaki H, Nakase T (2010b) Three anamorphic yeast species *Candida sanitii* sp. nov., *Candida sekii* sp. nov. and *Candida suwanaritii*, three novel yeasts in the *Saturnispora* clade isolated in Thailand. FEMS Yeast Res 10:114–122

Lopez LCS, Alves RR, Rios RI (2009) Micro-environmental factors and the endemism of bromeliad aquatic fauna. Hydrobiologia 625:151–156

McGinnis MR, Padhye AA, Ajello L (1974) Storage of stock cultures of filamentous fungi, yeasts, and some aerobic Actinomycetes in sterile distilled water. Appl Microbiol 28:218–222

Meyers SP, Ahearn DG, Miles PC (1971) Characterization of yeasts in Barataria Bay. LSU Coastal Stud Bull 6:7–17

Meyers SP, Ahearn DG, Alexander SK, Cook WL (1975) *Pichia spartinae* a dominant yeast of the *Spartina* salt marsh. Dev Ind Microbiol 16:262–267

Muyzer G (1999) DGGE/TGGE: a method for identifying genes from natural ecosystems. Curr Opin Microbiol 2:317–322

Mycobank (2016) Mycobank database. http://www.mycobank.org/quicksearch.aspx. Accessed on 15 Mar 2017

Nagahama T (2006) Yeast biodiversity in freshwater, marine, and deep-sea environments. In: Rosa CA, Gabor P (eds) Biodiversity and ecophysiology of yeasts. Springer, Berlin, pp 241–262

Odum EP (1993) Ecology and our endangered life-support systems. Sinauer Associates, Sunderland, MA

Pagnocca FC, Rodrigues A, Nagamoto N, Bacci M (2008) Yeasts and filamentous fungi carried by the gynes of leaf-cutting ants. Antonie Van Leeuwenhoek 94:517–526

Pires ACC, Cleary DFR, Almeida A, Cunha A, Dealtry S, Mendonça-Hagler L, Smalla K, Gomes NCM (2012) Denaturing gradient gel electrophoresis and barcoded pyrosequencing reveal unprecedented Archaeal diversity in mangroves sediments and rhizosphere samples. Appl Environ Microbiol 78:5520–5528

Polyakova AV, Chernov IY, Panikov NS (2001) Yeast diversity in hydromorphic soils with reference to a grass-*Sphagnum* wetland in western Siberia and a hummocky tundra region at Cape Barrow (Alaska). Microbiology 70:714–720

Richardson BA, Richardson MJ, Scatena FN, McDowell WH (2000) Effects of nutrient availability and other elevational changes on bromeliad populations and their invertebrate communities in a humid tropical forest in Puerto Rico. J Trop Ecol 16:167–188

Ruivo CCC (2005) Ocorrência de leveduras em espécies vegetais nativas da Mata Atlântica, Parque Estadual da Serra do Mar – Núcleo Picinguaba, São Paulo. Thesis UNESP Rio Claro. http://repositorio.unesp.br/handle/11449/103947. Accessed 12 Nov 2016

Ruivo CCC, Lachance M-A, Rosa CA, Bacci M, Pagnocca FC (2005) *Candida bromeliacearum* sp. nov. and *Candida ubatubensis* sp. nov., two yeast species isolated from the water tanks of *Canistropsis seidelii* (Bromeliaceae). Int J Syst Evol Microbiol 55:2213–2217

Ruivo CCC, Lachance MA, Rosa CA, Bacci M, Pagnocca FC (2006) *Candida heliconiae* sp. nov., *Candida picinguabensis* sp. nov. and *Candida saopaulonensis* sp. nov., three ascomycetous yeasts from *Heliconia velloziana* (Heliconiaceae). Int J Syst Evol Microbiol 56:1147–1151

Safar SVB, Gomes FCO, Marques AR, Lachance MA, Rosa CA (2013) *Kazachstania rupicola* sp. nov., a yeast species isolated from water tanks of a bromeliad in Brazil. Int J Syst Evol Microbiol 63:1165–1168

Schoch CL, Seifert KA, Huhndorf S, Robert V, Spouge JL, Lavesque CA, Chen W, Fungal barcoding consortium (2011) Nuclear ribosomal internal transcribed spacer (ITS) region as a universal DNA barcode marker for fungi. Proc Natl Acad Sci U S A 109:6241–6246

Soares CA, Maury M, Pagnocca FC, Araujo FA, Hagler AN, Mendonça-Hagler LC (1997) Yeast communities in dark muddy intertidal estuarine sediments of Rio de Janeiro, Brazil. J Gen Appl Microbiol 43:265–272

Sousa FMP, Morais PB, Lachance MA, Rosa CA (2014) *Hagleromyces* gen. nov., a yeast genus in the Saccharomycetaceae, and description of *Hagleromyces aurorensis* sp. nov., isolated from water tanks of bromeliads. Int J Syst Evol Microbiol 64:2915–2919

Species Fungorum (2016) CABI databases. www.speciesfungorum.org/names/names.asp. Accessed 15 Mar 2017

Starmer WT, Lachance MA (2011) Yeast ecology. In: Kurtzman CP, Fell JW, Boekhout T (eds) The yeasts, a taxonomic study, 5th edn. Elsevier, Amsterdam, pp 65–83

Statzell-Tallman A, Belloch C, Fell JW (2008) *Kwoniella mangroviensis* gen. nov., sp. nov. (Tremellales, Basidiomycota), a teleomorphic yeast from mangrove habitats in the Florida Everglades and Bahamas. FEMS Yeast Res 8:103–113

Statzell-Tallman A, Scorzetti G, Fell JW (2011) *Candida spencermartinsiae* sp. nov., *Candida taylorii* sp. nov. and *Pseudozyma abaconensis* sp. nov., novel yeasts from mangrove and coral reef ecosystems. Int J Syst Evol Microbiol 60:1978–1984

Taysi I, van Uden N (1964) Occurrence and population densities of yeast species in an estuarine-marine area. Limnol Oceanogr 9:42–45

Thormann MN, Rice AV, Beilman DW (2007) Yeasts in peatlands: a review of richness and roles in peat decomposition. Wetlands 27:761–773

Ueda-Nishimura K, Mikata K (1999) A new yeast genus, *Tetrapisispora* gen. nov.: *Tetrapisispora iriomotensis* sp. nov., *Tetrapisispora nanseiensis* sp. nov. and *Tetrapisispora arboricola* sp. nov., from the Nansei Islands, and reclassification of *Kluyveromyces phaffii* (van der Walt) van der Walt as *Tetrapisispora phaffii* comb. nov. Int J Syst Evol Microbiol 49:1915–1924

Van Uden N, Ahearn DG (1963) Occurrence and population densities of yeast species in a freshwater lake. Antonie Van Leeuwenhoek 29:308–312

Whittman PK (2000) The animal community associated with canopy bromeliads of the lowland Peruvian Amazon rain forest. Selbyana 21:48–51

Xu J (2006) Microbial ecology in the age of genomics and metagenomics: concepts, tools, and recent advances. Mol Ecol 5:1713–1731

Chapter 3
Yeasts in Forest Soils

Andrey Yurkov

Abstract Soil yeasts are common inhabitants of various soils, including those in forest biotopes. Historically, yeasts were studied mainly in vineyard, orchard and agricultural soils. Due to limited ecological surveys, yeasts represent yet a poorly known fraction of the microorganisms in forest soils. Our knowledge of soil yeasts is biased towards temperate and boreal forests, whereas data from Africa, Americas and Asia is scarce. Forest soils in the Southern hemisphere are strongly undersampled.

This chapter provides the first comprehensive review of yeasts in forest soils, their diversity, nutrition, traits and possible ecosystem services. Basidiomycetes are dominant in forest soils, but ascomycetes genera, including several fermenting yeasts, are also permanent residents in the soil. A particular focus in the chapter is dedicated to the review of yeast diversity after reclassification of previously polyphyletic yeast genera *Cryptococcus*, *Rhodotorula* and *Trichosporon*. Factors influencing distribution of soils yeasts are also discussed in this chapter.

Keywords Forest • Yeasts • Basidiomycetes • Cryptococcus • Trichosporon • Forest management

Contents

A. Yurkov (✉)
Leibniz Institute DSMZ-German Collection of Microorganisms and Cell Cultures, 38124 Braunschweig, Germany
e-mail: andrey.yurkov@dsmz.de

P. Buzzini et al. (eds.), *Yeasts in Natural Ecosystems: Diversity*,
DOI 10.1007/978-3-319-62683-3_3

3.1 Introduction

The role of forests for humankind is enormous. Trees are sources of food and energy and are used for building or the production of goods. Forests trap carbon dioxide from the atmosphere and release the oxygen that supports most of the life on the planet. There is no universally recognised definition of the forest, and this term is used to describe entities ranging from a terrestrial biome to a single tree plantation. The term can be employed to depict a biome (e.g. Taiga or Boreal forest biome), a vegetation type (e.g. coniferous forest) and a biotope (e.g. the Black Forest) but also a land use category. The tree density and height allow distinguishing forests from woodland, savannah, and scrubland vegetation types. Also, mangroves that grow in coastal saline or brackish water are usually referred to as forests. In this chapter, the forest is used as a synonym of the vegetation type without a regard to its properties such as age, density and management.

The Food and Agriculture Organization of the United Nations (FAO) has put forestry and sustainable forest management among its strategic goals. In collaboration with member countries, FAO carries out periodic global assessments of forest resources, which are made available through reports. The most recent report entitled "The Global Forest Resources Assessment 2015: How are the world's forests changing?" reviewed changes of forests over the last 25 years. According to this report, forests cover almost 4000 billion ha. But this area is continuously decreasing due to deforestation or forest conversion in another land use type from 31.6% of global land area in 1990 to 30.6% in 2015. Such net loss of some 129 million hectares (M Ha) of forest roughly corresponds to the size of South Africa. The annual forest loss, however, has decreased significantly over the last 5 years and is about 3.3 M Ha (33,000 km^2), which is about the size of the land area of Belgium or Netherlands. The largest forest area loss occurred in the tropics, particularly in South America, Africa and Southeast Asia. Although the rate of loss in those areas has decreased substantially in the past 5 years, as much as 8000 km^2 of tropical forest in Brazil was lost since August 2015 (Tollefson 2016). Sustainable forestry, reforestation and forest protection are employed to reduce net forest loss. The largest proportion of protected forest areas is located in tropics and subtropics (countries with more than 40% of protected forest area), i.e. Brazil, countries of Central America (Costa Rica, Panama, Guatemala), Thailand, Botswana, Cameroon and Mali. In Europe, Slovakia has a remarkably high percentage of protected forests. While the largest forest loss occurred in the tropics and subtropics, forest net gain, also due to regeneration of tree cover, is concentrated in the boreal climatic domain. On the country level, a substantial forest gain was achieved in Chile, China, India and USA. The bulk of the world's forests is natural forest, with reported natural forest area amounting to 93% of global forest area. However, primary forests (a forest undisturbed directly by human activities) accounted for 33% of the world's forests (1.3 billion ha), half of which are located in the tropics. Over half of the world's primary forests are found in only three countries, namely, Brazil, Canada and Russian Federation. The area of about 31 M

Ha (0.31 M km^2) of primary forest has been reported as modified or cleared since 1990. This corresponds to the size of the land area of Poland and almost as large as Germany (UN ESA 2006; FAO 2016).

Forests are the dominant terrestrial ecosystem of Earth and are distributed across the globe. Forests account for 75% of the gross primary productivity of the Earth's biosphere and contain 80% of the Earth's plant biomass (Pan et al. 2013). According to the FAO report (FAO 2016), the conservation of biodiversity is the primary objective for 13% of the world's forests and 150 M Ha (1.5 M km^2) of forest have been added to this area in the last 25 years. The area of forests designated for protection of soil and water has also increased and now represents 31% of the forest area. Forests support a large diversity of living organisms, including microscopic organisms such as fungi and yeasts (e.g. Phaff and Starmer 1987; Blackwell 2011; Starmer and Lachance 2011).

The present chapter is focused on yeasts residing in soils under forest vegetation. Yeasts inhabiting plant material above ground and agricultural soils are discussed in other chapters of this book. Major forest types are listed according to the classification of terrestrial biomes by the World Wide Fund for Nature (WWF Global 200, Olson and Dinerstein 1998) and a simplified Köppen climate classification (e.g. Belda et al. 2014; Rohli et al. 2015). Information about forests and soils reported in paragraphs 3.2 and 3.3 was also obtained from Encyclopedia of World Geography (McColl 2014), Encyclopedia of Soil Science (Chesworth 2008), the CABI Encyclopaedia of Forest Trees (Dixon et al. 2013), the WWF Global ecoregions database (www.worldwildlife.org/biomes) and regional maps. When possible, soil classification is given according to the FAO World Reference Base for Soil Resources (IUSS Working Group WRB 2014).

3.2 Distribution of Forests and Forest Soils

Forest vegetation can be found in all regions up to the tree line. Harsh environmental conditions, usually cold temperatures or lack of moisture, do not support sustainable forest growth beyond the tree line, e.g. at high altitudes (Alpine) and latitudes (Arctic and Antarctic). When forest growth is suppressed, trees may form bushes and scrublands, for example, krummholz (also deformed) and maquis. The latitudes 10° north and south of the equator are mostly covered in tropical rainforest, and the latitudes between 53°N and 67°N have boreal forest. Forests dominated by angiosperms (broadleaf forests) are often more species-rich than those dominated by gymnosperms (coniferous forests), although exceptions exist. In the two extremes, many tree species can found on a small area in a tropical rain forest, whereas large areas of taiga are occupied by a few dominant trees. Productivity and biomass of different forest types differ accordingly, although much of this biomass occurs, as a rule, below ground in the root systems and as partially decomposed plant residues.

Boreal forests or Taiga is the world's largest forest biome, which constitutes about a half of the whole forested area. Larches, the deciduous gymnosperms of the genus *Larix*, which have a circumpolar distribution in the Northern Hemisphere (i.e. in the Alpine region, Carpathian Mountains, in most of the Siberia, Alaska and Canada), cover most of this area. Fir (*Abies* spp.), pine (*Pinus* spp.) and spruce (*Picea* spp.) are the other most common conifers in the Northern Hemisphere. Small-leaved tree species such as alder (*Alnus* spp.), birch (*Betula* spp.) and poplar (*Populus* spp.) are sometimes found in mixed boreal forests. Sphagnum mosses frequently appear in the ground layer. Taiga is found in areas with cold, moist summers and cold winters. Because of the low temperatures, decomposition is fairly slow, and plant residues decompose slowly and remain on the soil surface for several seasons. Organic compounds released during decomposition are acid and leach minerals and clay particles, leaving the upper soil layer sandy in texture. Gelisols (Cryosols and Histosols) and Spodosols (Podzols) are typical boreal soils.

Temperate forests are coniferous or broadleaf forests that occur in the temperate zone and receive sufficient rainfall. Temperate forests cover a large part of the Earth, but temperate rainforests only occur in a few regions around the world, mostly in oceanic moist climates, i.e. the Pacific temperate rain forests in Western North America (Southeastern Alaska to Central California), the Valdivian and Magellanic temperate rainforests of southwestern South America (Southern Chile and adjacent Argentina), the Caspian temperate rainforests of Iran and Azerbaijan, the Colchan rainforests of the eastern Black Sea region (Turkey and Georgia), Asian mountain temperate rainforests (Japan, Korea and Taiwan), Knysna-Amatole coastal forests in South Africa, temperate rainforests of southeastern Australia (Tasmania and Victoria) and the New Zealand temperate rainforests (South Island's west coast). Pockets of rain forests also exist in Europe (e.g. Northern Iberia), North America (e.g. British Columbia) and Russian Far East (Ussuri, Outer Manchuria and Sakhalin).

Temperate deciduous forests or temperate broadleaf forests are found in areas with warm, moist summers and mild winters. When coniferous and deciduous trees grow together, this forest type is termed as "mixed forest". Depending on the region, these deciduous forests are typically composed by beech (*Fagus* spp.), birch (*Betula* spp.), elm (*Ulmus* spp.), hornbeam (*Carpinus betulus*) maple (*Acer* spp.), oak (*Quercus* spp.) and southern beech (*Nothofagus* spp., in Southern Hemisphere only). Temperate coniferous forests are diverse and can be formed by cedar (*Cedrus* spp.), cypress (*Cupressus* spp.), Douglas fir (*Pseudotsuga* spp.), fir (*Abies* spp.), juniper (*Juniperus* spp.), pine (*Pinus* spp.), podocarp (*Podocarpus* spp.), spruce (*Picea* spp.), red cedar (*Thuja* spp.), redwood (genera *Sequoia* and *Sequoiadendron* in North America, and *Metasequoia* in Asia) and yew (*Taxus* spp.). Trees of the genera *Araucaria*, *Fitzroya* (both in South America) and *Agatis* (New Zealand and Australia) make up temperate coniferous forests in the Southern Hemisphere. Some trees are restricted to a particular geographical region as, for example, Taiwan cypress (*Chamaecyparis taiwanensis*), camphor tree (*Cinnamomum camphora*), Japanese chinquapin (*Castanopsis cuspidata*) and Japanese stone oak (*Lithocarpus edulis*) are found in Asia only. In Australia and

Tasmania, endemic eucalypti and southern beech are accompanied by the deciduous trees coachwood (*Ceratopetalum apetalum*), pinkwood (*Eucryphia moorei*), southern sassafras (*Atherosperma moschatum*) as well as by a few conifers (mainly *Athrotaxis* spp.) and tree ferns (*Cyathea* spp. and *Dicksonia antarctica*). Typical soils of temperate forests are Alfisols and Spodosols along with some Histosols.

Tropical and subtropical forests occur in moist and warm climates. Subtropical forests have seasons, although winters are mild with infrequent or no frost. Seasonal droughts have great impact on these forests, and their strength varies with geographic location. Subtropical moist forests (e.g. monsoon forest, Afromontane forest) are found in Africa, Asia (Indomalaya ecoregion), Australasia, Central America and Southern America. They represent often a transition between the tropical and temperate forests, often following an elevation gradient. Vegetation is similar to those in tropical rainforests, although plant diversity is not as high. Conifers (genera *Araucaria*, *Cryptomeria*, *Podocarpus* and *Sequoia*), tree ferns and tree grasses (bamboos) also grow within subtropical climate regimes.

Dry forests (e.g. Caatinga, Cerrado) exist in the drier areas, north and south of the tropical rainforest belt in Australasia, Indochina, North America (Mexico), South America (pacific coast, Brazil and Bolivia) and Southern Africa (also on Madagascar). Similar dry forests are found in the Mediterranean climate, which is characterised by dry summers and mild, moist winters. Mediterranean climate zones are intermediate between subtropical and temperate climates. Mediterranean sclerophyll forests grow in the Mediterranean Sea region, North America (chaparral forest in California), South America (Matorral in Chile), South Africa (Fynbos) and Western and South Australia. These forests and woodlands are composed predominantly of evergreen sclerophyll trees (sometimes growing as scrubs), most of which are endemic species. Depending on the region, these deciduous forests are composed by acacia, eucalypts, maple, oak, olive tree, southern beech and several endemic trees, including boldo (*Peumus boldus*), ceiba (*Ceiba* spp.), Chilean wine palm (*Jubaea chilensis*), keule (*Gomortega keule*), *Melaleuca*, peumo (*Cryptocarya alba*), silver tree (*Leucadendron argenteum*) and Waboom tree (*Protea nitida*) among others.

Laurel forest (also laurisilva) is a type of humid subtropical forest, which is found in several separate areas. This type of vegetation contains remnants of archaic flora of the ancient supercontinent of Gondwana. The lauroid (also laurophyll) forest is characterised by broadleaf tree species with evergreen, glossy and elongated leaves, members of the families Lauraceae and Podocarpaceae as well as some Fagacea (e.g. genera *Castanopsis*, *Lithocarpus*, *Nothofagus* and *Quercus*). These forests are common in subtropical eastern Asia, Indochina, South America, Africa, New Zealand, Australia and New Caledonia. Laurisilva is also found in the Islands of Macaronesia in the Eastern Atlantic, in particular the Azores, Madeira and Canary Islands.

Tropical rainforests are dense and trees occupy several layers. These forests are found around the world, particularly in the Indo-Malayan Archipelagos, the Amazon Basin and the African Congo. A perpetually warm, wet climate promotes more explosive plant growth than in any other environment on Earth. The largest trees in the canopy layer include Brazil nut tree (*Bertholletia excelsa* in South America), *Bombax* (Asia and West Africa), *Cecropia* (South America), *Ceiba* (Central America, South America and West Africa), *Ficus* (Southeast Asia and Australasia), *Hevea* (originally South America), *Swietenia* (Central America and South America) and *Tectona* (Asia) among others.

Typical soils of tropical forests are the Oxisols (Ferralsols, Plinthosols and Nitisols), Ultisols (Plinthols, Planosols and Alisols) and some Alfisols (Albelluvisols, Luvisols and Planosols). Despite rich vegetation, tropical soils are not rich with humus. High temperatures and precipitation through the year support high decomposition rates of the soil organic matter. Humic substances acidify soils, release metals from minerals and turn the soil profile to reddish and ochre colours.

Mangroves are a type of woodland or swamp forest in coastal saline or brackish water in the tropics and subtropics. This vegetation occurs worldwide in the tropical and subtropical tidal areas, and the trees are salt tolerant. Due to harsh environmental conditions, a limited number of plants are able to colonise the coastal areas. By contrast with the tropical rainforest biome containing thousands of tree species, mangrove biotopes are naturally composed by a few plant species. Mangroves share some similarities with freshwater swamps as in both cases plant material can be turned into a peat deposits. Like in other forests, decomposition rates and chemistry of different mangrove tree species vary. Various plants, including members of the genera *Avicennia*, *Bruguiera*, *Rhizophora* and *Sonneratia* among others, form mangrove vegetation.

3.3 Forest Soils and Soil Types

Climate, a sum of temperatures, precipitation and their changes over time, defines the gradual sequence of natural landscapes from polar zones towards the equator. A population of plants and animals, which share adaptations to a certain climate, is called a biome. Terrestrial biomes, which can be found on a range of continents, also include forests. The influence of climatic and biotic factors on the formation of different soils was first suggested by Dokuchaev (Fairbridge 2008). He defined several natural zones considering above-ground vegetation, climate as well as hydrologic and geochemical conditions in soils. Later, this approach was adapted and developed by Russian geobotanists (reviewed by Aleksandrova 2012). Thus, in addition to the so-called zonal vegetation (and soils), intrazonal and azonal formations (also as biogeocenoses) have been recognised. Intrazonal formations reflect the dominant influence of a specific factor such as salinity, hydrology or parent rock material. While intrazonal soils are still defined by the climate, azonal formations are largely influenced by the factors, which are not restricted to a particular zone. For example, water currents erode, transfer and redeposit sand and clay material. Vegetation and soils, which are formed on that parent material, are often alike irrespective of the climate. Pine forests on sandy soils and altitudinal zonation of the vegetation are among of the prominent examples of azonal formations. To some extent, mangroves and forested wetlands represent an azonal vegetation. However, the particular type such as sphagnum swamps is restricted to the boreal climate and is, thus, an intrazonal formation. Relict vegetation growing as patches or pockets outside the main distribution range are called extrazonal.

It is important to document, however, that such views on the classification of the vegetation types are not universal but were mostly limited to the countries traditionally connected to the Russian school of soil science. Nevertheless, such an approach is a good starting point to understand the ecology of soil yeasts. Intuitively, the above- and below-ground parts of the forest ecosystem are tightly connected. The success of a tree species depends on soils properties, but soil properties are strongly influenced by the tree cover. Therefore, the information about vegetation properties and soil parameters are both crucial to understand the factors influencing yeasts. As the landscape approach was more common among soil scientists than plant ecologists, the early studies that systematically surveyed yeasts in soils have focused on a range of soils with different properties and vegetation even though approaches to the classification of soils varied (e.g. Starkey and Henrici 1927; Babjeva and Golovleva 1963; di Menna 1965).

Why is the information about soil type important? Although different soil classification systems do exist, they all rely on both properties and processes. While some basic properties can be easily measured in the field (e.g. temperature, pH and conductivity) or determined in the laboratory (e.g. total carbon and nitrogen), many other soil parameters require complex analyses and are laborious. Soil formation processes encompass chemical and physical factors changing organic and inorganic fractions and, thus, predict the range of the most of relevant parameters. Additionally, soil processes reflect history of the habitat and the factors that shaped yeast communities in the past. Thus, a few basic soil properties would potentially provide less information than the identification of the soil type. A description of soil texture (e.g. sandy soil, loam soil, etc.) cannot replace a proper soil classification either. Although certain trees require particular conditions to grow, the same type of forest can sometimes occur in different environmental conditions. Thus, these habitats need to be described with a combination of the soil type and vegetation. For example, birches are among the early succession species and antedate the climax-type spruce forests on acid soils in the boreal zone. However, primary birch forests occur in the northern part of the Western Siberian hemiboreal forest ecoregion (WWF code PA0444) as a part of the ancient Pliocene coniferous-broadleaf vegetation. Also, patches of birch forests grow as small groves within meadowed steppe vegetation of the Kazakh forest steppe ecoregion (WWF ecoregion PA0809). Although these plant communities are alike, the soil-forming factors would differ between the regions.

There are several different soil classifications existing. They have been developed for soils formed in dissimilar climates and, thus, weighted soil parameters, environmental factors and economic utility differently. Thus, none of them was able to cover properly the whole diversity of the soils. A common approach has been made by soil scientists to combine existing soil classifications into a single system presently known as the World Reference Base for Soil Resources, which replaced the previous FAO soil classification in 2006. Thus, a type of soils can be determined in the field according to the existing national resources (e.g. soil maps) and then translated into a common system, which will be understood by scientists worldwide.

3.4 Yeasts in Forest Soils

3.4.1 Methods to Study

Laboratory cultivation was the main tool to study soil-inhabiting microorganisms. Scientists have noticed that bacteria and filamentous fungi outnumber yeasts in most soils (e.g. Starkey and Henrici 1927; Phaff and Starmer 1987; Botha 2011). Low quantity of yeasts was interpreted as the evidence for minor importance of this group of fungi for soil functioning (e.g. Starkey and Henrici 1927; Phaff and Starmer 1987; Starmer and Lachance 2011). However, it is important to document that the plating of a soil suspension is not suitable for reliable enumeration of either bacteria or moulds. Thus, any discussion regarding the contribution of yeasts to soil microbial communities can be done using either the biomass or activity values. Yeast numbers are commonly provided as colony forming units, and these values can be transformed into biomass values, as demonstrated by Babjeva and Reshetova (1972) and recently by Botha (2011). Also, Botha (2011) emphasised that yeast cells are generally larger than those of bacteria, so even moderate yeast numbers would correspond to larger biomass values. Fungal-dominated soil microbial communities occur in less disturbed, late successional sites, often with acid soils that are of high organic matter content and low resource quality (Wardle et al. 2004; van der Heijden et al. 2008). Fungi, including yeasts, account for the major pool of microbial biomass in forest soils, and yeasts are more prominent in boreal soils, which are rich in raw organic matter (see below). Thus, the relative contribution of yeasts to the microbial pool in these soils can be as high as that of prokaryotes. Considering the growing body of evidence about soil yeast diversity, their adaptations and also interactions with plants and animals (reviewed by Botha (2011)), the views on soil yeasts as a minor component of the below-ground community should be regarded as inappropriate and obsolete.

Approaches that are used to isolate yeasts are not always suitable for yeast biodiversity assessments in soils. For example, enrichment techniques based on sugar concentration would favour a few fermenting ascomycetes, whereas most of the yeasts in forest soils are basidiomycetes (e.g. di Menna 1957). Cultivation temperatures are also important since many soil yeasts are mesophiles. Vegetation protects soil cover from overheating, so average soil temperatures remain moderate (often below 25–30 °C, Chesworth 2008) and temperature variation further decreases with depth. Cultivation at temperatures exceeding 25 °C is likely to reduce the number of species and shift the community structure towards a few thermophilic species (e.g. di Menna 1955). Due to the oligotrophic nature of soils (reviewed by Botha 2006, 2011), soil yeasts can be expected to be slow-growing species. Thus, a short incubation period does not favour their isolation. Considering the aforementioned limitations, soil yeasts can be fairly well cultured on artificial media, and the approaches to isolate yeasts more efficiently have been developed (e.g. Starkey and Henrici 1927; Miller and Webb 1954; Babjeva 1969; Arteaga-Reyes et al. 1977; Golubev 2000; Golubev and Golubeva 2004). A detailed overview of methods used to investigate yeast biodiversity is provided in the book chapter by Boundy-Mills (2006). Culture media, supplements and incubation techniques have been changing with the evolving knowledge of the taxonomic composition of soil yeast communities and the ecology of the dominant species. For example, dilute or nitrogen-depleted media can be employed to reflect the oligotrophic nature of the habitat (reviewed by Botha 2006). Additionally, several approaches have been made to culture yeasts on media using nutrient sources resembling presumably natural conditions such as soil-extract agars (e.g. di Menna 1957). Because soil yeasts receive nutrients from decomposing plant material, media based on plant-derived extracts such as leaf litter were tested (e.g. Golubev and Golubeva 2004). Acidification of culture media with organic and inorganic acids has been long used to suppress the growth of soil prokaryotes (e.g. Starkey and Henrici 1927; Miller and Webb 1954; di Menna 1957; Babjeva 1969). The main problem associated with the cultivation of yeasts includes development of moulds, which often overgrow yeasts and complicate their isolation. Thus, a few additives were used to isolate yeasts from soil, including propionic acid, biphenyl, dichloran, bromocresol purple, crystal violet, ox gall and Rose Bengal among others (e.g. Martin 1950; Miller and Webb 1954; Babjeva 1969; Boundy-Mills 2006). Although many of these supplements suppressed filamentous fungi, they also affected some soil yeasts (e.g. Babjeva 1969; Boundy-Mills 2006; Wehde 2011). Because basidiomycetes dominate in forest soils, the fungicide Benomyl, which suppresses ascomycetes, can be potentially used to isolate basidiomycete yeasts from soils (e.g. Summerbell 1988; Thorn et al. 1996; Baldrian 2008), though this compound supports also the development of Mucoromycotina moulds (Summerbell 1988). Tremelloid yeasts, which are common soil inhabitants, can be preferentially isolated on inositol- and D-glucuronate-containing agars (Golubev 2000).

Yeasts are reported among the most numerous fungal operational taxonomic units (OTU as a proxy for species) in culture-independent surveys (e.g. Buee et al. 2009; Voříšková and Baldrian 2013; Mašínová et al. 2017b). Although yeasts were among the first fungi routinely identified with rDNA sequencing (Kurtzman and Robnett 1998; Fell et al. 2000; Scorzetti et al. 2002; Schoch et al. 2012), identification of soil yeasts from environmental sequence libraries is yet inaccurate. It can result from incomplete reference databases, computational errors in sequence classification pipelines as well as the lack of the taxonomic resolution of the utilised DNA markers. Even the most prominent soil yeasts suffer from inappropriate identification. Besides wrong or obsolete taxonomic names, they often appear in publications classified to the level of large former polyphyletic genera (e.g. *Cryptococcus* sp., *Rhodotorula* sp. and *Trichosporon* sp.) or even higher taxonomic levels, i.e. as Microbotryomycetes and Tremellomycetes. It must be admitted that such level of taxonomic resolution is not sufficient for ecological studies and is below the one provided by previously employed physiological tests. With so little information in hand, readers can learn that yeasts inhabit forest soils and are sometimes as abundant as other filamentous fungi. However, culture-independent surveys often provide little information about the distribution of particular yeast species in soils. Although a cultivation bias does obviously exist (Boundy-Mills 2006), most of the available culture-independent experiments do not allow evaluation of the efficiency of cultivation-based approaches in species discovery. Most frequent OTUs detected from soils correspond to the frequently isolated soilborne yeasts (discussed in Yurkov et al. (2012a), Mašínová et al. (2017b)). The recent study that focused on soil yeasts and used an amplicon sequencing technique reported similar species richness values as those previously obtained with a cultivation (Mašínová et al. 2017b). The authors concluded that yeasts could be fairly well cultivated from temperate forest soils. In the same study, several OTUs defined with a conservative 97% threshold were matched to the same yeast species, and this cautions against overestimating diversity values with common identification pipelines.

3.4.2 Distribution and Numbers

Unlike in above-ground sources, soil yeasts are not numerous, and their numbers rarely exceed 10^3–10^4 cells g^{-1}, although counts reaching 10^5–10^6 cells occasionally occur (Phaff and Starmer 1987; Botha 2006). Soils rich with raw organic matter (OM) usually yield higher yeast colony numbers. Yeasts are more abundant in temperate and boreal soils, where OM decomposition rates are slow (e.g. Babjeva and Golovleva 1963; Babjeva and Chernov 1995; Chernov 2005). However, subtropical and tropical soils are insufficiently sampled to make any well-supported conclusion on this topic (see the discussion below). For example, hydromorphic soils (Gleysols) under swamp vegetation, which harbour large and diverse yeast communities (e.g. Babjeva and Golovleva 1963), are azonal soils and can be found in different climates (e.g. Jaiboon et al. 2016).

Quantity of yeast cells usually decreases with soil depth, and this trend was explained by the amount of available nutrients and soil OM (e.g. Danielson and Jurgensen 1973; Maksimova and Chernov 2004; Botha 2006; Starmer and Lachance 2011). Viable yeasts (*Lipomyces tetrasporus*) were observed in soil layers up to 100 cm (Vinovarova and Babjeva 1987), although soil yeasts become exceedingly rare below the top 20–30 cm (e.g. Phaff and Starmer 1987; Maksimova and Chernov 2004). A recent study investigated changes in yeast quality and community composition in three soil types, two of which were underneath boreal coniferous and temperate broadleaf forests, respectively (Glushakova et al. 2017). Unlike previous studies, soils were sampled up to 200 cm depth in this experiment. In agreement with previous observations, yeast quantity declined gradually from the surface to 100 cm depth but increased again at 120–160 cm depth and declined rapidly in deeper soil layers. The most pronounced effect was observed in soils with illuvial (B) horizon, i.e. Spodosol (spruce forest) and Alfisol (broadleaf forest). The authors didn't provide an explanation of the aforementioned effect but reported that the majority of isolated yeasts were allochthonous to soils. On one hand, yeast cells could be passively washed into deeper soil layers and finally deposited in the B (illuvial) horizon, where soluble constituents from the A and E horizons are accumulated. On the other hand, the leached organic material from the A horizon, which is deposited in the B horizon as a humus-rich horizon band, can support a larger yeast community than nutrient-depleted upper soil layers. Thus, the question whether the yeasts passively reside or actively propagate in the B horizon needs further studies.

Our knowledge of soil yeasts is biased towards temperate and boreal forests. Also, several regions received were studied intensively (Fig. 3.1). This map presents a simplified overview on the analysed forest soils. The data were obtained from the literature presented in Table 3.1 as well as from earlier reviews by Phaff and Starmer (1987) and by Spencer and Spencer (1997). Babjeva, Chernov and co-workers intensively surveyed boreal and temperate forests in the European part of Russia (e.g. Babjeva and Golovleva 1963; Babjeva and Chernov 1995; Maksimova and Chernov 2004; Yurkov et al. 2015). Temperate forests in Central Europe were studied in Austria (Wuczkowski and Prillinger 2004; Wuczkowski et al. 2005), Czech Republic (Mašínová et al. 2017a, b), Italy (França et al. 2016), Germany (Yurkov et al. 2011, 2012a, b, 2016b) and Slovakia (Sláviková and Vadkertiová 2000). Temperate forests in Northeast Asia were analysed by Babjeva and Reshetova (1996) and by Takashima et al. (2012). Data from North America is scarce. A few studies reported the isolation of yeasts from soils, but these results are difficult to evaluate due to obsolete identification approaches (e.g. Bouthilet 1951; Capriotti 1967), poor community assessment (numerous reposts dedicated to a few

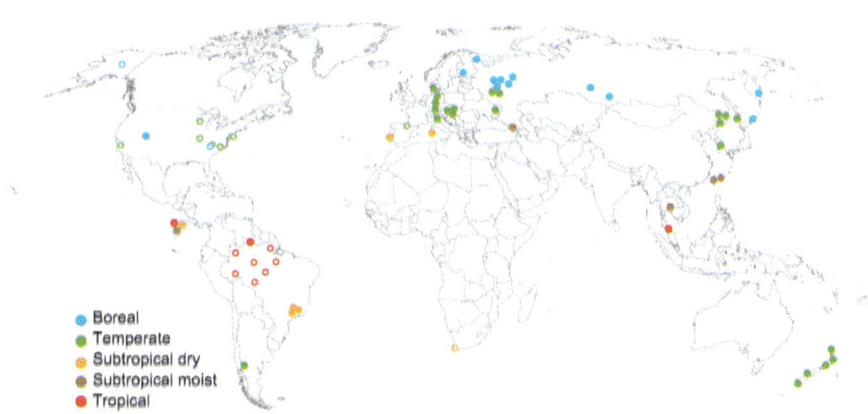

Fig. 3.1 Geographic distribution of the localities surveyed for soil yeasts. *Circles* are coloured according to the type of forest. *Empty circles* are the localities, where the information on soil yeast communities is limited

Table 3.1 Yeasts frequently isolated from forest soils during a number of surveys

Yeast (genus or species)	Selected species and names	Forest types and references (examples)
Basidiomycetous yeasts		
Apiotrichum dulcitum	*Trichosporon dulcitum*	Temperate deciduous (Wuczkowski and Prillinger 2004; Mestre et al. 2011; Yurkov et al. 2012a)
Apiotrichum porosum	*Trichosporon porosum*, also as *Trichosporon cutaneum**, *Trichosporon beigelii**	Boreal deciduous (Babjeva and Reshetova 1996). Temperate deciduous (di Menna 1960; Sláviková and Vadkertiová 2000; Takashima et al. 2012; Yurkov et al. 2012a; França et al. 2016)
Apiotrichum spp.	*Trichosporon laibachii*, *Trichosporon lignicola*, *Trichosporon loubieri*	Dry subtropical deciduous (Carvalho et al. 2013)
Bannozyma spp.	*Bensingtonia yamatoana*	Temperate deciduous (Mašínová et al. 2017a; Yurkov et al. 2016b)
Colacogloea spp.	*Rhodotorula diffluens*, *Rhodotorula foliorum*	Boreal deciduous (Golubtsova et al. 2007). Temperate deciduous (Mašínová et al. 2017a; Yurkov et al. 2016b)
Curvibasidium cygneicollum, *Curvibasidium* spp.	*Rhodotorula fujisanansis*, *Rhodotorula nothofagi*, *Curvibasidium pallidicorallinum*	Boreal deciduous (Yurkov et al. 2015). Temperate deciduous (Mašínová et al. 2017a). Dry subtropical deciduous (Yurkov et al. 2016a)
Cutaneotrichosporon curvatus	*Cryptococcus curvatus**, *Candida curvata**	Boreal deciduous (Babjeva and Reshetova 1996); Temperate deciduous (di Menna 1960, 1965)
Cutaneotrichosporon moniliiforme	*Trichosporon moniliiforme* (*Trichosporon beigelii**, *Trichosporon cutaneum**), *Trichosporon mucoides*	Boreal coniferous and deciduous (Vishniac 2006; Glushakova et al. 2015). Temperate deciduous (França et al. 2016; Yurkov et al. 2016b). Dry subtropical deciduous (Yurkov et al. 2016a). Tropical moist broadleaf (culture collections)
Cystobasidium spp.	*Rhodotorula laryngis*, *Rhodotorula minuta*, *Rhodotorula slooffiae*	Boreal coniferous and deciduous (Vishniac 2006; Golubtsova et al. 2007; Yurkov et al. 2015). Temperate deciduous (Takashima et al. 2012). Dry subtropical deciduous (Yurkov et al. 2016a). Tropical moist broadleaf (Vital et al. 2002)
Cystofilobasidium capitatum		Boreal coniferous and deciduous (Maksimova and Chernov 2004; Golubtsova et al. 2007). Temperate coniferous and deciduous (Wuczkowski et al. 2005; Sláviková and Vadkertiová 2000; Wuczkowski and Prillinger 2004; França et al. 2016; Yurkov et al. 2016b). Dry subtropical deciduous (Yurkov et al. 2016a)

(continued)

Table 3.1 (continued)

Yeast (genus or species)	Selected species and names	Forest types and references (examples)
Cystofilobasidium macerans	*Cryptococcus macerans*	Temperate coniferous and deciduous (Slávіková and Vadkertiová 2000; Wuczkowski and Prillinger 2004; Yurkov et al. 2012a)
Cystofilobasidium spp.	*Cystofilobasidium alribaticum*, *Cystofilobasidium intermedium*, *Cystofilobasidium infirmominiatum*	Boreal coniferous (Babjeva and Reshetova 1998). Temperate deciduous (Wuczkowski et al. 2005). Dry subtropical deciduous (Pontes et al. 2016; Yurkov et al. 2016a)
Dioszegia spp.	*Cryptococcus hungaricus*, *Dioszegia crocea*	Temperate deciduous (Wuczkowski and Prillinger 2004; França et al. 2016; Mašínová et al. 2017a; Yurkov et al. 2016b). Subtropical moist deciduous (Takashima et al. 2012)
Filobasidium magnum, *Filobasidium* spp.	*Cryptococcus magnus*, *Cryptococcus stepposus*, *Cryptococcus wieringae*, *Cryptococcus oeirensis*	Boreal coniferous and deciduous (Golubtsova et al. 2007; Vishniac 2006). Temperate deciduous (Wuczkowski et al. 2005; Yurkov et al. 2016b). Dry subtropical deciduous (Cardinali et al. 2012; Yurkov et al. 2016a)
Goffeauzyma gastrica	*Cryptococcus gastricus*	Boreal and temperate deciduous (di Menna 1960; Wuczkowski et al. 2005; Golubtsova et al. 2007; Yurkov et al. 2012a; França et al. 2016)
Hamamotoa spp.	*Hamamotoa cerberi*, *Hamamotoa telluris*	Temperate deciduous (Yurkov et al. 2016b)
Hannaella spp.	*Cryptococcus luteolus*, *Hannaella pagnoccae*	Subtropical deciduous (Vishniac 2006; Takashima et al. 2012; Landell et al. 2014)
Heterocephalacria spp.	*Cryptococcus arrabidensis*	Temperate deciduous (Takashima et al. 2012); Dry subtropical deciduous (Yurkov et al. 2016a)
Holtermanniella spp.	*Holtermanniella festucosa* (*Cryptococcus festucosus*), *Holtermanniella takashimae*, *Holtermanniella wattica* (*Cryptococcus watticus*)	Temperate deciduous (Takashima et al. 2012; França et al. 2016; Yurkov et al. 2012a). Dry subtropical deciduous (Yurkov et al. 2016a)
Krasilnikovozyma spp.	*Cryptococcus huempii*, *Mrakia curviuscula*	Temperate deciduous (França et al. 2016; Mašínová et al. 2017a). Dry subtropical deciduous (Yurkov et al. 2016a)
Leucosporidium scottii, *Leucosporidium* spp.	*Candida scottii*, *Candida* (*Rhodotorula*) *muscorum*, *Rhodotorula fragaria*, *Leucosporidium drummii*, *Leucosporidium golubevii*	Temperate coniferous and deciduous (di Menna 1965; Slávіková and Vadkertiová 2000; Wuczkowski and Prillinger 2004; Yurkov et al. 2012a; Mašínová et al. 2017a; França et al. 2016). Dry subtropical deciduous (Yurkov et al. 2016a)

(continued)

Table 3.1 (continued)

Yeast (genus or species)	Selected species and names	Forest types and references (examples)
Mrakia spp.	*Mrakia gelida, Mrakia frigida*	Boreal coniferous and deciduous (Maksimova and Chernov 2004). Temperate deciduous (Yurkov et al. 2016b)
Naganishia albida	*Cryptococcus albidus**	Boreal and temperate deciduous (di Menna 1960; Sláviková and Vadkertiová 2000; Golubev and Golubeva 2004; Yurkov et al. 2015). Tropical moist broadleaf (Vital et al. 2002)
Naganishia spp.	*Cryptococcus adeliensis, Cryptococcus albidosimilis, Cryptococcus diffluens*, Cryptococcus uzbekistanensis*	Boreal coniferous and deciduous (Vishniac 2006; Yurkov et al. 2015). Dry subtropical deciduous (Cardinali et al. 2012; Yurkov et al. 2016a)
Oberwinklerozyma spp.	*Rhodotorula silvestris, Rhodotorula yarrowii*	Temperate deciduous (Mašínová et al. 2017a)
Papiliotrema laurentii, Papiliotrema flavescens, Papiliotrema spp.	*Cryptococcus laurentii Cryptococcus flavescens, Cryptococcus terrestris*	Temperate coniferous and deciduous (Golubev and Golubeva 2004; Sláviková and Vadkertiová 2000; Wuczkowski and Prillinger 2004; Wuczkowski et al. 2005). Subtropical forests (Vital et al. 2002; Cardinali et al. 2012; Carvalho et al. 2013; Jaiboon et al. 2016; Yurkov et al. 2016a)
Piskurozyma spp.	*Bullera taiwanensis, Cryptococcus cylindricus, Cryptococcus filicatus, Piskurozyma tuonelana, Piskurozyma yama*	Temperate deciduous (Mašínová et al. 2017a; Yurkov et al. 2016b)
Rhodotorula spp. *Rhodosporidiobolus* spp.	*Rhodotorula mucilaginosa, Rhodotorula glutinis, Rhodosporidium babjevae Rhodotorula colostri*	Boreal coniferous and deciduous (Babjeva et al. 1973; Golubtsova et al. 2007). Temperate deciduous (Babjeva et al. 1973; Mestre et al. 2011; Sláviková and Vadkertiová 2000; Yurkov et al. 2016a, b; França et al. 2016). Subtropical forests (Ahansal et al. 2008; Carvalho et al. 2013; Jaiboon et al. 2016; Mok et al. 1984; Takashima et al. 2012; Yurkov et al. 2016a)
Saitozyma podzolica	*Cryptococcus podzolicus, Candida podzolica*	Boreal coniferous and deciduous (Babjeva and Reshetova 1975, 1996, 1998; Babjeva and Chernov 1995; Maksimova and Chernov 2004; Yurkov et al. 2015). Temperate coniferous and deciduous (Jensen 1963; Sláviková and Vadkertiová 2000; Golubev and Golubeva 2004; Wuczkowski and Prillinger 2004; Wuczkowski et al.

(continued)

Table 3.1 (continued)

Yeast (genus or species)	Selected species and names	Forest types and references (examples)
		2005; Mestre et al. 2011; Takashima et al. 2012; Yurkov et al. 2012a; França et al. 2016; Mašínová et al. 2017a). Subtropical forests (Vishniac 2006; Takashima et al. 2012)
Slooffia spp.	*Rhodotorula pilati*, *Sporobolomyces tsugae*	Boreal and temperate deciduous (Golubtsova et al. 2007; Yurkov et al. 2012b, 2016b)
Solicoccozyma aeria	*Cryptococcus aerius*, *Cryptococcus albidus* var. *aerius*	Boreal coniferous (Vishniac 2006); Temperate deciduous (Golubev and Golubeva 2004; Wuczkowski and Prillinger 2004; França et al. 2016). Dry subtropical deciduous (Cardinali et al. 2012; Yurkov et al. 2016a)
Solicoccozyma terrea	*Cryptococcus terreus*, *Cryptococcus elinovii*, *Cryptococcus himalayensis*, *Cryptococcus phenolicus*	Temperate coniferous and deciduous (di Menna 1960, 1965; Golubev and Golubeva 2004; Mestre et al. 2011; Yurkov et al. 2012a, b; França et al. 2016). Dry subtropical deciduous (Cardinali et al. 2012; Yurkov et al. 2016a)
Solicoccozyma terricola	*Cryptococcus terricola*, *Cryptococcus terricolus*, *Cryptococcus albidus* var. *terricolus*	Boreal coniferous and deciduous (Babjeva and Chernov 1995; Babjeva and Reshetova 1998; Maksimova and Chernov 2004; Yurkov et al. 2015). Temperate coniferous and deciduous (Wuczkowski and Prillinger 2004; Mestre et al. 2011; Takashima et al. 2012; Yurkov et al. 2012a; França et al. 2016; Mašínová et al. 2017a); Dry subtropical deciduous (Yurkov et al. 2016a)
Tausonia pullulans	*Guehomyces pullulans*, *Trichosporon pullulans*	Boreal coniferous and deciduous (Babjeva et al. 1973; Babjeva and Chernov 1995; Babjeva and Reshetova 1996, 1998; Maksimova and Chernov 2004; Vishniac 2006); Temperate coniferous and deciduous (di Menna 1965; Sláviková and Vadkertiová 2000; Wuczkowski and Prillinger 2004; Mestre et al. 2011; Takashima et al. 2012; Yurkov et al. 2012a)
Vanrija albida *Vanrija humicola* *Vanrija* spp.	*Cryptococcus ramirezgomesianus*, *Cryptococcus humicola* *Asterotremella* spp.	Temperate coniferous and deciduous (di Menna 1960, 1965; Sláviková and Vadkertiová 2000; Wuczkowski and Prillinger 2004; Mestre et al. 2011; Takashima et al. 2012; Yurkov et al. 2012a; França et al. 2016); Dry subtropical deciduous (Carvalho et al. 2013; Yurkov et al. 2016a)

(continued)

Table 3.1 (continued)

Yeast (genus or species)	Selected species and names	Forest types and references (examples)
Vishniacozyma victoriae *Vishniacozyma* spp.	*Cryptococcus victoriae* *Cryptococcus tephrensis*, *Cryptococcus heimaeyensis*, *Cryptococcus taibaiensis* *Cryptococcus laurentii**	Boreal coniferous and deciduous (Babjeva and Chernov 1995; Maksimova and Chernov 2004; Yurkov et al. 2015); Temperate coniferous and deciduous (Wuczkowski and Prillinger 2004; Takashima et al. 2012; Yurkov et al. 2012a; França et al. 2016; Mašínová et al. 2017a); Subtropical forests (Vishniac 2006; Takashima et al. 2012; Yurkov et al. 2016a)
Ascomycetous yeasts		
Barnettozyma californica, *Barnettozyma pratensis*, *Barnettozyma* spp.	*Williopsis californica*, *Williopsis pratensis*, *Barnettozyma vustinii*	Temperate deciduous (Bouthilet 1951; Wuczkowski and Prillinger 2004; Takashima et al. 2012; Yurkov et al. 2012a)
Candida maritima		Temperate deciduous (Mestre et al. 2011)
Candida parapsilosis	*Candida orthopsilosis*	Subtropical forests (Limtong et al. 2009; Vishniac 2006; Yurkov et al. 2016a)
Candida quercitrusa		Dry subtropical deciduous (Yurkov et al. 2016a)
Candida railenensis		Temperate deciduous (Mašínová et al. 2017a; Yurkov et al. 2016b)
Candida sake		Temperate deciduous (Wuczkowski and Prillinger 2004; Yurkov et al. 2016b)
Candida santamariae		Temperate deciduous (Yurkov et al. 2016b; França et al. 2016)
Candida solani		Temperate deciduous (Takashima et al. 2012)
Candida vartiovaarae		Temperate deciduous (Takashima et al. 2012; Yurkov et al. 2012a, b)
Candida spp.	*Candida albicans*, *Candida tropicalis*	Tropical moist broadleaf (Mok et al. 1984)
Cyberlindnera spp.	*Cyberlindnera* (*Pichia*) *misumaiensis*, *Cyberlindnera culbertsonii*, *Cyberlindnera rhizosphaerae*, *Cyberlindnera (Williopsis) saturnus*, *Cyberlindnera (Williopsis) suaveolens*, *Cyberlindnera subsufficiens*	Boreal coniferous and deciduous (Golubtsova et al. 2007); Temperate deciduous (Bouthilet 1951; Babjeva and Reshetova 1996; Wuczkowski and Prillinger 2004; Wuczkowski et al. 2005; Mestre et al. 2011; Sylvester et al. 2015; Yurkov et al. 2016b); Subtropical forests (Vital et al. 2002; Vishniac 2006; Takashima et al. 2012; Carvalho et al. 2013; Jaiboon et al. 2016)

(continued)

Table 3.1 (continued)

Yeast (genus or species)	Selected species and names	Forest types and references (examples)
Debaryomyces hansenii	*Candida famata*	Boreal coniferous and deciduous (Babjeva et al. 1973; Babjeva and Reshetova 1998; Sláviková and Vadkertiová 2000; Maksimova and Chernov 2004; Golubtsova et al. 2007); Temperate deciduous (Takashima et al. 2012; Yurkov et al. 2012a; França et al. 2016); Subtropical and tropical forests (Mok et al. 1984; Vital et al. 2002; Carvalho et al. 2013; Yurkov et al. 2016b)
Geotrichum spp.	*Geotrichum candidum*, *Geotrichum vulgare*	Boreal and temperate deciduous (Wuczkowski and Prillinger 2004; Golubtsova et al. 2007; Groenewald et al. 2012)
Hanseniaspora spp.	*Hanseniaspora guilliermondii*, *Hanseniaspora occidentalis*, *Hanseniaspora uvarum*	Boreal coniferous and deciduous (Maksimova and Chernov 2004); sub-tropical moist deciduous (Limtong et al. 2009)
Kazachstania piceae *Kazachstania* spp.	*Arxiozyma telluris*, *Saccharomyces exiguus*, *Kazachstania servazzii*, *Kazachstania siamensis*	Temperate deciduous (Wuczkowski and Prillinger 2004; Yurkov et al. 2012a, b); Subtropical forests (Vishniac 2006; Limtong et al. 2008; Cardinali et al. 2012; Carvalho et al. 2013); Tropical moist broadleaf (Vital et al. 2002)
Kodamaea ohmeri	*Pichia ohmeri*	Boreal deciduous (Babjeva et al. 1973); Dry subtropical deciduous (Carvalho et al. 2013)
Meyerozyma spp.	*Pichia* (*Candida*) *guilliermondii*, *Pichia caribbica*	Subtropical forests (Limtong et al. 2009; Jaiboon et al. 2016; Yurkov et al. 2016a)
Nadsonia starkeyi-henricii	*Schizoblastosporion starkeyi-henricii*	Temperate coniferous and deciduous (di Menna 1965; Babjeva and Chernov 1995; Yurkov et al. 2012a; França et al. 2016)
Pichia kudriavzevii	*Issatchenkia orientalis*, *Candida krusei*	Subtropical forests (Vishniac 2006; Carvalho et al. 2013); Tropical moist broadleaf (Vital et al. 2002)
Schwanniomyces spp.	*Schwanniomyces castellii*, *Schwanniomyces polymorphus*, *Schwanniomyces pseudopolymorphus*, *Schwanniomyces vanrijiae*	Boreal deciduous (Babjeva et al. 1973); Temperate coniferous and deciduous (Sláviková and Vadkertiová 2000; Wuczkowski and Prillinger 2004; Yurkov et al. 2012a); subtropical forests (Limtong et al. 2009; Carvalho et al. 2013; Jaiboon et al. 2016; Yurkov et al. 2016a)

(continued)

Table 3.1 (continued)

Yeast (genus or species)	Selected species and names	Forest types and references (examples)
Teunomyces kruisii	*Candida kruisii*	Temperate deciduous (Yurkov et al. 2016b)
Wickerhamomyces anomalus	*Pichia anomala*, *Hansenula anomala*	Temperate deciduous (Bouthilet 1951); subtropical moist deciduous (Limtong et al. 2009)

Note: Species identified solely based on morphological and physiological characters are marked with an asterisk

species) or the lack of information on the biotope (e.g. Bouthilet 1951; Miller et al. 1962; Capriotti 1967; Danielson and Jurgensen 1973; Sylvester et al. 2015). Many studies have focused on the description of new yeasts and did not provide any further information about other yeasts isolated from the same soils (e.g. taxonomic works by Capriotti, Phaff and Wickerham). A handful of publications report soil yeasts from subtropical and tropical forests worldwide (e.g. Mok et al. 1984; Vital et al. 2002; Vishniac 2006; Takashima et al. 2012). It is important to document that these forests received least attention despite their importance as the major biodiversity hotspot. Both large studies of yeasts from Amazon rainforests in Brazil had an applied focus and surveyed either species pathogenic to humans (Mok et al. 1984) or yeasts producing killer toxins (Vital et al. 2002). Likewise, Asian soils were mainly studied as the source of novel yeast species, but the information about the distribution of other species is scarce (e.g. Limtong et al. 2007, 2009; Landell et al. 2014; Jaiboon et al. 2016). Forest soils in the Southern hemisphere are strongly undersampled, although the temperate silver beech (*Nothofagus pumilio*) forest was studied in Patagonia (Mestre et al. 2011). Di Menna (1965) performed a broad survey of New Zealand soils, but the majority of them were under herbaceous vegetation. Furthermore, very few cultures isolated during this study were retained, so the majority of yeast names used at that time cannot be confidently translated into the currently taxonomy.

3.5 Nutrition and Traits

As it has been repeatedly emphasised before, soils represent a refuge for yeast propagules from above-ground sources (e.g. Miller et al. 1962; Phaff and Starmer 1987; Lachance and Starmer 1998; Starmer and Lachance 2011). Yeast cells enter the soil with fruits and leaves as well as with the activity of animals (e.g. Miller et al. 1962; Phaff and Starmer 1987; Yurkov et al. 2008). Soils are the largest storage of the dead OM that supports extremely rich microbial communities (e.g. van der Heijden et al. 2008; Parker 2010). Decomposition of the soil OM potentially allows yeasts from other sources to survive in soils. However, despite the astonishing 1550 gigatons of organic carbon stored in soils worldwide (Lal 2004), soils remain a nutrient-poor substrate because most of the carbon and nitrogen are stored in heterocyclic compounds (humic substances) and are not available for soil microorganisms. Sugars are also present in soils in the total concentration of about $10 \pm 5\%$ (Gunina and Kuzyakov 2015). They are represented mainly by hexoses (glucose, galactose, mannose and rhamnose) and pentoses (arabinose and xylose). The dominating sugar is glucose. The content of other prominent sugars such as galactose, mannose, arabinose and xylose is 1.5 to 2 times lower than glucose. Most of the soil sugars are of non-cellulose origin, and their turnover rates are very high, so the uptake of monosaccharides by microorganisms takes place in seconds to minutes (Gunina and Kuzyakov 2015). Thus, yeast species, which do not have physiological abilities to degrade complex molecules, will be outcompeted by faster soil microorganisms and become the source of sugars for other species, i.e. microbial recycling. Soil invertebrates and protists also graze on yeasts in the litter and topsoil and decrease their numbers (e.g. Heal 1963; Babjeva and Gorin 1987; Byzov et al. 1993; Botha 2006; Men'ko et al. 2006; Yurkov et al. 2008). These processes result in a frequently reported trend that forest soils contain more yeasts in the end of the vegetative period than in the beginning of the next year (e.g. Bouthilet 1951; Jensen 1963; Sláviková and Vadkertiová 2000; Golubtsova et al. 2007). Soils were long considered not suitable for yeasts, and the ability of yeasts to propagate in soils was repeatedly questioned (discussed by Danielson and Jurgensen 1973; Phaff and Starmer 1987). Yeast numbers in soils exceeding those on decaying plant material convinced researchers that yeasts do live and not reside in soils. Although the existence of allochthonous soil yeasts is taken now for granted, forest soils usually contain 10^2–10^3 yeast cells g^{-1} of soil, and these numbers are lower than those in phyllosphere and fresh litter (e.g. Fonseca and Inacio 2006). As outlined by Phaff and Starmer (1987), the repeated isolation of the same yeasts from soils and their absence in other sources above ground was employed as another argument to prove soil origin of several yeast species. Subsequent studies showed that these yeasts often possess adaptations that help them to live in soils. Unlike the typical saccharolytic phenotype often attributed to yeasts, basidiomycetous species are able to utilise a wide spectrum of carbon sources, including complex compounds (Fonseca 1992; Middelhoven 1993, 2004, 2006; Sampaio 1999). In his review on soil yeasts, Botha (2006) noted that most of the yeast species frequently encountered in soil are able to utilise hemicellulose-derived sugars L-arabinose, D-xylose and cellobiose (see also di Menna 1959; Sláviková and Vadkertiová 2000; Mestre et al. 2011). These carbohydrates are known to be products of the enzymatic hydrolysis of lignocellulosic plant materials by bacteria and moulds (Bisaria and Ghose 1981; Tomme et al. 1995). Some of the frequently encountered yeasts in soil were also found to assimilate intermediates of lignin degradation, i.e. ferulic, 4-hydroxybenzoic and vanillic acids (e.g. Henderson 1961; Botha 2006; Yurkov et al. 2016b). Also, this indicates that soil yeasts might contribute to the decomposition of woody material to soil organic matter and dissipation of nutrients within the soil.

Several adaptations might facilitate yeast capability for surviving in soil substrates (see also reviews by Botha 2006, 2011). For example, species frequently found in soil have been shown to be able to grow in media with low concentrations of nutrients (Vishniac 1983; Babjeva and Gorin 1987; Kimura et al. 1998). Oligotrophy is apparently thought to provide yeasts an advantage in competition with other soil microbes. Nitrogen oligotrophy is a widespread adaptation of yeast fungi, which enables them to colonise diverse substrates such as plant surface (Fonseca and Inacio 2006), tree fluxes (Golubev et al. 1977) and soils (Botha 2011). Yeasts were reported to grow at as much as 0.002% (di Menna 1959; Golubev et al. 1977) and even 0.0003% (Babjeva and Gorin 1987) of nitrogen in the media. Such ecological adaptation is very important since soils as a habitat are characterised by a strong limitation of available nitrogen (Date 1973; Reich et al. 2006). Furthermore, most soil nitrogen (some 96–98%) is bound within OM as complex insoluble polymers such as chitin, proteins and nucleic acids (van der Heijden et al. 2008). Interestingly, that typical soil yeasts which form the genus *Lipomyces* have the ability to assimilate nitrogen incorporated into heterocyclic compounds, such as imidazole, pyrimidine and pyrazine (LaRue and Spencer 1967; van der Walt 1992). Recent studies showed diversity of yeasts growing on imidazole is larger and includes both asco- and basidiomycetes (Cornelissen et al. 2003; Yurkov et al. 2011, 2016b). However, unlike typical oligotrophic organisms, many yeast species are able to grow in a wide range of nutrition levels (e.g. di Menna 1957; Yurkov et al. 2011).

The other adaptation frequently reported to be advantageous for soil is the capability of the production of extracellular polysaccharide capsules (EPS). The formation of these capsules is a known mechanism whereby microbes are able to sequester and concentrate nutrients while growing in low-nutrient environments or sustain low water activity and desiccation (di Menna 1959; Aksenov et al. 1972; Raspor and Zupan 2006). It was shown that semiarid soils, low in nutrients and moisture, were mostly populated by encapsulated anamorphic basidiomycetous yeasts (Spencer and Spencer 1997; Vishniac 2006). The ability of some of these soil yeasts to survive in sandy soils due to production of EPS has been demonstrated with soil yeast *Naganishia albida* (formerly *Cryptococcus albidus*, Vishniac 1995). Soil-borne *Naganishia* and *Solicoccozyma* species (*Cryptococcus diffluens* and *Cryptococcus terreus*, di Menna 1959) were viable after storage for 9 months in the dry stage. Due to this extraordinary ability to produce EPS, yeasts from the genera *Cryptococcus* and *Lipomyces* play a role in soil aggregation and stability, thereby impacting water-holding capacity, as well as improving soil fertility (reviewed by Botha 2011).

Within the ecosystem, organic carbon acting as nutrient source for soil-borne microbial decomposers ultimately originates from plants (Wardle et al. 2004). Assimilation of lignin and cellulose derivatives, oligotrophy and psychrotolerance imply that yeasts could play a role in the decomposition process, especially in cold regions or at low temperatures. These findings are supported with observation of Turchetti et al. (2008) that cold-adapted yeasts retain activity and produce enzymes at low temperatures. Babjeva and Golovleva (1963) reported successful isolation of several strains of *Na. albida* after incubation at −2 °C. Although yeasts were repeatedly isolated from decaying plant material, those studies were mainly addressed to the description of the novel species (e.g. Lee and Komagata 1980; Péter et al. 2003; Middelhoven 2006; Middelhoven and Kurtzman 2007). While filamentous fungi in the soils are commonly considered as primary degraders (Christensen 1989; Bridge and Spooner 2001; Baldrian 2016), it is especially difficult to prove that a certain yeast species is involved in degradation of soil OM. Our knowledge about possible roles of yeasts in decomposition of plant material is limited to a few studies, which demonstrated utilisation of several relevant compounds (Sampaio 1999; Middelhoven 2004, 2006). Another indirect evidence comes from the culture-independent study, which showed that the abundance of sequence reads corresponding to *Apiotrichum* (*Trichosporon*) yeasts was increasing during the decomposition of forest (oak) litter (Voříšková and Baldrian 2013).

Yeasts reflect changes in environmental conditions (pH, water activity, temperature) and nutrient availability. For example, after sampling different soil types, it was found that a positive correlation (at 1% level) exists between soil yeast population size and both organic carbon ($r = 0.884$) and organic nitrogen content ($r = 0.829$) of the soil (Moawad et al. 1986, reviewed by Botha 2006). Similarly, changes in the yeast community of forest soils correlated with soil moisture and, thus, followed seasonal changes (Sláviková and Vadkertiová 2000) and latitudinal changes of physicochemical environmental conditions (Chernov 2005; Vishniac 2006). However, abiotic soil parameters have little effect on soil yeast communities within the same type of forest. It has been shown that yeast quantity, diversity and community structure reflect forest properties, such as age and management history, but not the basic abiotic properties, including pH, nitrogen content and C/N ratio (Birkhofer et al. 2012; Yurkov et al. 2012a). Likewise, yeast communities in Mediterranean forest soils reflected the properties of the forest cover, which, in turn, is shaped by the local precipitation regime (Yurkov et al. 2016a).

As decomposers, yeasts rely on the above-ground nutrients from leaf litter and dead wood. It has been shown that forest litter differs in properties and composition and can affect soil microbial communities (Gunina and Kuzyakov 2015). Interestingly, forest litters with similar nutrient content but of different origin harbour different microbial communities (e.g. Urbanová et al. 2015). Tree species was the major factor explaining the occurrence of soil yeasts under boreal spruce, birch and alder forests (Maksimova and Chernov 2004). However, in contrast to the previous observation, the effect of tree species was not significant for soil yeasts in the temperate forest (Mašínová et al. 2017b). In the experiment, which studied the influence of the decomposing wood on soil yeasts, large wood logs provided stable conditions and promoted growth of a few soil-borne yeasts such as *Saitozyma podzolica* (*Cryptococcus podzolicus*) and *Kazachstania piceae* (Yurkov et al. 2012b). However, another typical soil yeast *Solicoccozyma terricola* (*Cryptococcus terricola*) showed no preference for decomposing plant material.

3.6 Diversity of Soil Yeasts

The current body of knowledge suggests that many yeast species are adapted to soil habitat. Some of them are widespread, while others were found in a certain type of forest. Species repeatedly reported from forest soils are summarised in Table 3.1. Ascomycetes genera *Lipomyces* and *Myxozyma* were not included in this table. These yeasts are permanent residents in the soil (Babjeva and Gorin 1987; Phaff and Starmer 1987; Lachance and Starmer 1998). Even though this table lists more than 130 yeasts, this number should be considered as approximate because many regions of the world are insufficiently sampled (Fig. 3.1). Not every yeast species isolated from soil is an indigenous soil inhabitant but which may originate from other sources other than soils (Table 3.1). For example, pigmented *Cystobasidium*, *Rhodotorula*, *Rhodosporidiobolus*, *Sporobolomyces* and *Vishniacozyma* yeasts from plant surfaces were frequently recovered from soils. Species of the Basidiomycete genera *Cystofilobasidium* and *Apiotrichum* as well as non-pigmented Microbotryomycetes (e.g. *Bannozyma*, *Colacogloea*, *Curvibasidium*, *Hamamotoa* and *Oberwinklerozyma*) are shared sometimes between topsoil and forest litter layers (Babjeva and Chernov 1995; Maksimova and Chernov 2004; Golubev and Golubeva 2004; Golubev and Tomashevskaya 2010; Mašínová et al. 2017a). Observation of fermenting Ascomycete yeasts frequently found on fruit surfaces, such as *Hanseniaspora*, suggests that they reside in soils (e.g. Phaff and Starmer 1987). However, the ability to ferment sugars does not predict well the transient habit of a yeast species since several autochthonous soil yeasts possess this trait, e.g. *Barnettozyma*, *Cyberlindnera*, *Kazachstania* and *Schwanniomyces*.

Yeast fungi occur in various soils worldwide and are also more prominent in the temperate zone and in tundra (Babjeva and Chernov 1995; Chernov 2005; Vishniac 2006; Botha 2011; Yurkov et al. 2012a). Basidiomycetes are dominant in forest soils, and yeasts of the former polyphyletic genus *Cryptococcus* are among most frequently reported species (e.g. Babjeva and Chernov 1995; Chernov 2005; Botha 2006; Yurkov et al. 2012a). As *Cryptococcus* was reclassified (Liu et al. 2015), soil-related species were accommodated in species belonging to various genera: *Goffeauzyma gastrica* (*Cryptococcus gastricus*), *Holtermanniella wattica* (*Cryptococcus watticus*), *Na. albida* (*Cr. albidus*), *Papiliotrema laurentii* and *Papiliotrema terrestris* (*Cryptococcus laurentii*, *Cryptococcus terrestris*), *Sa. podzolica* (*Cr. podzolicus*), *Solicoccozyma aeria*, *Sol. terricola* and *Solicoccozyma terrea* (e.g. *Cryptococcus aerius*, *Cr. terricola*, *Cr. terreus*) and *Vanrija humicola* and *Vanrija albida* (*Cryptococcus humicola*, *Cryptococcus ramirezgomesianus*). Other less frequent yeasts reported from soils include *Heterocephalacria arrabidensis* (*Cryptococcus arrabidensis*), *Krasilnikovozyma huempii* (*Cryptococcus huempii*) and *Piskurozyma cylindrica* (*Cryptococcus cylindricus*) among others. *Trichosporon* was another prominent yeast genus in soils and was reclassified (Liu et al. 2015) with soil-related species that are accommodated in the genera *Apiotrichum* (*Trichosporon dulcitum*, *Trichosporon laibachii*, *Trichosporon lignicola*,

Trichosporon loubieri and *Trichosporon porosum*) and *Cutaneotrichosporon* (*Cryptococcus curvatus*, *Trichosporon moniliiforme*).

Older studies report phenotypic species *Cr. albidus*, *Cr. laurentii*, *Trichosporon cutaneum* (also as *Trichosporon beigelii*) and *Trichosporon pullulans* from forest soils (Table 3.1). But reidentification of these yeast cultures with DNA-based tools has been performed in only a few cases. In boreal forests, *Cr. albidus* and *Cr. laurentii* identified with physiological tests corresponded to the genera *Filobasidium* and *Vishniacozyma* (Glushakova and Chernov 2010; Yurkov et al. 2015). Yeasts isolated by Babjeva and co-workers and identified as *Tr. cutaneum* and *Tr. beigelii* were found to be *Ap. porosum* (Kachalkin, personal communication). In the same time, *Na. albida* and *Tausonia pullulans* (formerly *Tr. pullulans*) are present in forest soils but not as dominant species (e.g. Yurkov et al. 2012a, 2015). More recently, *Ap. dulcitum* and *Ap. porosum* were found to be abundant in different forest types and climates (e.g. Wuczkowski and Prillinger 2004; Golubev and Tomashevskaya 2010; Yurkov et al. 2012a; Mašínová et al. 2017a). Interestingly, these two species have been described nearly a century ago. However, they were not recognised as widespread soil-borne yeasts before wide application of molecular techniques (discussed in Yurkov et al. (2012a)). Other former phenotypic species in question are *Cr. humicola* and *Cr. curvatus*, which were repeatedly reported in pioneering surveys. The former species belongs presently to the genus *Vanrija*, and the two closely related species *Va. albida* and *Vanrija musci* were isolated from forest soils. *Va. humicola* is a rather rare yeast in soils. Further studies are required to recollect and reidentify cultures resembling phenotypically *Cr. curvatus* to interpret the older reports of this species. Isolation of yeasts belonging to species of the genera *Saitozyma* (*Cr. podzolicus*), *Solicoccozyma* (e.g. *Cr. aerius*, *Cr. terricola*, *Cr. terreus*) from forest soils is consistent with more recent observations.

Many yeast species were isolated from soils. Although soil communities are frequently regarded as species-poor, low species richness in a single plot (alpha diversity) contrasts with the larger number of yeasts, which can be isolated from a forest or a region. Yeast distribution in soils is often fragmented with a few species only shared between sampling sites. For example, Vishniac (2006) reported nearly 40% of yeasts to be restricted to a single locality. Likewise, temperate forests in Germany (3 regions) had only *Ap. dulcitum* in common (Yurkov et al. 2012a). Three Mediterranean xerophyl forests sampled in a single locality had 8 out of 57 species shared between all three plots (Yurkov et al. 2016a). The dissimilarity in species composition between sites results in high diversity values on the regional level (e.g. Yurkov et al. 2011, 2016a). Recent studies showed that fairly well-analysed soils yield a large number of yet unknown yeasts. The proportion of potential novel taxa was estimated to exceed 30% in temperate beech and Mediterranean xerophyl forests (Yurkov et al. 2012a, 2016a). The same holds true for a few other temperate forests (Mestre et al. 2011; Takashima et al. 2012; Mašínová et al. 2017a, b) and is likely to be true for tropical biotopes.

3.7 Concluding Remarks

The term soil has a broad definition, and yeast researchers often considered weathered rocks, sediments and decaying plant material as soils. The litter layer in forests is a part of soil profile (O horizon), but it is not always simultaneously included in an ecological study (e.g. but see Danielson and Jurgensen (1973), Maksimova and Chernov (2004), Mašínová et al. (2017b)). This is one of the reasons why some yeasts cannot be unambiguously attributed to either soils or plant material. The soil litter is classified in three layers (L, F and H), which differ substantially in decomposition rate of plant material and quality of OM. It is likely that yeast communities reflect these changes through their composition, structure and activity in the same way as they respond to OM content in different soil types. As discussed above, the propagation, fate and transportation of yeast cells between and within soil horizons are still little understood and should be addressed in future studies.

Recent studies suggest that old natural forests harbour more diverse yeast population than managed ones (Yurkov et al. 2011, 2012a). Also, yeast communities respond to forest properties rather than to basic soil parameters even though the mechanisms underlying these effects remain unknown. Forest management is known to influence substrate-dependent taxa, such as bryophytes and sporocarp (fruiting body) forming fungi (Ódor et al. 2006). To what extent the distribution pattern of soil yeasts is following those of macro-fungi deserves a detailed investigation in the future. The question regarding the functional connection between soil yeasts and the above-ground habitats is especially interesting in the light of close phylogenetic relationships of tremellaceous mycoparasites (e.g. *Carcinomyces*, *Heterocephalacria*, *Rhynchogastrema*, *Syzygospora* and *Tremella*) and prominent soil-borne yeasts (Scorzetti et al. 2002; Millanes et al. 2011; Liu et al. 2015). These yeasts are known only from their saprobic asexual states and not outside the soil habitat. The question on whether or not the teleomorphs of soil yeasts depend on their hosts (lichens or other fungi) as well as the degree of the host specificity may shed new light on the ecology of below ground microorganisms.

Because yeasts respond to forest properties, they might be threatened by the habitat loss caused by the deforestation and forest fragmentation (e.g. di Menna 1960). This needs to be taken into consideration and focus future studies on natural and unmanaged forests. Although communities of yeasts in forest soils are often species-poor in a single sample, they are more diverse on the biotope level. Thus, it is important to remember that reliable sampling and cultivation efforts are both crucial for a biodiversity assessment. Yeast communities vary with forest age (e.g. Yurkov et al. 2012a), different succession stages (e.g. birch and spruce forests, Maksimova and Chernov 2004) and along altitudinal gradient (e.g. França et al. 2016). Long-term experiments addressed to the modelling of plant successions (e.g. litter and soil transplantation experiments) are important to understand the factors influencing stability, plasticity and resilience of soil yeasts communities.

Acknowledgements I am grateful to all people who taught me soil science, microbiology, yeast ecology and systematics. I owe much of my knowledge to Inna Babjeva, Ivan Chernov, Álvaro Fonseca and the teachers of the van Uden International Advanced Course on Molecular Ecology, Taxonomy and Identification of Yeasts. I would like to thank colleagues and project students from the Lomonosov Moscow State University, Ruhr-University Bochum and Universidade Nova de Lisboa for their assistance during the projects on soil yeasts. Aleksey Kachalkin, Tereza Mašínová and Petr Baldrian are acknowledged for sharing unpublished data, and Cletus P. Kurtzman is acknowledged for correcting the English text.

References

Ahansal L, Sassi AB, Martini A, Vaughan-Martini A, Walker G, Boussaid A (2008) Biodiversity of yeasts isolated from the indigenous forest of Argan (*Argania spinosa* (L.) Skeels) in Morocco. World J Microbiol Biotechnol 24:777–782

Aksenov SI, Babjeva IP, Golubev WI (1972) The investigation of desiccation–moistening cycles in *Cryptococcus albidus* var. *diffluens* by the NMR–Spin Echo technique. Biol Bull (former Izvestiya Rossiiskoi Akademii Nauk, Seriya Biologicheskaya) 4:545–558 (in Russian)

Aleksandrova VD (2012) Russian approaches to classification of vegetation. In: Whittaker RH (ed) Classification of plant communities. Springer, Heidelberg, pp 167–200

Arteaga-Reyes E, Echegaray-Aleman A, Garcia-Trejo A (1977) An improved method to isolate yeasts from soil. Soil Biol Biochem 9:367–369

Babjeva IP (1969) To the methods of enumeration of yeasts in soils. Biologiceskie nauki (former Nauchnye Doklady Vysshei Shkoly. Biologicheskie Nauki) 12:113–118 (in Russian)

Babjeva IP, Chernov IY (1995) Geographic aspects of yeast ecology. Physiol Gen Biol Rev 9:1–54

Babjeva IP Golovleva LA (1963) Yeast flora of the main soil types of the European part of the USSR. In: Microorganisms in agriculture. Moscow State Univ Press, pp 231–251 (in Russian)

Babjeva IP, Gorin SE (1987) Soil yeasts. Moscow State Univ Press, Moscow (in Russian)

Babjeva IP. Reshetova IS (1972) Quantity and biomass of yeast cells in soils. In: Current issues concerning the quantity, biomass and efficiency of soil organisms. Nauka, Leningrad, pp 71–79 (in Russian)

Babjeva IP, Reshetova IS (1975) A new yeast species *Candida podzolica* sp. nov. isolated from the soil. Microbiology 44:333–338 (in Russian)

Babjeva IP, Reshetova IS (1996) Taxonomic analysis of yeast fungi in the Russian Far East. Mikol Fitopatol 30(4):10–18

Babjeva IP, Reshetova IS (1998) Yeast resources in natural habitats at polar circle latitude. Food Technol Biotechnol 36:1–6

Babjeva IP, Golubev VI, Kartintsev AV, Gorin SE, Zaslavskaya PL (1973) Yeasts in the structure of forest and meadow biogeocenoses. Moscow Univ Soil Sci Bull (former Vestnik Moskovskogo Universiteta. Seriia 17, Pochvovedenie) 6:67–73 (in Russian)

Baldrian P (2008) Wood-inhabiting ligninolytic basidiomycetes in soils: ecology and constraints for applicability in bioremediation. Fungal Ecol 1:4–12

Baldrian P (2016) Forest microbiome: diversity, complexity and dynamics. FEMS Microbiol Rev 41:109–130

Belda M, Holtanová E, Halenka T, Kalvová J (2014) Climate classification revisited: from Köppen to Trewartha. Clim Res 59:1–13

Birkhofer K, Schöning I, Alt F, Herold N, Klarner B, Maraun M, Marhan S, Oelmann Y, Wubet T, Yurkov A, Begerow D, Berner D, Buscot F, Daniel R, Diekötter T, Ehnes RB, Erdmann G, Fischer C, Foesel B, Groh J, Gutknecht J, Kandeler E, Lang C, Lohaus G, Meyer A, Nacke H, Näther A, Overmann J, Polle A, Pollierer MM, Scheu S, Schloter M, Schulze ED, Schulze W, Weinert J, Weisser WW, Wolters V, Schrumpf M (2012) General relationships between abiotic

soil properties and soil biota across spatial scales and different land-use types. PLoS One 7: e43292
Bisaria VS, Ghose TK (1981) Biodegradation of cellulosic materials: substrates, microorganisms, enzymes and products. Enzym Microb Technol 3:90–104
Blackwell M (2011) The Fungi: 1, 2, 3... 5.1 million species? Am J Bot 98:426–438
Botha A (2006) Yeasts in soil. In: Rosa C, Peter G (eds) Biodiversity and ecophysiology of yeasts. Springer, Berlin, pp 221–240
Botha A (2011) The importance and ecology of yeasts in soil. Soil Biol Biochem 43:1–8
Boundy-Mills K (2006) Methods for investigating yeast biodiversity. In: Rosa C, Peter G (eds) Biodiversity and ecophysiology of yeasts. Springer, Berlin, pp 67–100
Bouthilet RJ (1951) A taxonomic study of soil yeasts. Mycopathol Mycol Appl 6:79–85
Bridge P, Spooner B (2001) Soil fungi: diversity and detection. Plant Soil 232:147–154
Buee M, Reich M, Murat C, Morin E, Nilsson RH, Uroz S, Martin F (2009) 454 Pyrosequencing analyses of forest soils reveal an unexpectedly high fungal diversity. New Phytol 184:449–456
Byzov BA, Thanh VN, Babjeva IP (1993) Interrelationships between yeasts and soil diplopods. Soil Biol Biochem 25:1119–1126
Capriotti A (1967) Yeasts from USA soils. Arch Microbiol 57:406–413
Cardinali G, Alli L, Corte L, Roscini L, Bagnetti A, Pelliccia C, Puddu G (2012) *Kazachstania ichnusensis* sp. nov., a diploid homothallic ascomycetous yeast from Sardinian lentisk rhizosphere. Int J Syst Evol Microbiol 62:722–727
Carvalho P, de Souza AC, Magalhatilde KT, Dias DR, Silva CF, Schwan RF (2013) Yeasts diversity in Brazilian Cerrado soils: study of the enzymatic activities. Afr J Microbiol Res 7: 4176–4190
Chernov IY (2005) The latitude-zonal and spatial-successional trends in the distribution of yeasts. Zh Obshch Biol 66:123–135 (in Russian)
Chesworth W (ed) (2008) Encyclopedia of soil science. Springer, Berlin
Christensen M (1989) A view of fungal ecology. Mycologia 81:1–19
Cornelissen S, Botha A, Conradie WJ, Wolfaardt GM (2003) Shifts in community composition provide a mechanism for maintenance of activity of soil yeasts in the presence of elevated copper levels. Can J Microbiol 49:425–432
Danielson RM, Jurgensen MF (1973) The propagule density of *Lipomyces* and other yeasts in forest soils. Mycopathol Mycol Appl 51:191–198
Date RA (1973) Nitrogen a major limitation in the productivity of natural communities crops and pastures in the pacific area. Soil Biol Biochem 5:5–18
Di Menna ME (1955) A search for pathogenic species of yeasts in New Zealand soils. Microbiology 12:54–62
Di Menna ME (1957) The isolation of yeasts from soil. Microbiology 17:678–688
Di Menna ME (1959) Some physiological characters of yeasts from soils and allied habitats. Microbiology 20:13–23
Di Menna ME (1960) Yeasts from soils under forest and under pasture. N Z J Agr Res 3:623–632
Di Menna ME (1965) Yeasts in New Zealand soils. N Z J Bot 3:194–203
Dixon C, Fyson GF, Pasiecznik N, Praciak A, Rushforth K, Sassen M, Sheil D, Correia CS, Teeling C, van Heist M (2013) CABI encyclopedia of forest trees. CABI Publishing, Oxford
Fairbridge RV (2008) History of soil science. In: Chesworth W (ed) Encyclopedia of soil science. Springer, Berlin, pp 307–312
FAO (2016) The global forest resources assessment 2015: how are the world's forests changing? 2nd edn. The Food and Agriculture Organization of the United Nations (FAO), Rome
Fell JW, Boekhout T, Fonseca A, Scorzetti G, Statzell-Tallman A (2000) Biodiversity and systematics of basidiomycetous yeasts as determined by large-subunit rDNA D1/D2 domain sequence analysis. Int J Syst Evol Microbiol 50:1351–1371
Fonseca A (1992) Utilization of tartaric acid and related compounds by yeasts: taxonomic implications. Can J Microbiol 38:1242–1251
Fonseca A, Inacio J (2006) Phylloplane yeasts. In: Rosa C, Peter G (eds) Biodiversity and ecophysiology of yeasts. Springer, Berlin, pp 263–301

França L, Sannino C, Turchetti B, Buzzini P, Margesin R (2016) Seasonal and altitudinal changes of culturable bacterial and yeast diversity in Alpine forest soils. Extremophiles 20:855–873
Glushakova AM, Chernov IY (2010) Seasonal dynamics of the structure of epiphytic yeast communities. Microbiology 79:830–839
Glushakova AM, Kachalkin AV, Chernov IY (2015) Effect of invasive herb species on the structure of soil yeast complexes in mixed forests exemplified by Impatiens parviflora DC. Microbiology 84:717–721
Glushakova AM, Kachalkin AV, Tiunov AV, Chernov IY (2017) Distribution of yeast complexes in the profiles of different soil types. Eurasian Soil Sci 50(7):820–825
Golubev VI (2000) Isolation of tremelloid yeasts on glucuronate medium. Microbiology 69: 490–493 (in Russian)
Golubev VI, Golubeva EV (2004) Yeast fungi in steppe and forest phytocenoses of the Prioksko-Terrasny Nature Reserve. Mikol Fitopatol 38:20–27 (in Russian)
Golubev WI, Tomashevskaya MA (2010) Yeast fungi in *Picea abies* (L.) Karst. needle litter. Microbiology 79:385–389
Golubev VI, Babjeva IP, Novik SN (1977) Yeast succession in birch sap flows. Ekologiia 5:21–26 (in Russian)
Golubtsova YV, Glushakova AM, Chernov IY (2007) The seasonal dynamics of yeast communities in the rhizosphere of soddy-podzolic soils. Eurasian Soil Sci 40:875–879
Groenewald M, Coutinho T, Smith MT, van der Walt JP (2012) Species reassignment of *Geotrichum bryndzae*, *Geotrichum phurueaensis*, *Geotrichum silvicola* and *Geotrichum vulgare* based on phylogenetic analyses and mating compatibility. Int J Syst Evol Microbiol 62: 3072–3080
Gunina A, Kuzyakov Y (2015) Sugars in soil and sweets for microorganisms: review of origin, content, composition and fate. Soil Biol Biochem 90:87–100
Heal OW (1963) Soil fungi as food for amoebae. Soil Org 289–297
Henderson ME (1961) Isolation, identification and growth of some soil hyphomycetes and yeast-like fungi which utilize aromatic compounds related to lignin. Microbiology 26:149–154
IUSS Working Group WRB (2014) World reference base for soil resources. World Soil Resources Report 105. The Food and Agriculture Organization of the United Nations (FAO), Rome 2015
Jaiboon K, Lertwattanasakul N, Limtong P, Limtong S (2016) Yeasts from peat in a tropical peat swamp forest in Thailand and their ability to produce ethanol, indole-3-acetic acid and extracellular enzymes. Mycol Prog 15:755–770
Jensen V (1963) Studies on the microflora of Danish beech forest soils. IV. Yeasts and yeast-like fungi. Zentralbl Bakteriol Mikrobiol Hyg A 117:41–65
Kimura Y, Nakano Y, Fujita K, Miyabe S, Imasaka S, Ishikawa Y, Sato M (1998) Isolation and characteristics of yeasts able to grow at low concentrations of nutrients. Yeast 14:233–238
Kurtzman CP, Robnett CJ (1998) Identification and phylogeny of ascomycetous yeasts from analysis of nuclear large subunit (26S) ribosomal DNA partial sequences. A van Leeuwenhoek 73:331–371
Lachance MA, Starmer WT (1998) Ecology and yeasts. In: Kurtzman CP, Fell J, Boekhout T (eds) The yeasts, a taxonomic study, 4th edn. Elsevier, Amsterdam, pp 21–30
Lal R (2004) Soil carbon sequestration impacts on global climate change and food security. Science 304:1623–1627
Landell MF, Brandão LR, Barbosa AC, Ramos JP, Safar SV, Gomes FC, Sousa FM, Morais PB, Broetto L, Leoncini O, Ribeiro JR, Fungsin B, Takashima M, Nakase T, Lee CF, Vainstein MH, Fell JW, Scorzetti G, Vishniac HS, Rosa CA, Valente P (2014) *Hannaella pagnoccae* sp. nov., a tremellaceous yeast species isolated from plants and soil. Int J Syst Evol Microbiol 64:1970–1977
LaRue TA, Spencer JFT (1967) The utilization of imidazoles by yeasts. Can J Microbiol 13:789–794
Lee JD, Komagata K (1980) *Pichia cellobiosa*, *Candida cariosilignicola*, and *Candida succiphila*, new species of methanol-assimilating yeasts. Int J Syst Evol Microbiol 30:514–519

Limtong S, Yongmanitchai W, Tun MM, Kawasaki H, Seki T (2007) *Kazachstania siamensis* sp. nov., an ascomycetous yeast species from forest soil in Thailand. Int J Syst Evol Microbiol 57:419–422

Limtong S, Yongmanitchai W, Kawasaki H, Fujiyama K (2009) *Wickerhamomyces edaphicus* sp. nov. and *Pichia jaroonii* sp. nov., two ascomycetous yeast species isolated from forest soil in Thailand. FEMS Yeast Res 9:504–510

Liu XZ, Wang QM, Göker M, Groenewald M, Kachalkin AV, Lumbsch HT, Millanes AM, Wedin M, Yurkov AM, Boekhout T, Bai FY (2015) Towards an integrated phylogenetic classification of the Tremellomycetes. Stud Mycol 81:85–147

Maksimova IA, Chernov IY (2004) Community structure of yeast fungi in forest biogeocenoses. Microbiology 73:474–481

Martin JP (1950) Use of acid, rose bengal, and streptomycin in the plate method for estimating soil fungi. Soil Sci 69:215–232

Mašínová T, Pontes A, Carvalho C, Sampaio JP, Baldrian P (2017a) *Libkindia masarykiana* gen. nov. et sp. nov., *Yurkovia mendeliana* gen. nov. et sp. nov., and *Leucosporidium krtinense* f.a. sp. nov., isolated from temperate forest soils. Int J Syst Evol Microbiol 67:902–908

Mašínová T, Bahnmann BD, Větrovský T, Tomšovský M, Merunková K, Baldrian P (2017b) Drivers of yeast community composition in the litter and soil of a temperate forest. FEMS Microbiol Ecol 93:fiw223

McColl RW (2014) Encyclopedia of world geography. Infobase Publishing

Men'ko EV, Chernov IY, Byzov BA (2006) Interrelationships between yeast fungi and collembolans in soil. Microbiology 75:814–822

Mestre MC, Rosa CA, Safar SV, Libkind D, Fontenla SB (2011) Yeast communities associated with the bulk-soil, rhizosphere and ectomycorrhizosphere of a *Nothofagus pumilio* forest in northwestern Patagonia, Argentina. FEMS Microbiol Ecol 78:531–541

Middelhoven WJ (1993) Catabolism of benzene compounds by ascomycetous and basidiomycetous yeasts and yeast like fungi. A van Leeuwenhoek 63:125–144

Middelhoven WJ (2004) *Trichosporon wieringae* sp. nov., an anamorphic basidiomycetous yeast from soil, and assimilation of some phenolic compounds, polysaccharides and other non-conventional carbon sources by saprophytic Trichosporon species. A van Leeuwenhoek 86:329–337

Middelhoven WJ (2006) Polysaccharides and phenolic compounds as substrate for yeasts isolated from rotten wood and description of *Cryptococcus fagi* sp. nov. A van Leeuwenhoek 90:57–67

Middelhoven WJ, Kurtzman CP (2007) Four novel yeasts from decaying organic matter: *Blastobotrys robertii* sp. nov., *Candida cretensis* sp. nov., *Candida scorzettiae* sp. nov. and *Candida vadensis* sp. nov. A van Leeuwenhoek 92:233–244

Millanes AM, Diederich P, Ekman S, Wedin M (2011) Phylogeny and character evolution in the jelly fungi (Tremellomycetes, Basidiomycota, Fungi). Mol Phylogenet Evol 61:12–28

Miller JJ, Webb NS (1954) Isolation of yeasts from soil with the aid of acid, Rose Bengal, and ox gall. Soil Sci 77:197–204

Miller MW, Phaff HJ, Snyder HE (1962) On the occurrence of various species of yeast in nature. Mycopathol Mycol Appl 16:1–18

Moawad H, Salem SH, El-Din SB, Khater T, Iskandar M (1986) Yeasts in soils of Egypt. Zentralbl Bakteriol Mikrobiol Hyg A 141:431–435

Mok WY, Luizao RC, da Silva MDSB, Teixeira MF, Muniz EG (1984) Ecology of pathogenic yeasts in Amazonian soil. Appl Environ Microbiol 47:390–394

Ódor P, Heilmann-Clausen J, Christensen M, Aude E, Van Dort KW, Piltaver A, Siller I, Veerkamp MT, Walleyn R, Standovár T, Van Hees AF, Kosece J, Matočeci N, Kraigherh H, Grebench T (2006) Diversity of dead wood inhabiting fungi and bryophytes in semi-natural beech forests in Europe. Biol Conserv 131:58–71

Olson DM, Dinerstein E (1998) The Global 200: a representation approach to conserving the Earth's most biologically valuable ecoregions. Conserv Biol 12:502–515

Pan Y, Birdsey RA, Phillips OL, Jackson RB (2013) The structure, distribution, and biomass of the world's forests. Annu Rev Ecol Evol Syst 44:593–622

Parker SS (2010) Buried treasure: soil biodiversity and conservation. Biodivers Conserv 19:3743–3756

Péter G, Tornai-Lehoczki J, Fülöp L, Dlauchy D (2003) Six new methanol assimilating yeast species from wood material. A van Leeuwenhoek 84:147–159

Phaff HJ, Starmer WT (1987) Yeasts associated with plants, insects and soil. In: Rose H, Harrison JS (eds) The yeasts, vol vol. 1. Academic Press, London, pp 123–180

Pontes A, Röhl O, Carvalho C, Maldonado C, Yurkov AM, Sampaio JP (2016) *Cystofilobasidium intermedium* sp. nov. and *Cystofilobasidium alribaticum* f.a. sp. nov., isolated from Mediterranean forest soils. Int J Syst Evol Microbiol 66(2):1058–1062

Raspor P, Zupan J (2006) Yeasts in extreme environments. In: Rosa C, Peter G (eds) Biodiversity and ecophysiology of yeasts. Springer, Berlin, pp 371–417

Reich PB, Hobbie SE, Lee T, Ellsworth DS, West JB, Tilman D, Knops JM, Naeem S, Trost J (2006) Nitrogen limitation constrains sustainability of ecosystem response to CO2. Nature 440:922–925

Rohli RV, Joyner TA, Reynolds SJ, Ballinger TJ (2015) Overlap of global Köppen–Geiger climates, biomes, and soil orders. Phys Geogr 36:158–175

Sampaio JP (1999) Utilization of low molecular weight aromatic compounds by heterobasidiomycetous yeasts: taxonomic implications. Can J Microbiol 45:491–512

Schoch CL, Seifert KA, Huhndorf S, Robert V, Spouge JL, Levesque CA, Chen W, Fungal Barcoding Consortium (2012) Nuclear ribosomal internal transcribed spacer (ITS) region as a universal DNA barcode marker for Fungi. Proc Natl Acad Sci U S A 109:6241–6246

Scorzetti G, Fell JW, Fonseca A, Statzell-Tallman A (2002) Systematics of basidiomycetous yeasts: a comparison of large subunit D1/D2 and internal transcribed spacer rDNA regions. FEMS Yeast Res 2:495–517

Sláviková E, Vadkertiová R (2000) The occurrence of yeasts in the forest soils. J Basic Microbiol 40:207–212

Spencer JFT, Spencer DM (1997) Ecology: where yeasts live. In: Spencer JFT, Spencer DM (eds) Yeasts in natural and artificial habitats. Springer, Berlin, pp 33–58

Starkey RL, Henrici AT (1927) The occurrence of yeasts in soil. Soil Sci 23:33–46

Starmer WT, Lachance MA (2011) Yeast ecology. In: Kurtzman CP, Fell JW, Boekhout T (eds) The yeasts: a taxonomic study, vol vol. 1. Elsevier, Amsterdam, pp 65–83

Summerbell RC (1988) Benomyl-tolerant microfungi associated with mycorrhizae of black spruce. Can J Bot 66:553–557

Sylvester K, Wang QM, James B, Mendez R, Hulfachor AB, Hittinger CT (2015) Temperature and host preferences drive the diversification of *Saccharomyces* and other yeasts: a survey and the discovery of eight new yeast species. FEMS Yeast Res 15:fov002

Takashima M, Sugita T, Van BH, Nakamura M, Endoh R, Ohkuma M (2012) Taxonomic richness of yeasts in Japan within subtropical and cool temperate areas. PLoS One 7:e50784

Thorn RG, Reddy CA, Harris D, Paul EA (1996) Isolation of saprophytic basidiomycetes from soil. Appl Environ Microbiol 62:4288–4292

Tollefson J (2016) Deforestation spikes in Brazilian Amazon. Nature 540:182

Tomme P, Warren RAJ, Gilkes NR (1995) Cellulose hydrolysis by bacteria and fungi. Adv Microb Physiol 37:1–81

Turchetti B, Buzzini P, Goretti M, Branda E, Diolaiuti G, D'Agata C, Smiraglia C, Vaughan-Martini A (2008) Psychrophilic yeasts in glacial environments of Alpine glaciers. FEMS Microbiol Ecol 63:73–83

UN ESA (2006) National trends in population, resources, environment and development 2005: country profiles. Department of Economic and Social Affairs of the United Nations, New York

Urbanová M, Šnajdr J, Baldrian P (2015) Composition of fungal and bacterial communities in forest litter and soil is largely determined by dominant trees. Soil Biol Biochem 84:53–64

van der Heijden MG, Bardgett RD, Van Straalen NM (2008) The unseen majority: soil microbes as drivers of plant diversity and productivity in terrestrial ecosystems. Ecol Lett 11:296–310
van der Walt JP (1992) The Lipomycetaceae, a model family for phylogenetic studies. A van Leeuwenhoek 62:247–250
Vinovarova ME, Babjeva IP (1987) Yeast fungi in steppe communities. Vestn Mosk Univ Ser Pochvoved 2:43–48 (in Russian)
Vishniac HS (1983) An enation system for the isolation of Antarctic yeasts inhibited by conventional media. Can J Microbiol 29:90–95
Vishniac HS (1995) Simulated in situ competitive ability and survival of a representative soil yeast, *Cryptococcus albidus*. Microb Ecol 30:309–320
Vishniac HS (2006) A multivariate analysis of soil yeasts isolated from a latitudinal gradient. Microb Ecol 52:90–103
Vital MJS, Abranches J, Hagler AN, Mendonça-Hagler LC (2002) Mycocinogenic yeasts isolated from Amazon soils of the Maracá ecological station, Roraima-Brazil. Braz J Microbiol 33:230–235
Voříšková J, Baldrian P (2013) Fungal community on decomposing leaf litter undergoes rapid successional changes. ISME J 7:477–486
Wardle DA, Bardgett RD, Klironomos JN, Setälä H, Van Der Putten WH, Wall DH (2004) Ecological linkages between aboveground and belowground biota. Science 304:1629–1633
Wehde T (2011) Untersuchungen zur saisonalen Dynamik der Biodiversität von bodenassoziierten Hefen im Natur-und Nutzwald. Diploma Thesis. Ruhr-University Bochum
Wuczkowski M, Prillinger H (2004) Molecular identification of yeasts from soils of the alluvial forest national park along the river Danube downstream of Vienna, Austria ("National park Donauauen"). Microbiol Res 159:263–275
Wuczkowski M, Metzger E, Sterflinger K, Prillinger H (2005) Diversity of yeasts isolated from litter and soil of different natural forest sites in Austria. Bodenkultur 56:201–208
Yurkov AM, Chernov IY, Tiunov AV (2008) Influence of *Lumbricus terrestris* earthworms on the structure of the yeast community of forest litter. Microbiology 77:107–111
Yurkov AM, Kemler M, Begerow D (2011) Species accumulation curves and incidence-based species richness estimators to appraise the diversity of cultivable yeasts from beech forest soils. PLoS One 6:e23671
Yurkov AM, Kemler M, Begerow D (2012a) Assessment of yeast diversity in soils under different management regimes. Fungal Ecol 5:24–35
Yurkov A, Wehde T, Kahl T, Begerow D (2012b) Aboveground deadwood deposition supports development of soil yeasts. Diversity 4:453–474
Yurkov A, Inácio J, Chernov IY, Fonseca A (2015) Yeast biogeography and the effects of species recognition approaches: the case study of widespread basidiomycetous species from birch forests in Russia. Curr Microbiol 70:587–601
Yurkov AM, Röhl O, Pontes A, Carvalho C, Maldonado C, Sampaio JP (2016a) Local climatic conditions constrain soil yeast diversity patterns in Mediterranean forests, woodlands and scrub biome. FEMS Yeast Res 16:fov103
Yurkov AM, Wehde T, Federici J, Schäfer AM, Ebinghaus M, Lotze-Engelhard S, Mittelbach M, Prior R, Richter C, Röhl O, Begerow D (2016b) Yeast diversity and species recovery rates from beech forest soils. Mycol Prog 15:845–859

Chapter 4
Yeasts in Agricultural and Managed Soils

Renáta Vadkertiová, Hana Dudášová, and Marta Balaščáková

Abstract All managed soils (agricultural soil, orchard soil, vineyard soil, pasture soil) are exposed to human intervention. These include regular tillage, crop or plant seeding, harvesting and the application of fertilizers, herbicides and pesticides. Yeasts are present in all types of managed soil; some of them are restricted to an individual season, soil horizon or locality, while others are present at all times in all soils. The abundance of yeasts depends on the availability of water, the type of soil and plant diversity. The composition and quantity of soil yeast communities are influenced by the yeasts originating from aerial parts of plants, which enter the soil during tillage or with decaying plant material. The size of the yeast population ranges from a few to several thousands of CFU per gram of soil. The diversity of ascomycetous yeasts present in agricultural soil includes fermentative species (e.g. *Candida* spp., *Metschnikowia* sp.), soil-related yeasts (e.g. *Cyberlindnera saturnus* and *Lipomyces* sp.), black yeasts (*Exophiala* sp., *Aureobasidium* sp.) and basidiomycetous yeasts, mainly species previously classified in the genera *Cryptococcus* and *Trichosporon*. Vineyard soils are inhabited by basidiomycetous yeasts (mainly *Naganishia* spp., *Sollicoccozyma* spp., *Filobasidium* sp.) and also by grape yeasts including *Aureobasidium pullulans*, *Metschnikowia* sp. and *Hanseniaspora uvarum*. In orchard soils, fruit-related yeasts *H'spora uvarum* and *Metschnikowia pulcherrima* are associated with the upper layer of soil. Species previously classified in the genera *Cryptococcus* and *Trichosporon* dominate the soil of citrus orchards. Grassland soils are mainly occupied by soil-related ascomycetous species *Schwanniomyces capriotti*, *Barnettozyma vustinii* and *Cyberlindnera suaveolens*.

Keywords Agricultural soil • Diversity of yeasts • Orchard soil • Vineyard soil • Pasture soil

Contents

R. Vadkertiová (✉) • H. Dudášová • M. Balaščáková
Culture Collection of Yeasts, Institute of Chemistry, SAS, Dúbravská cesta 9,
845 38 Bratislava, Slovakia
e-mail: renata.vadkertiova@savba.sk

P. Buzzini et al. (eds.), *Yeasts in Natural Ecosystems: Diversity*,
DOI 10.1007/978-3-319-62683-3_4

4.1 Introduction

Soil is a dynamic and multifunctional system and is the most important complex interface for the global exchange of matter and energy. It covers a multitude of diverse habitats of different scales, localities and properties and is utilized by humankind for housing, transportation, recreation and industry (Welbaum et al. 2004).

Soil has also a unique role in supplying humanity with food; agricultural, orchard and vineyard soils are managed with the aim of improving the quality and yield of the crops. As a result of the land use, soils are tilled and fertilized, and fruits, crops and plants are treated with pesticides.

Soil horizons in managed soil are also influenced by human intervention, whereby the topsoil is impacted the most significantly. This top layer is also the most important part of the soil, because it contains the highest content of soil organic matter (SOM), which can reach 1–6% in managed soils, and is comprised of living organisms (animals, plants, microorganisms), fresh residues and well-decomposed residues (humus) (Nielsen and Winding 2002; McCauley et al. 2005; Magdoff and van Es 2009).

Soil microorganisms play a crucial role in the decomposition of organic matter, mineralization, nutrient cycling (carbon, nitrogen, phosphorus and sulphur), and stabilization of soil aggregates. They also serve as a link between plant diversity and ecosystem functioning (Lemtiri et al. 2016; Zak et al. 2003). Microorganisms respond rapidly to environmental stress, resulting in fast changes in their diversity, abundance and activity. Therefore, they have been recognized as useful indicators of soil health (Nielsen and Winding 2002). Bacteria and fungi compose more than 90% of the soil microbial biomass. The amounts of culturable microorganisms in the managed soils range from 10^5 to 10^8 colony-forming units (CFU) g^{-1} soil for bacteria and from 10^4 to 10^6 CFU g^{-1} soil for fungi (Lawlor et al. 2000; Rinnan et al. 2009; López-Piñeiro et al. 2013). The quantity of the yeast population ranges from a few to several thousands of CFU g^{-1} soil (Sláviková and Vadkertiová 2003a; Yurkov et al. 2012a; López-Piñeiro et al. 2013).

Yeasts are present in all types of managed soil; some of them are restricted to an individual season, soil horizon or locality, while others are present at all times in all soils. The abundance of yeasts depends mainly on soil organic matter content and water availability. Other relevant factors include the type of soil, plant diversity, management regimes and the availability of overripe fruits. The composition and quantity of soil yeasts can be influenced by the yeasts associated with the aerial parts of plants (Botha 2011), which enter the soil during tillage or with decaying plant material. Plants also affect soil microbial communities (including yeasts) through their root system by the release of various substances (i.e. carbohydrates and organic acids such as amino, carboxylic, phenolic and long-chain aliphatic), which support or suppress the growth of specific microorganisms. This results in higher densities but in a lower diversity of the microbiota in the rhizosphere than in the surrounding bulk soil (Campbell et al. 1997; Berendsen et al. 2012).

Yeasts found in managed soils are able to degrade or transform various organic compounds, including pesticides and fertilizers. Some soil yeasts (e.g. *Filobasidium magnum*, *Naganishia albida* and *Lipomyces* spp.) produce extracellular polymeric substances, which protect them from unfavourable environmental conditions and contribute to the binding and formation of soil aggregates (Gulevskaya et al. 1982; Vishniac 1995; Botha 2006; Deng et al. 2015). Yeasts isolated from managed soils also produce substances, which promote the growth of both mycorrhizal fungi and plants, as well as compounds, which protect plants from fungal diseases (Nassar et al. 2005; Boby et al. 2008; Azcón et al. 2010).

4.2 Yeasts in Agricultural Soils

Agricultural soils utilized for the production of annual crops are exposed to the greatest human intervention among managed soils. The management practice includes regular tillage, crop seeding, harvesting and crop residue operations and the application of fertilizers, herbicides and pesticides (Shennan 2008; Sipilä et al. 2012; Janušauskaite et al. 2013). There are three different systems of tillage, which differ in their impact on soil parameters.

Conventional tillage consists of the primary deep operation, in which the soil is inverted with a mouldboard and/or a disc plough, and, simultaneously, crop residues, manures or fertilizers are incorporated into the tillage layer, down to a depth of 10–30 cm. This operation, when it is properly managed, improves soil aeration, mixes organic matter, controls weeds and reduces compaction of the soil, albeit temporarily. The secondary tillage aims at the preparation of a homogenous seedbed for crops. However, both primary and secondary operations result in topsoil which is bare of vegetation and, therefore, more susceptible to drying and, consequently, erosion by wind and water. Some agricultural practices involve crop rotation (to improve SOM) and the use of winter crops, which protect the soil from the negative effects of erosion (Altieri 1999; Welbaum et al. 2004; Shennan 2008; Magdoff and van Es 2009; Sipilä et al. 2012).

Conservational tillage refers to the minimal practice of tillage, in which the mechanical agitation of the soil is reduced significantly. A low or no surface disturbance is involved during seedbed preparation. In this system, plant residues (at least 30%, but also up to 100%) are left on the soil surface to prevent erosion and desiccation. Conservation tillage improves soil's physical, chemical and biological properties better than the conventional tillage (Janušauskaite et al. 2013; Busari et al. 2015).

The no-till system does not apply any tillage between harvest and sowing. A crop is sown directly into the soil, which was not tilled after the previous harvest. Weed control is achieved through the use of herbicides and erosion controlled by appropriate mulching and the retaining of stubble. However, this system can cause soil compaction (Shennan 2008; Magdoff and van Es 2009; Sipilä et al. 2012; Janušauskaite et al. 2013).

The main objective of agriculture is to cultivate a narrow spectrum of crop species as often as possible. As a result, the integrity of the microbial communities in the soil is regularly disturbed by tillage, crop rotation and the use of pesticides (Shennan 2008; Singh and Ryan 2015). The surface layer down to a depth of 15 cm is the zone with the highest content of organic carbon and nitrogen and harbours the most diverse microbiota. Conventional tillage reduces the three parameters over time (Wyland et al. 1996) and can result in a lower number of microorganisms in the surface layer of ploughed fields devoid of vegetation. This fact may also explain higher numbers of microorganisms present in the surface layer of no-till fields than in ploughed fields. However, the deeper layers of ploughed fields are more abundant in microorganisms than those of no-till fields (Sipilä et al. 2012). Tillage causes soil inversion and also moves the microorganisms (both bacteria and fungi) present in deeper layers to the surface layers and vice versa (McCauley et al. 2005; Magdoff and van Es 2009). Nevertheless, tillage affects positively soil aeration, which is required by the majority of organisms living there (Altieri 1999; Welbaum et al. 2004; Shennan 2008; Magdoff and van Es 2009; Sipilä et al. 2012; Janušauskaite et al. 2013).

The number of yeast cells present in agricultural soils ranges from a few to several thousands of cells g^{-1} of soil. The amounts of yeasts in the rhizosphere are higher; they were found to range from 1.2×10^4 to 2.2×10^5 cells g^{-1} (Babjeva and Belianin 1966). The quantity of yeasts present in soil in a region depends on the crop species, which are cultivated there. Among the crop fields examined, Lund (1954) reported the greatest amounts of yeasts in the surface layer of a barley field (3.7×10^3 cells g^{-1} of soil), whereas Sláviková and Vadkertiová (2003a) noted the highest number of yeasts in samples collected from a maize field (6.8×10^3 CFU g^{-1} soil), and Babjeva and Belianin (1966) found the highest quantity of yeasts in the soil beneath a beet (2.2×10^5 cells g^{-1}).

Different factors contribute to the variability in quantity and composition of yeast communities present in soil. Low water content in soil has been considered a factor which limits yeast abundance (Sláviková and Vadkertiová 2003a). The fields covered with oat and barley stubbles yielded only 10 yeast cells g^{-1} in soil which contained 3.2% water, whereas 370-fold higher yeast numbers were found, when the water content reached about 18% (Lund 1954).

Yeast abundance in soils decreases with the depth. A significant reduction in the yeast quantities and diversity was noted at a depth up to 20–30 cm (Lund 1954; Wuczkowski and Prillinger 2004). However, Vinovarova and Babjeva (1987) reported the yeast species *Lipomyces tetrasporus* down to a depth of 100 cm. A study based on the culture-independent approach of isolating DNA from soil discovered the presence of diverse yeast OTUs at a depth down to 70 cm. Moreover, some of the yeasts dominated the undisturbed rooted zone below the plough horizon, down to 40–50 cm (Moll et al. 2016).

The composition of soil microbial communities, including yeasts, is also influenced by plant diversity. Reduced plant species richness correlates with a decrease in microbial biomass present in the soil. Therefore, monoculture cropping, which is commonly used in agriculture, can negatively affect diversity and the amounts of soil microbiota (Spehn et al. 2000; Singh et al. 2009). The stage of crop growth affects the dynamic of the microbial community present in the soil. When a crop is in its rapid growth stage, it competes with microorganisms (mainly in the rhizosphere) for substrates and nutrients and may result in lower quantity of microbiota (Chiarini et al. 1998; Zhang et al. 2012). The age of plant also influences the yeast population. Over 2 months, a great fluctuation in yeast quantities was determined in the rhizosphere of various crops. At the beginning of July, when the plants were young, the number of yeasts reached 4.2×10^4 cells g^{-1} maximum, whereas at the end of July (blooming stage) increased to 2.2×10^5 cells g^{-1} and at the beginning of September (ripe plants) decreased to 3×10^4 cells g^{-1} soil. The highest fluctuation in yeast quantity was found in the soil beneath cabbages, whereas only insignificant changes were found in the soil under oats (Babjeva and Belianin 1966). Moreover, the rhizosphere of young maize plants (20 d) was inhabited only by fungi of the phylum Ascomycota, whereas the rhizosphere of senescent plants (90 d) harboured basidiomycetous yeasts (Gomes et al. 2003). Different amounts of yeasts over the growing period of sugar cane were reported. The yeast numbers reached up to 1.5×10^6 CFU g^{-1} soil at the early stage of the plant growth, whereas tenfold higher amounts were determined at the end of the growing season (de Azeredo et al. 1998).

The diversity of ascomycetous yeasts present in agricultural soil includes fermentative species, soil-related yeasts, black yeasts and yeast-like species. Until now, 23 ascomycetous species have been isolated from agricultural soils (Table 4.1); among them, *Candida parapsilosis* was the dominant species in a barley field, *Candida maltosa* formed about 46% of the total yeast population in a potato field and *Metschnikowia* (*Candida*) *pulcherrima* comprised more than 30% of the total yeast population in both a maize field and a sugar beet field (Lund 1954; Sláviková and Vadkertiová 2003a). Moreover, all these species have been reported as promoting the hyphal growth and germination of arbuscular mycorrhizal fungi (El-Mehalawy et al. 2004; Boby et al. 2008; Azcón et al. 2010). The soil-related yeasts *Barnettozyma* (*Hansenula*, *Williopsis*) *californica*, *Cyberlindnera* (*Hansenula*, *Lindnera*) *saturnus* and *Schwanniomyces occidentalis* were present only in low quantities (Babjeva and Belianin 1966; Wuczkowski and Prillinger 2004). Nevertheless, *Cyberlindnera* (*Hansenula*, *Lindnera*) *suaveolens* and the genus *Lipomyces* were found in higher amounts. The yeast *Cyb. suaveolens* was associated with barley and oat fields across the soil compartments, from the surface down to a depth of 20 cm, in both seasons examined (March and August), whereas the genus *Lipomyces* was the most abundant in the rhizosphere in September (Lund 1954; Babjeva and Belianin 1966; Wuczkowski and Prillinger 2004). Vustin and Babjeva (1981) observed *Barnettozyma* and *Cyberlindnera* yeasts (cited as

Table 4.1 Ascomycetous yeast species isolated from different types of managed soils

Species	Type of soil	Country of origin	Reference
Aureobasidium pullulans	Vineyard, agricultural	Dagestan	Kachalkin et al. (2015)
		Austria	Wuczkowski and Prillinger (2004)
		Spain	Sabate et al. (2002)
		Ukraine	Vinovarova and Babjeva (1987)
	Rhizosphere, grassland	New Mexico	Porras-Alfaro et al. (2011)
Barnettozyma (*Hansenula*, *Williopsis*) *californica*	Meadow	Germany	Yurkov et al. (2012a)
	Agricultural	Russia	Babjeva and Belianin (1966)
		Denmark	Lund (1954)
		Austria	Wuczkowski and Prillinger (2004)
	Orchard	Denmark	Lund (1954)
Barnettozyma vustinii	Meadow, pasture	Germany	Yurkov et al. (2009a)
Candida apis	Pasture	New Zealand	Parle and di Menna (1966)
Candida azyma	Rhizosphere	Brazil	de Azeredo et al. (1998)
Candida diddensiae-like	Rhizosphere	Brazil	de Azeredo et al. (1998)
Candida glabrata	Vineyard	Dagestan	Kachalkin et al. (2015)
Candida maltosa	Agricultural	Slovakia	Slávikova and Vadkertiová (2003a)
	Rhizosphere	Brazil	de Azeredo et al. (1998)
Candida parapsilosis	Agricultural	Denmark	Lund (1954)
	Rhizosphere		
Candida sake	Agricultural	Austria	Wuczkowski and Prillinger (2004)
	Pasture	Germany	Yurkov et al. (2012a, b)
Candida saitoana (*Torulopsis candida*)	Agricultural	Denmark	Lund (1954)
	Cultivated meadow		
Candida (*Torulopsis*) *magnoliae*	Agricultural	Denmark	Lund (1954)
	Cultivated meadow		

(continued)

Table 4.1 (continued)

Species	Type of soil	Country of origin	Reference
Candida spp.	Rhizosphere	Russia	Babjeva and Belianin (1966)
		Germany	Moll et al. (2016)
	Agricultural	Russia	Babjeva and Belianin (1966)
		Germany	Moll et al. (2016)
Candida vartiovaarae	Meadow, pasture	Germany	Yurkov et al. (2009b)
Clavispora (*Candida*) *lusitaniae*	Vineyard	New Zealand	Parle and di Menna (1966)
Clavispora reshetovae	Meadow, pasture	Germany	Yurkov et al. (2009b)
Cyberlindnera (*Lindnera*, *Williopsis*, *Hansenula*) *saturnus*	Agricultural	Austria	Wuczkowski and Prillinger (2004)
		United Arab Emirates	Nassar et al. (2005)
		Slovakia	Sláviková and Vadkertiová (2003a)
	Meadow, pasture	Germany	Yurkov et al. (2012a)
Cyberlindnera (*Hansenula*, *Lindnera*) *suaveolens*	Orchard	Denmark	Lund (1954)
	Agricultural		
	Meadow		
Debaryomyces hansenii (*Torulopsis famata*)	Rhizosphere	Brazil	de Azeredo et al. (1998)
		Russia	Babjeva and Belianin (1966)
Diutina (*Candida*) *rugosa*	Rhizosphere	China	Xiao et al. (2013)
Galactomyces sp.	Agricultural	Germany	Moll et al. (2016)
	Rhizosphere		
Hanseniaspora uvarum (*Kloeckera apiculata*)	Vineyard	Dagestan	Kachalkin et al. (2015)
		Spain	Sabate et al. (2002)
	Orchard	Denmark	Lund (1954)
Hanseniaspora valbyensis	Orchard	Denmark	Lund (1954)
Hanseniaspora vineae	Vineyard	South Africa	van der Walt and Tscheuschner (1957)
Kazachstania servazzii	Meadow	Germany	Yurkov et al. (2012b)
Kazachstania taianensis	Orchard	China	Chen et al. (2010)
Kodamaea ohmeri	Vineyard	Dagestan	Kachalkin et al. (2015)
Kregervanrija fluxuum (*Candida mycoderma*)	Orchard	Denmark	Lund (1954)
Kuraishia molischiana (*Torulopsis molischiana*)	Orchard	Denmark	Lund (1954)

(continued)

Table 4.1 (continued)

Species	Type of soil	Country of origin	Reference
Lachancea thermotolerans	Vineyard	Dagestan	Kachalkin et al. (2015)
Lipomyces tetrasporus	Agricultural	Ukraine	Vinovarova and Babjeva (1987)
Metschnikowia (Candida) pulcherrima	Vineyard	Dagestan	Kachalkin et al. (2015)
	Agricultural	Slovakia	Sláviková and Vadkertiová (2003a)
	Orchard	Denmark	Lund (1954)
Metschnikowia sinensis	Vineyard	Dagestan	Kachalkin et al. (2015)
Meyerozyma (*Candida*) *guilliermondii*	Rhizosphere	Brazil	de Azeredo et al. (1998)
	Vineyard	South Africa	Cornelissen et al. (2003)
Nakazawaea holstii (*Torulopsis ernobii*)	Rhizosphere	Russia	Babjeva and Belianin (1966)
Ogataea polymorpha (Hansenula angusta)	Orchard	Denmark	Lund (1954)
	Agricultural		
Pichia fermentans	Orchard	Denmark	Lund (1954)
Pichia kluyveri	Orchard	USA	Spencer and Gorin (1971)
Pichia membranifaciens (Candida valida)	Agricultural	Slovakia	Sláviková and Vadkertiová (2003a)
	Rhizosphere	United Arab Emirates	El-Tarabily (2004)
Pichia terricola	Vineyard	New Zealand	Parle and di Menna (1966)
Saccharomyces cerevisiae	Vineyard	New Zealand	Parle and di Menna (1966)
	Rhizosphere	China	Xiao et al. (2013)
		Brazil	de Azeredo et al. (1998)
Saccharomycopsis (*Arthroascus*) *fermentans*	Orchards	Taiwan	Lee et al. (1994)
Sampaiozyma (*Candida*) *ingeniosa*	Pasture	New Zealand	Parle and di Menna (1966)
Nadsonia (*Schizoblastosporion*) *starkeyi-henricii*	Mown pasture	Germany	Yurkov et al. (2012a)
Schwanniomyces capriottii (*Schwanniomyces castellii*)	Vineyard	Spain	Capriotti (1957)
	Meadow, pasture	Germany	Yurkov et al. (2012a)

(continued)

Table 4.1 (continued)

Species	Type of soil	Country of origin	Reference
Schwanniomyces (Debaryomyces) occidentalis	Vineyard	Dagestan	Kachalkin et al. (2015)
	Agricultural	Slovakia	Sláviková and Vadkertiová (2003a)
	Mown pasture	Germany	Yurkov et al. (2012a)
Torulaspora delbrueckii (*Saccharomyces rosei*, *Torulaspora nilssoni*)	Vineyard	New Zealand	Parle and di Menna (1966)
		Dagestan	Kachalkin et al. (2015)
	Rhizosphere	Brazil	de Azeredo et al. (1998)
	Agricultural	Sweden	Capriotti (1957)
Wickerhamomyces subpelliculosus (*Hansenula subpelliculosa*)	Cultivated meadow	Denmark	Lund (1954)
Zygosaccharomyces (*Saccharomyces*) *rouxii*	Rhizosphere	China	Xiao et al. (2013)

The species names follow the nomenclature of Kurtzman et al. (2011), Kurtzman and Robnett (2013) and Minter (2009)

Zygowilliopsis and *Williopsis*, respectively) associated with the rhizosphere of some agricultural plants. All these soil-related sporogenous species seem to be associated with the soil compartments down to a depth of 10–20 cm rather than with the litter layer. However, the black yeast *Aureobasidium pullulans* was isolated from the litter layer and from maximum depth of 5 cm. This typical phylloplane yeast exhibits a large spectrum of enzymatic activities and is probably involved in the decomposing of plant material in the litter layer (Buzzini and Martini 2002; Wuczkowski and Prillinger 2004; Molnárová et al. 2014).

A culture-independent survey reported four ascomycetous operational taxonomic units (OTUs) (*Candida* spp. and *Galactomyces* sp.). One of the *Candida* OTUs was exclusively related to a depth of 0–10 cm, whereas the other, together with the *Galactomyces* OTU, was associated with all soil compartments (0–70 cm) and, insignificantly, with the root compartment (Moll et al. 2016). In contrast to these yeasts, the black yeast *Exophiala* was a prominent yeast in the root compartment throughout the year. Different OTUs occupied the soil compartments down to a depth of 40–50 cm, reaching insignificant numbers at 60–70 cm (Moll et al. 2016). *Exophiala* OTUs were also associated with both bare soil and soil covered with winter barley (Klaubauf et al. 2010). All the *Exophiala* OTUs determined are probably linked to the environmental species *Exophiala xenobiotica* (which is able to degrade both aromatic compounds and dead plant material), rather than to human-related pathogens (de Hoog et al. 2006).

A higher diversity of species has been reported with basidiomycetous (Table 4.2) than with ascomycetous yeasts. Yeasts of the classes Tremellomycetes and Microbotryomycetes (cited in older literature as members of polyphyletic genera

Table 4.2 Basidiomycetous yeast species isolated from different types of managed soils

Species	Type of soil	Country of origin	Reference
Apiotrichum (Trichosporon) dulcitum	Agricultural	USA	Lynch and Thorn (2006)
		Austria	Klaubauf et al. (2010)
	Meadow, pasture	Germany	Yurkov et al. (2012a, b)
Apiotrichum laibachii (Trichosporon multisporum)	Meadow, pasture	Germany	Yurkov et al. (2012a)
Apiotrichum (Trichosporon) porosum	Meadow, pasture	Germany	Yurkov et al. (2012a)
Buckleyzyma (Rhodotorula) aurantiaca	Rhizosphere	Russia	Babjeva and Belianin (1966)
	Meadow	Denmark	Lund (1954)
Bullera alba (Bulleromyces albus)	Agricultural	Austria	Wuczkowski and Prillinger (2004)
		Germany	Moll et al. (2016)
Bullera hannae	Rhizosphere	Brazil	Gomes et al. (2003)
Bullera unica	Rhizosphere	Brazil	Gomes et al. (2003)
Bulleribasidium variabile-like (*Bullera variabilis*)	Rhizosphere	Brazil	de Azeredo et al. (1998)
Former (phenotypic, polyphyletic) *Cryptococcus* spp.	Rhizosphere	Korea	Hong et al. (2002)
	Agricultural	Austria	Wuczkowski and Prillinger (2004)
		Brazil	de Azeredo et al. (1998)
		Germany	Moll et al. (2016)
	Vineyard	South Africa	Cornelissen et al. (2003)
Cutaneotrichosporon (Cryptococcus) curvatus	Vineyard	New Zealand	Parle and di Menna (1966)
	Pasture		
Cutaneotrichosporon (Trichosporon) cutaneum	Agricultural	Slovakia	Sláviková and Vadkertiová (2003a)
	Orchard	USA	Spencer and Gorin (1971)
Cystobasidium minutum (Rhodotorula minuta)	Rhizosphere	Russia	Babjeva and Belianin (1966)
		Brazil	de Azeredo et al. (1998)
	Orchard	USA	Spencer and Gorin (1971)
	Vineyard	South Africa	Cornelissen et al. (2003)
Cystobasidium pallidum (Rhodotorula pallida)	Rhizosphere	Russia	Babjeva and Belianin (1966)
Cystobasidium (Rhodotorula) slooffiae	Vineyard	South Africa	Cornelissen et al. (2003)
Cystofilobasidium capitatum	Agricultural	Austria	Wuczkowski and Prillinger (2004)
		Slovakia	Sláviková and Vadkertiová (2003a, b)
Cystofilobasidium (Cryptococcus) macerans	Pasture	Germany	Yurkov et al. (2012a, b)
	Vineyard	New Zealand	Parle and di Menna (1966)
	Agricultural	USA	Lynch and Thorn (2006)
		Austria	Wuczkowski and Prillinger (2004)

(continued)

Table 4.2 (continued)

Species	Type of soil	Country of origin	Reference
Cystofilobasidium spp.	Rhizosphere	Germany	Moll et al. (2016)
Dioszegia sp.	Agricultural	Germany	Moll et al. (2016)
Exophiala spp.	Agricultural	Austria	Klaubauf et al. (2010)
		Germany	Moll et al. (2016)
	Grassland	Austria	Klaubauf et al. (2010)
	Rhizosphere	Germany	Moll et al. (2016)
Fellomyces horovitziae-like	Rhizosphere	Brazil	de Azeredo et al. (1998)
Filobasidium capsuligenum	Vineyard	Spain	Sabate et al. (2002)
Filobasidium magnum (*Cryptococcus magnus*)	Rhizosphere	New Mexico	Porras-Alfaro et al. (2011)
Goffeauzyma gastrica (*Cryptococcus gastricus*)	Meadow, pasture	Germany	Yurkov et al. (2012a, b)
Hannaella luteola (*Cryptococcus luteolus*)	Agricultural	USA	Lynch and Thorn (2006)
		Russia	Babjeva and Belianin (1966)
Hannaella (*Bullera*) *oryzae*	Rhizosphere	Brazil	Gomes et al. (2003)
Heterocephalacria (*Cryptococcus*) *arrabidensis*	Vineyard	South Africa	Cornelissen et al. (2003)
Holtermanniella festucosa	Pasture	Germany	Yurkov et al. (2012a)
Holtermanniella takashimae	Pasture	Germany	Yurkov et al. (2012a)
Holtermanniella wattica	Pasture	Germany	Yurkov et al. (2012a)
Kwoniella shandongensis	Orchard	China	Chen et al. (2012)
Leucosporidium drummii	Meadow	Germany	Yurkov et al. (2012b)
Leucosporidium golubevii	Pasture	Germany	Yurkov et al. (2012b)
Leucosporidium scottii	Rhizosphere	Brazil	de Azeredo et al. (1998)
	Vineyard	South Africa	Cornelissen et al. (2003)
	Pasture	New Zealand	Parle and di Menna (1966)
Microstroma bacarum	Agricultural	Austria	Wuczkowski and Prillinger (2004)
Mrakia sp.	Agricultural	Germany	Moll et al. (2016)
	Rhizosphere		
Naganishia (*Cryptococcus*) *adeliensis*	Pasture	Germany	Yurkov et al. (2012a, b)
Naganishia albida (*Cryptococcus albidus*)	Agricultural	Russia	Babjeva and Belianin (1966)
		Ukraine	Vinovarova and Babjeva (1987)
		Slovakia	Slávikova and Vadkertiová (2003a)
	Meadow	Denmark	Lund (1954)
	Rhizosphere	Brazil	de Azeredo et al. (1998)
	Vineyard	South Africa	Cornelissen et al. (2003)
	Pasture	New Zealand	Parle and di Menna (1966)
	Orchard	USA	Spencer and Gorin (1971)
Naganishia albida var. *kuetzingii*	Rhizosphere	New Mexico	Porras-Alfaro et al. (2011)

(continued)

Table 4.2 (continued)

Species	Type of soil	Country of origin	Reference
Naganishia (*Cryptococcus*) *bhutanensis*	Vineyard	South Africa	Cornelissen et al. (2003)
Naganishia (*Cryptococcus*) *diffluens*	Rhizosphere	Russia	Babjeva and Belianin (1966)
	Agricultural	Ukraine	Vinovarova and Babjeva (1987)
	Vineyard		
	Pasture	New Zealand	Parle and di Menna (1966)
	Orchard	USA	Spencer and Gorin (1971)
Naganishia (*Cryptococcus*) *friedmannii*	Rhizosphere	New Mexico	Porras-Alfaro et al. (2011)
Papiliotrema (*Cryptococcus*) *laurentii*	Agricultural	Slovakia	Sláviková and Vadkertiová (2003a)
		Ukraine	Vinovarova and Babjeva (1987)
	Mown pasture	Germany	Yurkov et al. (2012a)
	Rhizosphere	Brazil	de Azeredo et al. (1998)
	Vineyard	New Zealand	Parle and di Menna (1966)
	Pasture		
	Orchard	USA	Spencer and Gorin (1971)
Papiliotrema (*Cryptococcus*) *terrestris*	Vineyard	Dagestan	Kachalkin et al. (2015)
Former (phenotypic, polyphyletic) *Pseudozyma* sp.	Agricultural	Germany	Moll et al. (2016)
Rhodotorula babjevae	Pasture	Germany	Yurkov et al. (2012a)
Rhodotorula glutinis var. *glutinis*	Vineyard	New Zealand	Parle and di Menna (1966)
	Pasture	New Zealand	Parle and di Menna (1966)
	Agricultural	Russia	Babjeva and Belianin (1966)
		United Arab Emirates	Nassar et al. (2005)
	Orchard	Denmark	Lund (1954)
	Pasture	Germany	Yurkov et al. (2012a, b)
	Rhizosphere	Russia	Babjeva and Belianin (1966)
		United Arab Emirates	El-Tarabily (2004)
		Brazil	de Azeredo et al. (1998)
Rhodotorula graminis	Rhizosphere	Korea	Hong et al. (2002)
	Vineyard	South Africa	Cornelissen et al. (2003)
Rhodotorula mucilaginosa var. *mucilaginosa*	Rhizosphere	Argentina	Fracchia et al. (2003)
		Brazil	de Azeredo et al. (1998)
	Vineyard	Dagestan	Kachalkin et al. (2015)
	Orchard	USA	Spencer and Gorin (1971)
	Meadow	Denmark	Lund (1954)

(continued)

Table 4.2 (continued)

Species	Type of soil	Country of origin	Reference
Former (phenotypic, polyphyletic) *Rhodotorula* spp	Meadow, pasture	Germany	Yurkov et al. (2012a)
	Rhizosphere	China	Xiao et al. (2013)
		Brazil	de Azeredo et al. (1998)
	Agricultural	Germany	Moll et al. (2016)
Saitozyma podzolica (*Cryptococcus podzolicus*)	Agricultural	USA	Lynch and Thorn (2006)
	Vineyard	South Africa	Cornelissen et al. (2003)
	Rhizosphere	Korea	Hong et al. (2002)
Sampaiozyma (*Rhodotorula*) *ingeniosa*	Pasture	New Zealand	Parle and di Menna (1966)
Solicoccozyma aeria (*Cryptococcus aerius*)	Agricultural	Austria	Wuczkowski and Prillinger (2004)
	Meadow, pasture	Germany	Yurkov et al. (2012a, b)
	Rhizosphere	Russia	Babjeva and Belianin (1966)
	Grassland	Austria	Klaubauf et al. (2010)
Solicoccozyma phenolica (*Cryptococcus phenolicus*)	Vineyard	South Africa	Cornelissen et al. (2003)
Solicoccozyma (*Cryptococcus*) *terreus*	Grassland	New Zealand	di Menna (1960)
	Agricultural	USA	Lynch and Thorn (2006)
	Rhizosphere	Korea	Hong et al. (2002)
	Pasture	Germany	Yurkov et al. (2012a)
	Vineyard	New Zealand	Parle and di Menna (1966)
	Orchard	USA	Spencer and Gorin (1971)
Solicoccozyma (*Cryptococcus*) *terricola*	Agricultural	USA	Lynch and Thorn (2006)
		Austria	Klaubauf et al. (2010)
	Grassland	Germany	Yurkov et al. (2012a, b)
Sporobolomyces johnsonii (*Sporobolomyces holsaticus*)	Rhizosphere	Russia	Babjeva and Belianin (1966)
Sporobolomyces salmonicolor	Agricultural	Slovakia	Sláviková and Vadkertiová (2003a)
Former (phenotypic, polyphyletic) *Sporobolomyces* spp.	Rhizosphere	Korea	Hong et al. (2002)
	Agricultural	Germany	Moll et al. (2016)
Symmetrospora coprosmae/*oryzicola*	Agricultural	Austria	Wuczkowski and Prillinger (2004)
Tausonia (*Guehomyces*, *Trichosporon*) *pullulans*	Agricultural	Austria	Wuczkowski and Prillinger (2004)
			Klaubauf et al. (2010)
		Slovakia	Sláviková and Vadkertiová (2003a)
		USA	Lynch and Thorn (2006)
	Meadow, pasture	Germany	Yurkov et al. (2012a)

(continued)

Table 4.2 (continued)

Species	Type of soil	Country of origin	Reference
Tremella globispora	Vineyard	South Africa	Cornelissen et al. (2003)
Tremella mesenterica	Rhizosphere	Brazil	de Azeredo et al. (1998)
Trichosporon asahii	Rhizosphere	United Arab Emirates	El-Tarabily (2004)
Former (phenotypic, polyphyletic) *Trichosporon* spp.	Rhizosphere	Brazil	de Azeredo et al. (1998)
	Agricultural	Germany	Moll et al. (2016)
Vanrija (*Cryptococcus*) *humicola*	Vineyard	New Zealand	Parle and di Menna (1966)
Vishniacozyma (*Cryptococcus*) *tephrensis*	Agricultural	USA	Lynch and Thorn (2006)
	Mown pasture	Germany	Yurkov et al. (2012a)
	Grassland	Austria	Klaubauf et al. (2010)
Vishniacozyma (*Cryptococcus*) *victoriae*	Agricultural	Austria	Wuczkowski and Prillinger (2004)
	Pasture	Germany	Yurkov et al. (2012a)

The species names follow the nomenclature of Liu et al. (2015), Wang et al. (2015) and Kijpornyongpan and Aime (2017)

Cryptococcus and *Rhodotorula*) have been isolated the most frequently. Sláviková and Vadkertiová (2003a) noted that phenotypic *Papiliotrema* (*Cryptococcus*) *laurentii* was the main inhabitant of the maize field (35.6%) and was also found in all the fields examined (16.4–34.2%), followed by *Sporobolomyces* (*Sporidiobolus*) *salmonicolor* (2.7–17%), whereas Babjeva and Belianin (1966) reported the species *Na. albida* (formerly *Cryptococcus albidus*) and *Rhodotorula glutinis* as the dominant species of all the agricultural soils they examined. *Pa. laurentii*, together with *Fellomyces horovitziae*-like, were the most abundant inhabitants of sugar cane roots (de Azeredo et al. 1998). Among these species, *Rh. glutinis* and *Trichosporon asahii*, both isolated from the rhizosphere of sugar beet, promoted the growth of sugar beet plants and protected them from the soilborne plant pathogen *Rhizoctonia solani* (El-Tarabily 2004). Another red yeast, *Rhodotorula mucilaginosa*, positively affected the hyphal growth and germination of the arbuscular mycorrhizal fungi, *Glomus mosseae* and *Gigaspora rosea*, and exhibited antagonistic properties to *Cephalosporium maydis*, a causal agent of late wilt disease (Fracchia et al. 2003; El-Mehalawy et al. 2004; Sampedro et al. 2004; Boby et al. 2008).

Four species were associated with arable fields managed by conventional tillage. The species *Solicoccozyma terricola* (formerly *Cryptococcus terricola*) formed 5.3% of the total fungal clones in the soil covered with winter barley, whereas it comprised only 1% of the total fungal clones in the bare soil. Furthermore, at the same sampling site, there was *Apiotrichum* (*Trichosporon*) *dulcitum* (1% of the total fungal clones). The second sampling site without vegetation was inhabited with *Guehomyces* (*Tausonia*, *Trichosporon*) *pullulans* in the same proportion as with *Ap. dulcitum* (Klaubauf et al. 2010). All these species were also found, together with *Hannaella luteola* (*Cryptococcus luteolus*), *Cystofilobasidium* (*Cryptococcus*) *macerans*, *Saitozyma podzolica*, (*Cryptococcus podzolicus*), *Vishniacozyma* (*Cryptococcus*) *tephrensis* and *Solicoccozyma terrea* (*Cryptococcus terreus*), in maize fields managed by conventional tillage and no till (Lynch and Thorn 2006). *Solicoccozyma aeria* (*Cryptococcus aerius*) and *Ta. pullulans* were found down to a depth of 30–35 cm, with the exception of the litter layer, whereas *F. magnum* (former *Cryptococcus magnus*) and *Bullera alba* were exclusively found in the litter layer (Wuczkowski and Prillinger 2004).

The study on the fungal population present in maize field, managed by conventional tillage, over various seasons and soil compartments, ascertained a much larger spectrum of yeasts. The highest number of OTUs was classified in former polyphyletic genera *Cryptococcus* (10) and *Rhodotorula* (6). The individual "*Cryptococcus* OTU" and "*Trichosporon* OTU", not classified to monophyletic lineages, were prominent down to depths of 40–50 cm. A few yeast OTUs (cited as former polyphyletic *Cryptococcus* and *Rhodotorula*) were only rarely associated with the root compartment, whereas *Trichosporon* was not found there. Although yeasts have been said to be less abundant in soil than fungi, they formed a large proportion of fungal sequences (almost 70%) at a depth of 60–70 cm in September, shortly before harvest. Two OTUs (cited as former polyphyletic *Rhodotorula* and *Cryptococcus*) contributed most significantly to the yeast part of fungal population. The highest diversity of yeast OTUs was found at a depth of 40–50 cm in December, when the soil was left fallow (Moll et al. 2016). The cultivation-independent study showed that the rhizosphere of maize was inhabited by *Hann. luteola*, *F. magnum*, *Bullera unica*, *Sporidiobolus johnsonii*, *Hannaella oryzae* and *Bullera hannae* (Gomes et al. 2003).

It can be supposed that representatives of the yeast-like genera *Apiotrichum*, *Exophiala*, *Tausonia* and *Trichosporon*, which were found in higher quantities, probably also contribute to the binding of soil particles by the formation of long hyphae beneath the soil, whereas *Rhodotorula* spp. do it by the production of exopolymeric compounds (Cho et al. 2001; Breierová et al. 2008).

4.3 Yeasts in Orchard and Vineyard Soils

Orchards and vineyards are semi-permanent systems. Vines and fruit trees are cultivated in rows, which are bare of vegetation or, less frequently, covered with annual crops, grasses or mulches. Between the rows of vines and trees, cover crops are sown, which prevent the soil from drying. The fruit-producing period of a fruit tree ranges from 10 to 20 years, whereas for a vine crop, 20–40 years. Similar to agricultural soils, orchard and vineyard soils are managed with fertilizers, manures and crop residues. Tillage is practised to a maximum depth of 18 cm, leaving the plant roots undisturbed. Weeds, which compete with vine plants and fruit crops for moisture and nutrients, are controlled by tillage, herbicides and cover crops (Lieskovský and Kenderessy 2014; Hanson et al. 2017). In recent years, the no-till system with cover cropping has been recognized as a useful practice in increasing enzyme activity and the amounts of both organic carbon and microbial biomass in soil (Sipilä et al. 2012; Zhang et al. 2012). A positive effect on yeast amounts and the presence of fermentative yeasts have also been confirmed (González et al. 2007). However, another study on the microbial population associated with vineyard soils showed that the number of yeasts was not significantly affected by cover crops (natural grasses) (López-Piñeiro et al. 2013). Moreover, crops planted among vine plants in a dry and hot area negatively influenced the vigour and yields of the vines, which resulted in the reduced biodiversity of yeasts isolated from grapes, the inability of some musts to reach spontaneous fermentation and the development of fungal populations in the musts (Cordero-Bueso et al. 2011).

The abundance of yeasts in vineyard soil depends on the management regime, the season and soil water content. The quantity of yeasts ranged from 5.8×10^3 to 3.5×10^4 yeast CFU g^{-1} soil in sandy low-nutrient vineyards which contained about 15% water (Vreulink et al. 2007), whereas the number of yeasts reached only thousands of CFU g^{-1} of soil when the vineyard soil contained 2–5% water (López-Piñeiro et al. 2013). A significant seasonal fluctuation in the yeast population was also reported. The largest quantities of yeasts were isolated 1 week after the harvesting of grapes ($7–9 \times 10^4$ CFU g^{-1} of soil), when the soil was probably enriched with the yeasts originating from the grapes, whereas the lowest number was found 3 months before harvest, at an early stage of grape veraison (not exceeding 5×10^3 CFU g^{-1} of soil) (González et al. 2007).

The diversity of yeast populations in soil, similar to the diversity of yeasts associated with grapes, can be influenced by many factors, such as the locality, season, maturity of grapes, composition of the soil and application of agrochemicals (Vreulink et al. 2007). Basidiomycetous yeasts are mainly isolated from unripe grapes, whereas the species of the genera *Metschnikowia* and *Hanseniaspora* are more associated with latter states of fruit ripening (Sipiczki 2006, 2016; Vadkertiová et al. 2012). Although numerous investigations have been devoted to the yeast microbiota on grapes, only a limited number of studies have been focused on soil yeasts.

Parle and Di Menna (1966) found only basidiomycetous yeasts inhabiting the vineyard soil in New Zealand. *Na. albida* was the dominant species in the brown earth soils of the Barossa Valley, whereas *Cutaneotrichosporon* (*Cryptococcus*) *curvatus* was the most abundant in brown granular clay in New Zealand, and *Sol. terrea* was present in significant amounts in both localities (Parle and Di Menna 1966). Molecular methods discovered the basidiomycetous species *Filobasidium capsuligenum* (60% of the total yeast population) and ascomycetous yeasts *A. pullulans* (5%) and *Hanseniaspora uvarum* (5%), which were associated with soil in Spain. However, 20% of the yeast population remain unidentified (Sabate et al. 2002). A higher diversity of yeasts, mainly ascomycetous, was found in the soil of a vineyard in Dagestan (Kachalkin et al. 2015). Although *A. pullulans* has been recognized as a predominant epiphyte of the phylloplane environment (Fonseca and Inácio 2006), it dominated the yeast population (about 46% of the total yeast population) in this vineyard. *Metschnikowia sinensis*, a close relative of *M. pulcherrima*, comprised about 4% of the total yeast population, whereas *Candida glabrata*, *H'spora uvarum*, *M. pulcherrima* and *Rh. mucilaginosa* were found only in small amounts (less than 1%). All these species were also associated with grapes, but in more significant quantities (2–42%), and probably entered the soil with the fruits (Kachalkin et al. 2015).

The species previously classified in the former polyphyletic genera *Cryptococcus* and *Rhodotorula* were associated with the soil of a vineyard in South Africa. *Solicoccozyma* (*Cryptococcus*) *phenolicus* was present in more than 40% of the samples, followed by *Cystobasidium minutum* (*Rhodotorula minuta*), *Naganishia* (*Cryptococcus*) *bhutanensis* and *Meyerozyma* (*Candida*, *Pichia*) *guilliermondii*. However, the authors pointed out that the composition of the yeast populations could be influenced by the selective medium used for their isolation, i.e. thymine-mineral-vitamin agar (Cornelissen et al. 2003).

In orchard soils, the yeast quantities range from 4.5×10^2 to 3.4×10^5 CFU g^{-1} of soil, which can vary within orchards. Lund (1954) reported yeast amounts in the shrub berry orchard up to 2.4×10^5 cells g^{-1}, whereas the yeast quantities in other similar orchard reached only 1.4×10^3 cells g^{-1} in the same season. Yeasts are present in orchard soil in higher quantities during the period when the fruit is ripening and enter the soil with fallen overripe fruits (Lund 1956; Parle and Di Menna 1966).

A study on the microbiota present in a plum orchard (Węgierka Zwykła) soil showed yeast numbers peaked in July (2.8×10^5 CFU g^{-1}), when the plums were unripe. The yeast population was lower in October (1.7×10^5 CFU g^{-1}), a month after harvest (Tuszyński and Satora 2003).

Yeast quantity in the soil may be affected by neighbouring vegetation. A higher abundance of yeasts was found in the part of a plum orchard surrounded by forest than that located close to farm buildings (Tuszyński and Satora 2003). Different yeast species (identified using morphological and physiological methods) were associated with the soil beneath various plant species. *Rhodotorula* and *Candida* spp. were found in the soil in a plum orchard, ascomycetous yeasts were exclusively isolated from the soil beneath gooseberry and raspberry shrubs, and basidiomycetous yeasts were dominant in the soil of citrus orchards (Lund 1954; Spencer and Gorin 1971; Tuszyński and Satora 2003). In soil under berry shrubs, *Kregervanrija fluxuum* (*Candida mycoderma*) was found most frequently in the layer down to a depth of 30 cm. *Ogataea polymorpha* was isolated from a depth of 5–30 cm, whereas *Hanseniaspora valbyensis* was found in the upper layer. The soil-related yeasts *Barn. californica* and *Cyb. suaveolens* were found at a depth of 30 cm, while the fruit-related yeasts *H'spora uvarum* (*Kloeckera apiculata*) and *M. pulcherrima* were associated with the upper layer of soil (Lund 1954). The phenotypic (identified with assimilation tests alone) basidiomycetous yeast-like species *Trichosporon cutaneum* (*Apiotrichum*, *Cutaneotrichosporon*) dominated the soil in an orange orchard, followed by phenotypic *Cryptococcus diffluens* (*Naganishia* spp.) and

Sol. terrea. Both species were also associated with lemon and grape orchards. The highest diversity was found in the soil under decayed grapefruits. *Pa. laurentii* and *Pichia kluyveri* were predominant, present in similar quantities, whereas phenotypic *Tr. cutaneum* (see above) was less abundant. Yeasts of the species *Na. albida*, *Sol. terrea*, *Cyst. minutum* and *Vanrija humicola* (*Cryptococcus humicola*) were found in small amounts (Spencer and Gorin 1971).

Four yeast species associated with orchard soil have been discovered: *Saccharomycopsis* (*Arthroascus*) *fermentans* from the soil in a papaya and carambola orchard in Taiwan (Lee et al. 1994), *Kazachstania taianensis* and *Kwoniella shandongensis* from the orchards in China (Chen et al. 2010, 2012) and *Hanseniaspora vineae* from vineyard soil in South Africa (Cornelissen et al. 2003).

4.4 Yeasts in Grassland and Meadow Soils

The soils of meadows and grasslands are managed with significantly less intervention, compared to orchards and fields. The plants are allowed to grow for several weeks and subsequently are mowed and left to dry to produce hay. The biomass of fast-growing plants may also be removed by livestock, which recycles the plant nutrients by enriching the soil with manure. The soil under this system is termed managed or extensively managed. This soil usually contains more organic matter and is occupied by more species-rich vegetation (the number may reach more than 40 species/square meter) than those of other managed soils (Beeby and Brennan 2008). Moreover, herbaceous vegetation forms a dense cover above ground and an extensive root system below ground, both of which retain soil water and, thereby, support larger microbial biomass (Singh et al. 2009). Intensively managed soils are treated with fertilizers which accelerate the growth of plants and, thus, enables farmers to crop those plants several times during the year (Beeby and Brennan 2008; Yurkov et al. 2012a).

On average, the yeast quantity in soil varies between 10^3 and 10^4 CFU g^{-1} of soil (Lund 1954; Yurkov et al. 2012a). The highest number of yeasts, 3.6×10^4 CFU g^{-1} of soil, was found in the soil covered with wild plants and grazed by horses and cattle, whereas the lowest number (not exceeding 80 CFU g^{-1}) was found in soil of a cultivated meadow. In a meadow soil in late summer, there was found a higher abundance of yeasts at a depth of 15 cm than in the upper layer (Lund 1954). These types of soils are mainly occupied by soil-related species. *Schwanniomyces capriotti* (formerly *Schwanniomyces castellii*) and *Candida vartiovaarae* were found in significant quantities in all types of grasslands (intensively managed, managed and extensively managed) and in meadows (Yurkov et al. 2012a; Glushakova et al. 2015a, b, 2016). *Barnettozyma vustinii*, *Cyb. saturnus* and *Schw. capriotti* were restricted to the particular region in Germany (Yurkov et al. 2012a), whereas *Barn. californica* to a single region in Russia (Glushakova et al. 2015a, b, 2016) and *L. tetrasporus* to a pasture soil in Ukraine (Vinovarova and Babjeva 1987). In Denmark, *Cyb. suaveolens* was the main species of cultivated meadow and grassy soils. The osmotolerant species *Wickerhamomyces subpelliculosus* (*Hansenula subpelliculosa*) and *Candida* (*Torulopsis*) *magnoliae* were associated with depths of 15 and 20 cm, respectively, in both type of soils examined (Lund 1954).

A significantly higher proportion of yeasts belonging to the order Filobasidiales than that of the order Tremellales was found in an extensive grassland soil in Italy, whereas *Apiotrichum laibachii* (formerly *Trichosporon multisporum*), *Apiotrichum* (*Trichosporon*) *porosum* (order Trichosporonales) and *Sol. aeria* (order Filobasidiales) were the most abundant among basidiomycetous species in grassland soils in Germany and comprised about 10–15% of the total yeast community (Yurkov et al. 2012a; Ciccolini et al. 2015). The latter species, together with *Vishn. tephrensis*, were reported as the only yeast inhabitants of a grassy soil in Austria, where they formed, respectively, 4.3 and 1.1% of the total fungal population (Klaubauf et al. 2010). Former *Cryptococcus* species [*Na. albida*, *Naganishia* (*Cryptococcus*) *friedmannii* and *F. magnum*], together with ascomycetous *Aureobasidium* sp., were ascertained among the fungal community in the rhizosphere of grassland in New Mexico (Porras-Alfaro et al. 2011). Phenotypic (see discussion above) *Cr. albidus* was also associated with a pasture in Ukraine, followed by phenotypic *Cr. diffluens* and *Pa. laurentii* (Vinovarova and Babjeva 1987). *Ap. porosum*, *Cystofil. macerans*, *Rh. mucilaginosa* and *Sa. podzolica* were the predominant basidiomycetous species associated with meadows in Russia (Glushakova et al. 2015a, b, 2016).

4.5 Impact of Pesticides and Herbicides on Yeasts

The utilization of fertilizers, pesticides and herbicides has been an integral part of agriculture for a long time because they increase crop yields.

The latter two types of agrochemicals are used to protect or rid crops, fruit trees, vines and fruit shrubs of insect pests, fungal diseases and weeds (Aktar et al. 2009). They differ from each other in their chemical composition, physico-chemical properties and mode of action. Many pesticides are persistent in soil, but some of them are broken down or transformed by microorganisms. Some of these chemicals may reduce the total microbial biomass, change the proportions of individual groups of microorganisms (i.e. the ratio of bacteria to fungi) and impact on particular microbial species (Chowdhury et al. 2008). Yeasts present in soil may be resistant, sensitive or neutral to the presence of pesticides. The yeasts inhabiting aerial parts of plants regularly treated with pesticides enter the soil with plant material and may contribute to higher amounts of yeasts which are resistant to agrochemicals.

In general, yeasts are more tolerant to herbicides than fungicides; however, they show different responses to individual pesticides. A treatment of the bare soil in a vineyard with herbicide glyphosate did not affect the yeast microbiota inhabiting the grapes; a high diversity as well as the spontaneous fermentation capability of yeasts was maintained (Cordero-Bueso et al. 2011). A low inhibition of yeasts from agricultural soil was found with the herbicide metazachlor, whereas the herbicide lactofen significantly inhibited only these yeasts previously classified in the former genera *Cryptococcus* and *Trichosporon*. The fungicides prochloraz and penconazole inhibited the growth of all strains tested (Sláviková and Vadkertiová 2003b; Cordero-Bueso et al. 2014). A relative abundance of *Dioszegia* spp., *A. pullulans* and *Leucosporidium golubevii*, which inhabited wheat phyllosphere, was also lower in fungicide-treated (among them prochloraz) leaves than in control leaves (Karlsson et al. 2014). However, none of the fungicides tested (iprodione, fludioxonil + cyprodinil or pyrimethanil, sulphur) affected significantly the abundance of yeasts on grapes (Čadež et al. 2010; Cordero-Bueso et al. 2014).

The ability of yeasts to degrade pesticides has also been noted. The soil yeast *Lipomyces starkeyi* exhibited an ability to degrade the herbicide paraquat, whereas *Candida* sp. isolated from a sugar cane field, and *Rhodotorula* sp. from a sorghum field, degraded the insecticide lindane (Carr et al. 1985; Salam et al. 2013; Salam and Das 2014).

The utilization of fertilizers is essential for the production of healthy crops. Crops remove nutrients from the soil, and so it is necessary to supplement these nutrients with mineral or organic fertilizers. Fertilizers support the growth of roots and subsequently improve crop yields, SOM and microbial activity. However, crops take up only limited amounts of fertilizers (less than 50% for N fertilizers), and the rest is deposited in the soil, leached or released into the environment (Singh and Ryan 2015). The effect of fertilizers on microbiota present in soil has been published only in a limited number of studies. It was found that the long-term application of urea, ammonium and NPK (nitrogen, phosphorus and potassium) fertilizers can negatively affect the abundance of microbial biomass in soil as well as the composition of the microbial community (Allison and Martiny 2008; Geisseler and Scow 2014). The studies on the response of fungi to nitrogen input showed negative, positive or no effect on the diversity and structure of fungal communities (Porras-Alfaro et al. 2011; Pezzolla et al. 2015). Up to now, yeast communities have been mentioned only twice in the literature. Porras-Alfaro et al. (2011) compared the composition of fungal community inhabiting the rhizosphere and biological soil crust of semiarid grasslands. The experimental plots included both extensive and intensive grasslands, the latter treated twice a year with NH_4NO_3 fertilizer. No significant differences were found in the species richness and the community composition between the treated and the untreated plots. Among the fungal communities, the yeast community was represented only by a few species belonging to the genera *Aureobasidium* and *Naganishia*. However, a positive effect of an organic N fertilizer on microbiota associated with agricultural soil was reported by Pezzolla et al. (2015). The authors found that the addition of organic N fertilizer (pig slurry-derived digestate) resulted in an increased taxonomic richness of the fungal and yeast communities. The latter community was mainly represented by urease-positive species of the order Tremellales (cited as former polyphyletic *Cryptococcus* sp.), which was found to be the second most abundant group among basidiomycetous fungi. Moreover, fertilization has a more prolonger effect on the yeast and fungal taxa than on the bacterial communities.

4.6 Properties of Yeasts Inhabiting Managed Soils

The decomposition of organic material is one of the most important processes in soil. Managed soils contain not only crop residues, fertilizers and/or manures but also other anthropogenic substances, which may be utilized and transformed into other compounds by plants, microorganisms and other biota present there (Johnson et al. 2005; Winding et al. 2005). Yeasts also play an important role in the transformation of nitrogen, phosphorus and carbon sources.

Urea is the most widely used organic N fertilizer, which is also present in manure, organic matter and crop residues. This is utilized by plants directly, or is hydrolysed by plants or microorganisms, to ammonia and carbon dioxide (Johnson et al. 2005). Table 4.2 shows a large spectrum of basidiomycetous yeasts associated with managed soils. All these species, together with *A. pullulans* (Table 4.1) and *Exophiala* spp., are probably involved in mineralization of urea as they produce enzyme urease. Some of *Naganishia*, *Solicoccozyma*, and *Cystofilobasidium* spp. and the soil-related yeasts *Ta. pullulans*, *Barn. californica*, *Cyb. suaveolens* and *Cyb. saturnus* utilize nitrates (Kurtzman et al. 2011) and may be involved in the denitrification of N sources. The highest number of nitrate-positive species is associated with agricultural soils, whereas the lowest number of these species is found in the rhizosphere.

Phosphorus is the second most important mineral nutrient. It is utilized by plants for their metabolic processes and is required in larger quantities. As a result, supplementing soils with P fertilizers is required. However, the application of fertilizers results in a higher fixation of phosphorus in the soil and a decrease in its availability to plants (Khan et al. 2010; Sharma et al. 2013). Microorganisms, which are able to solubilize phosphate, play a crucial role in its accessibility to plants. Among the yeasts, only the isolates from the rhizospheric soil show this ability. *Debaryomyces hansenii* and *Diutina* (*Candida*) *rugosa* exhibited the greatest efficiency in solubilizing to P_2O_5, followed by *Rhodotorula* sp., *Saccharomyces cerevisiae* and *Zygosaccharomyces rouxii* (Narsian et al. 2010; Xiao et al. 2013).

Yeasts are not involved in the degradation of large polymeric substrates such as cellulose, hemicelluloses or lignin but contribute to their degradation in subsequent steps. Cellobiose is a product of cellulose degradation, which is a major component of plant cell walls. Most of ascomycetous species, and all basidiomycetous species associated with vineyard soils, produce β-glucosidase, which split cellobiose (Kurtzman et al. 2011). These species include grape-related genera *Metschnikowia* and *Hanseniaspora* and the species *A. pullulans*, *F. capsuligenum* and former *Cryptococcus* spp., which were found to be the most abundant in vineyard soils

(Parle and Di Menna 1966; Sabate et al. 2002; Molnárová et al. 2014; Kachalkin et al. 2015).

D-xylose is the main component of hemicellulose, another part of the cell wall. This and another pentose (arabinose), together with two hexoses (galactose and mannose), are among the most abundant sugars present in soil organic matter; glucose is the most abundant (Gunina and Kuzyakov 2015). Xylose is utilized by all ascomycetous yeasts found in agricultural soil as well as by all basidiomycetous species associated with all managed soils. Fewer yeast species assimilate arabinose and galactose than assimilate xylose (Sláviková and Vadkertiová 2003a; Kurtzman et al. 2011). The basidiomycetous yeasts, mainly of the genera *Apiotrichum*, *Naganishia* and *Solicoccozyma*, are also capable of utilizing low molecular weight aromatic compounds derived from lignin (Sampaio 1999). Starch is a heterogeneous polysaccharide which is produced by all higher plants as their major reserve carbohydrate. This is hydrolysed by glucoamylases and α-amylases to glucose. *Deb. hansenii*, *Nakazawaea* (*Pichia*) *holstii* and *Schwanniomyces* spp. are starch solubilizing ascomycetous species, whereas Tremellomycetes are the main representatives of starch solubilizing basidiomycetous yeasts (Kurtzman et al. 2011).

4.7 Concluding Remarks

Although more than 60 years have passed since the first study on yeasts present in managed soil (Lund 1954), knowledge on this topic is still limited. This may be due to the lower abundance and diversity of yeasts in comparison to bacteria and fungi. As a result, the role of yeasts in managed soils, similarly as in other soils, has been deemed insignificant. However, previous surveys of the literature (Botha 2006, 2011; Starmer and Lachance 2011), as well as the studies mentioned in this chapter, have advocated yeasts as important members of soil microbiota and provide examples of their roles in various soil processes.

Thanks to all the studies, which have been carried out, we are now able to estimate the number and diversity of yeasts present in managed soil. However, it should be mentioned that some limitations of the studies may have resulted in a lower diversity of yeast species being found.

The first is that some of the yeasts were identified on the base of their morphological and physiological properties, and, therefore, a much greater species richness may be "hidden" within "phenotypic" species solely identified with physiological parameters. This mainly concerns the species *Pa. laurentii*, *Na. albida*, *Cut. curvatus* and *Cut. cutaneum* (originally cited as *Cryptococcus* and *Trichosporon*).

The second limitation is the use of cultivation media, which may not meet the growth requirements of specific yeasts or may favour one particular group of yeasts over another, i.e. fermenting, and fast-growing yeasts can outgrow slow-growing yeasts in enrichment broths (Parle and di Menna 1966; Wawrik et al. 2005).

An additional limitation is the number of samples collected. Indirect molecular techniques have brought amazing results on the richness and diversity of yeast taxa across the soil layers and over the seasons, when the samples were taken over a 2-year period (Moll et al. 2016). However, these yeasts were classified only at genera level (including polyphyletic), and, therefore, no data on the individual species richness are available.

The last limitation is the low number of studies on the yeasts associated with all types of managed soils (agriculture, orchard, vineyard and pasture).

It has been clearly stated that the yeast populations inhabiting soils reflect the population associated with plants (Botha 2011). However, a very recent study also showed the opposite influence of the yeast microbiota, when the specific strain, *S. cerevisiae*, inoculated into the soil, was transported via roots and stems to the surface of grapes, where it was detected (Mandl et al. 2015). Moreover, another study noted the importance of soil microbiota for the microbial community associated with vine plants (Zarraonaindia et al. 2015). The authors demonstrated that soil bacteria are a source of the bacterial communities associated with grapevine organs and suggested an indirect, but significant, impact of the soil communities on the organoleptic properties and terroir of wines.

These new findings, as well as the limitations found within previous studies, give rise to the possibility of further investigations in which all types of managed soils will be examined, a larger spectrum of isolation media will be used, multiple and spatially diverse soil samples will be collected over various seasons (or, which is better, over a few years), the structure of the soil community will be compared with those found on the aerial parts of plants and the effects of human intervention (tillage, the application of pesticides and fertilizers) on soil yeasts will be ascertained.

Acknowledgements We would like to dedicate this chapter to the memory of Anna Kocková-Kratochvílová, the founder of the Culture Collection of Yeasts, a well-known researcher and an enthusiastic hunter of yeasts inhabiting different natural substrates. We would also like to remember Elena Sláviková, who carried on Anna Kocková-Kratochvílová's research and was the author of various studies on the diversity of yeasts, including the yeasts associated with agricultural soil. We would also like to express our thanks to P. Buzzini and A. Yurkov for their comments on and corrections of this chapter and Michael Hawkins for the language correction.

References

Aktar MW, Sengupta D, Chowdhury A (2009) Impact of pesticides use in agriculture: their benefits and hazards. Interdiscip Toxicol 2:1–12

Allison SD, Martiny JBH (2008) Resistance, resilience, and redundancy in microbial communities. Proc Natl Acad Sci U S A 105:11512–11519

Altieri MA (1999) The ecological role of biodiversity in agroecosystems. Agric Ecosyst Environ 74:19–31

Azcón R, Perálvarez MC, Roldán A, Barea JM (2010) Arbuscular mycorrhizal fungi, *Bacillus cereus*, and *Candida parapsilosis* from a multicontaminated soil alleviate metal toxicity in plants. Microb Ecol 59:668–677

Babjeva IP, Belianin AI (1966) Yeasts in the rhizosphere. Mikrobiologiya 35:712–720

Beeby A, Brennan AM (2008) First ecology: ecological principles and environmental issues, 3rd edn. Oxford University Press, Oxford

Berendsen RL, Pieterse CM, Bakker PA (2012) The rhizosphere microbiome and plant health. Trends Plant Sci 17:478–486

Boby VU, Balakrishna AN, Bagyaraj DJ (2008) Interaction between *Glomus mosseae* and soil yeasts on growth and nutrition of cowpea. Microbiol Res 163:693–700

Botha A (2006) Yeasts in soil. In: Rosa CA, Péter G (eds) Biodiversity and ecophysiology of yeasts. Springer, Berlin, pp 221–240

Botha A (2011) The importance and ecology of yeasts in soil. Soil Biol Biochem 43:1–8

Breierová E, Gregor T, Márová I, Čertík M, Kogan G (2008) Enhanced antioxidant formula based on a selenium-supplemented carotenoidproducing yeast biomass. Chem Biodivers 5:440–446

Busari MA, Kukal SS, Kaur A, Bhatt R, Dulazi AA (2015) Conservation tillage impacts on soil, crop and the environment. Int Soil Water Conserv Res 3:119–129

Buzzini P, Martini A (2002) Extracellular enzymatic activity profiles in yeast and yeast-like strains isolated from tropical environments. J Appl Microbiol 93:1020–1025

Čadež N, Zupan J, Raspor P (2010) The effect of fungicides on yeast communities associated with grape berries. FEMS Yeast Res 10:619–630

Campbell CD, Grayston SJ, Hirst DJ (1997) Use of rhizosphere carbon sources in sole carbon source tests to discriminate soil microbial communities. J Microbiol Methods 30:33–41

Capriotti A (1957) New blastomycetes isolated from soils of Spain I: *Schwanniomyces castellii* nov. spec. Arch Mikrobiol 26:434–438

Carr RJG, Bilton RF, Atkinson T (1985) Mechanism of biodegradation of paraquat by *Lipomyces starkeyi*. Appl Environ Microbiol 49:1290–1294

Chen R, Wei SC, Jiang YM, Wang QM, Bai FY (2010) *Kazachstania taianensis* sp. nov., a novel ascomycetous yeast species from orchard soil. Int J Syst Evol Microbiol 60:1473–1476

Chen R, Jiang YM, Wei SC, Wang QM (2012) *Kwoniella shandongensis* sp. nov., a basidiomycetous yeast isolated from soil and bark from an apple orchard. Int J Syst Evol Microbiol 62:2774–2777

Chiarini L, Bevivino A, Dalmastri C, Nacamulli C, Tabacchioni S (1998) Influence of plant development, cultivar and soil type on microbial colonization of maize roots. Appl Soil Ecol 8:11–18

Cho DH, Chae HJ, Kim EY (2001) Synthesis and characterization of a novel extracellular polysaccharide by *Rhodotorula glutinis*. Appl Biochem Biotechnol 95:183–193

Chowdhury A, Pradhan S, Monidipta S, Nilanjan S (2008) Impact of pesticides on soil microbiological parameters and possible bioremediation strategies. Indian J Microbiol 48:114–127

Ciccolini V, Bonari E, Pellegrino E (2015) Land-use intensity and soil properties shape the composition of fungal communities in Mediterranean peaty soils drained for agricultural purposes. Biol Fertil Soils 51:719–731

Cordero-Bueso G, Arroyo T, Serrano A, Valero E (2011) Influence of different floor management strategies of the vineyard on the natural yeast population associated with grape berries. Int J Food Microbiol 148:23–29

Cordero-Bueso G, Arroyo T, Valero E (2014) A long term field study of the effect of fungicides penconazole and sulfur on yeasts in the vineyard. Int J Food Microbiol 189:189–194

Cornelissen S, Botha A, Conradie WJ, Wolfaardt GM (2003) Shifts in community composition provide a mechanism for maintenance of activity of soil yeasts in the presence of elevated copper levels. Can J Microbiol 49:425–432

de Azeredo LA, Gomes EA, Mendonça-Hagler LC, Hagler AN (1998) Yeast communities associated with sugarcane in Campos, Rio de Janeiro, Brazil. Int Microbiol 1:205–208

de Hoog GS, Zeng JS, Harrak MJ, Sutton DA (2006) *Exophiala xenobiotica* sp. nov., an opportunistic black yeast inhabiting environments rich in hydrocarbons. A van Leeuwenhoek 90:257–268

Deng J, Orner EP, Chau JF, Anderson EM, Kadilak AL, Rubinstein RL, Bouchillon GM, Goodwin RA, Gage DJ, Shor M (2015) Synergistic effects of soil microstructure and bacterial EPS on drying rate in emulated soil micromodels. Soil Biol Biochem 83:116–124

di Menna ME (1960) Biological studies of some tussock-grassland soils. N Z J Agric Res 1:939–942

El-Mehalawy AA, Hassanein NM, Khater HM, El-Din EAK, Youssef YA (2004) Influence of maize root colonization by the rhizosphere actinomycetes and yeast fungi on plant growth and on the biological control of late wilt disease. Int J Agric Biol 6:599–605

El-Tarabily KA (2004) Suppression of *Rhizoctonia solani* diseases of sugar beet by antagonistic and plant growth-promoting yeasts. J Appl Microbiol 96:69–75

Fonseca Á, Inácio J (2006) Phylloplane yeasts. In: Rosa C, Péter G (eds) Biodiversity and ecophysiology of yeasts. Springer, Berlin, pp 263–301

Fracchia S, Godeas A, Scervino JM, Sampedro I, Ocampo JA, Garcıa-Romera I (2003) Interaction between the soil yeast *Rhodotorula mucilaginosa* and the arbuscular mycorrhizal fungi *Glomus mosseae* and *Gigaspora rosea*. Soil Biol Biochem 35:701–707

Geisseler D, Scow KM (2014) Long-term effects of mineral fertilizers on soil microorganisms – a review. Soil Biol Biochem 75:54–63

Glushakova AM, Kachalkin AV, Chernov IY (2015a) Soil yeast communities under the aggressive invasion of Sosnowsky' Hogweed (*Heracleum sosnowskyi*). Eurasian Soil Sci 48:201–207

Glushakova AM, Kachalkin AV, Chernov IY (2015b) Effect of invasive herb species on the structure of soil yeast complexes in mixed forests exemplified by *Impatiens parviflora* DC. Microbiology 84:717–721

Glushakova AM, Kachalkin AV, Chernov IY (2016) The influence of *Aster x salignus* Wiild. Invasion on the diversity of soil yeast communities. Eurasian Soil Sci 49:792–795

Gomes NCM, Fagbola O, Costa R, Rumjanek NG, Buchner A, Mendonca-Hagler L, Smalla K (2003) Dynamics of fungal communities in bulk and maize rhizosphere spoil in tropics. Appl Environ Microbiol 69:3758–3766

González AM, Piñeiro AL, Fernández MR (2007) Culturable microbial populations in a vineyard soil under different management regimes: influence on spontaneous must fermentation. In: Mendez-Vilas A (ed) Current research topics in applied microbiology and microbial biotechnology: Proceedings of the II International Conference on Environmental, Industrial and Applied Microbiology (BioMicroWorld 2007). World Scientific, Singapore, pp 12–15

Gulevskaya SA, Manukyan AR, Golubev WI (1982) Cytological studies of capsule formation in the yeast *Cryptococcus magnus* in the course of its growth. Mikrobiologiya 51:287–291 (in Russian)

Gunina A, Kuzyakov Y (2015) Sugars in soil and sweets for microorganisms: review of origin, content, composition and fate. Soil Biol Biochem 90:87–100

Hanson BD, Roncoroni J, Hembree KJ, Molinar R, Elmore CI (2017) Weed control in orchards and vineyards. In: Thomas B, Murray BG, Murphy DJ (eds) Encyclopedia of applied plant sciences, vol 3, 2nd edn. Elsevier, Oxford, pp 479–484

Hong SG, Lee KH, Bae KS (2002) Diversity of yeasts associated with natural environments in Korea. J Microbiol 40:55–62

Janušauskaite D, Kadžienė G, Auškalnienė O (2013) The effect of tillage system on soil microbiota in relation to soil structure. Pol J Environ Stud 22:1387–1391

Johnson C, Albrecht G, Ketterings Q, Beckman J, Stockin K (2005) Nitrogen Basics – the Nitrogen Cycle. Agronomy Fact Sheet Series, Fact Sheet 2, Cornell University Cooperative Extension, New York

Kachalkin AV, Abdullabekova DA, Magomedova ES, Magomedov GG, Chernov IY (2015) Yeasts of the vineyards in Dagestan and other regions. Microbiology 84:425–432

Karlsson I, Friberg H, Steinberg C, Persson P (2014) Fungicide effects on fungal community composition in the wheat phyllosphere. PLoS One 9:e111786

Khan MS, Zaidi A, Ahemad M, Oves M, Wani PA (2010) Plant growth promotion by phosphate solubilizing fungi – current perspective. Arch Agron Soil Sci 56:73–98

Kijpornyongpan T, Aime MC (2017) Taxonomic revisions in the Microstromatales: two new yeast species, two new genera, and validation of Jaminaea and two *Sympodiomycopsis* species. Mycol Prog 16:495–505

Klaubauf S, Inselsbacher E, Zechmeister-Boltenstern S, Wanek W, Gottsberger R, Strauss J, Gorfer M (2010) Molecular diversity of fungal communities in agricultural soils from Lower Austria. Fungal Divers 44:65–75

Kurtzman CP, Robnett CJ (2013) *Alloascoidea hylecoeti* gen. nov., comb. nov., *Alloascoidea africana* comb. nov., *Ascoidea tarda* sp. nov., and *Nadsonia starkeyi-henricii* comb. nov., new members of the Saccharomycotina (Ascomycota). FEMS Yeast Res 13:423–432

Kurtzman CP, Fell JW, Boekhout T (2011) The yeasts, a taxonomic study, 5th edn. Elsevier, Amsterdam

Lawlor K, Knight BP, Barbosa-Jefferson VL, Lane PW, Lilley AK, Paton GI, McGrath SP, O'Flaherty SM, Hirsch PR (2000) Comparison of methods to investigate microbial populations in soils under different agricultural management. FEMS Microbiol Ecol 33:129–137

Lee CF, Lee FL, Hsu WH, Phaff HJ (1994) *Arthroascus fermentans*, a new yeast species isolated from soil in Taiwan. Int J Syst Bacteriol 44:303–307

Lemtiri A, Degrune F, Barbieux S, Hiel MP, Chélin M, Parvin N, Vandenbol M, Francis F, Colinet G (2016) Crop residue management in arable cropping systems under temperate climate. Part 1: soil biological and chemical (phosphorus and nitrogen) properties. A review. Biotechnol Agron Soc Environ 20:236–244

Lieskovský J, Kenderessy P (2014) Modelling the effect of vegetation cover and different tillage practices on soil erosion in vineyards: a case study in Vráble (Slovakia) using WATEM/SEDEM. Land Degrad Dev 25:288–296

Liu XZ, Wang QM, Göker M, Groenewald M, Kachalkin AV, Lumbsch HT, Millanes AM, Wedin M, Yurkov AM, Boekhout T, Bai FY (2015) Towards and integrated phylogenetic classification of the Tremellomycetes. Stud Mycol 81:85–147

López-Piñeiro A, Muñoz A, Zamora E, Ramírez M (2013) Influence of the management regime and phenological state of the vines on the physicochemical properties and the seasonal fluctuations of the microorganisms in a vineyard soil under semi-arid conditions. Soil Tillage Res 126:119–126

Lund A (1954) Studies on the ecology of yeasts. Munksgaard, Copenhagen

Lund A (1956) Yeasts in nature. Wallerstein Lab Comm 19:221–236

Lynch MDJ, Thorn RG (2006) Diversity of basidiomycetes in Michigan agricultural soils. Appl Environ Microbiol 72:7050–7056

Magdoff F, van Es H (2009) Building soils for better crops: sustainable soil management, 3rd edn. SARE, Brentwood

Mandl K, Schieck J, Silhavy-Richter K, Schneider V, Schmidt HP (2015) Vines take up yeasts from soil and transport them through the vine to the stem and grapes. Ithaka J, ISSN 1663-0521, pp 349–355

McCauley A, Jones C, Jacobsen J (2005) Soil and water management module 1: basic soil properties. Montana State University Extension Service, Bozeman, pp 1–12

Minter DW (2009) *Cyberlindnera*, a replacement name for *Lindnera* Kurtzman et al., nom. illegit. Mocotaxon 110:473–476

Moll J, Hoppe B, König S, Wubet T, Buscot F, Krüger D (2016) Spatial distribution of fungal communities in an arable soil. PLoS One 11:e0148130

Molnárová J, Vadkertiová R, Stratilová E (2014) Extracellular enzymatic activities and physiological profiles of yeasts colonizing fruit trees. J Basic Microbiol 54:74–84

Narsian V, Samaha SMAA, Patel HH (2010) Rock phosphate dissolution by specific yeast. Indian J Microbiol 50:57–62

Nassar AH, El-Tarabily KA, Sivasithamparam K (2005) Promotion of plant growth by an auxin-producing isolate of the yeast *Williopsis saturnus* endophytic in maize (*Zea mays* L.) roots. Biol Fertil Soils 42:97–108

Nielsen MN, Winding A (2002) Microorganisms as indicators of soil health. National Environmental Research Institute, NERI Technical Report No. 388, Denmark

Parle JN, Di Menna ME (1966) The source of yeasts in New Zealand wines. N Z J Agric Res 9:98–107

Pezzolla D, Marconi G, Turchetti B, Zadra C, Agnelli A, Veronesi F, Onofri A, Benucci GMN, Buzzini P, Albertini E, Gigliotti G (2015) Influence of exogenous organic matter on prokaryotic and eukaryotic microbiota in an agricultural soil. A multidisciplinary approach. Soil Biol Biochem 82:9–20

Porras-Alfaro A, Herrera J, Natvig DO, Lipinski K, Sinsabaugh RL (2011) Diversity and distribution of soil fungal communities in a semiarid grassland. Mycologia 103:10–21

Rinnan R, Stark S, Tolvanen A (2009) Responses of vegetation and soil microbial communities to warming and simulated herbivory in a subarctic heath. J Ecol 97:788–800

Sabate J, Cano J, Esteve-Zarzoso B, Guillamón JM (2002) Isolation and identification of yeasts associated with vineyard and winery by RFLP analysis of ribosomal genes and mitochondrial DNA. Microbiol Res 157:267–274

Salam JA, Das N (2014) Lindane degradation by *Candida* VITJzN04, a newly isolated yeast strain from contaminated soil: kinetic study, enzyme analysis and biodegradation pathway. World J Microbiol Biotechnol 30:1301–1313

Salam JA, Lakshmi V, Das D, Das N (2013) Biodegradation of lindane using a novel yeast strain, *Rhodotorula* sp. VITJzN03 isolated from agricultural soil. World J Microbiol Biotechnol 29:475–487

Sampaio JP (1999) Utilization of low molecular weight aromatic compounds by heterobasidiomycetous yeasts: taxonomic implications. Can J Microbiol 45:491–512

Sampedro I, Aranda E, Scervino JM, Fracchia S, García-Romera I, Ocampo JA, Godeas A (2004) Improvement by soil yeasts of arbuscular mycorrhizal symbiosis of soybean (*Glycine max*) colonized by *Glomus mosseae*. Mycorrhiza 14:229–234

Sharma SB, Sayyed RZ, Trivedi MH, Gobi TA (2013) Phosphate solubilizing microbes: sustainable approach for managing phosphorus deficiency in agricultural soils. SpringerPlus 2:587

Shennan C (2008) Biotic interactions, ecological knowledge and agriculture. Philos Trans R Soc B 363:717–739

Singh B, Ryan J (2015) Managing fertilizers to enhance soil health. International Fertilizer Industry Association, Paris, pp 1–24

Singh BK, Dawson LA, Macdonald CA, Buckland SM (2009) Impact of biotic and abiotic interaction on soil microbial communities and functions: a field study. Appl Soil Ecol 41:239–248

Sipiczki M (2006) *Metschnikowia* strains isolated from botrytised grapes antagonize fungal and bacterial growth by iron depletion. Appl Environ Microbiol 72:6716–6724

Sipiczki M (2016) Overwintering of vineyard yeasts: survival of interacting yeast communities in grapes mummified on vines. Front Microbiol 7:1–17

Sipilä TP, Yrjälä K, Alakukku L, Palojärvi A (2012) Cross-site soil microbial communities under tillage regimes: fungistasis and microbial biomarkers. Appl Environ Microbiol 78:8191–8201

Sláviková E, Vadkertiová R (2003a) The diversity of yeasts in the agricultural soil. J Basic Microbiol 43:430–436

Sláviková E, Vadkertiová R (2003b) Effects of pesticides on yeasts isolated from agricultural soil. Z Naturforsch C 58:855–859

Spehn EM, Joshi J, Schmid B, Alphei J, Körner C (2000) Plant diversity and soil heterotrophic activity in experimental grassland ecosystems. Plant Soil 224:217–230

Spencer JFT, Gorin PAJ (1971) Yeasts isolated from soils of citrus orchards and citrus waste disposal areas in California and Florida: flavonoid utilization. Can J Microbiol 17:871–877

Starmer WT, Lachance MA (2011) Yeast ecology. In: Kurtzman CP, Fell J, Boekhout T (eds) The yeasts, a taxonomic study, 5th edn. Elsevier, Amsterdam, pp 65–83

Tuszyński T, Satora P (2003) Microbiological characteristics of the Węgierka Zwykła plum orchard in submontane region. Pol J Food Nutr Sci 12:43–48

Vadkertiová R, Molnárová J, Vránová D, Sláviková E (2012) Yeasts and yeast-like organisms associated with fruits and blossoms of different fruit trees. Can J Microbiol 58:1344–1352

van der Walt JP, Tscheuschner IT (1957) *Hanseniaspora vineae* sp. nov. Trans Br Mycol Soc 40:211–212

Vinovarova ME, Babjeva IP (1987) Yeast fungi in steppe community. Moscow Univ Soil Sci Bull (former Vestnik Moskovskogo Universiteta. Seriia 17, Pochvovedenie) 2:43–47 (in Russian)

Vishniac HS (1995) Simulated in situ competitive ability and survival of a representative soil yeast, *Cryptococcus albidus*. Microb Ecol 30:309–320

Vreulink JM, Esterhuyse A, Jacobs K, Botha A (2007) Soil properties that impact yeast and actinomycete numbers in sandy low nutrient soils. Can J Microbiol 53:1369–1374

Vustin MM, Babjeva IP (1981) Natural habitats of yeasts *Williopsis* Zender and *Zygowilliopsis* Kudriavzev. Microbiology 50:1088–1091 (in Russian)

Wang QM, Yurkov AM, Göker M, Lumbsch HT, Leavitt SD, Groenewald M, Theelen B, Liu XZ, Boekhout T, Bai FY (2015) Phylogenetic classification of yeasts and related taxa within Pucciniomycotina. Stud Mycol 81:149–189

Wawrik B, Kerkhof L, Kukor J, Zylstra G (2005) Effect of different carbon sources on community composition of bacterial enrichments from soil. Appl Environ Microbiol 71:6776–6783

Welbaum GE, Sturz AV, Dong Z, Nowak J (2004) Managing soil microorganisms to improve productivity of agro-ecosystems. Crit Rev Plant Sci 23:175–193

Winding A, Hund-Rinke K, Rutgers M (2005) The use of microorganisms in ecological soil classification and assessment concepts. Ecotoxicol Environ Saf 62:230–248

Wuczkowski M, Prillinger H (2004) Molecular identification of yeasts from soils of the alluvial forest national park along the river Danube downstream of Vienna, Austria ("Nationalpark Donauauen"). Microbiol Res 159:263–275

Wyland LJ, Jackson LE, Chaney WE, Klonsky K, Koike ST, Kimple B (1996) Winter cover crops in a vegetable cropping system: impacts on nitrate leaching, soil water, crop yield, pests and management costs. Agric Ecosyst Environ 59:1–17

Xiao C, Chi R, Pan X, Liu F, He J (2013) Rock phosphate solubilization by four yeast strains. Ann Microbiol 63:173–178

Yurkov A, Schäfer AM, Begerow D (2009a) *Barnettozyma vustinii*. Fungal Planet 38. Persoonia 23:188–189

Yurkov A, Schäfer AM, Begerow D (2009b) *Clavispora reshetovae*. Fungal Planet 35. Persoonia 23:182–183

Yurkov AM, Kemler M, Begerow D (2012a) Assessment of yeast diversity in soils under different management regimes. Fungal Ecol 5:24–35

Yurkov AM, Schäfer AM, Begerow D (2012b) *Leucosporidium drummii* sp. nov., a member of the Microbotryomycetes isolated from soil. Int J Syst Evol Microbiol 62:728–734

Zak DR, Holmes WE, White DC, Peacock AD, Tilman D (2003) Plant diversity, soil microbial communities, and ecosystem function: are there any links? Ecology 84:2042–2050

Zarraonaindia I, Owens SM, Weisenhorn P, West K, Hampton-Marcell J, Lax S, Bokulich NA, Mills DA, Martin G, Taghavi S, van der Lelie D, Gilbert JA (2015) The soil microbiome influences grapevine-associated microbiota. mBio 6:e02527–e02514

Zhang B, He H, Ding X, Zhang X, Zhang X, Yang X, Filley TR (2012) Soil microbial community dynamics over a maize (*Zea mays* L.) growing season under conventional- and no-tillage practices in a rainfed agroecosystem. Soil Tillage Res 124:153–160

Chapter 5
Yeast in Anthropogenic and Polluted Environments

Monika Novak Babič, Jerneja Zupančič, Nina Gunde-Cimerman, and Polona Zalar

Abstract In modern society, people spend most of their time indoors and are exposed to a variety of selected air- and waterborne microorganisms that survive indoors despite sanitation chemicals, hygiene measures and occasional high temperatures. Although public health has focussed on bacteria and viruses, fungi are increasingly recognised as opportunistic infective agents. Past reports on indoor mycobiota have emphasised airborne filamentous fungi, while yeast and tap water as their transmission vector have been little investigated. Recent studies of wet indoor niches, like in kitchens and bathrooms, and particularly extreme environments inside household appliances, have revealed a diversity of yeast from the genera *Debaryomyces*, *Meyerozyma*, *Pichia*, *Saccharomyces* and *Yarrowia*. While many of these are not considered pathogenic for humans, special concern should be given to commonly isolated opportunistic yeast from the genera *Candida*, *Naganishia* and *Rhodotorula* and the black yeast *Aureobasidium* and *Exophiala*. The main characteristics of these yeasts include production of extracellular polysaccharides, degrading of cleaning agents and tolerance to high temperatures, high salt concentrations, and alkaline pH. These selected and enriched yeast species can form biofilms on synthetic and metal materials, where they can come into daily contact with people and pose the risk of infection, especially with immunocompromised people.

Keywords Air • Black yeast • Household appliances • Tap water • Wet niches

Contents

M.N. Babič • J. Zupančič • N. Gunde-Cimerman • P. Zalar (✉)
Department of Biology, Biotechnical Faculty, University of Ljubljana, Jamnikarjeva 101, 1000 Ljubljana, Slovenia
e-mail: polona.zalar@bf.uni-lj.si

P. Buzzini et al. (eds.), *Yeasts in Natural Ecosystems: Diversity*,
DOI 10.1007/978-3-319-62683-3_5

5.1 Introduction

Nowadays, people in developed countries spend most of their time indoors, as they are mainly in their homes or their work places (Klepeis et al. 2001). Indoors, we are exposed to airborne and waterborne microorganisms that to some extent reflect the outdoor seasonal cycles, although they more closely parallel conditions and practices within a household (Adams et al. 2013a, b, 2015a, b). Due to the dramatically changed conditions in our indoor environments that have followed rapid economic growth and increased housing standards over recent decades, we are increasingly surrounding ourselves with microbial species that have been selected in new ways. Sanitation, hygiene measures and use of different chemicals have reduced the exposure of humans to microorganisms, which has also been accompanied by increased access of specific species through selection, as they can survive these measures (Levy 2001; Beumer and Kusumaningrum 2003). Indoor environments also contain a multitude of unusual substrates that can aid in the selection of certain species, such as biofilm-forming microbes, or those that degrade volatile organic compounds (Isola et al. 2013). Furthermore, the presence of habitats with temperatures above 40 °C that have become part of our homes over these recent decades, such as dishwashers, also selects for thermotolerant species, a trait that is crucial for human pathogenesis and that most fungi lack (Zalar et al. 2011; Döğen et al. 2013; Gümral et al. 2016; Zupančič et al. 2016).

In comparison to filamentous fungi, and in particular when considering the airborne species, for indoor yeast there is little data on their diversity and abundance (Glushakova et al. 2004). Recent reports of opportunistic pathogenic black yeast-like fungi (here referred to as black yeast, for simplicity) in what are believed to be aseptic in household appliances have stirred up a lot of public and media attention (Zalar et al. 2011). Follow-up investigations have indicated that tap water is the main source for this contamination (Novak Babič et al. 2016), with transfer via plastic and rubber materials to food and food preparation surfaces, upon which biofilms can be formed. They have also revealed that in particular wet areas within houses, such as in kitchens and bathrooms, different opportunistic pathogenic yeast species can be harboured (Matos et al. 2002; Hamada and Abe 2009; Zupančič et al. 2016), while remaining considerably less abundant in indoor air and in other drier parts of houses (Ejdys et al. 2009). Their frequent water-related indoor presence has prompted questions regarding these yeast species, in terms of their natural reservoirs, routes of entry into and dissemination within houses, and virulence towards humans as a host (Novak Babič et al. 2016).

5.2 Water

Water has long been recognised as a source of infective microorganisms and as a vector for waterborne and faecal-borne diseases (Gray 2014). Water cleaning practices in Europe started only after a major cholera outbreak in the nineteenth century and included filtration and application of chlorine. The provision of safe drinking water remains one of the main sanitary goals of modern societies worldwide (Gray 2014). Although the main groups of microorganisms targeted in the production of safe drinking water are protozoa and bacteria, fungi are also becoming increasingly recognised as potential infectious agents and to also form biofilms in water distribution systems. As with the other microorganisms, fungi can influence the taste and odour of water (Grabinska-Loniewska et al. 2007). Previous investigations on fungi in water have primarily focussed on filamentous, spore-forming fungi, which are known to be opportunistic pathogens that can cause skin, nail and eye infections, allergies and systemic infections. However, while yeast and yeast-like fungi have only been isolated rarely in water, this has mainly been due to isolation bias, which has favoured the growth of filamentous fungi (Hageskal et al. 2006; Pereira et al. 2010; Novak Babič et al. 2016).

5.2.1 Raw Water

Freshwater derived from glaciers, lakes, rivers and groundwater represents the raw water that is used to produce tap water. The location and type of a raw water source can have a significant effect on the final microbiological quality (Novak Babič et al. 2016). For instance, surface waters can be contaminated with soil and plant particles and with fertilisers used in agronomy, which results in high organic content and consequentially increased growth of filamentous fungi, due to their ability to degrade plant material (Hageskal et al. 2006; Pereira et al. 2010). To date, only a few yeast species have been reported for surface water, with the prevalence of the genera *Candida*, *Cystobasidium* and *Rhodotorula* and sporadic detection of yeast of the genera *Cutaneotrichosporon*, *Geotrichum*, *Naganishia*, *Hyphopichia*, *Saccharomyces* and *Yarrowia* (Table 5.1). Among the melanised black yeasts, only *Aureobasidium pullulans* has been reported for surface water (Hageskal et al. 2006).

Table 5.1 Yeast species isolated from water and water-related domestic environments

Yeast species	Water niche			Water-related domestic niche					
	Surface water	Groundwater	Tap water	Bathroom, sauna	Washing machine	Kitchen	Dishwasher	Dishwasher hot aerosol	Nebuliser
Ascomycota									
Aureobasidium spp.	–	–	–	++	–	–	–	–	–
Aureobasidium melanogenum	–	++	++	+	+	++	++	–	–
Aureobasidium pullulans	+	–	++	–	+	+	–	–	+
Candida albicans	++	++	+	–	–	+	–	–	+
Candida glabrata	++	++	+	–	–	–	–	–	–
Candida glaebosa	–	–	+	–	–	–	–	–	–
Candida intermedia	–	–	+	–	–	+	+	–	–
Candida orthopsilosis	–	–	–	–	–	+	–	–	–
Candida parapsilosis	++	++	++	–	++	++	++	+	++
Candida pararugosa	–	–	+	–	–	+	–	–	–
Candida pseudointermedia	–	–	+	–	–	–	–	–	–
Candida sake	++	–	–	–	–	–	–	–	+
Candida saitoana	–	–	+	–	–	–	–	–	–
Candida silvicola	++	–	–	–	–	–	–	–	–
Candida spp.	–	–	–	++	+	+	+	–	–
Candida tropicalis	++	++	–	–	–	+	–	–	+
Candida versatilis	–	–	+	–	–	–	–	–	–
Candida zeylanoides	–	–	–	–	–	+	+	–	+
Cladophialophora boppii	–	–	–	+	–	+	–	–	–

Clavispora lusitaniae	–	++	+	–	–	+	+	–	–
Debaryomyces hansenii	–	–	++	–	–	+	+	–	–
Exophiala alcalophila	–	–	+	–	+	–	–	–	–
Exophiala angulospora	–	+	+	–	–	–	–	–	–
Exophiala cancerae	–	–	+	–	–	–	–	–	–
Exophiala castellanii	–	+	+	–	–	–	–	–	–
Exophiala dermatitidis	–	+	+	+	–	++	++	+	–
Exophiala equina	–	–	+	–	+	–	–	–	–
Exophiala jeanselmei	–	–	+	–	–	–	–	–	+
Exophiala lecanii-corni	–	–	+	+	+	+	–	–	–
Exophiala mesophila	–	–	+	++	++	–	–	–	–
Exophiala oligosperma	–	–	+	+	–	–	–	–	+
Exophiala opportunistica	–	–	+	–	–	–	–	–	–
Exophiala phaeomuriformis	–	–	+	++	++	++	++	–	–
Exophiala pisciphila	–	+	+	–	–	–	–	–	–
Exophiala psychrophila	–	–	+	–	–	–	–	–	–
Exophiala salmonis	–	–	+	+	–	–	–	–	–
Exophiala spinifera	–	–	+	+	–	–	–	–	–
Exophiala xenobiotica	–	–	+	+	–	+	–	–	–
***Fonsecaea* sp.**	–	–	–	–	–	+	–	–	–
Geotrichum candidum	+	++	++	–	–	+	–	–	–
Hanseniaspora sp.	–	+	–	–	–	–	–	–	–
Hanseniaspora uvarum	–	–	–	–	–	+	–	–	–
Hyphopichia burtonii	+	–	–	–	–	+	–	–	–
Kazachstania exigua	–	–	–	–	–	–	–	–	+
Kluyveromyces lactis	–	–	+	–	–	–	–	–	–

(continued)

Table 5.1 (continued)

Yeast species	Water niche			Water-related domestic niche					
	Surface water	Groundwater	Tap water	Bathroom, sauna	Washing machine	Kitchen	Dishwasher	Dishwasher hot aerosol	Nebuliser
Kluyveromyces marxianus	–	–	++	–	–	+	–	–	–
Knufia epidermidis	–	–	–	–	–	+	–	–	–
Kwoniella europaea	–	–	–	–	–	+	–	–	–
Lodderomyces elongisporus	–	–	–	–	–	+	–	–	–
Magnusiomyces capitatus	–	–	–	–	–	–	++	–	–
Metschnikowia fructicola	–	–	–	–	–	+	–	–	–
Metschnikowia pulcherrima	–	–	–	–	–	+	–	–	–
Meyerozyma caribbica	–	–	+	–	–	–	–	–	–
Meyerozyma guilliermondii	–	+	+	–	+	++	+	–	++
Ochroconis constricta	–	–	–	–	–	++	–	–	–
Phialophora sp.	–	–	–	+	–	–	–	–	–
Pichia cactophila	–	–	–	–	–	+	–	–	–
Pichia fermentans	–	–	+	–	–	–	–	–	–
Pichia kluyveri	–	–	–	–	–	+	+	–	–
Pichia kudriavzevii	++	++	+	–	–	–	+	–	+
Pichia membranifaciens	–	–	+	–	–	+	–	–	–
Priceomyces carsonii	–	–	+	–	–	+	–	–	–
***Rhinocladiella* sp.**	–	–	–	+	–	–	–	–	–

Rhinocladiella similis	–	++	+	+	–	–	–	–	–
Saccharomyces cerevisiae	+	++	–	–	–	+	+	–	–
Saccharomycopsis capsularis	–	–	+	–	–	–	–	–	–
Saprochaete clavata	–	–	–	–	–	–	++	–	–
Wickerhamomyces anomalus	–	–	+	–	–	+	–	–	++
Wickerhamomyces onychis	–	–	–	–	–	+	–	–	–
Yarrowia lipolytica	+	++	+	–	–	+	–	–	+
Basidiomycota									
Apiotrichum montevideense	–	–	+	–	–	–	–	–	–
Cryptococcus spp.	–	–	–	++	+	–	–	–	+
Cutaneotrichosporon curvatus	+	–	–	–	–	–	–	–	–
Cystobasidium minutum	++	–	+	–	–	+	–	–	–
Cystobasidium slooffiae	–	–	+	–	++	+	+	–	–
Cystofilobasidium capitatum	–	–	+	–	–	–	–	–	–
Filobasidium uniguttulatum	–	–	–	–	–	–	–	–	+
Naganishia albida	+	–	–	–	–	+	+	–	+
Naganishia diffluens	–	–	–	–	–	+	–	–	–
Naganishia liquefaciens	–	–	–	–	–	+	+	–	–
Papiliotrema laurentii	–	+	+	–	–	–	–	–	–

(continued)

Table 5.1 (continued)

Yeast species	Water niche			Water-related domestic niche					
	Surface water	Groundwater	Tap water	Bathroom, sauna	Washing machine	Kitchen	Dishwasher	Dishwasher hot aerosol	Nebuliser
Rhodotorula diobovata	–	–	–	–	–	–	+	–	–
Rhodotorula glutinis	–	+	–	–	–	–	–	–	+
Rhodotorula kratochvilovae	–	–	–	–	–	–	+	–	–
Rhodotorula mucilaginosa	++	++	+	–	–	++	++	+	+
Rhodotorula spp.	–	–	++	++	++	–	–	–	+
Sporidiobolus salmonicolor	–	+	+	–	–	–	–	–	–
Sporobolomyces ruberrimus	–	–	–	–	+	–	–	–	–
Sporobolomyces sp.	–	–	+	–	–	–	–	–	–
Trichosporon coremiiforme	–	+	–	–	–	–	–	–	–
Triodiomyces crassus	–	–	+	–	–	+	–	–	–
Vishniacozyma carnescens	–	–	–	–	–	–	–	–	+

Legend: ++, most abundant yeast species per niche; **bold**, BSL-2 yeast. Data obtained from: Salonen and Ruokola (1969), Hinzelin and Block (1985), Frankova and Horecka (1995), Arvanitidou et al. (1999), Kinsey et al. (1999), Doggett (2000), Hamada and Fujita (2000), Göttlich et al. (2002), Hamada (2002), Matos et al. (2002), Nanbakhsh et al. (2004), de Hoog et al. (2006), Goksugur et al. (2006), Hageskal et al. (2006), Brandi et al. (2007), Grabinska-Loniewska et al. (2007), Yamaguchi et al. (2007), Hamada and Abe (2009), Lotrakul et al. (2009), Gattlen et al. (2010), Lian and de Hoog (2010), Pereira et al. (2010), Sammon et al. (2010), Miceli et al. (2011), Zalar et al. (2011), Ayanbimpe et al. (2012), Shaker and Sharif (2012), Adams et al. (2013a), Döğen et al. (2013), Heinrichs et al. (2013a, b), Isola et al. (2013), Jadhav et al. (2013), Kadaifciler et al. (2013), Stapleton et al. (2013), Biedunkiewicz et al. (2014), Samah et al. (2014), Novak Babič et al. (2015, 2016), Gümral et al. (2016), Peckham et al. (2016), Zupančič et al. (2016)

In comparison with surface water, groundwater is more oligotrophic and has higher concentrations of magnesium and calcium ions. The diversity of filamentous fungi is thus lower, while the diversity of yeast appears comparable to that of surface water (Novak Babič et al. 2016). The most abundant yeast species reported in groundwater belong to the genera *Candida*, *Clavispora*, *Geotrichum*, *Pichia*, *Saccharomyces* and *Yarrowia* and to the basidiomycetous phenotypic genus *Rhodotorula* (Table 5.1). The genera *Hanseniaspora*, *Meyerozyma*, *Papiliotrema*, *Trichosporon* and *Sporidiobolus* have been detected less frequently (Table 5.1). The most important difference observed between surface and groundwaters has been the greater presence of black yeasts in groundwater (Göttlich et al. 2002; Novak Babič et al. 2016). Groundwater harboured *Aureobasidium melanogenum*, a high diversity of *Exophiala* species, and *Rhinocladiella similis* (Table 5.1). These differences might have resulted from isolation bias, as slowly growing black yeast might have been overlooked or overgrown by filamentous species present in surface waters. This might also be caused by differences in the chemical compositions of the water, such as the higher concentrations of calcium and magnesium ions and nitrate in groundwater (Novak Babič et al. 2016).

5.2.2 Tap Water

Tap water intended for human consumption can be derived from either surface water or groundwater, which is taken through a series of physico-chemical cleaning processes that usually end with the addition of chlorine (Gray 2014), which removes or inactivates most microorganisms. These processes are regularly monitored and are subject to national directives and legislature (Gray 2014). Most recent directives do not include monitoring for fungi, despite alarming reports on the regular presence of emerging opportunistic fungal pathogens, such as the *Candida parapsilosis* species complex, that has indicated that tap water might be a vector for human infections (Yamaguchi et al. 2007; Biedunkiewicz et al. 2014; Novak Babič et al. 2016; Zupančič et al. 2016). Other *Candida* species have been isolated from tap water sporadically (Kadaifciler et al. 2013; Biedunkiewicz et al. 2014; Novak Babič et al. 2016; Zupančič et al. 2016), one of which is *Candida albicans*, the most common yeast human pathogen (see also Inácio and Daniel 2017). *C. albicans* was detected within biofilms formed in tap water distribution systems using a metagenomic approach (Heinrichs et al. 2013a) (Table 5.1). Of the non-*Candida* species, *Geotrichum candidum* has been most commonly reported (Frankova and Horecka 1995; Kinsey et al. 1999; Novak Babič et al. 2016; Zupančič et al. 2016), followed by yeast of the genera *Clavispora*, *Debaryomyces*, *Kluyveromyces*, *Meyerozyma*, *Pichia*, *Saccharomycopsis* and *Yarrowia* (Table 5.1). Among basidiomycetous yeast, the ubiquitous *Rhodotorula* species have prevailed, while members of the genera *Apiotrichum*, *Cystobasidium*, *Cystofilobasidium*, *Papiliotrema*, *Priceomyces*, *Sporobolomyces* and *Triodiomyces* have only been isolated sporadically (Table 5.1).

In agreement with the biodiversity that has been described for groundwater, tap water also contains a large spectrum of black yeast. Most studies have reported the oligophilic *A. melanogenum* and *A. pullulans*, followed by the waterborne *Rhin. similis*, and high diversity of *Exophiala* species (Kinsey et al. 1999; Göttlich et al. 2002; Heinrichs et al. 2013a, b; Biedunkiewicz et al. 2014; Novak Babič et al. 2016; Zupančič et al. 2016) (Table 5.1). Several *Exophiala* species are known as causative agents of waterborne diseases of cold-blooded animals, while *Exophiala jeanselmei*, *Exophiala dermatitidis* and *Exophiala phaeomuriformis* can cause diseases in immunocompromised people (Vicente et al. 2008).

5.3 Wet Areas in Houses

Tap water distribution networks can introduce outdoor yeast into indoor environments (Novak Babič et al. 2016). Water in houses is used for human consumption, food preparation, hygiene and leisure activities. The main wet niches within houses are found in kitchens and bathrooms, and in domestic saunas, and these are characterised by high humidity, presence of various salts and detergents, fluctuations in pH and temperature and frequent presence of natural and artificial carbon sources (Zhao et al. 2010; Zalar et al. 2011; Novak Babič et al. 2016; Zupančič et al. 2016).

5.3.1 Kitchens

Kitchens are usually heavily colonised by a wide diversity of microorganisms (Ojima et al. 2002; Sinclair and Gerba 2011; Flores et al. 2013). Humans are exposed to these microorganisms, and hence to potential infections, via the handling, preparation and consumption of food and indirectly by contact with contaminated surfaces (Scott 2000; Flores et al. 2013). The conditions in kitchens can favour fungal colonisation and propagation, due to the presence of water, the continuous flow of nutrients, and the temperature of the room. At the same time, the use of oxidative detergents and cleaning products, the rapid changes in pH, the mechanical forces and the presence of artificial materials (e.g. rubber, plastic, and metals) promote the selection of the most stress-resistant species. The kitchen sites that show the highest degree of contamination by fungi are kitchen drains (up to 78%) and dish-drying racks (60%), followed by surfaces of counters, rubber seals of kitchen drains, in kitchen sinks, and metal grids on taps (Zupančič et al. 2016).

Kitchen drains can harbour a larger and more stable fungal biomass than other sites in the kitchen. They are typically populated by *Exophiala* and *Candida*, and in particular *C. parapsilosis*, *Ex. dermatitidis* and *Ex. phaeomuriformis*. Metal or plastic dish-drying racks are usually located close to the kitchen sinks and are often constantly exposed to dripping water, which can promote the formation of a mixed bacterial–fungal biofilms. The primarily detected yeasts reported for dish-drying racks are *C. parapsilosis*, *Ex. phaeomuriformis*, *A. melanogenum* and *Ex. dermatitidis*. *C. albicans* is a well-known human pathogen and has been detected in kitchen drains and around the rubber seals, on filter grids in kitchen sinks and on dish-drying racks (Adams et al. 2013a; Zupančič et al. 2016).

The yeast mycobiota that can be found in kitchens differs considerably in terms of whether or not there is a dishwasher in the kitchen. In general, kitchens with dishwashers show higher occurrence of yeast species, although with less diversity, compared to those without dishwashers where situation is reversed. In kitchens with dishwashers, *Ex. dermatitidis* and other black yeasts prevail, while in those without dishwashers, the dominant black yeast is *Ex. phaeomuriformis*, the diversity of the black yeast is much lower, and *C. parapsilosis* also dominates (Zupančič et al. 2016).

5.3.2 Bathrooms and Domestic Saunas

Bathrooms and kitchens share many features, such as temperatures above 21 °C, high levels of humidity (including occasional condensation and accumulation of moisture on walls and floor) and exposure to detergents (Hamada and Abe 2009). In addition, the microbes present are exposed to personal hygiene products, such as soaps and shampoos. Although these cleaning agents successfully reduce the numbers of faecal bacteria and viruses, they can favour the growth of alkali-tolerant fungi that can degrade the components in the detergents (Hamada and Fujita 2000; Hamada and Abe 2009; Lian and de Hoog 2010). Bathroom floors and shower curtains can be contaminated with yeast from the genera *Candida*, *Cryptococcus* and *Rhodotorula* and with different black yeasts. In particular, these yeasts can form biofilms on wet or soapy grouting between ceramic bathroom tiles (Hamada and Fujita 2000). Among the black yeasts, *Aureobasidium* species prevail, followed by mesophilic *Exophiala* species, and the related *Cladophialophora boppii* and *Rhin. similis* (Table 5.1) (Hamada and Abe 2009; Lotrakul et al. 2009; Lian and de Hoog 2010). The highest numbers of black yeasts are found in bathroom sinks, filters and taps, which indicates that tap water might be the source of this contamination (Matos et al. 2002; de Hoog et al. 2006; Lian and de Hoog 2010). The growth of these black yeasts is favoured by sodium fatty acids, non-ionic surfactants used in soaps and shampoos, and alkaline conditions (Hamada and Abe 2009).

There are additional indoor wet niches that are related to leisure activities, which include domestic saunas and steam baths. In comparison to bathrooms, these are exposed to higher temperatures and humidity levels (Matos et al. 2002; Nanbakhsh et al. 2004). The wood of public Finnish saunas was shown to be colonised by black yeast from the genera *Phialophora* and *Rhinocladiella* (Salonen and Ruokola 1969), while wet ceilings and floors in Turkish steam baths harbour mainly species of the genus *Candida* (Nanbakhsh et al. 2004; Brandi et al. 2007) and the thermotolerant black yeast *Ex. dermatitidis* (Matos et al. 2002; Goksugur et al. 2006) (Table 5.1).

5.4 Extreme Domestic Environments

People increasingly use household appliances to facilitate their daily chores, with 80% to 100% of households in developed countries equipped with various ones. The most common of such appliances are refrigerators, washing machines and dishwashers (Berkholz et al. 2011). The conditions inside these appliances in particular during their functioning are considered extreme in comparison to normal indoor conditions. Dishwashers and washing machines operate at high temperatures, with specific niches exposed to alternate wet and dry periods and neutral to high pH (Zalar et al. 2011; Döğen et al. 2013; Novak Babič et al. 2015; Gümral et al. 2016; Zupančič et al. 2016). The insides of refrigerators on the other hand are exposed to constant low temperatures and elevated humidity (Godwin et al. 2007).

5.4.1 *Washing Machines*

The mycobiota of washing machines was investigated due to the potential transfer of dermatophyte fungi from clothes to these appliances and to humans (Shah et al. 1988; Novak Babič et al. 2015). Recent ecological trends that have resulted in lower temperatures of operation of washing machines and the use of biodegradable detergents have facilitated fungal colonisation of washing machines. This occurs in the form of persistent biofilms, where yeasts covered with extracellular polysaccharides prevail (Isola et al. 2013; Novak Babič et al. 2015). These extracellular polysaccharides enhance adhesion to different artificial materials and protect the cells against stress conditions, such as changes in pH and temperature (Cooper 2010). Washing machines have been shown to be populated mainly with *C. parapsilosis* species complex, the ubiquitous *Meyerozyma guilliermondii* and *Cystobasidium slooffiae* (Stapleton et al. 2013; Novak Babič et al. 2015) (Table 5.1). For their rubber seals, the lower shearing forces and rougher surfaces can result in significantly increased attachment in terms of the numbers and diversity of the yeast, in comparison to those seen for the plastic drawers for detergents, with a prevalence of the genera *Candida*, *Cystobasidium*, *Naganishia* and *Rhodotorula* (Gattlen et al. 2010; Novak Babič et al. 2015) (Table 5.1). Black yeasts from the genera *Aureobasidium* and *Exophiala* are prominent in the interiors of water supply connectors and plastic drawers for detergents (Hamada 2002; Isola et al. 2013; Novak Babič et al. 2015) (Table 5.1).

5.4.2 *Dishwashers*

Dishwashers are constantly exposed to contact with people, kitchen litter, and raw cooking ingredients. They were constructed to remove the need for people to wash dishes by hand and simultaneously with the intention to diminish or eliminate the associated high microbial burden besides. Zalar et al. (2011) demonstrated that although conditions in dishwashers prevent growth of most mesophilic microorganisms, they enrich the growth of sturdy and polyextremotolerant fungi. The term polyextremotolerant describes remarkable ability of specific fungi regarding their character to colonise a variety of different environments and endure a broad range of ecological conditions (Gostinčar et al. 2011). In case of dishwashers, such fungi can resist the frequent exchanges of wet and dry conditions, fluctuations in the temperature of up to 70 °C, addition of oxidative detergents, elevation of pH up to 12, high organic loads and NaCl concentrations and mechanical shearing of water ejectors.

The presence of *Candida* in dishwashers and their relation to human infections was first reported by Bennett (1998) and then by Nedret Koc et al. (2002). Since 2011 several studies have reported that geographically distant dishwashers across all continents are contaminated by similar set of fungi (Zalar et al. 2011; Döğen et al. 2013; Gümral et al. 2016; Zupančič et al. 2016). The dominant colonisers are the black yeast *Ex. dermatitidis* and *Ex. phaeomuriformis*, followed by *C. parapsilosis*, *Magnusiomyces capitatus*, *Rhodotorula mucilaginosa* and *Pichia guilliermondii* (now *Mey. guilliermondii*) (Zalar et al. 2011; Döğen et al. 2013; Gümral et al. 2016; Zupančič et al. 2016). Other species that have been sporadically detected in dishwashers are given in Table 5.1. The most heavily contaminated parts of dishwashers are the rubber seals, followed by the side nozzle, door and drain. The walls, sprinklers, cutlery rack, detergent dispenser and rinse aid dispenser are less affected. The hot aerosols that can be released from dishwashers when they are opened before the end of the cooling process, as also for the wastewater, can contain several opportunistic pathogenic yeast species (Table 5.1) (Zupančič et al. 2016).

5.4.3 Refrigerators

Microbiological spoilage of foods and food-borne diseases are common throughout the world, and these can be considerably diminished by correct food storage (Ojima et al. 2002). Increased demand for stable household food storage facilities resulted in the development of refrigerators for home use in the beginning of the twentieth century. Refrigerators and freezers that generally operate at temperatures from −18 °C to +12 °C slow down the metabolism of most mesophilic microbes, while promoting the growth of psychrotolerant and psychrophilic microbes. The moisture that is available due to evaporation and condensation, in addition to the food particles, can provide favourable conditions for their growth (Jeon et al. 2013). The majority of refrigerators also operate at higher temperatures than recommended (4.4 °C) (Godwin et al. 2007; Pal et al. 2008; Esfarjani et al. 2016).

In this context, it appears that the food is the main microbial transmission vector (Kennedy et al. 2005; James et al. 2008; Bharathirajan et al. 2012; Maktabi et al. 2013; Catellani et al. 2014). Yeasts dominate in the spoilage of refrigerated foods with low water activity and low pH. They have been isolated mainly from the plastic refrigerator vegetable compartments, rubber seals, walls and ice and water dispensers (NSF 2013; Catellani et al. 2014). *Candida* species have been isolated most frequently, with *Pichia kudriavzevii* (former *Candida krusei*) prevailing in the refrigerator air (Altunatmaz et al. 2012). All of the yeasts that have been detected in refrigerators to date are given in Table 5.2. Many yeasts are known to grow in natural environments with low temperatures, like polar and Alpine regions (reviewed in Chaps. 11 and 12 of this book). Considering the only publication (Altunatmaz et al. 2012), representatives of psychrotolerant generalist species occupy refrigerators, to the best of our knowledge, there have been no reports of yeast isolated from freezers, which have operating temperatures down to −18 °C.

5.4.4 Other Electric Appliances

5.4.4.1 Coffee and Tea Makers

Coffee and tea makers have heated aluminium water tanks, combined with water flow during beverage preparation, when the water temperature is increased to the boiling point and the pressure rises from 0.3 bars up to 5.0 bars in less than 3 min. The water that has not been used during the preparation of the beverage is also removed from the machine after its use. As the inlet tubes can contain residues of coffee, tea and sugar, these are coated with an antimicrobial layer (Iacobucci 2004). Additionally, the caffeine present in coffee and tea has well-known antibacterial properties. While the bacterial communities that can be found in coffee machines have been well described (Vilanova et al. 2015), to the best of our knowledge, nothing is known about the associated mycobiota. The only information available to date is related to the colonisation of coffee beans by different fungi (Falkowski et al. 2002) and yeast (Evangelista et al. 2014), as primarily *Pichia kluyveri*, *Wickerhamomyces anomalus* (former *Pichia anomala*), *Hanseniaspora uvarum*, *Saccharomyces cerevisiae*, *Debaryomyces hansenii*, *Meyerozyma caribbica* and *Torulaspora delbrueckii* (Masoud et al. 2004; Silva et al. 2008; Vilela et al. 2010; de Melo Pereira et al. 2014; Evangelista et al. 2015). To date, there have been no reports of the presence of these in coffee and tea makers.

5.4.4.2 Clothes Driers

Drying is the most reliable method of decontamination of textiles when it is carried out at temperatures of approx. 80 °C for 2 h (Scott and Bloomfield 1990). To date, it has been reported that in spite of these high temperatures, bacterial cross-contamination from laundry to driers can occur, which is particularly important in light of clinical outbreaks in hospital environments (Brunton 1995; Fijan and Turk 2012). At present, there is no information available relating to fungi inside driers.

Table 5.2 Yeast species from air-related domestic environments

Yeast species	Air-related domestic niche				
	Indoor air	Air conditioning	House dust	Refrigerator	Pets
Ascomycota					
Aureobasidium spp.	–	+	–	–	–
Aureobasidium pullulans	++	–	++	–	+
Candida catenulata	–	–	+	–	–
Candida haemulonis	–	–	+	–	–
Candida maltosa	–	–	+	–	–
Candida parapsilosis	++	–	–	–	–
Candida spp.	–	+	++	++	–
Candida tropicalis	–	–	+	–	–
Candida zeylanoides	+	–	–	–	–
Clavispora lusitaniae	+	–	–	–	–
Colwellia maris	+	–	–	–	–
Debaryomyces hansenii	+	–	+	–	–
Debaryomyces spp.	+	–	–	+	–
Diutina rugosa	–	–	+	–	–
***Exophiala* spp.**	–	+	–	–	+
Geotrichum candidum	++	–	–	–	–
Kluyveromyces spp.	+	–	–	–	–
Phaeococcomyces nigricans	–	–	–	–	+
Pichia fermentans	+	–	–	–	–
Pichia kudriavzevii	–	–	–	++	–
Pichia spp.	+	–	+	–	–
Saccharomyces cerevisiae	+	–	–	–	–
Saccharomyces spp.	–	–	–	+	–
Saccharomycopsis spp.	+	–	–	–	–
Saprochaete clavata	–	–	–	–	–
Scytalidium flavobrunneum	+	–	–	–	–
Yarrowia lipolytica	+	–	–	–	–
Wickerhamomyces onychis	–	–	+	–	–
Zygoascus spp.	+	–	–	–	–
Basidiomycota					
Cryptococcus spp.	+	–	++	–	–
Cryptococcus neoformans	–	–	–	–	+
Cryptococcus gattii	–	–	–	–	+
Cutaneotrichosporon cutaneum	+	–	–	–	–
Filobasidium oeirense	–	–	+	–	–
***Malassezia* spp.**	–	–	+	–	+
Malassezia restricta	–	–	–	–	++
Naganishia albida	+	–	++	–	–
Naganishia diffluens	+	–	++	–	–

(continued)

Table 5.2 (continued)

	Air-related domestic niche				
Yeast species	Indoor air	Air conditioning	House dust	Refrigerator	Pets
Rhodotorula spp.	–	+	–	–	+
Rhodotorula glutinis	++	–	+	–	+
Rhodotorula graminis	–	–	+	–	–
Rhodotorula mucilaginosa	++	–	++	–	+
Rhodotorula nothofagi	–	–	+	–	–
Sporobolomyces roseus	–	–	+	–	–
Tausonia pullulans	–	–	+	–	–
Trichosporon spp.	–	–	+	–	–
Vishniacozyma carnescens	–	–	–	–	+
Vishniacozyma tephrensis	–	–	–	–	+

Legend: ++, most abundant yeast species per niche; **bold**, BSL-2 yeast. Data obtained from: Davenport (1980), Nishimura and Miyaji (1982), Jay (1996), Górny and Dutkiewicz (2002), Brandt et al. (2004), Glushakova et al. (2004), Cheong and Neumeister-Kemp (2005), Dynowska et al. (2006), Ejdys (2009), Amend et al. (2010), Noris et al. (2011), Simoes et al. (2011), Altunatmaz et al. (2012), Khan and Karuppayil (2012), Adams et al. (2013b, 2015a), Catellani et al. (2014), Seyedmousavi et al. (2015)

5.4.4.3 Nebuliser Devices

Nebulisers are devices used to administer medication in the form of a mist that is inhaled into the lungs, as commonly used for treatment of patients with cystic fibrosis, asthma, chronic obstructive pulmonary disease and other respiratory disorders. Patients with cystic fibrosis represent a particularly sensitive and frequent group of users (Blau et al. 2007).

Nebulisers provide extreme environments due to the occasional high doses of different drugs (e.g. antibiotics, mucolytics), disinfection with 70% alcohol and dry periods when not in use. Nebulisers favour the selection and enrichment of certain opportunistic yeast and moulds. Approximately 25% of nebulisers tested were reported to be colonised by *Mey. guilliermondii* followed by *C. parapsilosis*, both of which are associated with the formation of biofilms inside such devices (Jadhav et al. 2013; Peckham et al. 2016). The other more sporadic contaminants that have been isolated from nebulisers are given in Table 5.1.

5.5 Indoor Air and Air-Conditioning Systems

5.5.1 Indoor Air

Bioaerosols consist of particles of biological origin (e.g. microorganisms, plants, animals) suspended in the air. Indoor aerosols originate from humans, pets, house dust, organic waste, food, houseplants, textiles, furniture and heating, ventilation and air conditioning (HVAC) systems. These are affected by variations in season, temperature, humidity and characteristics of the HVAC systems. Air-conditioning systems vary in design and can additionally include filters and humidifiers (McGrath et al. 1999), both of which can provide substrate for the proliferation of opportunistic microorganisms (Li et al. 2010).

Several studies have identified human activities as an important source for the presence of bacteria in indoor bioaerosols (Horner et al. 2004), and in particular for skin-related species (Hospodsky et al. 2012). Contrary to this situation for bacteria, the presence and diversity of fungi have mainly been connected to outdoor air and seasonal changes (Ejdys et al. 2009; Adams et al. 2013b), rather than to room use. No major differences have been reported for kitchens, bathrooms, bedrooms and living rooms (Ejdys et al. 2009; Adams et al. 2013b). Indoor air is dominated by moulds, while yeast did not exceed 10% of the mycobiota at normal humidities and temperatures (Ejdys et al. 2009), although they will increase to up to 20% under particularly damp conditions (Meklin et al. 2002). The genera and species detected in indoor air are given in Table 5.2. Some of these species have been described as aetiological factors of mycoses, and less frequently, of myco-allergies (Ejdys et al. 2009).

5.5.2 House Dust

House dust is composed of microorganisms and plant, pet and human remains. Similar to bioaerosols, the mycobiota detected for house dust shows a stronger correlation with outdoor air than with human activities (Adams et al. 2015a). The proportion of human-associated taxa can vary from 4% (Adams et al. 2015b) up to 17% (Qian et al. 2012) and is primarily represented by typical human skin commensals, like *Trichosporon* spp., *Rh. mucilaginosa*, *Candida* species, *Filobasidium* and *Malassezia* (Amend et al. 2010; Adams et al. 2013b, 2015a). The abundance of the last of these was mainly related to the amount of house dust and mattress dust, the use of cold-mist vaporisers and the general dampness (Flannigan et al. 2001). Among the black yeasts isolated, *A. pullulans* was particularly abundant. The indoor air study of 25 apartments in Moscow in Moscow region revealed plant-related/epiphytic species *Cryptococcus diffluens* (now *Naganishia diffluens*), *Rh. mucilaginosa*, *Sporobolomyces roseus*, *Cryptococcus albidus* (now *Naganishia albida*) and *Deb. hansenii* as the most common yeasts (Glushakova et al. 2004). A study by Amend et al. (2010) analysed dust from 72 locations across six continents, and this revealed that among the 45 fungal operational taxonomic units (OTUs) found with the highest frequency, only eight were attributed to yeast, which included mainly species of the genera *Aureobasidium*, *Cryptococcus* and *Filobasidium*. The diversity of the fungi in this study correlated primarily to the latitude, with greater diversity for temperate zones in comparison to tropical sites (Amend et al. 2010) (Table 5.2).

5.5.3 Air-Conditioning Systems

The ventilation in houses can be controlled by an independent HVAC system that supplies the outdoor air to the indoor areas, after it has passed through minimum efficiency reporting value (MERV) rating filters (Bennett et al. 2011; Adams et al. 2015a). HVAC systems are based on single-pass ventilation and do not include air recirculation (Adams et al. 2015b). Examination of HVAC systems has resulted in the isolation of mainly black yeast from the genera *Aureobasidium* and *Exophiala* and the ubiquitous yeast from the genera *Candida* and *Rhodotorula* (Nishimura and Miyaji 1982; Cheong and Neumeister-Kemp 2005; Noris et al. 2011; Simoes et al. 2011; Khan and Karuppayil 2012; Adams et al. 2015a) (Table 5.2).

5.6 Indoor Pets

Pets presumably facilitate the introduction of environmental air-, water-, and soil-borne microbes into the indoor environment. For the impact of dogs and cats on the microbial composition of house dust, a study was based on a high-density phylogenetic microarray that used 16S rRNA for bacteria and rRNA intergenic spacers for fungi. It revealed that the presence of dogs increases the bacteria diversity in house dust, which is paralleled by significantly decreased diversity of fungi (Fujimura et al. 2010). The majority of the fungi detected belonged to the subclasses Dothideomycetidae and Eurotiomycetidae and included many known filamentous allergenic fungi and the yeast *Malassezia restricta*. Other yeasts from the genera *Aureobasidium*, *Exophiala*, *Phaeococcomyces*, *Rhodotorula* and *Vishniacozyma* (formerly *Cryptococcus*) were also sporadically detected (Fujimura et al. 2010) (Table 5.2). It is also known that domestic animals are carriers of zoonoses, infectious diseases that can be naturally transmitted between vertebrate animals and man. Dogs and cats can be the carriers of *Malassezia* spp., while a wide variety of mammals, birds, reptiles and amphibians might be a source of *Cryptococcus neoformans* and *Cryptococcus gattii* (Seyedmousavi et al. 2015).

5.7 Hydrocarbon-Based Synthetic Materials

Aliphatic or aromatic hydrocarbons serve as raw materials to produce fibre, plastic, rubber and industrial chemicals (Naik et al. 2010). As such, they are widely present in modern domestic environments, particularly as parts of mattresses, plastic toys, synthetic carpeting, polyester fabrics, polyvinylchloride-based plumbing and silicone and rubber seals (Prenafeta-Boldú et al. 2006; Isola et al. 2013). These surfaces are usually not completely smooth, which allows microorganisms to form stable biofilms (Percival et al. 1999). Also, a wide spectrum of yeast can utilise hydrocarbons as their sole carbon sources and thus have the potential to be used for bioremediation purposes (Gargouri et al. 2015). The most common yeasts that are regularly isolated from hydrocarbon-based materials indoors are *C. albicans*; *Candida tropicalis*; *Deb. hansenii*; *Naganishia*, *Pichia*, *Rhodotorula* and *Vishniacozyma* species; *Trichosporon asahii*; and *Yarrowia lipolytica* (Margesin et al. 2003; Sabirova et al. 2011; Shumin et al. 2012; Gargouri et al. 2015), along with black yeast species from the genera *Cladophialophora* and *Exophiala* (Prenafeta-Boldú et al. 2006; Peng et al. 2008).

5.8 Concluding Remarks

Most studies on yeast ecology have overlooked the yeast that can be present in indoor environments, primarily due to their focus being on airborne filamentous fungi and their potential to cause pulmonary or allergic infections. Discoveries in recent years have changed focus from indoor air to tap water systems, with yeast contamination reported for many indoor wet areas, such as kitchens and bathrooms, and even in the extreme environments provided by household appliances connected to the tap water system, such as washing machines and dishwashers. These more recent studies have shown that the prevailing ecological and energy-preserving trends have resulted in lower temperatures for washing and the use of biodegradable detergents, which have considerably increased the occurrence of yeasts both indoors in general and in the aforementioned appliances. In contrast to bacteria, to date there are only few reports on the presence of yeast in other appliances, such as freezers, dryers and coffee machines. Indoors, we come into close contact with selected, and sometimes even enriched, species of stress-tolerant yeast on a daily basis. Many of the indoor species listed in Tables 5.1 and 5.2 have been recognised as potential human opportunistic pathogens and some even as emerging pathogens. Indoors, we are exposed to these yeasts via the air, water and physical contact with surfaces, many of which have the potential to cause cutaneous, subcutaneous, pulmonary, enteric and/or neurotropic infections. While this exposure might be less important for healthy individuals, there is the need to alert immunocompromised people and to stimulate further studies in the future.

References

Adams RI, Miletto M, Taylor JW, Bruns TD (2013a) The diversity and distribution of fungi on residential surfaces. PLoS One 8:e78866

Adams RI, Miletto M, Taylor JW, Bruns TD (2013b) Dispersal in microbes: fungi in indoor air are dominated by outdoor air and show dispersal limitation at short distances. ISME J 7:1262–1273

Adams RI, Bateman AC, Bik HM, Meadow JF (2015a) Microbiota of the indoor environment: a meta-analysis. Microbiome 3:1–18

Adams RI, Bhangar S, Pasut W, Arens EA, Taylor JW, Lindow SE, Nazaroff WW, Bruns TD (2015b) Chamber bioaerosol study: outdoor air and human occupants as sources of indoor airborne microbes. PLoS One 10:e0128022

Altunatmaz SS, Issa G, Aydin A (2012) Detection of airborne psychrotrophic bacteria and fungi in food storage refrigerators. Braz J Microbiol 43:1436–1443

Amend AS, Seifert KA, Samson R, Bruns TD (2010) Indoor fungal composition is geographically patterned and more diverse in temperate zones than in the tropics. Proc Natl Acad Sci U S A 107:13748–13753

Arvanitidou M, Kanellou K, Constantinides TC, Katsouyannopoulos V (1999) The occurrence of fungi in hospital and community potable waters. Lett Appl Microbiol 29:81–84

Ayanbimpe GM, Abbah VE, Ior CA (2012) Yeasts and yeast-like fungal contaminants of water used for domestic purposes in Jos, Nigeria. Microbiol Res 3:99–102

Bennett JE (1998) Candidiasis. In: Isselbacher KJ, Braunwald E, Wilson JD, Martin JB, Fauci AS, Kasper DL (eds) Harrison's principles of internal medicine. McGraw-Hill, New York, pp 860–861

Bennett DH, Apte MG, Wu X, Trout A, Faulkner D, Maddalena R, Sullivan D (2011) Indoor environmental quality and heating, ventilating, and air conditioning survey of small and medium size commercial buildings: field study: California Energy Commission, CEC-500-2011-043

Berkholz P, Kobersky V, Stamminger R (2011) Comparative analysis of global consumer behaviour in the context of different manual dishwashing methods. Int J Consum Stud 37:46–58

Beumer RR, Kusumaningrum HD (2003) Kitchen hygiene in daily life. Int Biodeterior Biodegrad 51:299–302

Bharathirajan R, Gopinathan R, Prakash M (2012) Microbial management of household cold storage exploratory study in Jeddah, Saudi Arabia. Int J Curr Microbiol App Sci 1:50–55

Biedunkiewicz A, Kowalska K, Schulz Ł, Stojek K, Dynowska M, Ejdys E, Sucharzewska E, Kubiak D (2014) Mycological monitoring of selected aquatic ecosystems in the context of epidemiological hazards. Drinking water. Ann Parasitol 60:191–198

Blau H, Mussaffi H, Mei Zahav M, Prais D, Livne M, Czitron BM, Cohen HA (2007) Microbial contamination of nebulizers in the home treatment of cystic fibrosis. Child Care Health Dev 33:491–495

Brandi G, Sisti M, Paparini A, Gianfranceschi G, Schiavano GF, De Santi M, Santoni D, Magini V, Romano-Spica V (2007) Swimming pools and fungi: an environmental epidemiology survey in Italian indoor swimming facilities. Int J Environ Health Res 17:197–206

Brandt ME, Kauffman CA, Pappas PG, Iqbal N, Arthington-Skaggs BA, Lee-Yang W, Smith MT (2004) Fungemia caused by *Zygoascus hellenicus* in an allogeneic stem cell transplant recipient. J Clin Microbiol 42:3363–3365

Brunton WA (1995) Infection and hospital laundry. Lancet 345:1574–1575

Catellani P, Miotti Scapin R, Alberghini L, Radu IL, Giaccone V (2014) Levels of microbial contamination of domestic refrigerators in Italy. Food Control 42:257–262

Cheong CD, Neumeister-Kemp HG (2005) Reducing airborne indoor fungi and fine particulates in carpeted Australian homes using intensive, high efficiency HEPA vacuuming. J Environ Health Res 4:3–16

Cooper RI (2010) Microbial biofilms: case reviews of bacterial and fungal pathogens persisting on biomaterials and environmental substrata. In: Mendez-Vilas A (ed) Current research, technology and education topics in applied microbiology and microbial biotechnology. Formatex, Badajoz, Spain, pp 807–817

Davenport RR (1980) Cold-tolerant yeasts and yeast-like microorganisms. In: Skinner FA, Passmore SM, Davenport RR (eds) Biology and activities of yeasts. Academic, New York, pp 215–230

de Hoog GS, Zeng JS, Harrak MJ, Sutton DA (2006) *Exophiala xenobiotica* sp. nov., an opportunistic black yeast inhabiting environments rich in hydrocarbons. A van Leeuwenhoek 90:257–268

de Melo Pereira GV, Soccol VT, Pandey A, Medeiros AB, Andrade Lara JM, Gollo AL, Soccol CR (2014) Isolation, selection and evaluation of yeasts for use in fermentation of coffee beans by the wet process. Int J Food Microbiol 188:60–66

Döğen A, Kaplan E, Oksüz Z, Serin MS, Ilkit M, de Hoog GS (2013) Dishwashers are a major source of human opportunistic yeast-like fungi in indoor environments in Mersin, Turkey. Med Mycol 5:493–498

Doggett MS (2000) Characterisation of fungal biofilms within a municipal water distribution system. Appl Environ Microbiol 66:1249–1251

Dynowska M, Rosłan M, Góralska K (2006) *Saccharomyces cerevisiae* in respiratory system, digestive system and on the skin in humans. Acta Mycol 41:139–144

Ejdys E, Michalak J, Szewczyk KM (2009) Yeast-like fungi isolated from indoor air in school buildings and the surrounding outdoor air. Acta Mycol 44:97–107

Esfarjani F, Khaksar R, Nasrabadi FM, Roustaee R, Alikhanian H, Khalaji N, Khaneghah AM, Hosseini H (2016) A preventative approach to promote food safety: bacterial contamination of domestic refrigerators. Br Food J 118:2076–2091

Evangelista SR, Miguel MG, Cordeiro Cde S, Silva CF, Pinheiro AC, Schwan RF (2014) Inoculation of starter cultures in a semi-dry coffee (*Coffea arabica*) fermentation process. Food Microbiol 44:87–95

Evangelista SR, Miguel MG, Silva CF, Pinheiro AC, Schwan RF (2015) Microbiological diversity associated with the spontaneous wet method of coffee fermentation. Int J Food Microbiol 210:102–112

Falkowski J, Jakubowska B, Janda K (2002) The evaluation of thermophilic fungi in raw coffee beans. Rocz Panstw Zakl Hig 53:277–282

Fijan S, Turk SŠ (2012) Hospital textiles, are they a possible vehicle for healthcare-associated infections? Int J Environ Res Public Health 9:3330–3343

Flannigan B, Samson RA, Miller JD (2001) Microorganisms in home and indoor work environments: diversity, health impact, investigation and control. CRC Press, London and New York

Flores GE, Bates ST, Caporaso JG, Lauber CL, Leff JW, Knight R, Fierer N (2013) Diversity, distribution and sources of bacteria in residential kitchens. Environ Microbiol 15:588–596

Frankova E, Horecka M (1995) Filamentous soil fungi and unidentified bacteria in drinking water from wells and water mains near Bratislava. Microbiol Res 150:311–313

Fujimura KE, Johnson CC, Ownby DR, Cox MJ, Brodie EL, Havstad SL, Zoratti EM, Woodcroft KJ, Bobbitt KR, Wegienka G, Boushey HA, Lynch SV (2010) Man's best friend? The effect of pet ownership on house dust microbial communities. J Allergy Clin Immunol 126:410–412

Gargouri B, Mhiri N, Karray F, Aloui F, Sayadi S (2015) Isolation and characterization of hydrocarbon-degrading yeast strains from petroleum contaminated industrial wastewater. Biomed Res Int. doi:10.1155/2015/929424

Gattlen J, Amberg C, Zinn M, Mauclaire L (2010) Biofilms isolated from washing machines from three continents and their tolerance to a standard detergent. Biofouling 26:873–882

Glushakova AM, Zheltikova TM, Chernov IY (2004) Groups and sources of yeasts in house dust. Microbiology 73:94–98

Godwin SL, Chen FC, Chambers E IV, Coppings R, Chambers D (2007) A comprehensive evaluation of temperatures within home refrigerators. Food Protect Trends 27:168–173

Goksugur N, Karabay O, Kocoglu E (2006) Mycological flora of the Hammams, traditional Turkish bath. Mycoses 49:411–414

Górny RL, Dutkiewicz J (2002) Bacterial and fungal aerosols in indoor environment in Central and Eastern European countries. Ann Agric Environ Med 9:17–23

Gostinčar C, Grube M, Gunde-Cimerman N (2011) Evolution of fungal pathogens in domestic environments? Fungal Biol 115:1008–1018

Göttlich E, van der Lubbe W, Lange B, Fiedler S, Melchert I, Reifenrath M, Flemming HC, de Hoog S (2002) Fungal flora in groundwater-derived public drinking water. Int J Hyg Environ Health 205:269–279

Grabinska-Loniewska A, Konillowicz-Kowalska T, Wardzynska G, Boryn K (2007) Occurrence of fungi in water distribution system. Pol J Environ Stud 16:539–547
Gray FN (2014) Pathogen control in drinking water. In: Percival LS, Yates VM, Williams D, Chalmers R, Gray N (eds) Microbiology of waterborne diseases. Elsevier, Oxford, pp 537–570
Gümral R, Özhak-Baysan B, Tümgör A, Saraçlı MA, Yıldıran ŞT, Ilkit M, Zupančič J, Novak Babič M, Gunde-Cimerman N, Zalar P, de Hoog GS (2016) Dishwashers provide a selective extreme environment for human-opportunistic yeast-like fungi. Fungal Divers 76:1–9
Hageskal G, Knutsen AK, Gaustad P, de Hoog GS, Skaar I (2006) The diversity and significance of mold species in Norwegian drinking water. Appl Environ Microbiol 72:7586–7593
Hamada N (2002) Fungal contamination in washing machines. Antibact Antifungal Agent 30:703–708
Hamada N, Abe N (2009) Physiological characteristics of 13 common fungal species in bathrooms. Mycoscience 50:421–429
Hamada N, Fujita T (2000) Growth rate of fungi in bathrooms – experimental survey. Mycoscience 41:297–301
Heinrichs G, Hübner I, Schmidt KC, de Hoog GS, Haase G (2013a) Analysis of black fungal biofilms occurring at domestic water taps (I): compositional analysis using Tag-encoded FLX amplicon pyrosequencing. Mycopathologia 175:387–397
Heinrichs G, Hübner I, Schmidt KC, de Hoog GS, Haase G (2013b) Analysis of black fungal biofilms occurring at domestic water taps (II): potential routes of entry. Mycopathologia 175:387–397
Hinzelin F, Block JC (1985) Yeasts and filamentous fungi in drinking water. Environ Technol Lett 6:101–106
Horner WE, Worthan AG, Morey PR (2004) Air- and dustborne mycoflora in houses free of water damage and fungal growth. Appl Environ Microbiol 70:6394–6400
Hospodsky D, Qian J, Nazaroff WW, Yamamoto N, Bibby K, Rismani-Yazdi H, Peccia J (2012) Human occupancy as a source of indoor airborne bacteria. PLoS One 7:e34867
Iacobucci E (2004) Machine for American style coffee for use on aircraft. United States Patent US006779435B1:22
Inácio J, Daniel HM (2017) Commensalism: the case of the human zymobiome. In: Buzzini P, Lachance MA, Yurkov AM (eds) Yeasts in natural ecosystems: ecology. Springer, Heidelberg, pp 211–228
Isola D, Selbmann L, de Hoog GS, Fenice M, Onofri S, Prenafeta-Boldú FX, Zucconi L (2013) Isolation and screening of black fungi as degraders of volatile aromatic hydrocarbons. Mycopathologia 175:369–379
Jadhav S, Sahasrabudhe T, Kalley V, Gandham N (2013) The microbial colonization profile of respiratory devices and the significance of the role of disinfection: a blinded study. J Clin Diagn Res 7:1021–1026
James SJ, Evans J, James C (2008) A review of the performance of domestic refrigerators. J Food Eng 87:2–10
Jay MJ (1996) Modern food microbiology. Chapman and Hall, New York
Jeon YS, Chun J, Kim BS (2013) Identification of household bacterial community and analysis of species shared with human microbiome. Curr Microbiol 67:557–563
Kadaifciler DG, Ökten S, Sen B (2013) Mycological contamination in dental unit waterlines in Istanbul, Turkey. Braz J Microbiol 44:977–981
Kennedy J, Jackson V, Blair IS, McDowell DA, Cowan C, Bolton DJ (2005) Food safety knowledge of consumers and the microbiological and temperature status of their refrigerators. J Food Prot 68:1421–1430
Khan HAA, Karuppayil SM (2012) Fungal pollution of indoor environments and its management. Saudi J Biol Sci 19:405–426
Kinsey GC, Paterson RR, Kelley J (1999) Methods for the determination of filamentous fungi in treated and untreated waters. J Appl Microbiol 85:214–224

Klepeis NE, Nelson WC, Ott WR, Robinson JP, Tsang AM, Switzer P, Behar JV, Hern SC, Engelmann WH (2001) The National Human Activity Pattern Survey (NHAPS): a resource for assessing exposure to environmental pollutants. J Expo Anal Environ Epidemiol 11:231–252

Levy SB (2001) Antibacterial household products: cause for concern. Emerg Infect Dis 7:512–515

Li A, Liu Z, Zhu X, Liu Y, Wangb Q (2010) The effect of air-conditioning parameters and deposition dust on microbial growth in supply air ducts. Energ Buildings 42:449–454

Lian X, de Hoog GS (2010) Indoor wet cells harbour melanized agents of cutaneous infection. Med Mycol 48:622–628

Lotrakul P, Deenarn P, Prasongsuk S, Punnapayak H (2009) Isolation of *Aureobasidium pullulans* from bathroom surfaces and their antifungal activity against some *Aspergilli*. Afr J Microbiol Res 3:253–257

Maktabi S, Jamnejad A, Faramarzian K (2013) Contamination of household refrigerators by *Listeria* species in Ahvaz, Iran. Jundishapur J Microbiol 6:301–305

Margesin R, Gander S, Zacke G, Gounot AM, Schinner F (2003) Hydrocarbon degradation and enzyme activities of cold-adapted bacteria and yeasts. Extremophiles 7:451–458

Masoud W, Cesar LB, Jespersen L, Jakobsen M (2004) Yeast involved in fermentation of Coffea arabica in East Africa determined by genotyping and by direct denaturing gradient gel electrophoresis. Yeast 21:549–556

Matos T, de Hoog GS, de Boer AG, de Crom I, Haase G (2002) High prevalence of the neurotrope *Exophiala dermatitidis* and related oligotrophic black yeasts in sauna facilities. Mycoses 45:373–377

McGrath JJ, Wong WC, Cooley JD, Straus DC (1999) Continually measured fungal profiles in sick building syndrome. Curr Microbiol 38:33–36

Meklin T, Husman T, Vepsäläinen A, Vahteristo M, Koivisto J, Halla-Aho J, Hyvärinen A, Moschandreas D, Nevalainen A (2002) Indoor air microbes and respiratory symptoms of children in moisture damaged and reference schools. Indoor Air 12:175–183

Miceli HM, Diaz AJ, Lee AS (2011) Emerging opportunistic yeast infections. Lancet Infect Dis 11:142–151

Naik SN, Goud VV, Rout KP, Dalai AK (2010) Production of first and second generation biofuels: a comprehensive review. Renew Sust Energ Rev 14:578–597

Nanbakhsh H, Diba K, Hazarti K (2004) Study of fungal contamination of indoor public swimming pools. Iran J Public Health 33:60–65

Nedret Koc A, Kocagöz S, Erdem F, Gündüz Z (2002) Outbreak of nosocomial fungemia caused by *Candida glabrata*. Mycoses 45:470–475

Nishimura K, Miyaji M (1982) Studies on a saprophyte of *Exophiala dermatitidis* isolated from a humidifier. Mycopathologia 77:173–181

Noris F, Siegel JA, Kinney KA (2011) Evaluation of HVAC filters as a sampling mechanism for indoor microbial communities. Atmos Environ 45:338–346

Novak Babič M, Zalar P, Ženko B, Džeroski S, Gunde-Cimerman N (2015) *Candida* and *Fusarium* species known as opportunistic human pathogens from customer-accessible parts of residential washing machines. Fungal Biol 119:95–113

Novak Babič M, Zalar P, Ženko B, Džeroski S, Gunde-Cimerman N (2016) Yeasts and yeast-like fungi in tap water and groundwater, and their transmission to household appliances. Fungal Ecol 20:30–39

NSF The Public Health and Safety Organization (2013) International Household germ study. https://www.nsf.org/newsroom_pdf/2013_germ_study_FOR-WEB-ONLY.pdf. Accessed 21 Nov 2016

Ojima M, Toshima Y, Koya E, Ara K, Kawai S, Ueda N (2002) Bacterial contamination of Japanese households and related concern about sanitation. Int J Environ Health Res 12:41–52

Pal A, Labuza TP, Diez-Gonzalez F (2008) Shelf life evaluation for ready-to-eat sliced uncured turkey breast and cured ham under probable storage conditions based on *Listeria monocytogenes* and psychrotroph growth. Int J Food Microbiol 126:49–56

Peckham D, Williams K, Wynne S, Denton M, Pollard K, Barton R (2016) Fungal contamination of nebuliser devices used by people with cystic fibrosis. J Cyst Fibros 15:74–77

Peng R, Xiong A, Xue Y, Fu XY, Gao F, Zhao W, Tian YS, Yao QH (2008) Microbial biodegradation of polyaromatic hydrocarbons. FEMS Microbiol Rev 32:927–955
Percival SL, Knapp JS, Wales DS, Edyvean RGJ (1999) The effect of turbulent flow and surface roughness on biofilm formation in drinking water. J Ind Microbiol Biotechnol 22:152–159
Pereira VJ, Basîlio MC, Fernandes D, Domingues M, Paiva JM, Benoliel MJ, Crespo MT, San Romão MV (2010) Occurrence of filamentous fungi and yeasts in three different drinking water sources. Water Res 43:3813–3819
Prenafeta-Boldú FX, Summerbell R, de Hoog GS (2006) Fungi growing on aromatic hydrocarbons: biotechnology's unexpected encounter with biohazard? FEMS Microbiol Rev 30:109–130
Qian J, Hospodsky D, Yamamoto N, Nazaroff WW, Peccia J (2012) Size-resolved emission rates of airborne bacteria and fungi in an occupied classroom. Indoor Air 22:339–351
Sabirova JS, Haddouche R, Van Bogaert I, Mulaa F, Verstraete W, Timmis K, Schmidt-Dannert C, Nicaud J, Soetaert W (2011) The 'LipoYeasts' project: using the oleaginous yeast *Yarrowia lipolytica* in combination with specific bacterial genes for the bioconversion of lipids, fats and oils into high-value products. Microb Biotechnol 4:47–54
Salonen A, Ruokola AL (1969) Mycoflora of the Finnish "sauna" (bath-house). Mycopathologia 38:327–336
Samah AM, Shimaa RH, Al-Wasify RS (2014) Relative diversity of filamentous fungi and yeasts in groundwater and their correlation to fecal pollution indicators and physicochemical parameters. Int J Curr Microbiol App Sci 3:905–919
Sammon BN, Harrower MK, Fabbro DL, Reed RH (2010) Incidence and distribution of microfungi in a treated municipal water supply system in sub-tropical Australia. Int J Environ Res Public Health 7:1597–1611
Scott E (2000) Relationship between cross-contamination and the transmission of foodborne pathogens in the home. Pediatr Infect Dis J 19:111–113
Scott E, Bloomfield SF (1990) The survival and transfer of microbial contamination via cloths, hands and utensils. J Appl Bacteriol 68:271–278
Seyedmousavi S, Guillot J, Tolooe A, Verweij PE, de Hoog GS (2015) Neglected fungal zoonoses: hidden threats to man and animals. Clin Microbiol Infect 21:416–425
Shah CP, Krajden S, Kane J, Summerbell RC (1988) *Tinea corporis* caused by *Microsporum canis*: report of a nosocomial outbreak. Eur J Epidemiol 4:33–38
Shaker BK, Sharif FM (2012) Isolation and identification of some fungi from Al-Sader Water Treatment Plant, Baghdad, Iraq. Al-Mustansiriyah J Sci 3:1–12
Shumin Y, Na N, Zhaoyanga X, Jinglia T (2012) Screening and identification of halotolerant yeast for hydrocarbon degrading and its properties studies. Afr J Microbiol 6:1819–1828
Silva CF, Batista LR, Abreu LM et al (2008) Succession of bacterial and fungal communities during natural coffee (*Coffea arabica*) fermentation. Food Microbiol 25:951–957
Simoes SA, Junior DPL, Hahn RC (2011) Fungal microbiota in air-conditioning installed in both adult and neonatal intensive treatment units and their impact in two university hospitals of the central western region, Mato Grosso, Brazil. Mycopathologia 172:109–116
Sinclair RG, Gerba CP (2011) Microbial contamination in kitchens and bathrooms of rural Cambodian village households. Lett Appl Microbiol 52:144–149
Stapleton K, Hill K, Day K, Perry JD, Dean JR (2013) The potential impact of washing machines on laundry malodour generation. Lett Appl Microbiol 56:299–306
Vicente AV, Attili-Angelis D, Pie MR, Queiroz-Telles F, Cruz LM, Najafzadeh MJ, de Hoog GS, Zhao J, Pizzirani-Kleiner A (2008) Environmental isolation of black yeast-like fungi involved in human infection. Stud Mycol 61:137–144
Vilanova C, Iglesias A, Porcar M (2015) The coffee-machine bacteriome: biodiversity and colonisation of the wasted coffee tray leach. Sci Rep 5:17163
Vilela DM, Pereira GV, Silva CF, Batista LR, Schwan RF (2010) Molecular ecology and polyphasic characterization of the microbiota associated with semi-dry processed coffee (Coffea arabica L.) Food Microbiol 27:1128–1135

Yamaguchi MU, Pontelllo Rampazo RC, Yamada-Ogatta SF, Vataru Nakamura C, Ueda-Nakamura T, Dias Filho BP (2007) Yeasts and filamentous fungi in bottled mineral water and tap water from municipal supplies. Braz Arch Biol Technol 50:1–9

Zalar P, Novak M, de Hoog GS, Gunde-Cimerman N (2011) Dishwashers – a man-made ecological niche accommodating human opportunistic fungal pathogens. Fungal Biol 115:997–1007

Zhao J, Zeng J, de Hoog GS, Attili-Angelis D, Prenafeta-Boldú FX (2010) Isolation and identification of black yeasts by enrichment on atmospheres of monoaromatic hydrocarbons. Microb Ecol 60:149–156

Zupančič J, Novak Babič M, Zalar P, Gunde-Cimerman N (2016) The black yeast *Exophiala dermatitidis* and other selected opportunistic human fungal pathogens spread from dishwashers to kitchens. PLoS One 11:e0148166

Chapter 6
Phylloplane Yeasts in Temperate Climates

Martin Kemler, Frederick Witfeld, Dominik Begerow, and Andrey Yurkov

Abstract Yeasts are integral parts of phylloplane communities of temperate regions, where ecosystems are not only influenced by short-term fluctuations in abiotic conditions, but additionally by cyclic seasonal changes. Phylloplane yeasts possess physiological adaptations, such as pigmentation and extracellular polysaccharides that enable them to resist harsh conditions encountered in these environments. Additionally, through production of plant hormone-like metabolites, they also might influence the behavior, fitness, and growth of their plant host. Here we review how the understanding of yeasts in this environment has improved in the last years due to discoveries in new habitats, new developments in taxonomy, but also the application of environmental sequencing and genomics. These new technologies, as well as traditional approaches, have made it clear that yeasts are not only occupying this environment to gain nutrients, but they are active participants that shape the structure of microbial communities by diverse interactions with other community members.

Keywords Yeast communities • Yeast taxonomy • *Sphagnum* • Biofilm • Yeast pigmentation

Contents

M. Kemler (✉) • F. Witfeld • D. Begerow
Ruhr-Universität Bochum, AG Geobotanik, ND03, Universitätsstr. 150, 44801 Bochum, Germany
e-mail: martin.kemler@rub.de

A. Yurkov
Leibniz Institute DSMZ-German Collection of Microorganisms and Cell Cultures, Inhoffenstraße 7B, 38124 Braunschweig, Germany

P. Buzzini et al. (eds.), *Yeasts in Natural Ecosystems: Diversity*,
DOI 10.1007/978-3-319-62683-3_6

6.1 Introduction

Plants and fungi have a common evolutionary history going back to at least the origin of land plants (Redecker et al. 2000). Since then many different kinds of interaction types were established between the two groups. Thereby, many of the interactions are very specific and often restricted to one of the three plant organs. From an ecological point of view, the rhizosphere describes the subterranean habitat associated with the root, whereas the habitats associated with the above-ground parts of plants (i.e., stem and leaves) define the phyllosphere. In a narrower sense, as applied most often in literature and also used in this review, the phyllosphere is used to describe the entirety of all leaves of a plant, whereas the caulosphere describes the stem parts. Due to their importance in agriculture and forestry, parasitic plant-fungus interactions in the phyllosphere are well studied. Recent years have also seen an increase of studies on endophytic fungi living within plant tissues without causing visible symptoms, as they show some potential benefits to the plant. Several excellent reviews on their diversity and potential role of these fungi have been published (e.g., Arnold 2007; Rodriguez et al. 2009).

In addition to the abovementioned definitions, the phylloplane describes only the habitats associated with the surface of leaves. The entire foliar plant surface including the epidermis, the cuticle, waxes, and other exudates is an important interface of plant-fungus interactions. It forms the barrier between the plant interior and the surrounding aerial environment, where physical contact between the interacting partners gets established (e.g., adhesion to the leaf or first plant defense mechanisms against potentially harmful fungi). Unless infecting via the open stomata, pathogenic fungi have to overcome the epidermis to establish a stable nutrient exchange with their host. In order to achieve this, different fungal groups evolved sophisticated mechanisms based on specific morphological structures like appressoria or physiological adaptations. *Diaporthe amygdali* (syn. *Fusicoccum amygdali*) for instance produces the terpenoid fusicoccin to achieve a permanent stomata opening of the host plant due to an increased potassium influx into the stomata (Turner 1973). Many other plant pathogenic fungi, including, *Botrytis cinerea*, *Magnaporthe grisea*, and *Aspergillus flavus*, use cell wall-degrading enzymes for plant invasion (Zhao et al. 2013). Some endophytic fungi seem to use similar mechanisms; however, the extent of this way of entry in this taxonomically diverse group is sparsely known (e.g., Cabral et al. 1993). Appressoria, for instance, are realized in *Apiognomonia errabunda* (formerly *Discula umbrinella*) (Viret and Petrini 1994) and the latent pathogen *Botryosphaeria dothidea* (Kim et al. 1999). Apart from these transient phylloplane fungi, which are on the plant surface only for attaching and subsequent infection, taxonomically diverse fungi have adapted to the environmental conditions of the leaf surface and are members of specific, resident phylloplane communities. Yeasts form an integral part of these.

Living conditions for microorganisms, including yeasts, in the phylloplane can be harsh. Nutrient availability is in general considered to be poor and fluctuating, and nutrients are made available to microorganisms either via plant exudates or through input from the external environment. Interestingly carbohydrate availability seems not to be a limiting factor in some phylloplane habitats, as studies creating artificial nutrient conditions on turf grass have shown. Additional sugar did not enhance yeast abundance, but yeast growth in this system was facilitated by the amount of organically available nitrogen (Nix-Stohr et al. 2008). Next to challenging nutrient acquisition, microorganisms are often exposed to challenging abiotic factors:

- Water, although not scarce in itself, has a high dynamic in the phylloplane. Whereas massively available during rain, it subsequently virtually vanishes from leaf surfaces. Phylloplane-associated yeasts have adapted to the dry conditions, by being extremely resistant to desiccation. Artificial storage of yeasts with different ecological strategies demonstrated that pigmented plant-associated yeasts (i.e., *Rhodotorula mucilaginosa* and *Sporobolomyces roseus*) exhibited some of the highest tolerance to desiccation in a diverse array of yeast taxa from different environments (Glushakova et al. 2015b). The ability to cope with dry conditions thereby is closely related to the ability to form extracellular polysaccharides (see paragraphs 6.5 and 6.6).
- High radiation is also a common feature in the phylloplane and exerts a strong selection pressure on microbes (Fonseca and Inácio 2006; Whipps et al. 2008). Phylloplane yeasts have evolved means to produce protective pigments against radiation, such as melanins (e.g., black yeast *Aureobasidium pullulans*), carotenoids (e.g., red yeasts *Rhodotorula* and *Sporobolomyces*), mycosporines (e.g., red yeasts *Cystobasidium* and *Erythrobasidium*, and nonpigmented *Filobasidium*) (Brandão et al. 2011; Libkind et al. 2011), and ubiquinone (Yurkov et al. 2008). Production of carotenoid pigments by members of Sporidiobolales (cited in the literature as *Rhodotorula*, *Rhodosporidium*, and *Sporobolomyces*) is a well-known trait (Yurkov et al. 2008; Mannazzu et al. 2015), which can aid yeasts to survive on surfaces exposed to UV radiation. Frequently observed in the phylloplane, yeasts in the genus *Vishniacozyma* (see paragraph 6.2) were traditionally not studied as producers of common carotenoid pigments (e.g., Brandão et al. 2011). However, extracts from *Vishniacozyma victoriae* strains contained torulene (14%), torularhodine (12.5%), β-carotene (6%), and a large amount (over 60%) of unidentified carotenoids (Yurkov et al. 2008). Red- and orange-colored *Cystofilobasidium* species produce common (torulene, torularhodine, and β-carotene) carotenoid pigments (Yurkov et al. 2008). And the orange pigmentation of *Dioszegia* species is due to the unusual carotenoid plectaniaxanthin, which has been also detected in *Papiliotrema flavescens* (Yurkov et al. 2008). Synthesis of mycosporines is common in plant-related Cystobasidiomycetes, i.e., members of the genera *Cystobasidium* (formerly *Rhodotorula minuta* clade), *Erythrobasidium*, *Sakaguchia*, and *Symmetrospora* (Brandão et al. 2011; Libkind et al. 2011). Orange-colored tremellaceous *Dioszegia* were also reported to be mycosporine-positive. Among nonpigmented (common carotenoids are not known) yeasts, the species

Filobasidium magnum, *Filobasidium stepposum*, *Filobasidium wieringae*, *Naganishia adeliensis*, *Naganishia diffluens*, and *Naganishia saitoi* produce mycosporines (Brandão et al. 2011).

In temperate regions, next to short-term fluctuations in abiotic factors, seasonal change involving resting periods in many plants, often associated with leaf shedding, is an additional challenge that members of phylloplane communities had to adapt to. Studies that evaluated the dynamics of yeast populations on different plant types showed in general a continuous increase in population size from spring to autumn and winter (reviewed by Fonseca and Inácio 2006). Newer studies on the seasonal dynamics of epiphytic yeasts on 25 different plant species followed the same trend and showed that the total yeast population increased from spring to autumn followed by a decline during winter months (Glushakova and Chernov 2007). The distribution of single species varied substantially during the year, and the peak relative abundances of many types of yeast did not coincide. The proportion of ascomycetous yeasts in the community increased from spring to autumn followed by a decline during the winter months. However, the abundance of single species varied substantially throughout the year and reached its maximum within a short period of time (Glushakova et al. 2007; Glushakova and Chernov 2010). For example, *Torulaspora delbrueckii* and *Saccharomyces paradoxus* were prominent on plant leaves during summer months, while *Candida oleophila* and *Kazachstania barnettii* were more prominent in autumn and winter, respectively (Glushakova et al. 2007; Glushakova and Chernov 2010). The only species constantly present on most of the studied plant species was *Debaryomyces hansenii* (Glushakova and Chernov 2010). The abundance of basidiomycetous species did not change throughout the year. The only significant shift was the increased proportion of *Vishniacozyma* yeasts (cited as "phenotypic" *Cryptococcus laurentii* (see paragraph 6.2) in Glushakova and Chernov 2010). Therefore, it remains unclear whether or not basidiomycetous yeasts undergo temporal changes in phylloplane communities. Other temporal shifts in yeast communities have been reviewed by Yurkov (2017), and yeasts in the phylloplane of tropical plants are discussed in Chap. 7 of this book.

6.2 Diversity of Phylloplane Yeasts in the Light of New Taxonomic Insight

Fonseca and Inácio (2006) provided an excellent overview on yeasts found on plant surfaces and the traditional methods used to study them. They also indicated the discrepancy between identification based on morphological and physiological approaches, and the taxonomic entities determined by DNA sequencing. Nucleotide sequence analyses nowadays have de facto replaced physiological tests for yeast identification. The latter were laborious and often not able to distinguish closely related species. Additionally, interspecific variability of widespread basidiomycetous yeasts with respect to the utilized carbon and nitrogen sources made species delimitation difficult. As a result, species diversity assessments to a large extent were dependent on the identification technique. Several widespread phylloplane-related yeasts have been shown taxonomically heterogeneous (e.g., the former species *Cryptococcus albidus*, *Cryptococcus laurentii*, *Rh. minuta*, *Rhodotorula glutinis*, *Sp. roseus*; discussed in Fonseca and Inácio 2006). In many instances, yeast species frequently reported in the literature before the late 1990s have not been identified in more recent studies. Therefore, the ubiquity of phenotypic phylloplane yeast species could result from inaccurate identification (Fonseca and Inácio 2006; Yurkov et al. 2015c). Despite the growing number of studies employing rDNA sequencing, only few studies have attempted to reassess identification of cultures originally identified using physiological tests or resampled previously surveyed habitats (Glushakova and Chernov 2010; Chernov et al. 2013; Yurkov et al. 2015c). Such studies make it possible to infer the identity of several frequent yeast species reported in older studies (see below).

The taxonomy of basidiomycetous yeasts has been thoroughly revised recently and now links asexual and sexual stages of many taxa (Liu et al. 2015; Wang et al. 2015a, b, c). Several polyphyletic basidiomycetous yeast genera (e.g., *Bullera*, *Cryptococcus*, *Rhodotorula*, *Sporobolomyces*, and *Trichosporon*) were reclassified and restricted to monophyletic lineages of the respective type species. Other clades were accommodated in new or reinstated genera (Liu et al. 2015; Wang et al. 2015a, b, c). Below we provide examples of re-identification of common phylloplane-related phenotypic species. We also list prominent epiphytic yeasts isolated in the temperate climate and identified by rDNA sequencing.

In the Filobasidiales, phenotypic *Cr. albidus* and *Cryptococcus diffluens* correspond to members of the genera *Filobasidium* (e.g., *F. magnum*, *Filobasidium oeirense*, *F. stepposum*, and *F. wieringae*) and *Naganishia* (e.g., *Naganishia albida*, *Naganishia albidosimilis*, and *Na. adeliensis*), respectively. Strains isolated from plants collected in birch forests in Russia previously assigned to phenotypic *Cr. albidus* were identified as *F. wieringae*, *F. magnum*, and *F. oeirensis* (Yurkov et al. 2015c). Likewise, *Na. albida*, *Na. adeliensis*, and *Na. albidosimilis* were identified among phenotypic *Cr. diffluens* isolates. Another morphologically similar species occasionally reported from plants in Russia is *F. stepposum* (Golubev et al. 2006). *Ficus* leaves collected from a greenhouse in Germany harbored *Na. albida*, *Na. albidosimilis*, and *F. magnum* (Solis et al. 2014). *F. magnum* was also a dominant species on *Sphagnum* mosses (Kachalkin and Yurkov 2012). *Na. albidosimilis* and *Na. adeliensis* were occasionally isolated from forest, meadow, and swamp plants predominantly collected in cold periods (Chernov et al. 2013).

In the Tremellales, phenotypic *Cr. laurentii* might correspond to members of the genus *Vishniacozyma* (e.g., *Vishniacozyma carnescens*, *Vishniacozyma heimaeyensis*, *Vishniacozyma tephrensis*, and *Vishn. victoriae*). Strains isolated from plants collected in birch forests in Russia previously assigned to phenotypic *Cr. laurentii* were identified as *Vishn. victoriae* (most frequent), *Vishn. carnescens*, *Vishn. heimaeyensis*, and *Vishn. tephrensis* (Yurkov et al. 2015c). These species were also found on wind-pollinated plants (*Vishn. victoriae* and *Vishn. tephrensis*; Glushakova et al. 2015a), various forest and meadow plants (*Vishn. victoriae*;

Glushakova and Chernov 2010; Yurkov et al. 2015c), *Sphagnum* moss (*Vishn. victoriae*; Kachalkin and Yurkov 2012), diverse tropical and temperate plants (*Vishn. victoriae*; Takashima et al. 2012), and European ash (*Vishn. victoriae*; Davydenko et al. 2013). Members of the genus *Vishniacozyma* have been repeatedly detected in culture-independent approaches using clone libraries, although identification to the species level is often lacking. One of the former taxonomic synonyms of *Cr. laurentii*, *Pa. flavescens* (formerly *Cryptococcus flavescens*), has been identified from different plants worldwide (Yurkov et al. 2015a). This study also suggested that *Pa. flavescens* represents a complex comprising several cryptic species (i.e., *Papiliotrema baii*, *Papiliotrema ruineniae*, and *Papiliotrema terrestris*). Orange- and red-pigmented Tremellomycetes yeasts of the genera *Cystofilobasidium* (e.g., formerly *Cryptococcus macerans*), *Dimennazyma* (formerly *Cryptococcus cisti-albidi*), *Dioszegia* (e.g., formerly *Cryptococcus hungaricus*), and *Genolevuria* (e.g., formerly *Cryptococcus amylolyticus* and *Cryptococcus armeniacus*) have also been isolated from different plant surfaces (e.g., Golubev and Golubeva 2004; Inácio et al. 2005; Landell et al. 2009; Glushakova and Chernov 2010). Other rare, or yet rarely reported from phylloplane, yeast genera are *Bullera*, *Derxomyces*, *Hannaella* (e.g., formerly *Cryptococcus luteolus*), and *Papiliotrema* (e.g., Golubev and Golubeva 2004; Takashima et al. 2012; Landell et al. 2014). Clinically relevant *Trichosporon cutaneum* (currently *Cutaneotrichosporon cutaneum*) was isolated from agricultural and wild plants in Russia (Kvasnikov et al. 1975) but not detected with DNA sequencing in recent studies (e.g., Glushakova and Chernov 2010).

In the Pucciniomycotina, red yeasts belonging to the genera *Rhodotorula* and *Sporobolomyces* have been repeatedly reported from plant surfaces (reviewed in Fonseca and Inácio 2006). However, as in the case of other fungal lineages, frequent isolation of several common phenotypic species, such as *Rh. glutinis* and *Rhodotorula graminis*, is questionable. Recent studies have shown that despite numerous reports, the number of known strains of these two yeast species is extremely low (Coelho et al. 2011; Yurkov et al. 2012). Most likely, *Rhodotorula babjevae* (formerly *Rhodosporidium babjevae*) and other Sporidiobolales (e.g., *Rhodosporidiobolus colostri*) have been mistaken for *Rh. glutinis* and *Rh. graminis* in older studies. Several strains isolated from plants collected in birch forests in Russia previously assigned to phenotypic *Rh. glutinis* were identified as *Rh. babjevae* (Yurkov et al. 2015c). Other prominent phylloplane-related yeasts isolated in the past, *Rh. mucilaginosa* and *Sp. roseus*, have been also isolated from plants in recent studies (Glushakova and Chernov 2010; Kachalkin and Yurkov 2012; Glushakova et al. 2014; Solis et al. 2014; Yurkov et al. 2015c). Members of the *Rh. minuta* clade have been recently reclassified in the genus *Cystobasidium* (Cystobasidiomycetes, Pucciniomycotina) in order to reduce polyphyly of the genus *Rhodotorula* (Yurkov et al. 2015b). Older reports of phenotypic *Rh. minuta* are likely to refer to different species of the genus *Cystobasidium*, such as *Cystobasidium benthicum*, *Cystobasidium calyptogenae*, *Cystobasidium lysiniphylum*, *Cystobasidium minutum*, *Cystobasidium pinicola*, and *Cystobasidium slooffiae* (Fonseca and Inácio 2006; Glushakova and Chernov 2010; Kachalkin and

Yurkov 2012; Takashima et al. 2012; Glushakova et al. 2014, 2015a; Solis et al. 2014). Additionally, red yeasts of the closely related genus *Occultifur* (Cystobasidiomycetes, Pucciniomycotina) could be mistaken for phenotypic *Rh. minuta* (e.g., Takashima et al. 2012; Kurtzman and Robnett 2015). Further yeast genera in the Pucciniomycotina with recent taxonomic changes that have been occasionally reported from plants are *Ballistosporomyces*, *Bensingtonia*, *Kondoa*, and *Ruinenia* in the Agaricostilbomycetes; *Buckleyzyma*, *Erythrobasidium*, and *Symmetrospora* in the Cystobasidiomycetes; and *Curvibasidium*, *Leucosporidium*, *Oberwinklerozyma*, and *Phenoliferia* in the Microbotryomycetes (e.g., Golubev and Golubeva 2004; Takashima et al. 2012; Wang et al. 2012; Yurkov et al. 2015c).

Taxonomic changes have also occurred in potential phylloplane ascomycetous yeasts. The clinically relevant *Candida krusei* (currently *Pichia kudriavzevii*) and *Candida tropicalis* were isolated from agricultural and wild plants in Russia (Kvasnikov et al. 1975). However, later studies, which utilized DNA sequencing, have so far been unable to detect these species on plants (e.g., Glushakova and Chernov 2010). Among the dominating ascomycetous yeasts, *C. oleophila*, *Deb. hansenii*, *Metschnikowia pulcherrima*, and *Meyerozyma guilliermondii* (Glushakova et al. 2007, 2015a; Glushakova and Chernov 2010) on various plants, only *Deb. hansenii* has been identified in older studies. Asexual yeast forms of the plant parasites *Taphrina* (formerly *Lalaria*) have been reported from healthy plants that are not known as hosts for these fungi (Inácio et al. 2005; Fonseca and Inácio 2006; Fonseca and Rodriguez 2011). Because saprobic cultures of *Taphrina* often show a narrow spectrum of assimilated carbon sources (Fonseca and Inácio 2006; Fonseca and Rodriguez 2011) and grow slowly in culture, their observation in culture-dependent assays might be difficult (see also paragraph 6.4). To date, a few isolations of *Taphrina* cultures from plant surfaces are known from berries and flowers collected in boreal forests in Russia (Babjeva and Reshetova 1998; Babjeva et al. 1999; Maksimova and Chernov 2004; Yurkov et al. 2004), in *Silene* flowers in the USA (Golonka and Vilgalys 2013), and in forest soils in Austria (Wuczkowski et al. 2005) and from the maize phyllosphere in Germany (see PYCC database: http://pycc.bio-aware.com/). Yeasts as parasites of plants have been thoroughly reviewed by Begerow et al. (2017).

6.3 Yeasts in the Phylloplane of Bryophytes and Lichens

The moss phyllosphere is widespread and abundant, as mosses grow in different climates, where they can be found in many ecosystems with sufficient water availability. *Sphagnum* mosses dominate bogs and swamps in boreal ecosystems, thereby constituting a significant part of plant biomass in this region (reviewed in Turetsky 2003). Additionally, annual productivity of mosses can be higher than the foliage production of coniferous trees (Oechel and Van Cleve 1986). Yeasts in peat (i.e., dead but not decomposed *Sphagnum* material) have been reviewed by

Thormann et al. (2007). Despite the ecological significance of these mosses, there are only few studies about yeasts on living *Sphagnum* moss.

Yeast communities of *Sphagnum* mosses have been studied in boreal swamps and swamp forests in Russia (Kachalkin et al. 2008; Kachalkin and Yurkov 2012). Unlike vascular plants, which release a variety of sugars, alcohols, and amino acids, the *Sphagnum* phyllosphere is enriched with a large spectrum of complex organic acids and phenolic compounds (Rasmussen et al. 1995; Rydin et al. 2006; Montenegro et al. 2009). In addition to the specific chemical composition of exudates, *Sphagnum* species shape microbial communities through increased water holding capacity, lower temperature, acidification, and nitrogen immobilization in the environment (van Breemen 1995; Painter 1998; Maksimova and Chernov 2004; Kachalkin et al. 2008). Kachalkin and Yurkov (2012) suggested that nitrogen availability could be the key factor limiting development of yeasts on *Sphagnum* mosses. Plant surfaces in general are known to contain diverse carbon sources, while availability of nitrogen is limited (Fonseca and Inácio 2006). In the *Sphagnum* phyllosphere, potentially available nitrogen is bound to phenolic compounds (Painter 1998) and thereby is not available to yeasts. Neither insulation (temperature and ultraviolet radiation) nor humidity (water availability) affected yeast numbers and species richness values on *Sphagnum* (Kachalkin and Yurkov 2012). Interestingly though and despite nitrogen limitation, yeast abundances on mosses were higher than those on vascular plants in the same region (Babjeva et al. 1999; Maksimova and Chernov 2004).

Compared to the phylloplane of vascular plants, *Sphagnum* mosses provide rather different environments for yeasts. Consequently, yeast communities developing on *Sphagnum* mosses differ considerably from those found on vascular plants (e.g., Kachalkin et al. 2008; Kachalkin and Yurkov 2012). Although yeast communities on *Sphagnum* mosses comprised species also found in the phylloplane of vascular plants (e.g., *F. magnum*, *F. wieringae*, *Rh. mucilaginosa*), additionally they shared species with nearby boreal soils (e.g., *Saitozyma podzolica*, *Schwanniomyces* spp., *Nadsonia starkeyi-henricii*) (Kachalkin et al. 2008; Kachalkin and Yurkov 2012). Compared to phylloplane communities on vascular plants, *Sphagnum* was characterized by a low proportion of phylloplane-related Tremellales yeasts (e.g., *Vishniacozyma*, *Dioszegia*). Additionally, an increased diversity of ascomycetous yeasts was recovered. The authors suggested that this elevated occurrence on *Sphagnum* could be explained by the higher water holding capacity of the moss, thereby protecting yeasts from drought. The few available reports additionally suggest that the diversity of psychrophilic yeasts in the moss layer is higher than on leaves of boreal vascular plants. Specifically, the yeasts *Aureobasidium subglaciale*, *Leucosporidium scottii*, *Oberwinklerozyma silvestris* (formerly *Rhodotorula silvestris*), *Phenoliferia psychrophenolica* (formerly *Rhodotorula psychrophenolica*), and *Sterigmatosporidium polymorphum* were isolated from moss cover collected from a boreal spruce forest and a swamp (Babjeva et al. 1999; Maksimova and Chernov 2004; Kachalkin 2010).

Yeast communities on lichens have been studied even more scarcely than on mosses. Studies on yeasts on lichens in boreal spruce forest (Babjeva et al. 1999) demonstrated that this substrate is a suitable habitat for yeasts, as total yeast counts on lichens were as high as on vascular plants (Babjeva et al. 1999). In regions where other vegetations are nonexistent, lichens substitute vascular plants as the substrate for yeasts (e.g., Vasileva-Tonkova et al. 2014; Duarte et al. 2016; Santiago et al. 2017). Yeasts found in the phylloplane of plants in temperate climates (e.g., *Na. albida*, *Vishn. victoriae*, and *Rh. mucilaginosa*, *Sp. roseus*) have been recovered from these environments, but also species so far restricted to extreme environments (e.g., *Naganishia antarctica*, *Naganishia friedmannii*) have been reported (Vishniac 2006; Buzzini et al. 2012).

6.4 Phylloplane Yeast Diversity Using High-Throughput DNA Sequencing

The increased interest in microbial communities in recent years has gone hand in hand with technological advances in high-throughput or next-generation DNA sequencing (NGS). Although microbial diversity was assumed to be high before, only by the application of these technologies has it been possible to detect so far unprecedented species diversity in microbial communities, including fungal phyllosphere communities. Microbiota thereby describes the entirety of microbial organisms in a given habitat, whereas microbiome is the entire genomic content of a given microbial community. However, the two terms have been used interchangeably, whereby microbiome is the preferred term. In accordance to microbiome, the entire fungal community in a given habitat is increasingly referred to as mycobiome.

Microbial diversity studies using marker-assisted NGS methods have several advantages compared to culture-dependent approaches. Less labor-intensive workflows, for instance, enable the high-throughput assessments of many more ecosystems. A further advantage is the possibility to obtain information for potentially non-cultivable microorganisms. Despite being a powerful tool, there are also some issues with NGS methods in general as well as specific to assessing yeast diversity in the phyllosphere. DNA is a rather stable molecule and this makes it difficult to distinguish between living or dead organisms. Additionally, it is not possible to distinguish between actively growing members of phylloplane communities and resting or dispersal structures (e.g., sclerotia or spores) that are not actively growing on the plant surface. To circumvent these problems, approaches using RNA instead of DNA have been employed to differentiate between active and total fungal community (Womack et al. 2015). Primer biases resulting in skewed diversity patterns that not necessarily reflect real diversity patterns are known technical problems (Tedersoo et al. 2015). Further problems arise when identifying organisms via a single molecular marker. Although studies are often performed using the internal transcribed spacer region of the rDNA, which is the fungal barcode, inherent problems with marker variability as well as sequencing length can prevent identification down to the species level. However, it is often possible to identify the families (e.g., Kemler et al. 2013) or genera (e.g., Bálint et al. 2015) that operational taxonomic units (OTUs; i.e., clusters of sequences with predefined similarity) belong to. A phyllosphere specific problem is the impossibility to distinguish endophytic and epiphytic mycobiota, due to the lack of NGS studies having applied surface sterilization. A specific problem for assessing yeast diversity arises due to their polyphyletic origin and associated primer binding site heterogeneity. Therefore, no NGS study in the phyllosphere has been exclusive on yeasts so far. However, from culture-dependent endophytic studies using surface sterilization (e.g., Unterseher and Schnittler 2010) as well as culture-dependent phylloplane studies (e.g., Inácio et al. 2002), it can be extrapolated that yeasts and yeast-like fungi are adapted specifically to the leaf surface, and only few occur also in the interior of leaves. Still, until some of the abovementioned problems have been solved and species identification is fully feasible using NGS, some of the conclusions drawn below remain to a certain extent speculative.

A large proportion of identified phyllosphere fungi of common tree genera in temperate regions (i.e., *Fagus*, *Populus*, *Quercus*) recovered by NGS studies (Jumpponen and Jones 2009; Cordier et al. 2012b; Bálint et al. 2015) are yeasts or dimorphic fungi (Fig. 6.1) showing high taxon diversity (Fig. 6.2). The most

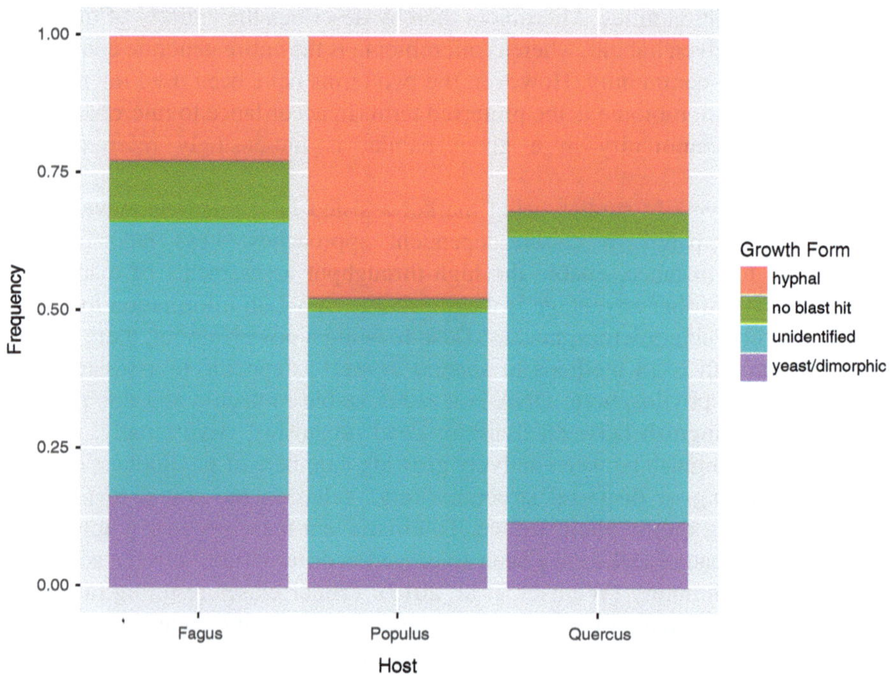

Fig. 6.1 Proportion of yeast and dimorphic taxa recovered from the reanalysis of three NGS studies conducted on phylloplane fungal communities of trees from temperate regions. Proportions are calculated based on OTU abundances in each study

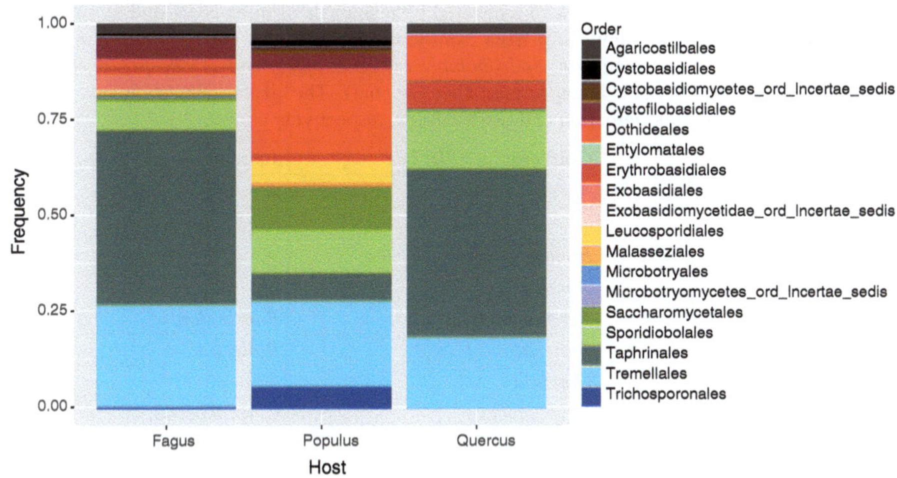

Fig. 6.2 Proportion of fungal orders within yeast and dimorphic taxa (Fig. 6.1) recovered from the reanalysis of three NGS studies conducted on phylloplane fungal communities of trees from temperate regions

common yeast and yeast-like genera recovered by NGS in the phyllosphere of different hosts belong to the genera *Aureobasidium*, *Cryptococcus*, *Dioszegia*, *Erythrobasidium*, *Rhodotorula*, *Taphrina* (including its anamorph *Lalaria*), and *Sporobolomyces* (Table 6.1). The dominance of *Taphrina* (e.g., Cordier et al. 2012b) thereby is surprising, as culture-dependent methods indicated that mostly basidiomycetous yeasts inhabit the phyllosphere (Fonseca and Inácio 2006; however, see Inácio et al. (2002)).

Identification of yeasts by NGS is dependent on taxonomic assignments through reference databases. As taxonomic updates to reference databases are laborious and time-consuming, the interpretation of culture-independent studies can be difficult to some extent. Therefore, OTUs are often reported as members of the polyphyletic phenotypic genera *Candida*, *Cryptococcus*, *Rhodotorula*, and *Trichosporon* and not assigned to the particular phylogenetic lineage or clade (Fig. 6.3). Due to newer taxonomic insights (e.g., Liu et al. 2015; Wang et al. 2015b), certain traditional genera are known to be polyphyletic, and in some cases its members are dispersed within a whole class of fungi (i.e., *Candida* species can be found in the whole of Saccharomycetes, *Cryptococcus* in Tremellomycetes, and *Rhodotorula* in Microbotryomycetes and Cystobasidiomycetes; see Fig. 6.3). To highlight the problem of incorrect taxonomic assignments based on erroneous database entries, we manually identified some phylloplane yeasts recovered in NGS (see paragraph 6.2) that were classified as *Cryptococcus* OTUs by automatic approaches: *Vishn. victoriae*, *Vishn. tephrensis*, *F. magnum* (and possibly *F. oeirensis*), *Papiliotrema pseudoalba*, *Bullera alba*, *Hannaella luteola*, and *Genolevuria* sp. (published in Jumpponen and Jones 2009), *Vishniacozyma* spp. (*Vishn. victoriae*, *Vishn. heimaeyensis*, *Vishn. carnescens*, *Vishn. tephrensis*, and *Vishniacozyma tibetensis*),

Table 6.1 Yeast and dimorphic genera recovered by reanalysis of three NGS studies conducted on phylloplane fungal communities of trees from temperate regions. The genera *Bensingtonia*, *Bullera*, *Cryptococcus*, *Rhodotorula*, *Sporobolomyces*, *Tilletiopsis*, and *Trichosporon* comprise distantly related phylogenetic lineages (see also Fig. 6.3). The genera *Rhynchogastrema* and *Tausonia* are cited as *Bandoniozyma* and *Guehomyces*, respectively

Bálint et al. (2015)	Cordier et al. (2012)	Jumpponen and Jones (2009)
Aureobasidium	*Aureobasidium*	*Aureobasidium*
–	–	*Auriculibuller*
–	*Bandoniozyma*	–
–	*Bannoa*	–
Bensingtonia	*Bensingtonia*	*Bensingtonia*
Bullera	*Bullera*	*Bullera*
Candida	*Candida*	–
Cryptococcus	*Cryptococcus*	*Cryptococcus*
–	*Cystobasidium*	–
–	*Cystofilobasidium*	–
–	*Debaryomyces*	–
–	*Derxomyces*	–
Dioszegia	*Dioszegia*	*Dioszegia*
–	–	*Entyloma*
Erythrobasidium	*Erythrobasidium*	*Erythrobasidium*
–	*Exobasidium*	–
–	*Fellomyces*	–
–	*Guehomyces*	–
Hannaella	*Hannaella*	*Hannaella*
Itersonilia	*Itersonilia*	–
Kondoa	*Kondoa*	*Kondoa*
Kurtzmanomyces	*Kurtzmanomyces*	–
–	*Lachancea*	–
Leucosporidium	*Leucosporidium*	–
Malassezia	*Malassezia*	–
–	*Metschnikowia*	*Metschnikowia*
Meyerozyma	–	–
–	*Microbotryozyma*	–
Mrakia	*Mrakia*	–
–	*Phaffia*	–
Pichia	–	–
Rhodotorula	*Rhodotorula*	*Rhodotorula*
–	*Sakaguchia*	–
Sporobolomyces	*Sporobolomyces*	*Sporobolomyces*
Taphrina	*Taphrina*	*Taphrina*
–	*Tilletiopsis*	*Tilletiopsis*
Tremella	*Tremella*	*Tremella*
Trichosporon	*Trichosporon*	–
–	*Udeniomyces*	–
Zygoascus	–	–

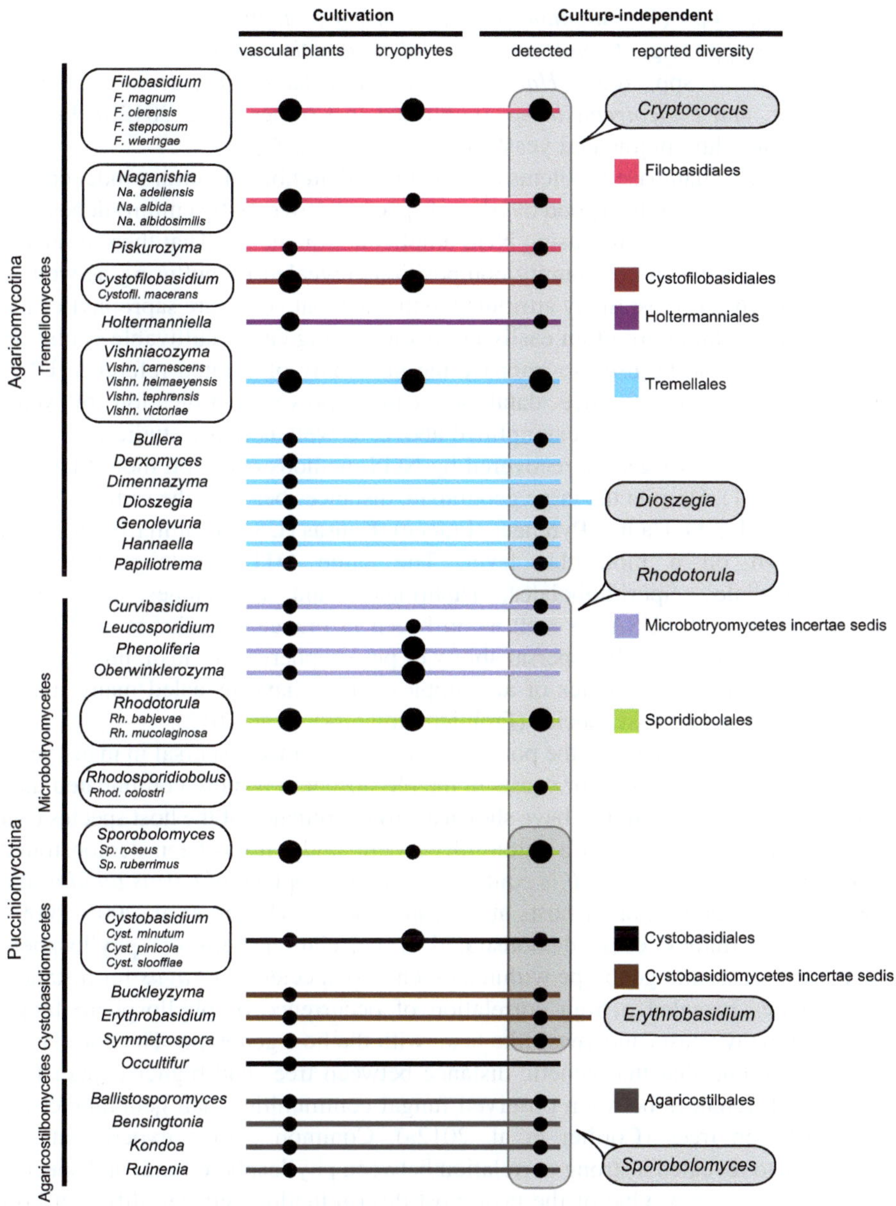

Fig. 6.3 Yeast species frequently isolated from the phylloplane that are also often recovered by culture-independent methods. The figure also indicates newer taxonomic revisions of yeast taxa that are often not incorporated into fungal databases used for taxon assignment in culture-independent methods

Filobasidium spp. (*F. magnum*, *F. oeirensis*, *F. stepposum*, *F. wieringae*), *Genolevuria* spp. (e.g., *Genolevuria amylolytica*), *Dioszegia* spp., *Piskurozyma* spp., *Hannaella* spp. (e.g., *Hann. luteola*), *Pa. flavescens*, *Holtermanniella takashimae*, and *Phaeotremella* sp. (published in Cordier et al. 2012b). Further examples including pigmented yeasts are shown in Fig. 6.3.

Despite these database problems and the limited number of culture-independent studies, there seems to be a good overlap of species recovered by culture-dependent isolation studies and studies using NGS amplicon sequencing. That there is such a good overlap between community compositions revealed by culturing as well as NGS approaches is most likely attributed to the fact that yeasts are saprophytic, and therefore these fungi are often easily cultivable. Being easily cultivable is also an advantage that can be used to improve the taxonomy of curated reference DNA databases. Curated reference databases can improve identification of yeast sequences from previous, as exemplified above, as well as future studies.

Many of the yeast genera recovered by NGS methods occur on more than one host (Table 6.1). Despite the large geographic distances between the sampling sites (i.e., Alaska (USA), Hautes-Pyrénées (France), Kansas (USA)), similarity has even been shown on a molecular level. The same OTUs, belonging to the Erythrobasidiales, Sporidiobolales, Taphrinales, and Tremellales, have been found on oak in the USA, as well as on beech in France (Cordier et al. 2012a). These results imply a high dispersal ability of phyllosphere yeast taxa, findings that are supported by NGS studies of air samples. These have revealed many fungal classes that contain yeast species (Fröhlich-Nowoisky et al. 2012).

However, even so there is the potential for long-distance dispersal in many yeast taxa, community assembly for yeasts in the phyllosphere is more complicated than only dispersal ability. Studies have shown a strong influence of the host species (but see Yurkov et al. 2015c). In a culture-dependent study, it has been demonstrated that *Dimennazyma cisti-albidi* is confined to the phylloplane of *Cistus albidus* and could not be found on other hosts at the same locality (Inácio et al. 2005, 2009). Some NGS studies even demonstrated a correlation between phyllosphere inhabiting yeasts and genotype within a given host species. On *Fagus* for instance, NGS studies revealed a high correlation of occurrence of phyllosphere fungi, including many yeasts and yeast-like taxa, with the host genotype. The correlation was so profound that that genetic distance between trees had higher explanatory value for differences between observed fungal communities than geographic distance between trees (Cordier et al. 2012a). Common garden experiments on *Populus* also revealed a strong correlation between phyllosphere fungi and specific host plant genotypes. One of the taxa most discriminating between different host genotypes thereby was *Sporobolomyces*, a yeast genus commonly found in the phylloplane of trees (Bálint et al. 2013).

Despite the knowledge that abiotic factors and host genotype exhibit an influence on microbial species composition, mechanistic insights into how these factors shape the microbial community are sparse. Using a systems biological approach incorporating NGS methods and scale-free network modeling, Agler et al. (2016) demonstrated that different microbial phyllosphere taxa respond in different ways to different factors. Additionally, direct influence on a taxon by the factors that shape microbial community composition is limited, and many effects are mediated via species interactions. Interestingly, several yeast species exhibited a strong effect on bacterial species, whereby the abundance of *Dioszegia*, *Udeniomyces*, and *Leucosporidium* (referred to as *Leucosporidiella*) negatively correlated with bacterial abundance. The effect most likely was mediated through direct competition for space and through the excretion of metabolites. Especially *Dioszegia* had a strong effect on community composition. By using network modeling, the authors demonstrated that this yeast was connected strongly with many other microbial species. Therefore, abiotic factors and host genotype that had an effect on *Dioszegia*, through its strong interaction with other species, showed a strong effect on many of the other microbes in the community (Agler et al. 2016).

6.5 Insight from Phylloplane Yeast Genomics

The yeast *Saccharomyces cerevisiae* was the first eukaryotic organism having its genome sequenced (Goffeau et al. 1996). Since then many yeast genomes have been sequenced, including several species that are common in the phylloplane. These genome-sequencing projects highlight some of the mechanisms and physiological pathways that enable these species to cope with the environmental factors encountered in the phylloplane and adapt to this specific niche.

The ubiquitous members of the black yeast *A. pullulans* s.l. are fungi with great ecological amplitude that show tolerance for many extreme environments (Gostinčar et al. 2010) and also occur frequently in the phylloplane (Fonseca and Inácio 2006). Genome sequencing of specimens from different habitats revealed, among other things, genes involved in melanin production as well as production of extracellular enzymes and polysaccharides (Gostinčar et al. 2014). Melanins are polyphenolic pigments that are produced by many fungi for coping with various environmental stresses, including protection against radiation, low water activity, and low nutrient availability (Cordero and Casadevall 2017). The sequenced *Aureobasidium* specimens contained homologues of genes involved in the biosynthesis of 1,8-dihydroxynaphthalene-melanin that have also been found in other melanin-pigmented fungi. The study also supported the proposed pathway for production of the extracellular polysaccharide pullulan by confirming the presence of all the necessary enzymes involved in its synthesis (Duan et al. 2008).

The genome sequence of *Phaffia rhodozyma*, a yeast specifically associated with trees (fluxes and inside tree-associated fungi) in the Southern and Northern Hemisphere, unraveled the genetic machinery of several photoprotective and antioxidant molecules. Most likely due to the high radiation stress this yeast is exposed to in the phyllosphere, it evolved biosynthetic pathways for synthesizing the carotenoid astaxanthin as well as the sunscreen molecule mycosporine-glutaminol-glucoside. Additionally, genes encoding several antioxidants, including catalases and superoxide dismutases, occurred in the genome (Bellora et al. 2016). These enzymes are particularly well suited to degrade H_2O_2, a reactive oxygen species often involved in plant defense against pathogens (Shetty et al. 2008). However, whether these antioxidants are involved in plant interaction and the actual roles they play in the biology of *Ph. rhodozyma* is so far unknown.

Rhodotorula toruloides (formerly *Rhodosporidium toruloides*) is a further red yeast containing large amount of carotenoids, most likely as radiation protection, with several strains having a sequenced genome (Kumar et al. 2012; Zhu et al. 2012; Morin et al. 2014; Hu and Ji 2016). Genome analyses revealed genes encoding protein complexes and a range of transcription factors involved in regulating protein synthesis, as well as fatty acid synthesis in *Rh. toruloides* that might be regulated by nitrogen availability (Zhu et al. 2012). Nitrogen deficiency leads to a downregulation of protein synthesis, but an upregulation of the fatty acid synthesis, which provides the precursors for carotenoid pigments. Thereby, nitrogen deficiency in the phylloplane could be a factor providing an advantage to red yeasts and explain their high abundance in this environment. Another *Rh. toruloides* (also cited as *Rh. glutinis* ATCC 204091) genome analysis highlighted the gene repertoire available to synthesize carotenoid pigments including torulene, torularhodin, γ-carotene, and β-carotene (Paul et al. 2014).

Pathways for the production of carotenoids were also identified in the genome of different *Taphrina* species (Cissé et al. 2013; Tsai et al. 2014). Although *Taphrina* is well known as a plant parasite causing tumors in different host plants, for parts of its life cycle, it occurs as yeast (formerly known as *Lalaria*). These have been isolated from the phylloplane (Inácio et al. 2002) and also form a large part of the fungal community recovered by NGS methods (see paragraph 6.3). As in other phylloplane yeasts, the carotenoids are interpreted as protection against radiation and other stress factors occurring in this environment.

An endophytic stem isolate of *Rh. graminis* showing strong plant growth promotion has recently been sequenced (Firrincieli et al. 2015). Species of *Rhodotorula* are also frequently reported as foliar epiphytes (Fonseca and Inácio 2006). *Rh. graminis* seems to promote plant growth in a different way than many other beneficial microorganisms that produce indole-3-acetic acid (IAA; see below) via indole-3-pyruvate from tryptophan precursors excreted by the plant (Hardoim et al. 2008). The genome analysis of *Rh. graminis* revealed that genes for standard IAA production were lacking. The genome does contain, however, three genes encoding enzymes involved in the transversion of tryptophan into IAA via tryptamine (Firrincieli et al. 2015). Additional genes identified in the *Rh. graminis* genome that could be involved in plant growth stimulation were putative hydrolases and putative proteins involved in the production of gibberellins. The genome analysis also revealed genes potentially involved in the production of volatile components (VOCs; see below) that could be important in the interaction with other microorganisms. Further, the putative genes CAP59 and CAP10 were observed in the genome. These genes have homologues in *Cryptococcus neoformans*, where they are involved in the production of capsule polysaccharides that are vital for the production of biofilms and can serve as protection against desiccation (e.g., Vishniac et al. 1997).

6.6 Yeast Interactions with Other Community Members

The NGS high-throughput diversity assessments confirm observations based on culture-dependent techniques that phylloplane yeasts occur in complex microbial communities (Fig. 6.1, see paragraph 6.3). It is therefore not surprising that they have evolved different strategies of interaction and communicate with each other, as well as with other community members. To achieve this interaction, they have to leak soluble or volatile compounds into the environment. Genome sequencing has shown that yeasts, like other fungi, produce a large amount of secreted proteins, which are potential triggers of interaction with other microbes. A well-understood form of interaction involving secreted proteins is the mating reaction that many haploid sporidia of smut fungi perform to form infective structures (see also Mittelbach and Vannette 2017). Thereby, pheromones secreted into the environment are perceived by a compatible partner via pheromone receptors, and subsequent mating is initiated by triggering cascades of physiological reactions (Kahmann and Kämper 2004).

Another group of signaling compounds used by microorganisms are volatile organic compounds (VOCs), which are commonly produced for interaction between organisms. Over 250 VOCs have already been described from fungi (Bennett et al. 2012). VOCs are comprised of many different chemical classes, the most well known being 1-octen-3-ol, causing the typical mushroom smell (Combet et al. 2006). In fungal interactions VOCs have numerous functions, including the induction of conidiogenesis in other members of the same species (Nemčovič et al. 2008), as well as serving as pheromones (Eastwood et al. 2013). Fungal VOCs also play a pivotal role in microbe insect interactions (Davis et al. 2013). The phylloplane dimorphic fungus *A. pullulans*, for instance, attracts a substantial amount of eusocial wasps via VOCs. In this interaction yeast growths most likely indicate an available food source for the wasps, and the insect in return acts as vector for the yeast (Davis et al. 2012). Yeast-mediated attraction of other insects is known from many groups, including bark beetles, mushroom-feeding beetles, drosophilids, and nitidulids, among others (see also Chap. 13 of this book). However, it has been also emphasized that microbial VOCs involved in insect interactions are functionally redundant (Davis et al. 2013). Additionally, the specificity of attraction varies largely, and several studies showed low specificity of insect attraction, i.e., distantly related yeasts equally attracted the same group of insects. For example, endemic flower-inhabiting *Metschnikowia hawaiiensis* (Saccharomycotina, Metschnikowiaceae), three closely related species from the *M. pulcherrima* species complex, and the basidiomycete *Vishn. tephrensis* all attracted arthropods in trap experiments (Andreadis et al. 2015). Remarkably, most arthropods did not distinguish between VOCs of the two widespread phylloplane species *M. pulcherrima* (Ascomycota) and *Vishn. tephrensis* (Basidiomycota) and were equally attracted to agar cultures with these yeasts. Only the numbers of Chironomidae (chironomids or lake flies) and Sarcophagidae (flesh flies) were higher in traps with *M. pulcherrima* and *Vishn. tephrensis*, respectively (Andreadis et al. 2015). In another study focusing on springtail feeding behavior (Men'ko et al. 2006), the phylloplane-related yeasts *Sp. roseus*, *Rh. babjevae* (cited as *Rh. glutinis*), and *M. pulcherrima* were investigated. Although observed effects were often strain-specific and varied between two springtail species, all phylloplane yeasts were eaten by either of the two springtails. Throughout the experiment, the collembolans also changed their preferences, but phylloplane yeasts were generally preferred over soil-borne species (Men'ko et al. 2006). Unfortunately, nothing is known about the influence of VOCs in the interaction of yeasts with other microorganisms. However, as genome sequencing indicates that they are able to produce VOCs (see paragraph 6.5), it can be speculated that phylloplane yeast species are able to use them for intraspecific as well as interspecific microbial interactions. Further investigations into yeast VOCs and their influence on other microbes and macroorganisms in the phyllosphere therefore promise to be fruitful in order to understand interactions in epiphytic microbial communities as well as the dispersal ways of individual species.

Secreted compounds are also essential in the formation of biofilms, an important way in which single cell microbes interact with each other. Biofilms describe microbial organisms that are embedded in a self-synthesized extracellular matrix (Blankenship and Mitchell 2006). Microbes within a biofilm are protected from adverse conditions acting on individual cells. This protection on one hand is provided by the extracellular matrix created by the cells, but also by changed physiological conditions that occur with increasing cell densities coupled with quorum sensing (Fanning and Mitchell 2012). Biofilm formation is a well-known phenomenon in bacterial communities, but is also known for yeasts. Most studies thereby have concentrated on clinically relevant species, such as *Candida*, *Cryptococcus*, *Cutaneotrichosporon*, and *Trichosporon* species (Fanning and Mitchell 2012). *Candida* biofilms are initiated by the attachment of yeast cells to the substrate, subsequent yeast proliferation that cumulates in the formation of hyphal growth (Blankenship and Mitchell 2006). Several phylloplane yeasts are known to have similar properties. Species producing polysaccharide capsules are abundant in this habitat (reviewed in Fonseca and Inácio 2006), and genome analysis (see paragraph 6.5) has recovered the genes involved in the production of extracellular matrices. Additionally, many of these yeasts are capable of producing hyphae or pseudo-hyphae on culture media, but their growth characteristics and morphology on leaves are largely unknown since direct observation with microscopic techniques (e.g., FISH or SEM) is scarce (e.g., Fonseca and Inácio 2006; Inácio et al. 2009). However, despite all indications that they are able, whether phyllosphere yeasts actually form biofilms is unknown. As with VOCs the behavior of yeasts in biofilms seems to be a promising avenue for further studies in order to understand microbial interactions in the phylloplane.

Phylloplane yeasts also exhibit antagonistic behavior toward other microorganisms. Especially the effects of yeasts as biological control agents on plant pathogens are a dynamic research area. By excretion of so far unidentified metabolites, *Moesziomyces* (formerly *Pseudozyma*) *aphidis*, for instance, directly inhibits growth of several severe bacterial and fungal plant pathogens (Buxdorf et al. 2013; Barda et al. 2014). Switching to a hyphal stage enabled the same *Moesz. aphidis* isolate to parasitize the fungal pathogen *Podosphaera xanthii*, thereby reducing the infection of cucumber by the powdery mildew (Gafni et al. 2015). But *Moesz. aphidis* does not only suppress growth of pathogenic microbes by excretion of metabolites or ectoparasitism. Additionally it has been shown to increase plant resistance by triggering plant defense response pathways (Buxdorf et al. 2013; Barda et al. 2014). Antagonistic interactions of yeasts with other microbial organisms including production of killer toxins, glycolipids, and chelating agents (pulcherrimine of *M. pulcherrima*) have been discussed by Klassen et al. (2017).

6.7 Yeasts as Plant Growth Promoters

Indole-3-acetic acid (IAA), a member of the auxin class, is the most common phytohormone occurring in plants. It mainly affects cell stretching, cell division, tissue development, apical dominance, abscission, and responses to light and gravity (Aloni et al. 2006; Teale et al. 2006; Scarpella et al. 2010; Tian et al. 2014). Consequently it has great importance for plant growth and development processes, most often in combination with other phytohormones like cytokinin or gibberellin. Although it occurs in the whole plant body, its main function is in developing and growing plant tissues, e.g., the coleoptile, shoot and root tips, juvenile leaves, and active cambium cells.

The ability to produce IAA is not restricted to plants and is a widespread feature found in bacteria and fungi, including yeasts (Spaepen et al. 2007; Reineke et al. 2008; Ali et al. 2009; Sun et al. 2014). A broad range of yeasts, including genera from Ascomycota and Basidiomycota, occurring in the plant phyllosphere have the capability of IAA biosynthesis (Xin et al. 2009; Limtong and Koowadjanakul 2012; Sun et al. 2014; Streletskii et al. 2016). The IAA biosynthesis of yeasts depends on multiple factors such as pH value of the substrate, temperature, utilized carbon and nitrogen sources, and tryptophan availability (Spaepen et al. 2007; Nutaratat et al. 2015).

Since both plants and several microorganisms are producing and using IAA as a signal molecule, IAA is also involved in plant-yeast interactions, and some studies have focused on the effects that IAA-producing yeasts have on plant growth. Advantageous effects of IAA-producing yeasts have been reported from *Nicotiana tabacum* (Fu et al. 2016) and *Zea mays* (Nassar et al. 2005). However, these studies were conducted with root-associated yeast taxa. Again, not much is known about actual growth-promoting capabilities of phylloplane-associated yeasts, except for many taxa having the ability to produce IAA. A study of 124 yeast strains showed that ability to produce IAA is present in the majority (92%) of yeast isolates of both plant and soil origin (Streletskii et al. 2016). Earlier reports on IAA synthesis used photometric detection technique, which fails to detect the compound in low concentrations (Streletskii et al. 2016). This explained lower numbers of yeasts reported for IAA production, i.e., 16–50% depending on the substrate and study (Limtong and Koowadjanakul 2012; Limtong et al. 2014; Nutaratat et al. 2014). Reports available to date suggest that although IAA synthesis is a frequent trait among yeasts, it is strain-dependent, and studied strains showed tenfold to 100-fold variation in amounts of these compounds (e.g., Limtong and Koowadjanakul 2012; Streletskii et al. 2016). High amount of IAA (above 1000 μg/g) has been detected in the following phylloplane-related yeasts: *Deb. hansenii* and *M. pulcherrima* in Ascomycota and *Cystofilobasidium capitatum*, *F. magnum*, *Rh. babjevae*, *Rh. mucilaginosa*, and *Sp. roseus* in Basidiomycota (Streletskii et al. 2016 and references therein). It has been suggested that ascomycetous yeasts produce on average more IAA in culture than basidiomycetous species, and yeasts from tropical regions show higher IAA yield than those from temperate regions (Streletskii et al. 2016).

6.8 Concluding Remarks

Assessment of whole microbial communities using massive parallel DNA marker sequencing, whole-genome sequencing of microbes, and other molecular high-throughput technologies have changed the way we understand the ecology and evolution of plants and their associated microbial partners. These methods also moved phylloplane-associated microbial communities in temperate climates, where yeasts form an integral part of, in the focus of different research disciplines. It is apparent that these methods will keep increasing our knowledge about the yeasts in this environment and that these technologies will become standard repertoire for phylloplane yeast research in the future. Going hand in hand with these technologies should be standard isolation methods of yeasts from the phylloplane and the long-term deposition in an acknowledged repository. This will not only make them available for future research, but due to the extreme environment that they adapted to, many of these yeasts have acquired characteristics that make them desirable biotechnological and biocontrol organisms. Living cultures are the source of the reference data, which is used to identify organisms in culture-independent studies. Future research of phylloplane yeasts and their allies should devote resources to the improvement of the databases and identification tools. Since questions about functional diversity, redundancy, and resilience of microbial communities have recently received interest in ecology, direct observation of yeast species on plant surfaces and studies addressing community dynamics will provide a valuable contribution to this topic.

References

Agler MT, Ruhe J, Kroll S, Morhenn C, Kim ST, Weigel D, Kemen EM (2016) Microbial hub taxa link host and abiotic factors to plant microbiome variation. PLoS Biol 14:e1002352

Ali B, Sabri AN, Ljung K, Hasnain S (2009) Auxin production by plant associated bacteria: impact on endogenous IAA content and growth of *Triticum aestivum* L. Lett Appl Microbiol 48:542–547

Aloni R, Aloni E, Langhans M, Ullrich CI (2006) Role of cytokinin and auxin in shaping root architecture: regulating vascular differentiation, lateral root initiation, root apical dominance and root gravitropism. Ann Bot 97:883–893

Andreadis SS, Witzgall P, Becher PG (2015) Survey of arthropod assemblages responding to live yeasts in an organic apple orchard. Front Ecol Evol 3:121

Arnold AE (2007) Understanding the diversity of foliar endophytic fungi: progress, challenges, and frontiers. Fungal Biol Rev 21:51–66

Babjeva IP, Reshetova I (1998) Yeast resources in natural habitats at polar circle latitude. Food Technol Biotechnol 36:1–6

Babjeva IP, Kartintseva AA, Maksimova IA, Chernov IY (1999) Yeasts in the spruce forests of the Central Forest Reserve. Vestn Mosk Univ Ser Pochvoved 4:45–49

Bálint M, Tiffin P, Hallström B, O'Hara RB, Olson MS, Fankhauser JD, Piepenbring M, Schmitt I (2013) Host genotype shapes the foliar fungal microbiome of balsam poplar (*Populus balsamifera*). PLoS One 8:e53987

Bálint M, Bartha L, O'Hara RB, Olson MS, Otte J, Pfenninger M, Robertson AL, Tiffin P, Schmitt I (2015) Relocation, high-latitude warming and host genetic identity shape the foliar fungal microbiome of poplars. Mol Ecol 24:235–248

Barda O, Shalev O, Alster S, Buxdorf K, Gafni A, Levy M (2014) *Pseudozyma aphidis* induces salicylic-acid-independent resistance to *Clavibacter michiganensis* in tomato plants. Plant Dis 99:621–626

Begerow D, Kemler M, Feige A, Yurkov AM (2017) Parasitism in yeasts. In: Buzzini P, Lachance MA, Yurkov AM (eds) Yeasts in natural ecosystems: ecology. Springer, Heidelberg, pp 179–210

Bellora N, Moliné M, David-Palma M, Coelho MA, Hittinger CT, Sampaio JP, Gonçalves P, Libkind D (2016) Comparative genomics provides new insights into the diversity, physiology, and sexuality of the only industrially exploited tremellomycete: *Phaffia rhodozyma*. BMC Genomics 17:901

Bennett JW, Hung R, Lee S, Padhi S (2012) Fungal and bacterial volatile organic compounds: an overview and their role as ecological signaling agents. In: Hock B (ed) Fungal associations. Springer, Berlin, pp 373–393

Blankenship JR, Mitchell AP (2006) How to build a biofilm: a fungal perspective. Curr Opin Microbiol 9:588–594

Brandão LR, Libkind D, Vaz ABM, Espírito Santo LC, Moliné M, de García V, van Broock M, Rosa CA (2011) Yeasts from an oligotrophic lake in Patagonia (Argentina): diversity, distribution and synthesis of photoprotective compounds and extracellular enzymes. FEMS Microbiol Ecol 76:1–13

Buxdorf K, Rahat I, Gafni A, Levy M (2013) The epiphytic fungus *Pseudozyma aphidis* induces jasmonic acid- and salicylic acid/nonexpressor of PR1-independent local and systemic resistance. Plant Physiol 161:2014–2022

Buzzini P, Branda E, Goretti M, Turchetti B (2012) Psychrophilic yeasts from worldwide glacial habitats: diversity, adaptation strategies and biotechnological potential. FEMS Microbiol Ecol 82:217–241

Cabral D, Stone JK, Carroll GC (1993) The internal mycobiota of *Juncus* spp.: microscopic and cultural observations of infection patterns. Mycol Res 97:367–376

Chernov IY, Glushakova AM, Kachalkin AV (2013) Annotated list of yeasts from Moscow region. Mikol Fitopatol 47:103–115

Cissé OH, Almeida JM, Fonseca A, Kumar AA, Salojärvi J, Overmyer K, Hauser PM, Pagni M (2013) Genome sequencing of the plant pathogen *Taphrina deformans*, the causal agent of peach leaf curl. MBio 4:e00055-13

Coelho MA, Gonçalves P, Sampaio JP (2011) Evidence for maintenance of sex determinants but not of sexual stages in red yeasts, a group of early diverged basidiomycetes. BMC Evol Biol 11:249

Combet E, Eastwood DC, Burton KS, Henderson J (2006) Eight-carbon volatiles in mushrooms and fungi: properties, analysis, and biosynthesis. Mycoscience 47:317–326
Cordero RJB, Casadevall A (2017) Functions of fungal melanin beyond virulence. Fungal Biol Rev 31(2):99–112
Cordier T, Robin C, Capdevielle X, Desprez-Loustau ML, Vacher C (2012a) Spatial variability of phyllosphere fungal assemblages: genetic distance predominates over geographic distance in a European beech stand (*Fagus sylvatica*). Fungal Ecol 5:509–520
Cordier T, Robin C, Capdevielle X, Fabreguettes O, Desprez-Loustau ML, Vacher C (2012b) The composition of phyllosphere fungal assemblages of European beech (*Fagus sylvatica*) varies significantly along an elevation gradient. New Phytol 196:510–519
Davis TS, Boundy-Mills K, Landolt PJ (2012) Volatile emissions from an epiphytic fungus are semiochemicals for eusocial wasps. Microb Ecol 64:1056–1063
Davis TS, Crippen TL, Hofstetter RW, Tomberlin JK (2013) Microbial volatile emissions as insect semiochemicals. J Chem Ecol 39:840–859
Davydenko K, Vasaitis R, Stenlid J, Menkis A (2013) Fungi in foliage and shoots of *Fraxinus excelsior* in eastern Ukraine: a first report on *Hymenoscyphus pseudoalbidus*. For Pathol 43:462–467
Duan X, Chi Z, Wang L, Wang X (2008) Influence of different sugars on pullulan production and activities of α-phosphoglucose mutase, UDPG-pyrophosphorylase and glucosyltransferase involved in pullulan synthesis in *Aureobasidium pullulans* Y68. Carbohydr Polym 73:587–593
Duarte AWF, Passarini MRZ, Delforno TP, Pellizzari FM, Cipro CVZ, Montone RC, Petry MV, Putzke J, Rosa LH, Sette LD (2016) Yeasts from macroalgae and lichens that inhabit the South Shetland Islands, Antarctica. Environ Microbiol Rep 8:874–885
Eastwood DC, Herman B, Noble R, Dobrovin-Pennington A, Sreenivasaprasad S, Burton KS (2013) Environmental regulation of reproductive phase change in *Agaricus bisporus* by 1-octen-3-ol, temperature and CO_2. Fungal Genet Biol 55:54–66
Fanning S, Mitchell AP (2012) Fungal biofilms. PLoS Pathog 8:e1002585
Firrincieli A, Otillar R, Salamov A, Schmutz J, Khan Z, Redman RS, Fleck ND, Lindquist E, Grigoriev IV, Doty SL (2015) Genome sequence of the plant growth promoting endophytic yeast *Rhodotorula graminis* WP1. Front Microbiol 6:978
Fonseca Á, Inácio J (2006) Phylloplane yeasts. In: Rosa C, Gábor P (eds) Biodiversity and ecophysiology of yeasts. Springer, Berlin, pp 263–301
Fonseca Á, Rodriguez M (2011) *Taprhina* fries (1832). In: Kurtzman CP, Fell JW, Boekhout T (eds) The yeasts: a taxonomic study, vol 2, 5th edn. Elsevier, Amsterdam, pp 823–858
Fröhlich-Nowoisky J, Burrows SM, Xie Z, Engling G, Solomon PA, Fraser MP, Mayol-Bracero OL, Artaxo P, Begerow D, Conrad R, Andreae MO, Després VR, Pöschl U (2012) Biogeography in the air: fungal diversity over land and oceans. Biogeosciences 9:1125–1136
Fu SF, Sun PF, Lu HY, Wei JY, Xiao HS, Fang WT, Cheng BY, Chou JY (2016) Plant growth-promoting traits of yeasts isolated from the phyllosphere and rhizosphere of *Drosera spatulata* Lab. Fungal Biol 120:433–448
Gafni A, Calderon CE, Harris R, Buxdorf K, Dafa-Berger A, Zeilinger-Reichert E, Levy M (2015) Biological control of the cucurbit powdery mildew pathogen *Podosphaera xanthii* by means of the epiphytic fungus *Pseudozyma aphidis* and parasitism as a mode of action. Front Plant Sci 6:132
Glushakova AM, Chernov IY (2007) Seasonal dynamic of the numbers of epiphytic yeasts. Microbiology 76:668–674
Glushakova AM, Chernov IY (2010) Seasonal dynamics of the structure of epiphytic yeast communities. Microbiology 79:830–839
Glushakova AM, Yurkov AM, Chernov IY (2007) Massive isolation of anamorphous ascomycete yeasts *Candida oleophila* from plant phyllosphere. Microbiology 76:799–803
Glushakova AM, Kachalkin AV, Chernov IY (2014) Yeasts in the flowers of entomophilic plants of the Moscow region. Microbiology 83:125–134

Glushakova AM, Kachalkin AV, Zheltikova TM, Chernov IY (2015a) Yeasts associated with wind-pollinated plants – leading pollen allergens in Central Russia. Microbiology 84:722–725

Glushakova AM, Kachalkin AV, Zheltikova TM, Chernov IY (2015b) Resistance of various yeast ecological groups to prolonged storage in dry state. Microbiology 84:379–385

Goffeau A, Barrell BG, Bussey H, Davis RW, Dujon B, Feldmann H, Galibert F, Hoheisel JD, Jacq C, Johnston M, Louis EJ, Mewes HW, Murakami Y, Philippsen P, Tettelin H, Oliver SG (1996) Life with 6000 Genes. Science 274:546–567

Golonka AM, Vilgalys R (2013) Nectar inhabiting yeasts in Virginian populations of *Silene latifolia* (Caryophyllaceae) and coflowering species. Am Midl Nat 169:235–258

Golubev WI, Golubeva EW (2004) Yeast fungi in steppe and forest phytocenoses of the Prioksko-terrasny biosphere reserve. Mikol Fitopatol 38:20–27

Golubev WI, Sampaio JP, Golubeva EW (2006) *Cryptococcus stepposus*, a new filobasidiaceous yeast species found in the Prioksko-terrasny biosphere reserve in Russia. Mycol Res 110:957–961

Gostinčar C, Grube M, de Hoog S, Zalar P, Gunde-Cimerman N (2010) Extremotolerance in fungi: evolution on the edge. FEMS Microbiol Ecol 71:2–11

Gostinčar C, Ohm RA, Kogej T, Sonjak S, Turk M, Zajc J, Zalar P, Grube M, Sun H, Han J, Sharma A, Chiniquy J, Ngan CY, Lipzen A, Barry K, Grigoriev IV, Gunde-Cimerman N (2014) Genome sequencing of four *Aureobasidium pullulans* varieties: biotechnological potential, stress tolerance, and description of new species. BMC Genomics 15:549

Hardoim PR, van Overbeek LS, Elsas JD (2008) Properties of bacterial endophytes and their proposed role in plant growth. Trends Microbiol 16:463–471

Hu J, Ji L (2016) Draft genome sequences of *Rhodosporidium toruloides* strains ATCC 10788 and ATCC 10657 with compatible mating types. Genome Announc 4:e00098-16

Inácio J, Pereira P, Carvalho M, Fonseca Á, Amaral-Collaço MT, Spencer-Martins I (2002) Estimation and diversity of phylloplane mycobiota on selected plants in a Mediterranean-type ecosystem in Portugal. Microb Ecol 44:344–353

Inácio J, Portugal L, Spencer-Martins I, Fonseca Á (2005) Phylloplane yeasts from Portugal: seven novel anamorphic species in the Tremellales lineage of the Hymenomycetes (Basidiomycota) producing orange-coloured colonies. FEMS Yeast Res 5:1167–1183

Inácio J, Ludwig W, Spencer-Martins I, Fonseca Á (2009) Assessment of phylloplane yeasts on selected Mediterranean plants by FISH with group- and species-specific oligonucleotide probes. FEMS Microbiol Ecol 71:61–72

Jumpponen A, Jones KL (2009) Massively parallel 454 sequencing indicates hyperdiverse fungal communities in temperate *Quercus macrocarpa* phyllosphere. New Phytol 184:438–448

Kachalkin AV (2010) New data on the distribution of certain psychrophilic yeasts in Moscow oblast. Microbiology 79:840–844

Kachalkin AV, Yurkov AM (2012) Yeast communities in *Sphagnum* phyllosphere along the temperature-moisture ecocline in the boreal forest-swamp ecosystem and description of *Candida sphagnicola* sp. nov. A van Leeuwenhoek 102:29–43

Kachalkin AV, Glushakova AM, Yurkov AM, Chernov IY (2008) Characterization of yeast groupings in the phyllosphere of *Sphagnum* mosses. Microbiology 77:474–481

Kahmann R, Kämper J (2004) *Ustilago maydis*: how its biology relates to pathogenic development. New Phytol 164:31–42

Kemler M, Garnas J, Wingfield MJ, Gryzenhout M, Pillay K-A, Slippers B (2013) Ion Torrent PGM as tool for fungal community analysis: a case study of endophytes in *Eucalyptus grandis* reveals high taxonomic diversity. PLoS One 8:e81718

Kim KW, Park EW, Ahn KK (1999) Pre-penetration behavior of *Botryosphaeria dothidea* on apple fruits. Plant Pathol J 15:223–227

Klassen R, Schaffrath R, Buzzini P, Ganter PF (2017) Antagonistic interactions and killer yeasts. In: Buzzini P, Lachance MA, Yurkov AM (eds) Yeasts in natural ecosystems: ecology. Springer, Heidelberg, pp 229–275

Kumar S, Kushwaha H, Bachhawat AK, Raghava GPS, Ganesan K (2012) Genome sequence of the oleaginous red yeast *Rhodosporidium toruloides* MTCC 457. Eukaryot Cell 11:1083–1084

Kurtzman CP, Robnett CJ (2015) *Occultifur kilbournensis* f.a. sp. nov., a new member of the Cystobasidiales associated with maize (*Zea mays*) cultivation. A van Leeuwenhoek 107:1323–1329

Kvasnikov EI, Nagornaia SS, Shchelokova IF (1975) Yeast flora of plant rhizosphere and phyllosphere. Microbiology 44:339–346

Landell MF, Inácio J, Fonseca A, Vainstein MH, Valente P (2009) *Cryptococcus bromeliarum* sp. nov., an orange-coloured basidiomycetous yeast isolated from bromeliads in Brazil. Int J Syst Evol Microbiol 59:910–913

Landell MF, Brandão LR, Barbosa AC, Ramos JP, Safar SV, Gomes FC, Sousa FM, Morais PB, Broetto L, Leoncini O, Ribeiro JR, Fungsin B, Takashima M, Nakase T, Lee CF, Vainstein MH, Fell JW, Scorzetti G, Vishniac HS, Rosa CA, Valente P (2014) *Hannaella pagnoccae* sp. nov., a tremellaceous yeast species isolated from plants and soil. Int J Syst Evol Microbiol 64:1970–1977

Libkind D, Moliné M, Sommaruga R, Sampaio JP, van Broock M (2011) Phylogenetic distribution of fungal mycosporines within the Pucciniomycotina (Basidiomycota). Yeast 28:619–627

Limtong S, Koowadjanakul N (2012) Yeasts from phylloplane and their capability to produce indole-3-acetic acid. World J Microbiol Biotechnol 28:3323–3335

Limtong S, Kaewwichian R, Yongmanitchai W, Kawasaki H (2014) Diversity of culturable yeasts in phylloplane of sugarcane in Thailand and their capability to produce indole-3-acetic acid. World J Microbiol Biotechnol 30:1785–1796

Liu XZ, Wang QM, Göker M, Groenewald M, Kachalkin AV, Lumbsch HT, Millanes AM, Wedin M, Yurkov AM, Boekhout T, Bai FY (2015) Towards an integrated phylogenetic classification of the Tremellomycetes. Stud Mycol 81:85–147

Maksimova IA, Chernov IY (2004) Community structure of yeast fungi in forest biogeocenoses. Microbiology 73:474–481

Mannazzu I, Landolfo S, da Silva TL, Buzzini P (2015) Red yeasts and carotenoid production: outlining a future for non-conventional yeasts of biotechnological interest. World J Microbiol Biotechnol 31:1665–1673

Men'ko EV, Chernov IY, Byzov BA (2006) Interrelationships between yeast fungi and collembolans in soil. Microbiology 75:708–715

Mittelbach M, Vannette RL (2017) Mutualism in yeasts. In: Buzzini P, Lachance MA, Yurkov AM (eds) Yeasts in natural ecosystems: ecology. Springer, Heidelberg, pp 155–178

Montenegro G, Portaluppi MC, Salas FA, Díaz MF (2009) Biological properties of the Chilean native moss *Sphagnum magellanicum*. Biol Res 42:233–237

Morin N, Calcas X, Devillers H, Durrens P, Sherman DJ, Nicaud J-M, Neuvéglise C (2014) Draft genome sequence of *Rhodosporidium toruloides* CECT1137, an oleaginous yeast of biotechnological interest. Genome Announc 2:e00641-14

Nassar AH, El-Tarabily KA, Sivasithamparam K (2005) Promotion of plant growth by an auxin-producing isolate of the yeast *Williopsis saturnus* endophytic in maize (*Zea mays* L.) roots. Biol Fertil Soils 42:97–108

Nemčovič M, Jakubíková L, Víden I, Farkaš V (2008) Induction of conidiation by endogenous volatile compounds in *Trichoderma* spp. FEMS Microbiol Lett 284:231–236

Nix-Stohr S, Burpee LL, Buck JW (2008) The influence of exogenous nutrients on the abundance of yeasts on the phylloplane of turfgrass. Microb Ecol 55:15–20

Nutaratat P, Srisuk N, Arunrattiyakorn P, Limtong S (2014) Plant growth-promoting traits of epiphytic and endophytic yeasts isolated from rice and sugar cane leaves in Thailand. Fungal Biol 118:683–694

Nutaratat P, Amsri W, Srisuk N, Arunrattiyakorn P, Limtong S (2015) Indole-3-acetic acid production by newly isolated red yeast *Rhodosporidium paludigenum*. J Gen Appl Microbiol 61:1–9

Oechel WC, Van Cleve K (1986) The role of bryophytes in nutrient cycling in the Taiga. In: Van Cleve K, Chapin FS, Flanagan PW, Viereck LA, Dyrness CT (eds) Forest ecosystems in the Alaskan Taiga. Springer, New York, pp 121–137

Painter TJ (1998) Carbohydrate polymers in food preservation: an integrated view of the Maillard reaction with special reference to discoveries of preserved foods in *Sphagnum*-dominated peat bogs. Carbohydr Polym 36:335–347

Paul D, Magbanua Z, Arick M, French T, Bridges SM, Burgess SC, Lawrence ML (2014) Genome sequence of the oleaginous yeast *Rhodotorula glutinis* ATCC 204091. Genome Announc 13: e00046-14

Rasmussen S, Wolff C, Rudolph H (1995) Compartmentalization of phenolic constituents in *Sphagnum*. Phytochemistry 38:35–39

Redecker D, Kodner R, Graham LE (2000) Glomalean fungi from the Ordovician. Science 289:1920–1921

Reineke G, Heinze B, Schirawski J, Buettner H, Kahmann R, Basse CW (2008) Indole-3-acetic acid (IAA) biosynthesis in the smut fungus *Ustilago maydis* and its relevance for increased IAA levels in infected tissue and host tumour formation. Mol Plant Pathol 9:339–355

Rodriguez RJ, White JF, Arnold AE, Redman RS (2009) Fungal endophytes: diversity and functional roles. New Phytol 182:314–330

Rydin H, Gunnarsson U, Sundberg S (2006) The role of *Sphagnum* in peatland development and persistence. In: Wieder RK, Vitt DH (eds) Boreal peatland ecosystems. Springer, Berlin, pp 47–65

Santiago IF, Rosa CA, Rosa LH (2017) Endophytic symbiont yeasts associated with the Antarctic angiosperms *Deschampsia antarctica* and *Colobanthus quitensis*. Polar Biol 40:177–183

Scarpella E, Barkoulas M, Tsiantis M (2010) Control of leaf and vein development by auxin. Cold Spring Harb Perspect Biol 2:a001511

Shetty NP, Jørgensen HJL, Jensen JD, Collinge DB, Shetty HS (2008) Roles of reactive oxygen species in interactions between plants and pathogens. Eur J Plant Pathol 121:267–280

Solis MJL, Yurkov A, dela Cruz TE, Unterseher M (2014) Leaf-inhabiting endophytic yeasts are abundant but unevenly distributed in three *Ficus* species from botanical garden greenhouses in Germany. Mycol Prog 14:1019

Spaepen S, Vanderleyden J, Remans R (2007) Indole-3-acetic acid in microbial and microorganism-plant signaling. FEMS Microbiol Rev 31:425–448

Streletskii RA, Kachalkin AV, Glushakova AM, Demin VV, Chernov IY (2016) Quantitative determination of indole-3-acetic acid in yeasts using high performance liquid chromatography – tandem mass spectrometry. Microbiology 85:727–736

Sun PF, Fang WT, Shin LY, Wei JY, Fu SF, Chou JY (2014) Indole-3-acetic acid-producing yeasts in the phyllosphere of the carnivorous plant *Drosera indica* L. PLoS One 9:e114196

Takashima M, Sugita T, Van BH, Nakamura M, Endoh R, Ohkuma M (2012) Taxonomic richness of yeasts in Japan within subtropical and cool temperate areas. PLoS One 7:e50784

Teale WD, Paponov IA, Palme K (2006) Auxin in action: signalling, transport and the control of plant growth and development. Nat Rev Mol Cell Biol 7:847–859

Tedersoo L, Anslan S, Bahram M, Põlme S, Riit T, Liiv I, Kõljalg U, Kisand V, Nilsson H, Hildebrand F, Bork P, Abarenkov K (2015) Shotgun metagenomes and multiple primer pair-barcode combinations of amplicons reveal biases in metabarcoding analyses of fungi. MycoKeys 10:1–43

Thormann MN, Rice AV, Beilman DW (2007) Yeasts in peatlands: a review of richness and roles in peat decomposition. Wetlands 27:761–773

Tian H, De Smet I, Ding Z (2014) Shaping a root system: regulating lateral versus primary root growth. Trends Plant Sci 19:426–431

Tsai IJ, Tanaka E, Masuya H, Tanaka R, Hirooka Y, Endoh R, Sahashi N, Kikuchi T (2014) Comparative genomics of *Taphrin*a fungi causing varying degrees of tumorous deformity in plants. Genome Biol Evol 6:861–872

Turetsky MR (2003) The role of bryophytes in carbon and nitrogen cycling. Bryologist 106:395–409

Turner NC (1973) Action of fusicoccin on the potassium balance of guard cells of *Phaseolus vulgaris*. Am J Bot 60(7):717–725

Unterseher M, Schnittler M (2010) Species richness analysis and ITS rDNA phylogeny revealed the majority of cultivable foliar endophytes from beech *Fagus sylvatica*. Fungal Ecol 3:366–378

van Breemen N (1995) How *Sphagnum* bogs down other plants. Trends Ecol Evol 10:270–275

Vasileva-Tonkova E, Romanovskaya V, Gladka G, Gouliamova D, Tomova I, Stoilova-Disheva-M, Tashyrev O (2014) Ecophysiological properties of cultivable heterotrophic bacteria and yeasts dominating in phytocenoses of Galindez Island, Maritime Antarctica. World J Microbiol Biotechnol 30:1387–1398

Viret O, Petrini O (1994) Colonization of beech leaves (*Fagus sylvatica*) by the endophyte *Discula umbrinella* (teleomorph: *Apiognomonia errabunda*). Mycol Res 98:423–432

Vishniac HS (2006) Yeast biodiversity in the Antarctic. In: Péter G, Rosa C (eds) Biodiversity and ecophysiology of yeasts. Springer, Berlin, pp 419–440

Vishniac HS, Anderson JA, Filonow AB (1997) Assimilation of volatiles from ripe apples by *Sporidiobolus salmonicolor* and *Tilletiopsis washingtonensis*. A van Leeuwenhoek 72:201–207

Wang QM, Boekhout T, Bai FY (2012) *Bensingtonia rectispora* sp. nov. and *Bensingtonia bomiensis* sp. nov., ballistoconidium-forming yeast species from Tibetan plant leaves. Int J Syst Evol Microbiol 62:2039–2044

Wang QM, Groenewald M, Takashima M, Theelen B, Han P-J, Liu X-Z, Boekhout T, Bai F-Y (2015a) Phylogeny of yeasts and related filamentous fungi within Pucciniomycotina determined from multigene sequence analyses. Stud Mycol 81:27–53

Wang QM, Yurkov AM, Göker M, Lumbsch HT, Leavitt SD, Groenewald M, Theelen B, Liu XZ, Boekhout T, Bai FY (2015b) Phylogenetic classification of yeasts and related taxa within Pucciniomycotina. Stud Mycol 81:149–189

Wang QM, Begerow D, Groenewald M, Liu XZ, Theelen B, Bai FY, Boekhout T (2015c) Multigene phylogeny and taxonomic revision of yeasts and related fungi in the Ustilaginomycotina. Stud Mycol 81:55–83

Whipps JM, Hand P, Pink D, Bending GD (2008) Phyllosphere microbiology with special reference to diversity and plant genotype. J Appl Microbiol 105:1744–1755

Womack AM, Artaxo PE, Ishida FY, Mueller RC, Saleska SR, Wiedemann KT, Bohannan BJM, Green JL (2015) Characterization of active and total fungal communities in the atmosphere over the Amazon rainforest. Biogeosciences 12:6337–6349

Wuczkowski M, Metzger E, Sterflinger K, Prillinger H (2005) Diversity of yeasts isolated from litter and soil of different natural forest sites in Austria. Die Bodenkultur 56:201–2008

Xin G, Glawe D, Doty SL (2009) Characterization of three endophytic, indole-3-acetic acid-producing yeasts occurring in *Populus* trees. Mycol Res 113:973–980

Yurkov AM (2017) Temporal and geographic patterns in yeast distribution. In: Buzzini P, Lachance MA, Yurkov AM (eds) Yeasts in natural ecosystems: ecology. Springer, Heidelberg, pp 101–130

Yurkov AM, Maksimova IA, Chernov IY (2004) The comparative analysis of yeast communities in birch forests of the European part of Russia and Western Siberia. Mikol Fitopatol 38:71–79

Yurkov AM, Vustin MM, Tyaglov BV, Maksimova IA, Sineokiy SP (2008) Pigmented basidiomycetous yeasts are a promising source of carotenoids and ubiquinone Q10. Microbiology 77:1–6

Yurkov AM, Krüger D, Begerow D, Arnold N, Tarkka MT (2012) Basidiomycetous yeasts from Boletales fruiting bodies and their interactions with the mycoparasite *Sepedonium chrysospermum* and the host fungus *Paxillus*. Microb Ecol 63:295–303

Yurkov A, Guerreiro MA, Sharma L, Carvalho C, Fonseca Á (2015a) Multigene assessment of the species boundaries and sexual status of the basidiomycetous yeasts *Cryptococcus flavescens* and *C. terrestris* (Tremellales). PLoS One 10:e0120400

Yurkov AM, Kachalkin AV, Daniel HM, Groenewald M, Libkind D, de Garcia V, Zalar P, Gouliamova DE, Boekhout T, Begerow D (2015b) Two yeast species *Cystobasidium psychroaquaticum* f.a. sp. nov. and *Cystobasidium rietchieii* f.a. sp. nov. isolated from natural environments, and the transfer of *Rhodotorula minuta* clade members to the genus *Cystobasidium*. A van Leeuwenhoek 107:173–185

Yurkov AM, Inácio J, Chernov IY, Fonseca Á (2015c) Yeast biogeography and the effects of species recognition approaches: the case study of widespread basidiomycetous species from birch forests in Russia. Curr Microbiol 70:587–601

Zhao Z, Liu H, Wang C, J-R X (2013) Comparative analysis of fungal genomes reveals different plant cell wall degrading capacity in fungi. BMC Genomics 14:274

Zhu Z, Zhang S, Liu H, Shen H, Lin X, Yang F, Zhou YJ, Jin G, Ye M, Zou H, Zou H, Zhao ZK (2012) A multi-omic map of the lipid-producing yeast *Rhodosporidium toruloides*. Nat Commun 3:1112

Chapter 7
Phylloplane Yeasts in Tropical Climates

Savitree Limtong and Rujikan Nasanit

Abstract Phylloplane, which refers to the surfaces of aboveground parts of plants, includes mainly leaves and has been recognized as an important habitat for microorganisms. The growth of phylloplane microorganisms is dependent on organic and inorganic substances on the leaf surfaces that are either secreted by the plant or originate from external sources. Several factors that structure the microbial communities in phylloplane such as leaf age, plant species, growing conditions, environmental factors, geography and competing microorganisms have been reported. To date, the diversity of microorganisms in the phylloplane has been studied intensively. However, only a small number of articles have focused on yeasts. In this chapter, we review the methodologies, both culture-dependent and culture-independent methods, which have been utilized for assessment of leaf-associated yeasts. The diversity of phylloplane yeasts in the tropical regions assessed by those techniques is discussed. In the course of investigations, phylloplane yeasts in tropical countries have been shown to be highly diverse. Many novel yeast species have been discovered by both culture-dependent and culture-independent methods. These studies suggest that extensive studies of phylloplane yeasts from tropical regions will lead to recovering a so far underappreciated diversity.

Keywords Phylloplane • Yeasts • Tropical country

Contents

S. Limtong (✉)
Department of Microbiology, Faculty of Science, Kasetsart University, Jatujak, Bangkok 10900, Thailand
e-mail: fscistl@ku.ac.th

R. Nasanit
Department of Biotechnology, Faculty of Engineering and Industrial Technology, Silpakorn University, Sanamchandra Palace Campus, Amphoe Muang, Nakhon Pathom 73000, Thailand

P. Buzzini et al. (eds.), *Yeasts in Natural Ecosystems: Diversity*,
DOI 10.1007/978-3-319-62683-3_7

7.1 Introduction

The aboveground parts of plants are usually referred to as phylloplane or phyllosphere (Phaff and Starmer 1987; Fonseca and Inácio 2006). In their review on phylloplane yeasts, Fonseca and Inácio (2006) used the term phylloplane, which was proposed by Last and Price (1969) to describe leaf surfaces. The term phyllosphere also includes additionally internal leaf tissues (reviewed by Fonseca and Inácio 2006). The aboveground parts include mainly leaves although surfaces of stems, flowers and fruits are sometimes also considered phylloplane. The phylloplane has been recognized as an important habitat for epiphytic microorganisms, including bacteria, yeasts and fungi that are capable of surviving, growing and reproducing in this constantly changing environment (Phaff and Starmer 1987; Andrews and Harris 2000; Lindow and Brandl 2003; Fonseca and Inácio 2006). The growth of microorganisms in the phylloplane depends mainly on plant metabolites that are secreted on the leaf surface or materials from external sources that are deposited on the leaf surface. Plant exudates are mainly simple sugars (e.g. glucose, fructose and sucrose), although amino acids, organic acids and sugar alcohols are also released (Fiala et al. 1900; Xin et al. 2009). In contrast, external sources often provide inorganic nutrients (Xin et al. 2009).

Epiphytic microorganisms have been reported to actively participate in the release of nutrients by their host plant. For instance, they can produce plant hormones, which promote cell wall lysis and release of mono- and oligosaccharides from plant cells (Fry 1989; Brandl and Lindow 1998; Lindow and Brandl 2003). Furthermore, it has been reported that epiphytes may produce biosurfactants facilitating wettability, thereby further enhancing leaching of organic substances (Bunster et al. 1989; Schreiber et al. 2005). The availability of nutrients in the phylloplane, which partly shape microbial communities, is dependent on leaf age, plant species and growing conditions. Additionally, it has been reported that environmental factors (e.g. climate condition, seasonal fluctuation, radiation, pollution and nitrogen fertilization) and competing microorganisms further affect the phyllosphere, including phylloplane, microorganisms (Irvine et al. 1978; Maksimova et al. 2009; Glushakova and Chernov 2010; Jumpponen and Jones 2010; Kachalkin and Yurkov 2012; Zimmerman and Vitousek 2012; Voriskova et al. 2014; Yurkov et al. 2015b). Due to the many different factors involved, microorganisms in the phylloplane are often distributed irregularly (Inácio et al. 2005, 2010; Fonseca and Inácio 2006; Remus-Emsermann et al. 2012; Vorholt 2012). For instance, Finkel et al. (2011) demonstrated that microbial phylloplane communities are similar within the same location, whereas trees of the same species growing in different climate host distinct microbial communities.

Tropical regions contain important phylloplane habitat, as, for instance, the leaf surface of tropical rain forests is nearly 140×10^6 km^2 (Morris and Kinkel 2002). Yeasts form an important part of microbial communities in the phylloplane of tropical regions. Therefore, in this chapter, we review their diversity and importance, whereas phylloplane yeast diversity in temperate regions is reviewed in Chap. 6 of this book.

7.2 Culture-Dependent Methods for Assessment of Phylloplane Yeasts

The diversity of yeasts in their natural habitats is assessed with both culture-dependent and culture-independent methods. Culture-dependent methods are based on isolation and purification of yeasts using appropriate culture medium and conditions. Subsequently, pure cultures are identified based on morphological, physiological or molecular characteristics. Most of the current knowledge of yeast communities in the phylloplane has been obtained using culture-dependent methods.

Yeasts from the phylloplane have been isolated using different methods, including the spore-fall method (Nakase and Takashima 1993; Nakase et al. 2001; Wang et al. 2011; Liu et al. 2012; Toome et al. 2013; reviewed by Boundy-Mills 2006), plating of leaf washings (de Azeredo et al. 1998; Fonseca and Inácio 2006; Slavikova et al. 2007; Inácio et al. 2008; Kurtzman et al. 2011a; Kachalkin and Yurkov 2012; Surussawadee et al. 2014) and enrichment cultures (Limtong and Koowadjanakul 2012; Kaewwichian et al. 2013a; Limtong et al. 2014; Limtong and Kaewwichian 2015). The spore-fall method is often used to isolate ballistoconidia-forming phylloplane yeasts. Thereby, leaves are cut into small pieces and attached to the inner lid of a Petri dish containing an appropriate culture medium (e.g. Toome et al. 2013). Petri dishes are inspected every day by the eye for the presence of yeast colonies. Yeast isolation by plating of leaf washings is the most commonly used method to study phylloplane yeasts. Yeast cells are detached from leaf surfaces using washing solutions such as water (Fonseca and Inácio 2006; Slavikova et al. 2007; Kachalkin and Yurkov 2012), saline solution (Surussawadee et al. 2014), saline solution with Triton X-100 (Weber et al. 2008) or Tween 40 (de Azeredo et al. 1998), Ringer's solution (Fonseca and Inácio 2006; Jager et al. 2001) and Tween 20 solution (Inácio et al. 2008). A leaf sample is aseptically put in a washing solution and shaken at an appropriate temperature for a certain period of time with the aid of rotary shaker (de Azeredo et al. 1998; Slavikova et al. 2007; Surussawadee et al. 2014, 2015), vortex (Kachalkin and Yurkov 2012) and sonicator (de Azeredo et al. 1998; Jager et al. 2001; Fonseca and Inácio 2006). An aliquot of the washing solution is spread on the surface of a culture medium such as malt extract agar (Weber et al. 2008), acidified malt extract agar (Slavikova et al. 2007), acidified glucose-peptone-yeast extract (acidified GPY) agar (Kachalkin and

Yurkov 2012), yeast extract-malt extract (YM) agar (de Azeredo et al. 1998; Surussawadee et al. 2014), acidified YM agar (Inácio et al. 2008; Landell et al. 2010), yeast nitrogen base with glucose (de Azeredo et al. 1998) and wort agar (Jager et al. 2001). Acidified media (reviewed by Boundy-Mills 2006) and media supplemented with antibiotics, such as chloramphenicol (Jager et al. 2001; Fonseca and Inácio 2006; Inácio et al. 2008; Landell et al. 2010; Surussawadee et al. 2014; reviewed by Boundy-Mills 2006), oxytetracycline (reviewed by Boundy-Mills 2006), streptomycin (Slavikova et al. 2007), chloramphenicol with rose bengal, streptomycin sulfate with penicillin (Weber et al. 2008) or streptomycin with benzylpenicillin (Golubev and Sampaio 2009) are used in order to inhibit bacterial growth. Compounds to inhibit fungal growth are also sometimes added to the medium (e.g. calcium propionate, sodium propionate rose bengal, biphenyl, eugenol, oligomycin, ox gall; Surussawadee et al. 2014, 2015; reviewed by Boundy-Mills 2006). Incubation at low temperature has been used to delay fungal growth (reviewed by Fonseca and Inácio 2006). The enrichment method is employed for selective isolation of specific yeast groups, for instance, ascomycetous methanol-assimilating yeasts (Dlauchy et al. 2003; Péter et al. 2007; Koowadjanakul et al. 2011; Limtong et al. 2013). The isolation of methylotrophic yeasts, a group of yeast that utilize methanol as the sole source of carbon and energy (Ogata et al. 1969; Kato et al. 1974), often includes two to three consecutive enrichments of leaf samples in methanol-yeast nitrogen base (YNB) broth in an Erlenmeyer flask on a rotary shaker. Enriched cultures are streaked on 0.5% v/v methanol-YNB agar and incubated at room temperature until yeast colonies appear (Koowadjanakul et al. 2011; Limtong et al. 2013). Methylotrophic yeasts can also be isolated using a serial dilution of methanol enrichment culture and plating on rose bengal agar containing chloramphenicol (Péter et al. 2007). Enrichment methods have been successfully used to assess yeast diversity in the phylloplane in several studies performed in Thailand (Limtong and Koowadjanakul 2012; Kaewwichian et al. 2013a; Limtong et al. 2014; Limtong and Kaewwichian 2015). These studies utilized yeast extract-malt extract (YM) broth as the enrichment medium and incubation with a rotary shaker at 30 ± 2 °C for 2 days. Subsequently, an inoculating loop full of the enriched culture was streaked on YM agar supplemented with 250 mg/l sodium propionate and 200 mg/l chloramphenicol.

Phenotypic characteristics (i.e. morphological features and selected physiological tests) to identify phylloplane yeasts following taxonomic keys were utilized in older studies (de Azeredo et al. 1998). However, in recent years, most of yeast identification has been based on molecular taxonomy and phylogenetic analyses. The molecular identification has mainly been performed by analysing the D1/D2 region in the large subunit (LSU) of the ribosomal RNA gene sequence (Kurtzman and Robnett 1998; Fell et al. 2000; Limtong and Koowadjanakul 2012; Limtong et al. 2014; Limtong and Kaewwichian 2015). The internal transcribed spacer (ITS) region, which is the standard barcode for the kingdom Fungi (Schocha et al. 2012), is another gene sequence used for rapid yeast identification (Hamamoto et al. 2002; Leaw et al. 2006; Kurtzman 2014). Sequences of the ITS region generally result in similar species resolution as those of the D1/D2 region (Kurtzman 2014). In both ascomycetous and basidiomycetous yeasts, strains of a species in general differ in the sequences of the D1/D2 and the ITS regions by no more than 1% sequence similarity (Kurtzman and Robnett 1998; Fell et al. 2000; Scorzetti et al. 2002; Sugita et al. 2002). However, in some instances, the resolution provided by the ITS region is superior to the D1/D2 region or vice versa (Kurtzman et al. 2011b). The intergenic spacer (IGS) is another rRNA gene that has been used for identification of some closely related species (Diaz et al. 2000; Sugita et al. 2002; Diaz and Fell 2005; Fell et al. 2007; Kurtzman et al. 2011b). When the sequence of the D1/D2 region does not provide a clear separation of closely related species, identification has been performed by sequence analyses of protein-coding genes (Kurtzman et al. 2011a, b), such as actin (ACT; Daniel and Meyer 2003), translation elongation factor-1α (TEF1α; Kurtzman et al. 2008) or cytochrome oxidase II (COX II; Belloch et al. 2000; Kurtzman and Robnett 2003). In some case, analyses of multiple gene sequences are required (Kurtzman and Robnett 2003; Kurtzman et al. 2008; Sugita et al. 2001; Yurkov et al. 2015a).

7.3 Yeasts in the Tropical Phylloplane by Culture-Dependent Methods

Bacteria are the most abundant phylloplane microorganisms; however, yeasts and yeast-like fungi are also active phylloplane colonizers (Andrews and Harris 2000; Lindow and Brandl 2003). The phylloplanes of dicotyledonous and monocotyledonous plants in temperate and tropical regions are colonized by both basidiomycetous and ascomycetous yeasts (Nakase and Suzuki 1985; de Azeredo et al. 1998; Nakase et al. 2001; Inácio et al. 2005; Fonseca and Inácio 2006; Slavikova et al. 2007; Glushakova and Chernov 2010; Landell et al. 2010). In tropical regions, only limited information is available on phylloplane yeast diversity. Most knowledge is based on culture-dependent techniques (Nakase and Suzuki 1985; de Azeredo et al. 1998; Nakase et al. 2001; Limtong and Koowadjanakul 2012; Limtong et al. 2014; Limtong and Kaewwichian 2015), with only few culture-independent surveys (see paragraphs 7.4 and 7.5).

Yeast communities associated with sugarcane leaves during different phases of plant development were studied near Campos, in Rio de Janeiro, Brazil, by de Azeredo et al. (1998). Yeasts were isolated by plating of leaf washings on yeast nitrogen base agar with 0.5% glucose and on yeast extract-malt extract agar, both supplemented with chloramphenicol. However, no significant differences in yeast counts between the two culture media could be found. Yeast cultures in this study were identified using phenotypic characteristics. A total of 103 strains were obtained from 23 leaf samples. Most of the yeast species belonged to Basidiomycota (95 strains, 92.2%), whereas Ascomycota were rare (8 strains, 7.8%). Species in Basidiomycota, many of which have been transferred to other genera since then, comprised *Naganishia albida* (formerly *Cryptococcus albidus*),

Vanrija humicola (*Cryptococcus humiculus*), *Papiliotrema* (*Cryptococcus*) *laurentii*, *Cystofilobasidium* (*Cryptococcus*) *macerans*, *Cystofilobasidium infirmominiatum*, *Cystobasidium minutum* (*Rhodotorula minuta*), *Dioszegia hungarica* (*Cryptococcus hungaricus*), *Rhodotorula* (*Rhodosporidium*) *toruloides*, *Rhodotorula glutinis*, *Rhodotorula mucilaginosa*, *Sporobolomyces roseus*, *Apiotrichum* (*Trichosporon*) *dulcitum*-like, *Cutaneotrichosporon* (*Trichosporon*) *cutaneum*-like, *Fellomyces horovitziae*-like, *Sporidiobolus pararoseus*-like, *Tausonia* (*Trichosporon*) *pullulans*-like, and *Tremella aurantia*-like. Species in Ascomycota were *Candida azyma*, *Debaryomyces hansenii*, *Saccharomyces cerevisiae* and *Zyogoascus hellenicus*.

The spore-fall method was used to isolate ballistoconidia-forming yeasts from dead leaves and stems of rice in Japan, as well as from rice and various other plant species in Thailand (Nakase et al. 2001). In this study, 136 ballistoconidia-forming yeast strains were obtained from plant samples collected in Thailand in 1987 and 1990. They were identified on the basis of morphological, physiological and biochemical characteristics. When conventional identification was doubtful, chemotaxonomic, molecular phylogenetic and DNA-DNA hybridization techniques were employed. The study yielded 21 yeast species, viz. *Bullera alba*, *Bullera penniseticola*, *Bulleribasidium variabile* (*Bullera variabilis*), *Dioszegia* (*Bullera*) *crocea*, *Hannaella* (*Bullera*) *sinensis*, *Gjaerumia* (*Tilletiopsis*) *penniseti*, *Kockovaella imperatae*, *Kockovaella sacchari*, *Kockovaella thailandica*, *Kondoa* (*Bensingtonia*) *thailandica*, *Phragmotaenium derxii* (*Tilletiopsis derxii*), *Phragmotaenium flavum* (*Tilletiopsis flava*), *Phragmotaenium* (*Tilletiopsis*) *oryzicola*, *Rhodosporidiobolus* (*Sporidiobolus*) *ruineniae*, *Rhodosporidiobolus* (*Sporobolomyces*) *nylandii*, *Rhodosporidiobolus* (*Sporobolomyces*) *poonsookiae*, *Sporobolomyces blumeae*, *Sporobolomyces japonicus* (formerly *Sporobolomyces shibatanus*), *Sporobolomyces salmonicolor* and *Symmetrospora vermiculata* (*Sporobolomyces vermiculatus*).

Investigation of phylloplane yeasts of diverse plant species in Thailand also followed the enrichment isolation method combined with molecular identification on the basis of the D1/D2 region of the LSU rRNA gene sequences (Limtong and Koowadjanakul 2012). A total of 114 yeast strains and 10 strains of yeast-like fungi were obtained from 91 (out of 97) leaf samples. Ninety-eight strains were identified to 37 known yeast species in 17 genera belonging to Ascomycota, viz. *Candida* (*amphixiae*) *amphicis*, *Candida apicola*, *Candida etchellsii*, *Candida glabrata*, *Candida jaroonii*, *Candida maltosa*, *Candida metapsilosis*, *Candida morakotiae*, *Candida nivariensis*, *Candida parapsilosis*, *Candida potacharoeniae*, *Candida sorboxylosa*, *Candida stigmatis*, *Candida tropicalis*, *Candida trypodendroni*, *Clavispora* (*Candida*) *lusitaniae*, *Cyberlindnera rhodanensis*, *Debaryomyces nepalensis*, *Diutina* (*Candida*) *rugosa*, *Hanseniaspora guilliermondii*, *Hanseniaspora opuntiae*, *Hanseniaspora thailandica*, *Hyphopichia burtoni*, *Kazachstania siamensis*, *Kluyveromyces marxianus*, *Kodamaea ohmeri*, *Lachancea thermotolerans*, *Metschnikowia koreensis*, *Pichia kudriavzevii*, *Pichia manshurica*, *Starmerella meliponinorum*, *Torulaspora delbrueckii*, *Torulaspora pretoriensis* and *Wickerhamomyces edaphicus*, and to Basidiomycota, viz. *Rhod. ruineniae* and *Trichosporon asahii*. Three strains represented the two novel species *Candida sirachaensis* and *Candida sakaeoensis* (Limtong et al. 2012). Ten strains of yeast-like fungi were identified as *Aureobasidium pullulans*. Species in the phylum Ascomycota accounted altogether for 98% of the 98 strains. The most frequent species was *C. tropicalis* with an isolation frequency of 14%.

The diversity of cultivable yeasts in the phylloplane of sugarcane in Thailand was reported by Limtong et al. (2014). Of the 158 strains identified from 94 sugar-cane leaf samples in this study, 144 strains belonged to 24 known species in the Ascomycota, viz. *Candida akabanensis*, *Candida dendronema*, *Candida mesorugosa*, *Candida michaelii*, *C. nivariensis*, *Candida orthopsilosis*, *Candida quercitrusa*, *C. tropicalis*, *Cyberlindnera fabianii*, *Cyb. rhodanensis*, *Deb. nepalensis*, *Diut. rugosa*, *Hannaella* aff. *coprosmae* (*coprosmaensis*), *H'spora guilliermondii*, *K. marxianus*, *Lach. thermotolerans*, *Lodderomyces elongisporus*, *M. koreensis*, *Meyerozyma caribbica*, *Millerozyma koratensis*, *P. kudriavzevii*, *Suhomyces* (*Candida*) *xylopsoci*, *T'spora delbrueckii* and *W. edaphicus*, and 12 species in the Basidiomycota, viz. *Filobasidium magnum* (*Pa. laurentii*), *Kwoniella heveanensis*, *Papiliotrema* (*Cryptococcus*) *flavescens*, *Papiliotrema* (*Cryptococcus*) *rajasthanensis*, *Rhodosporidiobolus fluvialis* (*Rhodosporidium fluviale*), *Rhod. nylandii*, *Rhod. ruineniae*, *Rhodotorula paludigena* (*Rhodosporidium paludigenum*), *Rh. mucilaginosa*, *Rhodotorula sesimbrana*, *Rhodotorula taiwanensis* and *Sporobolomyces carnicolor*. An additional seven strains were identical or similar to four yet undescribed species from public databases, and seven other strains represented four novel species. A total of 69% of the isolated strains were ascomycetous yeasts, and 31% were basidiomycetous yeast. The most prevalent species was *Mey. caribbica* (23% frequency of occurrence) followed by *Rh. taiwanensis* (11%) and *C. tropicalis* (10%).

Limtong and Kaewwichian (2015) reported the diversity of cultivable yeasts in the phylloplane of rice in Thailand. In this investigation, 156 strains from 85 leaf samples were identified as 25 species in 13 genera in the phylum Ascomycota and 9 species in 5 genera in the phylum Basidiomycota, respectively. The species in the phylum Ascomycota were *C. glabrata*, *C. jaroonii*, *Candida membranifaciens*, *Candida pseudolambica*, *Candida ruelliae*, *Candida entomaea* (*Candida terebra*), *C. tropicalis*, *Cl. lusitaniae*, *Cyb. fabianii*, *Cyb. rhodanensis*, *Deb. nepalensis*, *Diutina catenulata*, *Diut. rugosa*, *Kod. ohmeri*, *Lach. thermotolerans*, *M. koreensis*, *Metschnikowia lopburiensis*, *Mey. caribbica*, *Meyerozyma guilliermondii*, *Mill. koratensis*, *P. kudriavzevii*, *Wickerhamomyces ciferrii*, *Yamadazyma mexicana* (*Yamadazyma mexicanum*) and *Yarrowia lipolytica*. The species placed in Basidiomycota were *Pa. flavescens*, *F. magnum*, *Pa. rajasthanensis*, *Moesziomyces antarcticus* (*Pseudozyma antarctica*), *Moesziomyces* (*Pseudozyma*) *aphidis*, *Rh. taiwanensis*, *Sp. blumeae* and *Tr. asahii*. The most prevalent species was *Rh. taiwanensis* (23% frequency of occurrence) followed by *C. tropicalis* (16%). Thereby, 64% of the strains obtained by the enrichment isolation technique at 30 ± 2 °C were ascomycete yeasts, and only 36% represented basidiomycetous yeasts.

Comparison of the prevalent yeast species in the rice phylloplane with the prevalent species in the sugarcane phylloplane revealed that *Rh. taiwanensis*, *C. tropicalis*, *Cyb. fabianii*, *Cyb. rhodanensis* and *Mey. caribbica* were common species found in the phylloplane of both plant species. Many yeast species found in the rice phylloplane had also been detected in the phylloplane of diverse other plant species in Thailand. These included *C. glabrata*, *C. jaroonii*, *Diut. rugosa*, *Cl. lusitaniae*, *Pa. flavescens*, *Pa. rajasthanensis*, *Deb. nepalensis*, *Kod. ohmeri*, *M. koreensis*, *Mey. guilliermondii*, *Mill. koratensis*, *P. kudriavzevii*, *Sp. blumeae* and *Tr. asahii* (Nakase et al. 2001; Limtong and Koowadjanakul 2012; Limtong et al. 2014; Limtong and Kaewwichian 2015). The distribution of the above-mentioned species on many different host plants therefore might be an indication that plant species have little control on the occurrence of yeast species in the phylloplane.

In the last few years, we have investigated yeasts in the sugarcane phylloplane in Thailand by using dilution plating of leaf washings and not enrichment techniques as in the aforementioned studies. We obtained and identified molecular strains and found that 87 strains (50%) represented known yeast species, including 61 strains (35%) of basidiomycetous and 26 strains (15%) of ascomycetous yeasts, respectively. Interestingly, the majority of taxa are potentially new species (Limtong et al. 2014). In the course of investigations of phylloplane yeast in tropical countries, several novel yeast species were discovered and described from Thailand, as showed in Table 7.1.

The results of the investigations of phylloplane yeast diversity by using enrichment isolation are in contrast with the investigations that used other isolation methods. Whereas using enrichment protocols resulted in most species being Ascomycota, basidiomycetous yeasts dominated in the phylloplane of both tropical and temperate climates using other isolation methods (de Azeredo et al. 1998; Nakase et al. 2001; Fonseca and Inácio 2006; Slavikova et al. 2007; Glushakova and Chernov 2010). In our studies, the ascomycetous yeasts were most often isolated from rice (Limtong and Kaewwichian 2015) and sugarcane phylloplane (Limtong et al. 2014) in Thailand when we applied enrichment techniques. Contrary, plating of leaf washings obtained from sugarcane leaves in Brazil (de Azeredo et al. 1998) and from rice, sugarcane and corn leaves in Thailand (Limtong et al. 2014) showed that the prevalent species were basidiomycetous yeasts. These results indicate that the enrichment technique we applied might support the growth of small populations of ascomycetous yeasts leading to a greater representation in the results. The difference in the results may be influenced by oxygen requirements and cultivation temperature employed for isolation. Many basidiomycetous yeasts do not grow well at elevated temperatures used in enrichment cultivation technique. Our studies using the enrichment method however were performed at a culturing temperature of 30 ± 2 °C, where we recovered mostly Ascomycota. In contrast in many investigations using leaf washings, which resulted in mainly Basidiomycota, culture plates were incubated at 25 °C or even lower temperatures (de Azeredo et al. 1998; Maksimova and Chernov 2004; Slavikova et al. 2007), as in at least one instance the ballistoconidium-fall method has shown that more yeast species are recovered on YM plates incubated at 23 °C than at 30 °C (Nakase et al. 2001).

Table 7.1 Novel yeast species proposed based on the strains isolated from phylloplane in tropical regions

Species	Plant leaf	Location	Isolation method	Reference
Phylum Basidiomycota				
Bullera koratensis	*Lagerstroemia calyculata* and *Calamus viminalis*	Nakhon Ratchasima, Thailand	Ballistoconidium-fall	Fungsin et al. (2006)
Bullera lagerstroemia	*Lagerstroemia calyculata*	Nakhon Ratchasima, Thailand	Ballistoconidium-fall	Fungsin et al. (2006)
Bullera penniseticola	*Pennisetum* sp.	Bangkok, Thailand	Ballistoconidium-fall	Takashima and Nakase (1998)
Bullera sakaeratica	*Setaria pallide-fusca*	Nakhon Ratchasima, Thailand	Ballistoconidium-fall	Fungsin et al. (2003b)
Bullera siamensis	*Urena lobata* var. *sinuata*	Nakhon Ratchasima, Thailand	Ballistoconidium-fall	Fungsin et al. (2003a), Liu et al. (2015)
Bulleribasidium (*Bullera*) *panici*	*Panicum multinodes*	Nakhon Ratchasima, Thailand	Ballistoconidium-fall	Fungsin et al. (2003a), Liu et al. (2015)
Carcinomyces (*Bullera*) *arundinariae*	*Arundinaria pusilla*	Nakhon Ratchasima, Thailand	Ballistoconidium-fall	Fungsin et al. (2002b), Liu et al. (2015)
Carlosrosaea (*Bullera*) *vrieseae*	*Vriesea friburgensis* and *Tillandsia gardneri*	Brazil	Plating of leaf washings	Landell et al. (2015), Liu et al. (2015)
Farysia (*Farysizyma*) *itapuensis*	*Vriesea friburgensis* and *Dyckia* sp.	Rio Grande do Sul State, Brazil	Plating of leaf washings	Inacio et al. (2008), Wang et al. (2015b)
Genolevuria (*Cryptococcus*) *bromeliarum*	*Vriesea procera*, *Vriesea friburgensis* and *Tillandsia gardneri*	Rio Grande do Sul State, Brazil	Plating of leaf washings	Landell et al. (2009), Liu et al. (2015)
Gjaerumia (*Tilletiopsis*) *penniseti*	*Pennisetum pedicellatum*	Bangkok, Thailand	Ballistoconidium-fall	Takashima and Nakase (2001), Wang et al. (2015b)
Hannaella pagnoccae	*Vriesea procera*, *Vriesea friburgensis*, *Tillandsia gardneri*, *Saccharum* sp., *Arundinaria pusilla*, *Vitis vinifera* and *Cratoxylum maingayi*	Brazil, Thailand	Plating of leaf washings	Landell et al. (2014)

(continued)

Table 7.1 (continued)

Species	Plant leaf	Location	Isolation method	Reference
Hannaella phetchabunensis	Corn leaves (*Zea mays*)	Phetchabun, Thailand	Plating of leaf washings	Kaewwichian et al. (2014)
Hannaella phyllophila	Sugarcane (*Saccharum officinarum*), morning glory (*Merremia* sp.) and cassava (*Manihot esculenta*)	Lopburi, Nong Khai and Sakon Nakhon, Thailand	Plating of leaf washings	Surussawadee et al. (2015)
Hannaella siamensis	Rice leaves (*Oryza sativa*)	Nakhon Pathom, Suphan Buri and Nonthaburi, Thailand	Plating of leaf washings	Kaewwichian et al. (2014)
Kalmanozyma (*Pseudozyma*) *vetiver*	Vetiver grass (*Vetiveria zizanioides*)	Ratchaburi, Thailand	Plating of leaf washings	Chamnanpa et al. (2013), Wang et al. (2015b)
Holtermanniella takashimae	*Nothofagus pumilio*	Patagonia, Argentina	No data	Wuczkowski et al. (2011)
Kockovaella barringtoniae	*Barringtonia* sp.	Nakhon Ratchasima, Thailand	Ballistoconidium-fall	Fungsin et al. (2002a)
Kockovaella imperatae	Cogon (*Imperata cylindrica*)	Bangkok, Thailand	Ballistoconidium-fall	Nakase et al. (1991)
Kockovaella sacchari	Sugarcane (*Saccharum officinarum*)	Bangkok, Thailand	Ballistoconidium-fall	Takashima and Nakase (1998)
Kockovaella thailandica	Cogon (*Imperata cylindrica*), rice (*Oryza sativa*) and nadi blue grass (*Dichanthium caricosum*)	Bangkok and Nakhon Ratchasima, Thailand	Ballistoconidium-fall	Nakase et al. (1991)
Kondoa (*Bensingtonia*) *thailandica*	*Cratoxylum maingayi*, grape (*Vitis vinifera*), Burmese ironwood (*Xylia xylocarpa*), yellow bristle grass (*Setaria pallide-fusca*), *Dendrolobium triangulare*, fingerroot (*Boesenbergia pandurata*) and mission grass (*Pennisetum polystachyon*)	Nakhon Ratchasima, Thailand	Ballistoconidium-fall	Fungsin et al. (2001), Wang et al. (2015a)
Meredithblackwellia eburnea	Unidentified fern	Pakaraima Mountains, Western Guyana	Ballistoconidium-fall	Toome et al. (2013)

Occultifur tropicalis	Sugarcane (*Saccharum officinarum*)	Lopburi, Thailand	Plating of leaf washings	Khunnamwong et al. (2015)
Papiliotrema siamense	Sugarcane (*Saccharum officinarum*)	Chai Nat, Thailand	Plating of leaf washings	Surussawadee et al. (2014)
Phragmotaenium (Tilletiopsis) derxii	Rice (*Oryza sativa*)	Bangkok, Thailand	Ballistoconidium-fall	Takashima and Nakase (2001), Wang et al. (2015b)
Phragmotaenium (Tilletiopsis) oryzicola	Rice (*Oryza sativa*)	Bangkok, Thailand	Ballistoconidium-fall	Takashima and Nakase (2001), Wang et al. (2015b)
Pseudobensingtonia (Bensingtonia) musae	Banana (*Musa paradisiaca*)	Bangkok, Thailand	Ballistoconidium-fall	Takashima et al. (1995), Wang et al. (2015c)
Sporobolomyces blumeae	*Blumea* sp.	Bangkok, Thailand	Ballistoconidium-fall	Takashima and Nakase (2000)
Rhodosporidiobolus (Sporobolomyces) nylandii	Rice (*Oryza sativa*), perennial grass, wild sugarcane (*Saccharum spontaneum*) and common reed (*Phragmites karka*)	Bangkok, Thailand	Ballistoconidium-fall	Takashima and Nakase (2000), Wang et al. (2015c)
Rhodosporidiobolus (Sporobolomyces) poonsookiae	Silk reed (*Neyraudia reynaudiana*), mango (*Mangifera indica*), cogon grass/lalang grass (*Imperata cylindrica*), eucalyptus (*Eucalyptus* sp.) and tropic ageratum (*Ageratum conyzoides*)	Bangkok, Thailand	Ballistoconidium-fall	Takashima and Nakase (2000), Wang et al. (2015c)
Symmetrospora (Sporobolomyces) vermiculata	Perennial grass, cogon grass/lalang grass (*Imperata cylindrica*), rice (*Oryza sativa*), annual or perennial grass, hairy fountain grass (*Pennisetum pedicellatum*)	Bangkok, Thailand	Ballistoconidium-fall	Takashima and Nakase (2000), Wang et al. (2015c)

(continued)

Table 7.1 (continued)

Species	Plant leaf	Location	Isolation method	Reference
Phylum Ascomycota				
Candida aechmeae	*Aechmea recurvata*	Rio Grande do Sul State, Brazil	Plating of leaf washings	Landell et al. (2010)
Candida chumphonensis	Cassava (*Manihot esculenta*)	Chumphon, Thailand	Three consecutive methanol enrichments	Koowadjanakul et al. (2011)
Candida mattranensis	Cassava (*Manihot esculenta*)	Chumphon, Thailand	Three consecutive methanol enrichments	Koowadjanakul et al. (2011)
Candida morakotiae	Unknown plant	Narathiwat, Thailand	Enrichment technique	Nakase et al. (2009)
Candida phyllophila	Grape (*Vitis vinifera*)	Kanchanaburi, Thailand	Enrichment technique	Limtong and Kaewwichian (2013)
Candida sakaeoensis	Burma padauk tree (*Pterocarpus indicus*)	Sa Kaeo, Thailand	Enrichment	Limtong et al. (2012)
Candida sirachaensis	Copper pod tree (*Peltophorum pterocarpum*)	Chonburi, Thailand	Enrichment	Limtong et al. (2012)
Candida vitiphila	Grape (*Vitis vinifera*)	Kanchanaburi, Thailand	Enrichment	Limtong and Kaewwichian (2013)
Candida wancherniae	Unknown plant	Narathiwat, Thailand	Enrichment	Nakase et al. (2009)
Hannaella siamensis	Rice (*Oryza sativa*)	Nakhon Pathom, Nonthaburi Province and Suphan Buri, Thailand	Plating of leaf washings	Kaewwichian et al. (2015)
Hannaella phetchabunensis	Corn (*Zea mays*)	Phetchabun, Thailand	Plating of leaf washings	Kaewwichian et al. (2015)
Metschnikowia lopburiensis	Rice (*Oryza sativa*)	Lopburi, Thailand	Enrichment	Kaewwichian et al. (2012)
Metschnikowia saccharicola	Sugarcane (*Saccharum officinarum*)	Nakhon Ratchasima and Suphan Buri, Thailand	Enrichment	Kaewwichian et al. (2012)
Nakazawaea siamensis	Sugarcane (*Saccharum officinarum*)	Nakhon Ratchasima, Thailand	Enrichment	Kaewwichian and Limtong (2014)

Ogataea kanchanaburiensis	Mango (*Mangifera indica*)	Kanchanaburi, Thailand	Three consecutive methanol enrichments	Limtong et al. (2013)
Ogataea mangiferae	Mango (*Mangifera indica*)	Minas Gerais state, Brazil	Two-step methanol enrichment	Santos et al. (2015)
Ogataea phyllophila	Cassava (*Manihot esculenta*)	Chumphon, Thailand	Three consecutive methanol enrichments	Koowadjanakul et al. (2011)
Ogataea wangdongensis	Grapes (*Vitis vinifera*)	Kanchanaburi, Thailand	Three consecutive methanol enrichments	Limtong et al. (2013)
Wickerhamiella siamensis	Sugarcane (*Saccharum officinarum*)	Kaeng Khoi District, Saraburi Province, Thailand	Plating of leaf washings	Khunnamwong et al. (2014)
Wickerhamomyces siamensis	Sugarcane (*Saccharum officinarum*)	Lopburi, Thailand	Enrichment	Kaewwichian et al. (2013a)
Yamadazyma epiphylla	Rice (*Oryza sativa*)	Nonthaburi, Thailand	Plating of leaf washings	Jindamorakot et al. (2015)
Yamadazyma phyllophila	Corn (*Zea mays*) and bean (*Phaseolus vulgaris*)	Nakhon Ratchasima and Kanchanaburi, Thailand	Enrichment	Kaewwichian et al. (2013b)
Yamadazyma siamensis	Sugarcane (*Saccharum officinarum*)	Phetchabun, Thailand	Enrichment	Kaewwichian et al. (2013b)

7.4 Culture-Independent Methods for Assessment of Leaf-Associated Fungi

Environmental microbial diversity has been widely evaluated by culture-dependent methods. One of the major obstacles of these techniques is the presence of potentially uncultivable or slow-growing strains (on commonly used media and culture conditions) in the microbial community. This bias may lead to the underestimation of biodiversity and the inability to detect important community members (Amann et al. 1995; Takaku et al. 2006). It has been estimated that 99% of bacteria and 95% of fungi presented in nature have not yet been cultivated (Amann et al. 1995; Kaeberlein et al. 2002; Takaku et al. 2006). To circumvent these problems, culture-independent molecular approaches have increasingly been developed to investigate microbial communities in diverse environments to more accurately elucidate microbial diversity. Over the past decade, several PCR-based techniques have been widely employed to assess fungal communities (including yeasts) on plant leaves on the basis of LSU and ITS sequences. Total genomic DNA extracted directly from natural samples is subjected to PCR amplification and separation to identify a single organism using techniques such as denaturing gradient gel electrophoresis (DGGE) (e.g. Duong et al. 2006), temperature gradient gel electrophoresis (TGGE) (e.g. Sole et al. 2008), automated ribosomal DNA restriction analysis (ARDRA) or restriction fragment length polymorphism (RFLP) (e.g. Gao et al. 2005; Zhou et al. 2010; Nasanit et al. 2015a; Tantirungkij et al. 2015), terminal restriction fragment length polymorphism (T-RFLP) (e.g. Drage et al. 2014), ribosomal intergenic spacer analysis (RISA) and automated ribosomal intergenic spacer analysis (ARISA) (e.g. Szink et al. 2016). As with culture-dependent approaches, each molecular technique has advantages and disadvantages. For example, DGGE and TGGE are reliable, reproducible, rapid and economical techniques, which are able to separate different DNA fragments of the same size based on the different melting temperature of each nucleotide sequence. The drawbacks of these methods are that single DNA bands may still contain more than one DNA sequence type (Gelsomino et al. 1999) and that good resolution of the bands in a profile requires DNA fragments to be short (<500 bp). Additionally, it is estimated that DGGE can only detect 1–2% of the microbial population and that PCR fragments might represent mainly dominant species present in an environmental sample (MacNaughton et al. 1999). Duong et al. (2006) studied fungal communities on living leaves of *Magnolia liliifera* from Thailand by PCR-DGGE and cultivation methods. Their studies implicated that DGGE detected taxa that were not recovered from culture studies; moreover, other predominant taxa isolated by traditional method were not detected by DGGE technique. It might be possible that these fungi are present in small numbers in leaf tissues, thereby complicating detection by DGGE due to insufficient amount of DNA template. ARDRA and RFLP are tools used to study microbial diversity that rely on DNA polymorphisms of digested PCR amplicons that are separated with agarose or non-denaturing polyacrylamide gels, respectively. For instance, RFLP banding patterns were used to distinguish the LSU rDNA clones obtained by PCR cloning analysis in studying yeast diversity within rice leaves (Nasanit et al. 2015a; Tantirungkij et al. 2015). Although the DNA polymorphisms are useful for detecting structure in microbial communities from environmental samples without cultivation technique, DNA banding patterns in diverse communities are often too complex to analyse (Tiedje et al. 1999). Moreover, the RFLP technique has certain limitations for yeast identification: (1) there is no available database for RFLP profiles of LSU and therefore the presence of false-positive and false-negative results cannot be excluded and (2) fragments differing in 20–50 bp cannot be sufficiently distinguished. Therefore, the PCR products of each clone library showing distinct RFLP profiles should be always selected for DNA sequence-based identification. T-RFLP is a technique that has enabled significantly increased throughput compared to gel-based community profiling of RFLP technique by labelling one primer with a fluorescent dye, such as TET (4,7,2′,7′-tetrachloro-6-carboxyfluorescein) or 6-FAM (phosphoramidite fluorochrome 5-carboxyfluorescein). This allows detection of only the labelled terminal restriction fragments and simplifies the banding pattern for complex community analysis (Liu et al. 1997). Nonetheless, it is impossible to identify microbial species from the T-RFLP peaks. RISA and ARISA provide ribosomal-based fingerprinting of the microbial community from the intergenic spacer (IGS) region based on the lengths of the PCR amplicons, which are separated using polyacrylamide gel electrophoresis. The silver staining used for these methods tends to be somewhat insensitive, and resolution can be low for the RISA method. ARISA was developed to address these limitations by labelling the forward primer with a fluorescent dye, which allows automatic detection of labelled sequences. This increases the sensitivity of the method and additionally makes detection more time efficient. However, this technique still has some limitation from PCR biases (e.g. Fisher and Triplett 1999). Although these PCR-based techniques have their advantages for phylloplane fungal community assessments, PCR limitations can bias diversity studies because of, for instance, different affinities of primers to templates, different copy numbers of target genes, hybridization efficiency, primer specificity and G + C content in target genes (Kirk et al. 2004).

Another technique developed to monitor DNA polymorphisms using gel electrophoresis is single-strand conformation polymorphism (SSCP). It can distinguish single-strand nucleotide sequences of identical length based on their different conformations induced by differences in the nucleotide sequences (Orita et al. 1989; Schwieger and Tebbe 1998). Nonetheless, the detection sensitivity decreases when amplicon sizes exceed 500 bp. SSCP techniques have been optimized for characterizing the molecular signature of soil fungal communities by adding capillary electrophoresis (CE-SSCP) (Zinger et al. 2008). CE-SSCP is a simple, reproducible and time-efficient technique, which enables the analysis of large sample numbers (Hong et al. 2007). The method has also been successfully implemented to investigate the fungal assemblages of plant leaves (Cordier et al. 2012).

DNA hybridization-based methods have been developed for the visual assessment of fungal communities in the phyllosphere. For example, Inácio et al. (2010) optimized fluorescence in situ hybridization (FISH) method for phylloplane yeast community assessment. FISH probes were designed on the basis of a comparative analysis of hundreds of fungal 18S and 26S rRNA gene sequences and labelled at their 5′-end with fluorochrome Cy3. However, yeast cells were virtually impossible to visualize directly on the surface of the whole fixed leaves due to the autofluorescence and/or the nonspecific binding of the probes to leaf tissue. Therefore, the detection of yeast cells in leaf washings is preferable to the in vivo technique. Nevertheless, this study was able to demonstrate a log-normal species distribution, a result that has implications for the future quantification of phylloplane yeasts based on the washing and plating of bulk leaf samples. Based on the results of Inácio et al., previous studies most likely overestimated the size of respective yeast populations (Inácio et al. 2010).

High-throughput methods have also been developed to understand microbial diversity in the phyllosphere. Oligo-DNA custom macroarray techniques, for instance, have been developed to detect and monitor phylloplane microorganisms, i.e. major pathogenic and non-pathogenic fungi and bacteria, of apple trees (He et al. 2012). The 40 bp oligo-DNAs of fungal ITS rDNA and bacterial SSU rDNA of target microorganisms were fixed on a nylon membrane and hybridized with digoxigenin-labelled cRNA probes prepared for each species. This technique is a strong tool to monitor the major microbial species and identify key species on phyllosphere samples. Furthermore, it was possible to detect fungal species that were not detected by culture-dependent method. Until now, high-throughput sequencing methods have been developed for direct sampling and increasingly been employed in determining the fungal community composition from environmental samples including plant leaves (Jumpponen and Jones 2009, 2010; Cordier et al. 2012; Zimmerman and Vitousek 2012; Abdelfattah et al. 2015). These methods provide the possibility of investigating hundreds of thousands of sequences simultaneously. The bioinformatic analysis of sequence data had been a challenge of these sequencing technologies (Pop and Salzberg 2008). However, bioinformatics pipelines (i.e. QIIME) have been developed recently for performing microbiome analysis from raw DNA sequencing data (Kuczynski et al. 2012).

7.5 Yeasts in the Tropical Phylloplane by Culture-Independent Methods

To the best of our knowledge, only our research group has employed culture-independent approaches for investigation of phylloplane yeasts in the tropical region. We assessed yeast diversity in the phylloplane of rice (*Oryza sativa*) (Nasanit et al. 2015a), sugarcane (*Saccharum officinarum* L.) (Nasanit et al. 2015b) and corn (*Zea mays* L.) (Nasanit et al. 2016) in Thailand by PCR-RFLP with sequence-based analysis of the D1/D2 region of the LSU rDNA. Basidiomycetous species were the major component, and yeasts in the order Ustilaginales were predominant on leaf surfaces of all three plant species. These results are in accordance with a previous study in Portugal by Inácio et al. (2010), who reported that the abundant yeasts on phylloplane of Montpellier maple (*Acer monspessulanum*), Portuguese oak (*Quercus faginea*) and grey-leaved cistus (*Cistus albidus*) were basidiomycetes. However, the prevalent yeast species were different from our studies, which may be due to the different plant species and climate. By comparing the communities of phylloplane yeasts on corn, rice and sugarcane leaves, only the species *Moesz. aphidis* was in common. *Pseudozyma hubeiensis* pro tem. and *Moesz. antarcticus* were the most common yeasts found in the corn phylloplane; *Moesz. antarcticus* and *Papiliotrema japonica* were prevalent in the rice phylloplane, while a phylotype closely related to *Ustilago maydis* was commonly detected on sugarcane phylloplane. Additionally, the phylotype richness showed significant difference among these plant species. The Shannon index (H') values revealed that yeast communities of the corn phylloplane ($H' = 3.60$) appear to be more diverse than those of rice ($H' = 2.33$) and sugarcane ($H' = 2.19$). This suggests that the plant species may directly influence parameters of yeast community assembly. Microbial community structure on leaf surfaces naturally depends on many factors, including the availability of nutrients, which in turn varies with leaf age and plant species, and environmental factors such as growing conditions, geography and climatic regions. Additionally, biotic factors including competing microorganisms can affect microbial communities in the plant phylloplane (Irvine et al. 1978; Maksimova et al. 2009; Glushakova and Chernov 2010; Jumpponen and Jones 2010; Finkel et al. 2011; Kachalkin and Yurkov 2012; Voriskova et al. 2014; Yurkov et al. 2015b). Similar results were obtained by a study on fungal diversity on *Metrosideros polymorpha* leaves in the tropical region across Hawaiian landscapes using high-throughput barcoded amplicon pyrosequencing (Zimmerman and Vitousek 2012). Next to very high levels of fungal diversity (> 4200 OTUs), the variation of fungal communities among sites was strongly correlated with environmental factors such as temperature and rainfall. High-throughput DNA sequencing could also demonstrate high diversity of fungal OTUs on narrower geographic scales. Across four plots of a tropical plantation stand of *Shorea leprosula* in Central Kalimantan, Indonesia, fungal OTU compositions of leaves were different and even between plots 15 m apart (Izuno et al. 2016).

Recently, the endophytic yeast diversity in rice leaves collected in Thailand has been analysed using culture-independent methods (Tantirungkij et al. 2015). Semi-nested PCR technique was used to increase the detection sensitivity and specificity of the D1/D2 region from endophytic yeasts. Similarly, to leaf surfaces, the endophytic yeast community in rice leaves was dominated by phylotypes belonging to the Basidiomycota. The most prevalent yeasts were those located in the genus *Vishniacozyma* (formerly *Cryptococcus victoriae* clade, Tremellales) followed by *Deb. hansenii* and *Moesz. antarcticus*. In comparison to the species reported as rice epiphytic yeasts in Thailand by the enrichment (Limtong and Kaewwichian 2015) and culture-independent methods (Nasanit et al. 2015a), it could be noted that the phylotypes related to *C. metapsilosis*, *Deb. hansenii*, *Debaryomyces vindobonensis*, *Tr. asahii*, *Ustilago abaconensis*, *Vishniacozyma foliicola* and *Vishn. victoriae* live endophytically on rice since they have not been detected on rice leaf surfaces. *Moesz. antarcticus* and *Mey. guilliermondii* detected on rice phylloplane are probably adapted to endophytic proliferation since their DNA was present in the leaf tissue. Epiphytic yeasts, such as *M. koreensis*, *Pa. flavescens* and *Sp. blumeae*, were detected by both enrichment and culture-independent methods in the rice phylloplane and not within rice leaf tissues.

During recent studies, we were able to obtain numerous D1/D2 sequences that may represent novel yeast taxa from corn, rice and sugarcane leaves. Most of them were classified in the subphylum Ustilaginomycotina. The detection of numerous rare or novel D1/D2 sequences and the occurrence of various species within tropical plant species by culture-independent techniques from these studies suggest that extensive studies of yeasts from tropical plant phylloplane would enable the discovery of novel species. Therefore, these will lead to a better understanding of their diversity. Furthermore a better knowledge of yeast biodiversity would facilitate development of novel applications for the agricultural, industrial and pharmaceutical industries from these microorganisms.

7.6 Concluding Remarks

Phylloplane is known to be colonized by a large number of microorganisms. The investigations of yeast biodiversity in the phylloplane are currently increasing. However, only a small number of articles have focused on phylloplane yeasts in tropical regions. Many methodologies have been used to assess the phylloplane yeasts. Using different culture-dependent methods might have facilitated differences observed in yeast communities. For instance, the enrichment technique is likely to support the growth of small populations of ascomycetous yeasts, whereas basidiomycetous yeasts are prevalent when leaf washings are plated out. Culture-dependent methods however may underestimate microbial diversity due to the presence of potentially uncultivable or slow-growing strains in the microbial community in commonly used media and culture conditions. To overcome these problems, during the last 30 years, molecular techniques have been employed as very valuable research tools in phylloplane yeast diversity studies. Nevertheless, these techniques have their own limitations, e.g. PCR bias cannot be ruled out. In the course of investigations of phylloplane yeasts in tropical countries by both culture-dependent and culture-independent methods, many factors that shape these yeast communities, such as plant species, geography and environmental factors, have been elucidated. Moreover, several novel yeast species from phylloplane in tropical countries obtained by culture-dependent methods have recently been described, and it is expected that many other new species descriptions will follow. Indications for this unknown diversity also come from the numerous D1/D2 DNA sequences obtained by culture-independent methods that are related to yeast-containing phylogenetic lineages, but where no cultured yeast specimen is known. These results emphasize the importance of further research to get better understanding of phylloplane yeast biodiversity in tropical regions.

References

Abdelfattah A, Li Destri Nicosia MG, Cacciola SO, Droby S, Schena L (2015) Metabarcoding analysis of fungal diversity in the phyllosphere and carposphere of olive (*Olea europaea*). PLoS One 10:e0131069

Amann RI, Ludwig W, Schleifer KH (1995) Phylogenetic identification and in situ detection of individual microbial cells without cultivation. Microbiol Rev 59:143–169

Andrews JH, Harris RE (2000) The ecology and biogeography of microorganisms on plant surfaces. Annu Rev Phytopathol 38:145–180

Belloch C, Querol A, Garcia MD, Barrio E (2000) Phylogeny of the genus *Kluyveromyces* inferred from the mitochondrial cytochrome c oxidase II gene. Int J Syst Evol Microbiol 50:405–416

Boundy-Mills K (2006) Methods for investigating yeast biodiversity. In: Rosa C, Péter G (eds) Biodiversity and ecophysiology of yeasts. Springer, Berlin, pp 67–100

Brandl MT, Lindow SE (1998) Contribution of indole-3-acetic acid production to the epiphytic fitness of *Erwinia herbicola*. Appl Environ Microbiol 64:3256–3263

Bunster L, Fokkema NJ, Schippers B (1989) Effect of surface active *Pseudomonas* spp. on leaf wettability. Appl Environ Microbiol 55:1340–1345

Chamnanpa T, Limtong P, Srisuk N, Limtong S (2013) *Pseudozyma vetiver* sp. nov., a novel anamorphic ustilaginomycetous yeast species isolated from the phylloplane in Thailand. A van Leeuwenhoek 104:637–644

Cordier T, Robin C, Capdevielle X, Desprez-Loustau ML, Vacher C (2012) Spatial variability of phyllosphere fungal assemblages: genetic distance predominates over geographic distance in a European beech stand (*Fagus sylvatica*). Fungal Ecol 5:509–520

Daniel HM, Meyer W (2003) Evaluation of ribosomal RNA and actin gene sequences for the identification of ascomycetous yeasts. Int J Food Microbiol 86:71–78

de Azeredo LAI, Gomes EAT, Mendonca-Hagler LC, Hagler AN (1998) Yeast communities associated with sugarcane in Campos, Rio de Janeiro, Brazil. Int Microbiol 1:205–208

de Jager ES, Wehner FC, Korsten L (2001) Microbial ecology of the mango phylloplane. Microb Ecol 42:201–207

Diaz MR, Boekhout T, Theelen B, Fell JW (2000) Molecular sequence analyses of the intergenic spacer (IGS) associated with rDNA of the two varieties of the pathogenic yeast, *Cryptococcus neoformans*. Syst Appl Microbiol 23:535–545

Dlauchy D, Tornai-Lehoczki J, Fulop L, Péter G (2003) *Pichia (Komagataella) pseudopastoris* sp. nov., a new yeast species from Hungary. A van Leeuwenhoek 83:327–332

Drage S, Mitter B, Trols C, Muchugi A, Jamnadass RH, Sessitsch A, Hadacek F (2014) Antimicrobial drimane sesquiterpenes and their effect on endophyte communities in the medical tree *Warburgia ugandensis*. Front Microbiol 5:13

Duong LM, Jeewon R, Lumyong S, Hyde KD (2006) DGGE coupled with ribosomal DNA gene phylogenies reveal uncharacterized fungal phylotypes. Fungal Divers 23:121–138

Fell JW, Boekhout T, Fonseca A, Scorzetti G, Statzell-Tallman A (2000) Biodiversity and systematics of basidiomycetous yeasts as determined by large-subunit rDNA D1/D2 domain sequence analysis. Int J Syst Evol Microbiol 50:1351–1371

Fell JW, Scorzetti G, Statzell-Tallman A, Mills KB (2007) Molecular diversity and intragenomic variability in the yeast genus *Xanthophyllomyces*: the origin of *Phaffia rhodozyma*? FEMS Yeast Res 7:1399–1408

Fiala V, Glad C, Martin M, Jolivet E, Derridj S (1900) Occurrence of soluble carbohydrates on the phylloplane of maize (*Zea mays* L.): variations in relation to leaf heterogeneity and position on the plant. New Phytol 115:609–615

Finkel OM, Burch AY, Lindow SE, Post AF, Belkin S (2011) Geographical location determines the population structure in phyllosphere microbial communities of a salt-excreting desert tree. Appl Environ Microbiol 77:7647–7655

Fisher MM, Triplett EW (1999) Automated approach for ribosomal intergenic spacer analysis of microbial diversity and its application to freshwater bacterial communities. Appl Environ Microbiol 65:4630–4636

Fonseca A, Inácio J (2006) Phylloplane yeasts. In: Rosa C, Péter G (eds) Biodiversity and ecophysiology of yeasts. Springer, Berlin, pp 263–301

Fry SC (1989) Cellulases, hemicelluloses and auxin-stimulated growth: a possible relationship. Physiol Plant 75:532–536

Fungsin B, Hamamoto M, Arunpairojana V, Sukhumavasi J, Atthasampunna P, Nakase T (2001) *Bensingtonia thailandica* sp. nov., a novel basidiomycetous yeast species isolated from plant leaves in Thailand. Int J Syst Evol Microbiol 51:1209–1213

Fungsin B, Hamamoto M, Arunpairojana V, Sukhumavasi J, Atthasampunna P, Nakase T (2002a) *Kockovaella barringtoniae* sp. nov., a new basidiomycetous yeast species isolated from a plant leaf collected in a tropical rain forest in Thailand. Int J Syst Evol Microbiol 52:281–284

Fungsin B, Takashima M, Artjariyasripong S, Arunpaiojana V, Hamamoto M, Nakase T (2002b) *Bullera arundinariae* sp. nov., a new yeast species of ballistoconidium-forming yeast, isolated from a plant in Thailand. Microbiol Cult Coll 18:83–90

Fungsin B, Takashima M, Bai FY, Artjariyasripong S, Nakase T (2003a) *Bullera panici* sp. nov. and *Bullera siamensis* sp. nov., two new yeasts in the *Bullera variabilis* cluster isolated in Thailand. Microbiol Cult Coll 19:23–32

Fungsin B, Takashima M, Artjariyasripong S, Potacharoen W, Nakase T (2003b) *Bullera sakaeratica* sp. nov., a new species of ballistoconidium-forming yeast found in Thailand. Microbiol Cult Coll 19:33–39

Fungsin B, Takashima M, Sugita T, Atjariyasripong S, Potacharoen W, Tanticharoen M, Nakase T (2006) *Bullera koratensis* sp. nov. and *Bullera lagerstroemiae* sp. nov., two new ballistoconidium-forming yeast species in the Trichosporonales-clade isolated from plant leaves in Thailand. J Gen Appl Microbiol 52:73–81

Gao X-X, Zhou H, Xu D-Y, Yu C-H, Chen Y-Q, Qu L-H (2005) High diversity of endophytic fungi from the pharmaceutical plant, *Heterosmilax japonica* Kunth revealed by cultivation-independent approach. FEMS Microbiol Lett 249:255–266

Gelsomino A, Keijzer-Wolters AC, Cacco G, van Elsas JD (1999) Assessment of bacterial community structure in soil by polymerase chain reaction and denaturing gradient gel electrophoresis. J Microbiol Methods 38:1–15

Glushakova AM, Chernov IY (2010) Seasonal dynamics of the structure of epiphytic yeast communities. Microbiology 79:830–839

Golubev W, Sampaio J (2009) New filobasidiaceous yeasts found in the phylloplane of a fern. J Gen Appl Microbiol 55:441–446

Hamamoto M, Nagahama T, Tamura M (2002) Systematic study of basidiomycetous yeasts – evaluation of the ITS regions of rDNA to delimit species of the genus *Rhodosporidium*. FEMS Yeast Res 2:409–413

He YH, Isono S, Shibuya M, Tsuji M, Purushothama CRA, Tanaka K, Sano T (2012) Oligo-DNA custom macroarray for monitoring major pathogenic and non-pathogenic fungi and bacteria in the phyllosphere of apple trees. PLoS One 7:e34249

Hong H, Pruden A, Reardon KF (2007) Comparison of CE-SSCP and DGGE for monitoring a complex microbial community remediating mine drainage. J Microbiol Methods 69:52–64

Inácio J, Portugal L, Spencer-Martins I, Fonseca A (2005) Phylloplane yeasts from Portugal: seven novel anamorphic species in the Tremellales lineage of the Hymenomycetes (Basidiomycota) producing orange-coloured colonies. FEMS Yeast Res 5:1167–1183

Inácio J, Landell MF, Valente P, Wang PH, Wang YT, Yang SH, Manson JS, Lachance MA, Rosa CA, Fonseca A (2008) *Farysizyma* gen. nov., an anamorphic genus in the Ustilaginales to accommodate three novel epiphytic basidiomycetous yeast species from America, Europe and Asia. FEMS Yeast Res 8:499–508

Inácio J, Ludwig W, Spencer-Martins I, Fonseca A (2010) Assessment of phylloplane yeasts on selected Mediterranean plants by FISH with group- and species-specific oligonucleotide probes. FEMS Microbiol Ecol 71:61–72

Irvine JA, Dix NJ, Warren RC (1978) Inhibitory substances in *Acer platanoides* leaves: seasonal activity and effects on growth of phylloplane fungi. Trans Br Mycol Soc 70:363–371

Izuno A, Tanabe AS, Toju H, Yamasaki M, Indrioko S, Isagi Y (2016) Structure of phyllosphere fungal communities in a tropical dipterocarp plantation: a massively parallel next-generation sequencing analysis. Mycoscience 57:171–180

Jindamorakot S, Am-In S, Kaewwichian R, Limtong S (2015) *Yamadazyma insecticola* f.a., sp. nov. and *Yamadazyma epiphylla* f.a., sp. nov., two novel yeast species. Int J Syst Evol Microbiol 65:1290–1296

Jumpponen A, Jones KL (2009) Massively parallel 454 sequencing indicates hyperdiverse fungal communities in temperate *Quercus macrocarpa* phyllosphere. New Phytol 184:438–448

Jumpponen A, Jones KL (2010) Seasonally dynamic fungal communities in the *Quercus macrocarpa* phyllosphere differ between urban and nonurban environments. New Phytol 186:496–513

Kachalkin AV, Yurkov AM (2012) Yeast communities in *Sphagnum* phyllosphere along the temperature-moisture ecocline in the boreal forest-swamp ecosystem and description of *Candida sphagnicola* sp. nov. A van Leeuwenhoek 102:29–43

Kaeberlein T, Lewis K, Epstein SS (2002) Isolating "uncultivable" microorganisms in pure culture in a simulated natural environment. Science 296:1127–1129

Kaewwichian R, Limtong S (2014) *Nakazawaea siamensis* f.a., sp. nov., a novel yeast species isolated from phylloplane in Thailand. Int J Syst Evol Microbiol 64:266–270

Kaewwichian R, Yongmanitchai W, Kawasaki H, Limtong S (2012) *Metschnikowia saccharicola* sp. nov. and *Metschnikowia lopburiensis* sp. nov., two novel yeast species isolated from phylloplane in Thailand. A van Leeuwenhoek 102:743–751

Kaewwichian R, Kawasaki H, Limtong S (2013a) *Wickerhamomyces siamensis* sp. nov., a novel yeast species isolated from phylloplane in Thailand. Int J Syst Evol Microbiol 63:1568–1573

Kaewwichian R, Yongmanitchai W, Kawasaki H, Wang PH, Yang SH, Limtong S (2013b) *Yamadazyma siamensis* sp. nov., *Yamadazyma phyllophila* sp. nov. and *Yamadazyma paraphyllophila* sp. nov., three novel yeast species isolated from phylloplane in Thailand and Taiwan. A van Leeuwenhoek 103:777–788

Kaewwichian R, Jindamorakot S, Am-In S, Sipiczki M, Limtong S (2015) *Hannaella siamensis* sp. nov. and *Hannaella phetchabunensis* sp. nov., two new anamorphic basidiomycetous yeast species isolated from plants. Int J Syst Evol Microbiol 65:1297–1303

Kato K, Kurimura Y, Makiguchi N, Asai Y (1974) Determination of strongly methanol assimilating yeasts. J Gen Appl Microbiol 20:123–127

Khunnamwong P, Surussawadee J, Jindamorakot S, Limtong S (2014) *Wickerhamiella siamensis* f.a., sp. nov., a novel endophytic and epiphytic yeast species isolated from sugarcane leaf in Thailand. Int J Syst Evol Microbiol 64:3849–3855

Khunnamwong P, Surussawadee J, Jindamorakot S, Ribeiro JRA, Hagler AN, Limtong S (2015) Occultifur tropicalis f.a., sp. nov., a novel Cystobasidiomycetous yeast species isolated from tropical regions. Int J Syst Evol Microbiol 65:1578–1582

Kirk JL, Beaudette LA, Hart M, Moutoglis P, Klironomos JN, Lee H, Trevors JT (2004) Methods of studying soil microbial diversity. J Microbiol Methods 58:169–188

Koowadjanakul N, Jindamorakot S, Yongmanitchai W, Limtong S (2011) *Ogataea phyllophila* sp. nov., *Candida chumphonensis* sp. nov. and *Candida mattranensis* sp. nov., three methylotrophic yeast species from phylloplane in Thailand. A van Leeuwenhoek 100:207–217

Kuczynski J, Stombaugh J, Walters WA, Gonzalez A, Caporaso JG, Knight R (2012) Using QIIME to analyze 16S rRNA gene sequences from microbial communities. Curr Protoc Microbiol 27:1E.5.1–1E.5.20

Kurtzman CP (2014) Use of gene sequence analyses and genome comparisons for yeast systematics. Int J Syst Evol Microbiol 64:325–332

Kurtzman CP, Robnett CJ (1998) Identification and phylogeny of ascomycetous yeasts from analysis of nuclear large subunit (26S) ribosomal DNA partial sequences. A van Leeuwenhoek 73:331–371
Kurtzman CP, Robnett CJ (2003) Phylogenetic relationships among yeasts of the "*Saccharomyces* complex" determined from multigene sequence analyses. FEMS Yeast Res 3:417–432
Kurtzman CP, Robnett CJ, Basehoar-Powers E (2008) Phylogenetic relationships among species of *Pichia*, *Issatchenkia* and *Williopsis* determined from multigene phylogenetic analysis, and the proposal of *Barnettozyma* gen. nov., *Lindnera* gen. nov. and *Wickerhamomyces* gen. nov. FEMS Yeast Res 8:939–954
Kurtzman CP, Jack WF, Boekhout T, Robert V (2011a) Methods for isolation, phenotypic characterization and maintenance of yeasts. In: Kurtzman CP, Fell JW, Boekhout T (eds) The yeasts, a taxonomic study, vol 2, 5th edn. Elsevier, Amsterdam, pp 87–110
Kurtzman CP, Jack WF, Boekhout T (2011b) Gene sequence analyses and other DNA-based methods for yeast species recognition. In: Kurtzman CP, Fell JW, Boekhout T (eds) The yeasts, a taxonomic study, vol 2, 5th edn. Elsevier, Amsterdam, pp 137–144
Landell MF, Inácio J, Fonseca A, Vainstein MH, Valente P (2009) *Cryptococcus bromeliarum* sp. nov., an orange-coloured basidiomycetous yeast isolated from bromeliads in Brazil. Int J Syst Evol Microbiol 59:910–913
Landell MF, Billodre R, Ramos JP, Leoncini O, Vainstein MH, Valente P (2010) *Candida aechmeae* sp. nov. and *Candida vrieseae* sp. nov., novel yeast species isolated from the phylloplane of bromeliads in Southern Brazil. Int J Syst Evol Microbiol 60:244–248
Landell MF, Brandao LR, Barbosa AC, Ramos JP, Safar SV, Gomes FC, Sousa FM, Morais PB, Broetto L, Leoncini O, Ribeiro JR, Fungsin B, Takashima M, Nakase T, Lee CF, Vainstein MH, Fell JW, Scorzetti G, Vishniac HS, Rosa CA, Valente P (2014) *Hannaella pagnoccae* sp. nov., a tremellaceous yeast species isolated from plants and soil. Int J Syst Evol Microbiol 64:1970–1977
Landell MF, Brandão LR, Gomes FC, Félix CR, Santos AR, Pagani DM, Ramos JP, Broetto L, Mott T, Vainstein MH, Valente P, Rosa CA (2015) *Bullera vrieseae* sp. nov., a tremellaceous yeast species isolated from bromeliads. Int J Syst Evol Microbiol 65:2466–2471
Last FT, Price D (1969) Yeasts associated with living plants and their environs. In: Rose AH, Harrison JS (eds) The yeasts, vol 1, 1st edn. Academic Press, New York, pp 183–217
Leaw SN, Chang HC, Sun HF, Barton R, Bouchara JP, Chang TC (2006) Identification of medically important yeast species by sequence analysis of the internal transcribed spacer regions. J Clin Microbiol 44:693–699
Limtong S, Kaewwichian R (2013) *Candida phyllophila* sp. nov. and *Candida vitiphila* sp. nov., two novel yeast species from grape phylloplane in Thailand. J Gen Appl Microbiol 59:191–197
Limtong S, Kaewwichian R (2015) The diversity of culturable yeasts in the phylloplane of rice in Thailand. Ann Microbiol 65:667–675
Limtong S, Koowadjanakul N (2012) Yeasts from phylloplanes and their capability to produce indole-3-acetic acid. World J Mcrobiol Biotechnol 28:3323–3335
Limtong S, Koowadjanakul N, Jindamorakot S, Yongmanitchai W, Nakase T (2012) *Candida sirachaensis* sp. nov. and *Candida sakaeoensis* sp. nov. two anamorphic yeast species from phylloplane in Thailand. A van Leeuwenhoek 102:221–229
Limtong S, Kaewwichian R, Groenewald M (2013) *Ogataea kanchanaburiensis* sp. nov. and *Ogataea wangdongensis* sp. nov., two novel methylotrophic yeast species from phylloplane in Thailand. A van Leeuwenhoek 103:551–558
Limtong S, Kaewwichian R, Yongmanitchai W, Kawasaki H (2014) Diversity of culturable yeasts in phylloplane of sugarcane in Thailand and their capability to produce indole-3-acetic acid. World J Microbiol Biotechnol 30:1785–1796
Lindow SE, Brandl MT (2003) Microbiology of the phyllosphere. Appl Environ Microbiol 69:1875–1883

Liu WT, Marsh TL, Cheng H, Forney LJ (1997) Characterization of microbial diversity by determining terminal restriction fragment length polymorphisms of genes encoding 16S rRNA. Appl Environ Microbiol 63:4516–4522

Liu XZ, Wang QM, Boekhout T, Bai FY (2012) *Derxomyces amylogenes* sp. nov., *Derxomyces bambusicola* sp. nov. and *Derxomyces corylopsis* sp. nov., three ballistoconidium-forming yeast species isolated from subtropical plant leaves. Int J Syst Evol Microbiol 62:996–1001

Liu XZ, Wang QM, Göker M, Groenewald M, Kachalkin AV, Lumbsch HT, Millanes AM, Wedin M, Yurkov AM, Boekhout T, Bai FY (2015) Towards an integrated phylogenetic classification of the Tremellomycetes. Stud Mycol 81:85–147

MacNaughton SJ, Stephen JR, Venosa AD, Davis GA, Chang Y-J, White DC (1999) Microbial population changes during bioremediation of an experimental oil spill. Appl Environ Microbiol 65:3566–3574

Maksimova IA, Chernov IY (2004) Community structure of yeast fungi in forest biogeocenoses. Microbiology 73:474–481

Maksimova IA, Yurkov AM, Chernov IY (2009) Spatial structure of epiphytic yeast communities on fruits of *Sorbus aucuparia* L. Biol Bull 36:613–618

Morris CE, Kinkel LL (2002) Fifty years of phyllosphere microbiology: significant contributions to research in related fields. In: Lindow SE, Hechter-Poinar EI, Elliot VJ (eds) Pyllosphere microbiology. APS Press, Minnesota, pp 365–375

Nakase T, Suzuki M (1985) Ballistospore-forming yeasts found on the surface of the Japanese rice plants *Oryza sativa* L. J Gen Appl Microbiol 31:457–474

Nakase T, Takashima M (1993) A simple procedure for the high frequency isolation of new taxa of ballistosporous yeasts living on the surfaces of plants. RIKEN Rev 3:33–34

Nakase T, Itoh M, Takematsu A, Mikata K, Banno I, Yamada Y (1991) *Kockovaella*, a new ballistospore-forming anamorphic yeast genus. J Gen Appl Microbiol 37:175–197

Nakase T, Takashima M, Itoh M, Fungsin B, Potacharoen W, Atthasampunna P, Komagata K (2001) Ballistoconidium-forming yeasts found in the phyllosphere of Thailand. Microbiol Cult Coll 17:23–33

Nakase T, Jindamorakot S, Ninomiya S, Imanishi Y, Kawasaki H (2009) *Candida wancherniae* sp. nov. and *Candida morakotiae* sp. nov., two novel ascomycetous anamorphic yeast species found in Thailand. J Gen Appl Microbiol 55:93–100

Nasanit R, Krataithong K, Tantirungkij M, Limtong S (2015a) Assessment of epiphytic yeast diversity in rice (*Oryza sativa*) phyllosphere in Thailand by a culture-independent approach. A van Leeuwenhoek 107:1475–1149

Nasanit R, Tangwong-o-thai A, Tantirungkij M, Limtong S (2015b) The assessment of epiphytic yeast diversity in sugarcane phyllosphere in Thailand by culture-independent method. Fungal Biol 119:1145–1157

Nasanit R, Jaibangyang S, Tantirungkij M, Limtong S (2016) Yeast diversity and novel yeast D1/D2 sequences from corn phylloplane obtained by a culture-independent approach. A van Leeuwenhoek 109:1615–1634

Ogata K, Nishikawa H, Ohsugi M (1969) A yeast capable of utilizing methanol. Agric Biol Chem 33:13–19

Orita M, Iwahana H, Kanazawa H, Hayashi K, Sekiya T (1989) Detection of polymorphisms of human DNA by gel electrophoresis as single-strand conformation polymorphisms. Proc Natl Acad Sci U S A 86:2766–2770

Péter G, Tornai-Lehoczki J, Dlauchy D (2007) *Ogataea allantospora* sp. nov., an ascomycetous yeast species from phylloplane. A van Leeuwenhoek 92:443–448

Phaff HJ, Starmer WT (1987) Yeasts associated with plants, insects and soil. In: Rose AH, Harrison JS (eds) The yeasts, 2nd edn. Academic Press, London, pp 123–180

Pop M, Salzberg SL (2008) Bioinformatics challenges of new sequencing technology. Trends Genet 24:142–149

Remus-Emsermann MNP, Tecon R, Kowalchuk GA, Leveau JHJ (2012) Variation in local carrying capacity and the individual fate of bacterial colonizers in the phyllosphere. ISME J 6:756–765

Santos ARO, Faria ES, Lachance MA, Rosa CA (2015) *Ogataea mangiferae* sp. nov., a methylotrophic yeast isolated from mango leaves. Int J Syst Evol Microbiol 65:1855–1859

Schocha CL, Seifertb KA, Huhndorfc S, Robertd V, Spougea JL, Levesqueb CA, Chenb W, Fungal Barcoding Consortiuma (2012) Nuclear ribosomal internal transcribed spacer (ITS) region as a universal DNA barcode marker for Fungi. PNAS 109:6241–6246

Schreiber L, Krimm U, Knoll D, Sayed M, Auling G, Kroppenstedt RM (2005) Plant–microbe interactions: identification of epiphytic bacteria and their ability to alter leaf surface permeability. New Phytol 166:589–594

Schwieger F, Tebbe CC (1998) A new approach to utilize PCR-single-strand-conformation polymorphism for 16S rRNA gene-based microbial community analysis. Appl Environ Microbiol 64:4870–4876

Scorzetti G, Fell JW, Fonseca A, Statzell-Tallman A (2002) Systematics of basidiomycetous yeasts: a comparison of large subunit D1/D2 and internal transcribed spacer rDNA regions. FEMS Yeast Res 2:495–517

Slavikova E, Vadkertiova R, Vranova D (2007) Yeasts colonizing the leaf surfaces. J Basic Microbiol 47:344–350

Sole M, Chatzinotas A, Sridhar KR, Harms H, Krauss G (2008) Improved coverage of fungal diversity in polluted groundwaters by semi-nested PCR. Sci Total Environ 406:324–330

Sugita T, Ikeda R, Shinoda T (2001) Diversity among strains of *Cryptococcus neoformans* var. *gattii* as revealed by sequence analysis of multiple genes and chemotype analysis of capsular polysaccharide. Microbiol Immunol 45:757–768

Sugita T, Nakajima M, Ikeda R, Matsushima T, Shinoda T (2002) Sequence analysis of the ribosomal DNA intergenic spacer 1 regions of *Trichosporon* species. J Clin Microbiol 40:1826–1830

Surussawadee J, Khunnamwong P, Srisuk N, Limtong S (2014) *Papiliotrema siamense* f.a., sp. nov., a novel yeast species isolated from plant leaf in Thailand. Int J Syst Evol Microbiol 64:3058–3062

Surussawadee J, Jindamorakot S, Nakase T, Lee CF, Limtong S (2015) *Hannaella phyllophila* sp. nov., a novel basidiomycetous yeast species associated with plants in Thailand and Taiwan. Int J Syst Evol Microbiol 65:2135–2140

Szink I, Davis EL, Ricks KD, Koide RT (2016) New evidence for broad trophic status of leaf endophytic fungi of *Quercus gambelii*. Fungal Ecol 22:2–9

Takaku H, Kodaira S, Kimoto A, Nashimoto M, Takagi M (2006) Microbial communities in the garbage composting with rice hull as an amendment revealed by culture-dependent and -independent approaches. J Biosci Bioeng 101:42–50

Takashima M, Nakase T (1998) *Bullera penniseticola* sp. nov, and *Kockovaella sacchari* sp. nov., two new yeast species isolated from plants in Thailand. Int J Syst Evol Microbiol 48:1025–1030

Takashima M, Nakase T (2000) Four new species of the genus *Sporobolomyces* isolated from leaves in Thailand. Mycoscience 41:357–369

Takashima M, Nakase T (2001) *Tilletiopsis derxii*, *Tilletiopsis oryzicola* and *Tilletiopsis penniseti*, three new species of the ustilagionomycetous anamorphic genus *Tilletiopsis* isolated from leaves in Thailand. A van Leeuwenhoek 80:43–56

Takashima M, Suh SO, Nakase T (1995) *Bensingtonia musae* sp. nov. isolated from a dead leaf of *Musa paradisiaca* and its phylogenetic relationship among basidiomycetous yeasts. J Gen Appl Microbiol 41:143–151

Tantirungkij M, Nasanit R, Limtong S (2015) Assessment of endophytic yeast diversity in rice leaves by a culture-independent approach. A van Leeuwenhoek 108:633–647

Tiedje JM, Asuming-Brempong S, Nusslein K, Marsh TL, Flynn SJ (1999) Opening the black box of soil microbial diversity. Appl Soil Ecol 13:109–122

Toome M, Roberson RW, Aime MC (2013) *Meredithblackwellia eburnea* gen. et sp. nov., Kriegeriaceae fam. nov. and Kriegeriales ord. nov.-toward resolving higher-level classification in Microbotryomycetes. Mycologia 105:486–495

Vorholt JA (2012) Microbial life in the phyllosphere. Nat Rev Microbiol 10:828–840

Voriskova J, Brabcova V, Cajthaml T, Baldrian P (2014) Seasonal dynamics of fungal communities in a temperate oak forest soil. New Phytol 201:269–278

Wang QM, Boekhout T, Bai FY (2011) *Cryptococcus foliicola* sp. nov. and *Cryptococcus taibaiensis* sp. nov., novel basidiomycetous yeast species from plant leaves. J Gen Appl Microbiol 57:285–291

Wang QM, Groenewald M, Takashima M, Theelen B, Han PJ, Liu XZ, Boekhout T, Bai FY (2015a) Phylogeny of yeasts and related filamentous fungi within Pucciniomycotina determined from multigene sequence analyses. Stud Mycol 81:27–53

Wang QM, Begerow D, Groenewald M, Liu XZ, Theelen B, Bai FY, Boekhout T (2015b) Multigene phylogeny and taxonomic revision of yeasts and related fungi in the Ustilaginomycotina. Stud Mycol 81:55–83

Wang QM, Yurkov AM, Göker M, Lumbsch HT, Leavitt SD, Groenewald M, Theelen B, Liu XZ, Boekhout T, Bai FY (2015c) Phylogenetic classification of yeasts and related taxa within Pucciniomycotina. Stud Mycol 81:149–189

Weber RW, Becerra J, Silva MJ, Davoli P (2008) An unusual Xanthophyllomyces strain from leaves of *Eucalyptus globulus* in Chile. Mycol Res 112:861–867

Wuczkowski M, Passoth V, Turchetti B, Andersson AC, Olstorpe M, Laitila A, Theelen B, van Broock M, Buzzini P, Prillinger H, Sterflinger K, Schnürer J, Boekhout T, Libkind D (2011) Description of *Holtermanniella* gen. nov., including *Holtermanniella takashimae* sp. nov. and four new combinations, and proposal of the order Holtermanniales to accommodate tremellomycetous yeasts of the *Holtermannia* clade. Int J Syst Evol Microbiol 61:680–689

Xin G, Glawe D, Doty SL (2009) Characterization of three endophytic, indole-3-acetic acid-producing yeasts occurring in Populus trees. Mycol Res 113:973–980

Yurkov AM, Kachalkin AV, Daniel HM, Groenewald M, Libkind D, de Garcia V, Zalar P, Gouliamova DE, Boekhout T, Begerow D (2015a) Two yeast species *Cystobasidium psychroaquaticum* f.a. sp. nov. and *Cystobasidium rietchieii* f.a. sp. nov. isolated from natural environments, and the transfer of *Rhodotorula minuta* clade members to the genus *Cystobasidium*. A van Leeuwenhoek 107:173–185

Yurkov A, Inácio J, Chernov I, Fonseca A (2015b) Yeast biogeography and the effects of species recognition approaches: the case study of widespread basidiomycetous species from birch forests in Russia. Curr Microbiol 70:587–601

Zhou ZX, Jiang H, Yang C, Yang MZ, Zhang HB (2010) Microbial community on healthy and diseased leaves of an invasive plant *Eupatorium adenophorum* in Southwest China. J Microbiol 48:139–145

Zimmerman NB, Vitousek PM (2012) Fungal endophyte communities reflect environmental structuring across a Hawaiian landscape. Proc Natl Acad Sci U S A 109:13022–13027

Zinger L, Gury J, Alibeu O, Rioux D, Gielly L, Sage L, Pompanon F, Geremia RA (2008) CE-SSCP and CE-FLA, simple and high-throughput alternatives for fungal diversity studies. J Microbiol Methods 72:42–53

Chapter 8
Yeasts in Cacti and Tropical Fruit

Philip F. Ganter, Paula B. Morais, and Carlos A. Rosa

Abstract Yeast communities associated with necrotic tissues of cacti and ripe tropical fruits are excellent models for studying the ecology and evolution of these fungi. The availability of carbon sources, chemistry of the substrate, presence of antimicrobial compounds, mutualisms with insect vectors, competition, predation, and habitat heterogeneity in space and time all influence cactophilic yeast occurrence. Four yeast species are prevalent in necrotic cactus tissues: *Pichia cactophila*, *Candida* (*Ogataea*) *sonorensis*, *Clavispora opuntiae*, and *Sporopachydermia cereana* species complex. Other species are limited to determined geographic areas or host plants. Species isolated only from necrotic cactus tissues are considered as cactophilic and are vectored by insects that feed and breed in these substrates, mainly *Drosophila* spp. Ripe fruit yeast communities are composed mainly of species with a limited physiological nutritional profile during the early stages of fruit deterioration. Ascomycetous yeast species are most frequently associated with ripe fruits, such as *Hanseniaspora uvarum, Pichia kudriavzevii, Meyerozyma guilliermondii*, *Aureobasidium pullulans*, *Wickerhamiella* spp., and others. Basidiomycetous species occur mainly in the later stages of fruit deterioration. Necrotic cactus tissues and ripe tropical fruits also represent rich sources for the discovery of new yeast species.

Keywords Cactophilic yeasts • Ripe tropical fruits • Yeast interactions • Biogeography • Diversity

---

P.F. Ganter
Department of Biological Sciences, College of Life & Physical Sciences, Tennessee State University, 3500 John A Merritt Blvd, Nashville, TN 37209, USA
e-mail: pganter@tnstate.edu

P.B. Morais
Laboratório de Microbiologia Ambiental e Biotecnologia, Universidade Federal do Tocantins, Av. NS15, ALCNO14, Bloco II, Sala 04, 77020-120 Palmas, Tocantins, Brazil
e-mail: moraispb@mail.uft.edu.br

C.A. Rosa (✉)
Departamento de Microbiologia, Instituto de Ciências Biológicas, C.P. 486, Universidade Federal deMinas Gerais, CEP 31270-901 Belo Horizonte, MG, Brazil
e-mail: carlrosa@icb.ufmg.br

P. Buzzini et al. (eds.), *Yeasts in Natural Ecosystems: Diversity*,
DOI 10.1007/978-3-319-62683-3_8

Contents

8.1 Introduction

Yeast colonizes a wide variety of natural substrates, and the ecological success of these microorganisms depends on several biotic and abiotic factors: availability of carbon sources, chemistry of the substrate, presence of antimicrobial compounds, mutualism with insect vectors, competition, predation, and habitat heterogeneity in space and time. Cactus necroses and ripe tropical fruits are two excellent models for studying the ecology and evolution of yeast communities. The decaying tissues of cacti provide a habitat for yeasts and their associated vectors. Yeast communities are found in flowers, fruits, and stem and cladode rots of cacti. In addition, the rot juices can soak the soil below the cactus, representing a fourth substrate for yeast growth and insect colonization. Several yeast species can colonize these substrates. Some of them are found only in cacti and are considered cactophilic. Four dominant cactophilic yeasts have been reported worldwide associated with cactus rots: *Pichia cactophila, Ogataea (Candida) sonorensis, Sporopachydermia cereana* species complex, and *Clavispora opuntiae*. Other species have limited geographical distributions and are restricted to specific cactus hosts. According to Starmer et al. (2006), differences in host plants had a larger influence on the diversity and the composition of these yeast communities than geographic separation. The yeasts are vectored among the host plants by insects, mainly flies in the genus *Drosophila*. The interactions among the yeast communities, the host cactus plants, and the insect vectors (cactophilic *Drosophila)* are rich sources of studies on evolution and adaptation of these microorganisms. In this chapter, we will update what we know about the yeasts associated with cacti, describing the cactophilic yeasts, their habitats, vectors, and phylogeny. In addition, we include a brief history of how we came to study the cactophilic habitat.

Tropical fruits are also an excellent model to study the diversity and ecology of yeast. Several species of yeasts were described from studies of the microorganisms colonizing fruits in tropical habitats. Colonization of fruits by yeast communities has a pattern that may be successional. Species having restricted nutritional profiles dominate the early stages of the fruit's deterioration. Species with a broad nutritional profile colonize the final stages. Killer-producing strains may be prevalent during the first days of fruit spoiling, when the pH is low (see also Klassen et al. 2017). In this chapter, we also report the diversity of the fruit yeast communities and biotic and abiotic factors affecting these yeasts.

8.2 The Cactophilic Habitat

Cacti are a monophyletic lineage within the Caryophyllales, a diverse group of angiosperms with an unresolved phylogeny. They are related to the Portulacaceae or Anacampserotaceae, both clades of succulent plants (Nyffeler and Eggli 2010; Yang et al. 2015). Cacti are almost all stem succulents with leaves modified into spines that grow from unique clusters of buds called aureoles, which also produce flowers, branches, and roots. Cacti are native to the Nearctic and Neotropical biogeographic regions (with one exception) and occur from southern Ontario to Patagonia. They are associated with xeric habitats but are also found as part of mesophytic assemblages on well-drained soils in the tropical and temperate regions.

There are four yeast habitats associated with cacti. Like many plants, yeasts may grow in cactus flower nectaries (Lachance et al. 2001b; Lachance 2013) and in fruit that has been damaged sufficiently to break open the integument (Starmer et al. 1987b). The third habitat is related to their succulence. When a stem's integument is breached, it may be healed and a callus formed or it may be invaded by pectinolytic microbes, mostly bacteria, that result in stem tissue breakdown (Fogleman and Foster 1989). These areas of softened tissue, often reduced to a liquid, are known as necroses or, more commonly, rots (Figs. 8.1 and 8.2). Stem rots host more than bacteria and have been exploited by yeasts and numerous insects (Starmer 1982; Starmer et al. 1982, 1991; Starmer and Phaff 1983; Fogleman and Starmer 1985; Rosa et al. 1994). Only a few insects can initiate a rot by penetrating the cactus cuticle. Many insects are recruited to the rot by their attraction to volatiles produced by microbial activity in the rot. Initially, the volatiles produced are mainly due to bacterial activity with yeasts contributing only after they have been brought to the rot by an insect (Fogleman 1982; Fogleman and Abril 1990; Foster and Fogleman 1993).

The final habitat is unusual and is not as well studied. The rots in some large cacti may persist for long periods of time and may leak fluids for more than a year. The leakage from *Carnegiea gigantea*, giant saguaro, or *Pachycereus pringlei*, cardon, may be so persistent that it becomes the site for complex insect mating behaviors (Mangan 1979; Myles 1986). Such a rot can soak the soil below the cactus with rot juices, and evaporation may concentrate plant secondary chemicals in the soaked soil up to 30 times the concentration in healthy stem tissue (Meyer and Fogleman 1987). This rot-soaked soil represents a fourth habitat for fly and yeast growth. The flies reared from Saguaro stem rots are mostly *Drosophila nigrospiracula,* and the flies from soaked-soils are almost all *Drosophila mettleri*.

Fig. 8.1 Different necrotic stems of columnar cacti. Clockwise from the *upper left: Stenocereus hystrix* necrosis on Curaçao; *Carnegiea gigantea* with multiple necroses leaking fluids near Tucson, AZ, USA; *Echinopsis chiloensis* necrosis in central Chile; unidentified columnar cactus in Brazil

Fig. 8.2 *Opuntia* sp. plantation and necrotic cladodes of *Opuntia stricta*

However, this yeast community from this fourth habitat is similar to the yeast community in the stem rot (Fogleman et al. 1982).

Not all cacti act as hosts for stem rot communities and some may do so only rarely. Some cacti never reach a size that can accommodate a rot and its associated insects and microbes. In some cases, the climate may be too harsh for rots or, perhaps, the vectors. In 2000, a search through a dense patch of *Eulychnia iquiquensis* in the Atacama Desert north of Chañaral, Chile, growing on an exposed bluff in the Parque Nacional Pan de Azucar produced no stem necroses but many stems on which damaged tissue had dried so quickly that the stems appeared to be living and succulent but were, in fact, dead and thoroughly desiccated. Stem rots were found in the same species of cactus in other locales in Chile (Ganter, unpublished data). In many cases, it is not clear why a cactus hosts no rots. A search for stem rots in *Melocactus intortus* (Turk's cap cactus), at a site in Antigua in 1990, found no rots in over 150 plants inspected, many of them with visible damage, although nearby *Opuntia* species and *Pilosocereus royenii* contained rots (Ganter, unpublished data). However, Lachance et al. (2001a) report two strains from the *Sporop. cereana* species complex taken from *Meloc. intortus* in the British Virgin Islands (other yeast species were also present: Starmer, unpublished data). Such occasional hosts probably have little impact on local cactophilic yeast community structure, species persistence, or evolutionary history. Conversely, yeasts described from stem rots can sometimes be found in the necrotic tissue of other succulents. *Dipodascus australiensis* has been collected only from three *Opuntia* rots in Australia and one *Euphorbia ingens* rot in South Africa (de Hoog et al. 1986).

Cacti are both useful and attractive plants that have been introduced in many locations worldwide, including Australia, Hawaii, both sides of the Mediterranean Sea, and South Africa. There are reports of wild populations on the Indian subcontinent (Pareek et al. 2003). The anthropogenic spread of cacti has undoubtedly had an impact on the biogeography of yeasts associated with cacti. A clear example of this influence is linked to the control of cactus populations, a textbook example of biological control. In some areas, introduced cacti have become pests or have impeded agriculture, and efforts have been made to eradicate cacti in both native and non-native locations. When several imported *Opuntia* species invaded rangelands in Queensland and New South Wales, the Australian government imported cactus-feeding insects and unhealthy cactus tissue from the Americas (Dodd 1940). After several attempts to establish populations of different insects failed to control targeted *Opuntia* species, the moth *Cactoblastis cactorum* was introduced from a collection made near Concordia, Entre Rios, Argentina, in 1925. *Cact. cactorum* is a moth that lays its eggs only on *Opuntia*. The caterpillars penetrate the cactus' tough cuticle and feed on the stem tissue. In 10 years, the moth-cactus system stabilized at a cactus density low enough to control the problem (Murray 1982). Subsequently, the moth was introduced from Australia to islands in the Caribbean region for the control of *Opuntia* there, although *Opuntia* was native to the islands (Simmonds and Bennett 1966). Since the first Caribbean introduction in 1957, the moth has dispersed or been introduced throughout the Caribbean and into southern Florida (Simmonds and Bennett 1966; Soberón 2002). The moth is now viewed as a threat to native *Opuntia* species in both Florida (Pemberton 1995; Pemberton and Cordo 2001b; Stiling et al. 2004) and Mexico (Soberon et al. 2001; Vigueras and Portillo 2001) and is itself the target of biological pest control programs (Pemberton and Cordo 2001a).

The biogeography of *Cl. opuntiae*, a yeast that often co-occurs with *Cact. cactorum* (Lachance 1990), has been influenced by biological control programs aimed at its host (Lachance et al. 2000). Comparison of sequences and restriction-site maps of portions of the ribosomal DNA transcriptional unit of strains collected from localities where the moth was native and from localities where the moth was introduced or not known to be present found a clear geographic signal. The spacer map characteristic of strains from Argentina, the original source of the introduced moth populations and the presumptive source of the *Cl. opuntiae* found with the introduced moth populations, was found in areas where the moth had been introduced or had invaded through natural dispersal. The "Argentina" spacer map did not dominate in *Cl. opuntiae* strains collected from areas where the moth had not been introduced. A 1986 collection from *Opuntia* on Big Pine Key, Florida, USA, and on Playa Linda, Canaveral National Seashore, Florida, USA, sites separated by over 450 km, found no evidence of the moth and no *Cl. opuntiae* (Starmer et al. 1988a). A 1990 collection from the same sites found moth and yeast present at both locales (Ganter et al. 1993). The *Cl. opuntiae* strains had the "Argentine" spacer map (Lachance et al. 2000).

Cacti stems, flowers, and fruits are moist, nutrient-rich patches in what may be a very dry habitat. These patches are discrete, separated by inhospitable territory, and connected by the activities of animals that visit or live in the rots, vector yeasts among rots, and, potentially, between rots and other yeast habitats. Observations or collections of naturally occurring rots have recorded many different arthropods emerging from it or living in the rots (Starmer et al. 1988b). Castrezana and Markow (2001) observed 34 different arthropods (mites, various Diptera, ants, beetles from 5 families, wasps, a termite, a dermapteran, a moth, a jumping spider, and a pseudoscorpion in a sample of 36 rots) (12 each from saguaro, senita, and organ pipe cacti). Rots appear where stems or fruits are damaged, but, with the exception of the rots caused by cactus moths (there are more species than *Cact. cactorum*), the insects, bacteria, and yeasts play no known role in making the breach in the cactus' integument that can initiate a rot.

Among the most common insects reared from cactus rots are *Drosophila*, and the relationship between cactophilic *Drosophila* and cactophilic yeasts has been stressed in studies of cactophilic yeasts. It is possible that many cactus-associated insects act as yeast vectors. There may even be a role for vertebrates as yeast vectors. Birds, reptiles, and mammals all visit rots, but most researchers have stressed vertebrate's role in starting rots, especially fruit rots, and have considered them only occasional visitors. Most of the research effort on the vectors of cactophilic yeasts has been focused on the role *Drosophila* play. This disparity must partly arise from the well-known relationship between *Drosophila* and yeasts in general but also partly from the observation that *Drosophila* are often the first insects to arrive on new rots (Fogleman and Danielson 2001). Starmer et al. (1988b) found that, of the many insects feeding on a saguaro rot, only *Drosophila* and nereid flies were effective vectors of yeasts. The other insects carried molds. Arrival order is important in yeast community dynamics (Gurevitch 1984), and early arrival is an advantage in predicting maximal yeast population size.

In addition to the four different habitats, cacti promote yeast biodiversity because they differ widely in stem and, to a lesser extent, fruit chemistry, and these differences can be important factors in yeast distributions (variation in nectar composition has not been studied extensively) (Starmer 1982; Starmer and Phaff 1983; Fogleman and Starmer 1985; Fogleman and Abril 1990). Many cacti accumulate secondary metabolites in somatic tissues. In cacti that host rots, these metabolites can be mucopolysaccharides, alkaloids, toxic medium-chain fatty acids, sterol diols, and triterpene glycosides (Kircher 1982). They can range from 1 to 40% of stem tissue dry weight and have well-known influences on *Drosophila* oviposition, growth, and development (Fogleman and Armstrong 1989; Fogleman and Danielson 2001). In general, *Opuntia* stems are characterized by mucopolysaccharides, and most columnar cacti have secondary plant metabolites that are often toxic to animals and, sometimes, yeast. In addition, *Opuntia* stems are relatively rich in simple sugars, while sugars in most columnar cacti are bound into complex carbohydrates or triterpene glucosides in cacti rich in these saponins. Thus, in columnar cacti, most of the sugars become available only after they are enzymatically freed from more complex molecules (Fogleman and Abril 1990).

The role that the diversity of stem tissue chemistry plays in yeast community ecology is not completely understood. Stem rot communities from different species of cacti are diverse and vary greatly, as referenced above. But differences in stem chemistry may have both direct and indirect effects on yeast distributions. Some species of the genus *Pichia, Phaffomyces,* and *Starmera* are unable to grow on media made from cereoid cacti stems containing triterpene glycosides, while congeners are not (Starmer et al. 1980). At this point, no other single stem constituent has been found to have such a marked effect on growth as triterpene glycosides, saponins, that can constitute from 25 to 40% of the dry weight of stem tissue of cacti in the subtribe Stenocereinae. Stem chemistry is known to affect the distribution of cactophilic *Drosophila* (Heed and Kircher 1965; Fogleman et al. 1986; Fogleman and Heed 1989), and chemical diversity may indirectly affect yeast distributions through its effect on the yeast's vector's distribution.

More recent data further support the idea that stem chemistry is an important factor in cactophilic yeast diversity. In 2004, W. T. Starmer isolated yeasts from stem rots of two cacti from different subtribes: *Carn. gigantea* (saguaro, subtribe Pachycereinae) growing in Saguaro National Monument, Tucson, AZ, USA, and *Stenocereus thurberi* (organ pipe cactus, subtribe Stenocereinae) growing in Organ Pipe Cactus National Monument, Ajo, AZ, USA. Strains from eight yeast species were isolated. All were cactophilic species. When the strains were inoculated onto standard lab medium [YM (yeast extract-malt extract) agar] supplemented with triterpene glycosides extracted from *Stenocereus gummosus*, not all were able to grow (Table 8.1; Starmer, unpublished data). As expected, *Starmera amethionina* (formerly *Pichia amethionina*) and *Pichia heedii* were sensitive to the saponins and were collected mostly from *Carn. gigantea*, which lacks triterpene glycosides. Three of 18 *P. heedii* strains were collected from *Sten. thurberi*, although all *P. heedii* strains were sensitive to *Sten. gummosus* triterpene glycosides. Starmer et al. (1980) also found that *P. heedii*, while generally confined to cacti in the Pachycereinae, could grow on *Sten. thurberi* tissue but not on *Sten. gummosus* tissue. The reason for this anomaly is not known, but there are notable differences

Table 8.1 Growth of yeast strains isolated from *Carnegiea gigantea* (saguaro, subtribe Pachycereinae) in Saguaro National Monument, Tucson, AZ, USA, and *Stenocereus thurberi* (organ pipe cactus, subtribe Stenocereinae) in Organ Pipe National Monument, Ajo, AZ, USA, in 2004 on YM media supplemented with *Stenocereus gummosus* triterpene glycosides

Source	Saguaro			Organ pipe		
Growth category[a]	ttg-r	ttg-ws	ttg-s	ttg-r	ttg-ws	ttg-s
Yeast						
Starmera amethionina			15			
Pichia heedii			15			3
Pichia cactophila	1	11		10		
Sporopachydermia cereana complex		5		6	1	
Magnusiomyces starmeri		4	1	8	1	
Pichia deserticola	2		2	3	2	1
Phaffomyces thermotolerans	3					
Candida (Ogataea) sonorensis	1					
Total	7	20	33	27	4	4

[a]ttg-r = triterpene glycoside resistant (strong growth), ttg-ws = somewhat sensitive to triterpene glycoside (weak growth), ttg-s = sensitive to triterpene glycoside (no growth)

between *Sten. thurberi* and *Sten. gummosus* stem chemistry (Kircher 1980, 1982). But it is the distribution of sensitive and resistant strains in three common species that reveals the impact of stem chemistry on cactophilic diversity.

Starmer et al. (1980) proposed that host plant shifts between cacti with different secondary chemical compositions were important in speciation in several cactophilic clades. Table 8.1 suggests that host plant secondary chemicals are either important in maintaining within-species variation in cactophilic yeasts or that the strains of three species in these represent complexes of cryptic species. *P. cactophila, Sporop. cereana* species complex, and *Dipodascus* (*Magnusiomyces*) *starmeri* were all collected on both hosts, and all were variable in their sensitivity to triterpene glycosides. The pattern in all three species was for sensitive or weakly sensitive strains to come from *Carn. gigantea*, with no triterpene glycosides, and resistant strains to come from *Sten. thurberi*, which does have them. The presence of cryptic species is already known for the *Sporop. cereana* species complex (Lachance et al. 2001a) and *P. cactophila* (Ganter et al. 2010).

Although we have defined four distinct habitats associated with cacti, these distinctions must be reflected in yeast distributions if the distinctions are meaningful. Ganter (2011) summarized data from his and W. T. Starmer's cactophilic yeast collections totaling 8244 isolates. Excluding unidentified strains, 38 species of yeasts were collected. Of those, 25 were unambiguously cactophilic, 10 were "probably" cactophilic, and only 3 were listed as widespread, occurring in cactophilic and non-cactophilic habitats. When some species never collected by Starmer or Ganter and species recently described are added (de Hoog et al. 1986; Freitas et al. 2015), the number of definitely cactophilic species described from stem necroses now totals 33 (Table 8.2). The undescribed new species collected in Australia and South, Central, and North America have their D1/D2 sequences

Table 8.2 Cactophilic yeast species, cactus host, and geographical distribution

Species	Cactus host	Geographical distribution
Candida (Ogataea) sonorensis	Columnar cacti and *Opuntia* spp.	North, Central, and South America and Australia
Clavispora opuntiae	*Opuntia* spp., columnar cacti, and tunnels of cactus-feeding moth	North, Central, and South America, Hawaii, and Australia
Dipodascus australiensis	*Opuntia inermis*	Australia
Dipodascus sp. A	*Opuntia stricta*	Australia
Dipodascus sp. B	*Opuntia* spp., *Pilosocereus gounellei*, *Pilosocereus pachycereus*, *Cereus pernambucensis*	Brazil and Peru
Dipodascus sp. C	*Cereus repandus*, *Opuntia caracassana*, *Stenocereus griseus*, *Cephaloscereus urbanianus*	Curaçao and Saint Martin
Kodamaea (Candida) restingae	Flowers and fruits of *Pilosocereus arrabidae* and *Cereus pernambucensis*	Brazil
Kodamaea nitidulidarum	Flowers and fruits of *Pilosocereus arrabidae* and *Cereus pernambucensis*	Brazil
Kluyveromyces sp.	*Cereus saddianus*, *Micranthocereus dolichospermaticus*, and *Pilosocereus arrabidae*	Brazil
Kurtzmaniella cleridarum	Cactus flowers and associated beetles of *Opuntia* and *Echinocereus* spp.	USA
Magnusiomyces starmeri	Columnar cacti and *Opuntia* spp.	USA and Mexico
Myxozyma mucilagina	Columnar cacti	Mexico and Brazil
Phaffomyces (Candida) coquimbonensis	*Echinopsis chiloensis* and *Opuntia stricta*	Chile and Australia
Phaffomyces (Candida) orba	*Opuntia* sp.	Australia
Phaffomyces antillensis	*Cephalocereus royenii* and *Melocactus intortus*	West Indies
Phaffomyces opuntiae	*Opuntia* spp.	Australia
Phaffomyces thermotolerans	Cardon and Senita cacti of the subtribe *Pachycereinae*	Mexico
Pichia barkeri	*Opuntia stricta*	Mexico, Caribbean islands, and Australia
Pichia cactophila	Columnar cacti and *Opuntia* spp.	North, Central, and South America and Australia
Pichia cephalocereana	*Cephalocereus roye*	Caribbean island of Montserrat
Pichia deserticola	Columnar cacti and *Opuntia* spp.	USA, Mexico, and Caribbean islands
Pichia eremophila	*Opuntia* spp.	USA

(continued)

Table 8.2 (continued)

Species	Cactus host	Geographical distribution
Pichia heedii	*Lophocereus schotti* and *Carnegiea gigantea*	USA and Mexico
Pichia insulana	*Stenocereus griseus, Solanum hystrix, Cereus repandus, Opuntia* spp., and *Pilosocereus royenii*	Curaçao, Hispaniola, British Virgin Islands, Coastal Venezuela, and Cayman Islands
Pichia pseudocactophila	Cardon cactus and its near relatives	USA, Mexico
Pichia sp.	*Cereus saddianus* and *Micranthocereus dolichospermaticus*	Brazil
Sporopachydermia cereana complex	Columnar cacti and *Opuntia* spp.	USA, Mexico, Hawai'i, and Haiti
*Sporopachydermia australis**	Columnar cacti and *Opuntia* spp.	Venezuela, Argentina, and Brazil
*Sporopachydermia brasiliensis**	*Pilosocereus arrabidae, Cereus pernambucensis,*	Brazil
*Sporopachydermia centralis**	*Opuntia* spp.	Honduras and Mexico
*Sporopachydermia obscura**	Columnar cacti and *Opuntia* spp.	Mexico, Dominican Republic, Venezuela, Argentina, and Brazil
*Sporopachydermia opuntiana**	Columnar cacti and *Opuntia* spp.	USA, Central America, Peru, and Australia
Starmera pachycereana	*Pachycereus pringlei*	Mexico
*Sporopachydermia trichocereana**	*Trichocereus pasacana*	Argentina
Starmera amethionina	Predominantly on cacti of the subtribe *Stenocereinae* and in *Opuntia stricta* in Australia	USA, Mexico, and Australia
Starmera caribaea	Different species of columnar and *Opuntia* spp.	Caribbean islands, USA to Venezuela
Starmera pachycereana	Predominantly on cacti of the subtribe *Pachycereinae*	USA and Mexico
Starmera pilosocereana	*Pilosocereus arrabidae*	Coastal region of Brazil
Starmera sp. A	*Opuntia tomentosa*	Australia
Starmera sp. B	*Stenocereus griseus, Cereus repandus, Opuntia* sp.	Curaçao
Tortispora caseinolytica	Different species of columnar cacti and *Opuntia* spp.	USA, Mexico, Caribbean islands, Argentina, Hawai'i
Tortispora cuajiniquilana	Unidentified columnar cactus	Costa Rica
Tortispora ganteri	*Stenocereus gummosus* and *Myrtillocactus cochal*	Mexico
Tortispora mauiana	*Opuntia ficus-indica*	Hawaii

(continued)

Table 8.2 (continued)

Species	Cactus host	Geographical distribution
Tortispora phaffii	*Opuntia bonaerensis*, *Cereus saddianus*, and *Pilosocereus arrabidae*	Argentina and Brazil
Tortispora starmeri	*Stenocereus thurberi*	USA
Wickerhamiella cacticola	Flowers and fruits of *Pilosocereus arrabidae* and *Cereus pernambucensis*	Brazil

[a]New species informally reported in the literature as belonging to the *Sporop. cereana* complex (Lachance et al. 2001a; Ganter 2011) but still lacking a formal description

deposited in GenBank or have distinct and consistent physiologies and have been isolated multiple times (Rosa, unpublished data; Ganter, unpublished data). Of the 8244 strains in the combined collection data from Ganter and Starmer, there were 4 species (*P. cactophila, C. sonorensis, Sporop. cereana* species complex, and *Cl. opuntiae*) that constituted at least 5% of the total, and, combined, those four species represented 60% of the total.

Cactus flowers also present a yeast habitat with species specific to the overall cactophilic habitat. Three species associated with columnar cactus flowers and fruit in South America plus a new genus described from a large collection of yeast strains isolated from nitidulid beetles in the genus *Carpophilus* found in the flowers of various cacti in Arizona, USA, (Lachance et al. 1998; Rosa et al. 1999; Lachance and Starmer 2008) have been described from this habitat. If we add these species to the total of cactophilic species, the count of cactophilic yeast species rises to 37. As most regions containing cacti are either under-collected or never collected, the number of cactophilic species is most probably underestimated in Table 8.2.

Differences between cactophilic and non-cactophilic habitats and among cactophilic habitats are also evident in proportional overlap between yeast communities from three of the four habitats (cactus flowers, cactus fruits, and cactus stem rots) and neighboring yeast communities from a non-cactus hosts (Ganter 2011). For each overlap calculated, only yeasts collected from each habitat during the same collection trip were compared, and for the comparisons between cactus habitats, the strains came from the same locales as well. This was done to be sure that neither space nor time could reduce the degree of overlap. The neighboring yeast communities consisted of tree fluxes collected in Arizona within 30 km and 2 days of the cactus samples (from *Opuntia phaeacantha* and *Carn. gigantea*) and a collection from *Ipomoea* spp. flowers in Peru that grew within 40 km of the cactophilic communities from *Opuntia berteri* and *Corryocactus brevistylus*, to which the *Ipomoea* yeasts were compared. The comparisons were done for yeasts collected in Peru, Brazil, Texas, Florida, and Arizona so a wide spectrum of cactus communities was represented. However, in each location sampled, only a subset of comparisons was possible. The cactus may not have been flowering or no fruit may have been ripened (green fruit is much like stem tissue), and neighboring communities were normally not sampled. But there was sufficient data to make eight comparisons between yeast communities from cactus habitats and six comparisons between cactus communities and neighboring communities.

There was no overlap between *Ipomoea* flower yeasts (from Peru) and neighboring cactus-yeast communities and very little overlap (mean proportional overlap = 0.03, $n = 3$) between tree flux communities in Arizona with neighboring cactus stem rot communities. Comparisons between different cactus habitats averaged an overlap of 0.16 ($n = 8$) with the highest degree of overlap between cactus stem and cactus flower rot communities from Brazil (0.38). Although these are low overlap values for habitats that might possibly occur on the same plant, the hypothesis of distinct yeast communities requires that within-habitat overlap be higher than between-habitat overlap. Stem rot community overlap values in collections from various species of *Opuntia* in Texas collected in different years and from locations separated by at least 100 km were indeed higher (mean = 0.57, $n = 8$). Collections from various species of columnar cacti taken years apart and often thousands of kilometers apart were also higher (mean proportional overlap = 0.42, $n = 9$). It appears that the flower, stem, and fruit habitats on a cactus harbor different communities of yeasts. The differences among communities are persistent and stable (Starmer et al. 1988a; Rosa et al. 1995; Latham 1998).

8.3 A Brief History of the Study of Cactophilic Yeast

Studies of cactophilic yeasts lie at the intersection of two long-term, multiple-contributor research programs. This history is an intersection of ideas and people, and we will attempt to sketch its outline in terms of both. To understand the potential for further study of cactophilic yeasts, we believe it is useful to understand where it is placed within each of those larger programs. Research into the cactus-yeast-*Drosophila* system differs in significant ways from studies of the relationship between *Saccharomyces* and *Drosophila*. The systems differ. The cactus-yeast-*Drosophila* system is less influenced by human activity and is both a more diverse and a more clearly delimited system. The questions addressed by those interested in each system have been influenced by a somewhat different history, and we will focus on cactus yeasts, although some of those mentioned here play significant roles in *Saccharomyces cerevisiae* research.

Two of the three components of the cactus-yeast-*Drosophila* system are closely related to important model organisms. It is hard to pinpoint the time and place of the adoption of *S. cerevisiae* as a model for research. *S. cerevisiae* has influenced human culture for millennia. Even before its role in baking and brewing was understood, it was the focus of attempts to improve bread, wine, and beer, research that is ongoing today. It emerged as a model for the eukaryotic cell, a role it still enjoys (Botstein et al. 1997; Botstein and Fink 2011). PubMed lists over 100,000 publications that mention *S. cerevisiae*.

But why use a yeast as a model? Doubtless, many factors and incidents lead to its choice, but one of the pivotal moments must be the publication of "Die Einheit in der Biochemie" (Unity in Biochemistry) by Albert Kluyver and Hendrick Donker in 1926 (Bennett and Phaff 1993). Kluyver was a very early advocate for the adoption of model organisms (Woods 1957). As he famously said "From elephant to butyric acid bacterium – it is all the same" (Bennett and Phaff 1993), linking bacteria to animals through common cellular biochemical pathways. The idea of using a small number of model organisms must certainly depend on a fundamental belief that all living things, no matter how diverse in detail, are linked by a unity based on a common ancestry and fundamental physical law. Importantly, Kluyver, an early advocate of this view, also held a chair in microbiology at the Technical University in Delft, the Netherlands. Delft was already a center for yeast research, and Kluyver worked with yeasts, publishing research into yeast fermentation of various sugars. Through his publications and lectures and the students trained in his laboratory, Kluyver's influence was felt in the fields of physiology, biochemistry, and microbiology (Woods 1957; Bennett and Phaff 1993).

A scientist's influence on his or her field is often felt through the students he or she mentors, and this means of exerting influence is especially important to this narrative. In 1932, Herman J. Phaff, a scion of a family engaged in the brewing industry, entered the Technical University at Delft as a first-year college student. He wrote his undergraduate thesis under Kluyver's direction (Phaff 1986). After graduation, Phaff accepted an invitation to visit the University of California's Division of Fruit Products in Berkeley, CA, in 1939, on the eve of the outbreak of war in Europe (Phaff 1986).

At Berkeley, Phaff earned his doctorate (1943), worked with Emil Mrak, was put on the faculty, and held this post until his retirement, although the department moved from Berkeley to Davis. At first, Phaff's research centered on various yeast biochemical pathways, an interest he never lost. By 1948, he and Mrak were sampling yeasts from natural substrates (including shrimp and the guts of rabbits), and one of his graduate students, Moshe Shifrine, sampled yeasts found associated with beetles. These were Phaff's first forays into yeast ecology and they broke new ground for the discipline. After a suggestion by T. Dobzhansky, Phaff turned his interest to the yeasts in *Drosophila* and to their larval and adult habitats. Eventually, pressed by the need to improve the accuracy of assigning yeast strains to their correct species, yeast systematics based on DNA analysis was added as a third important research focus (Phaff 1986).

For those of us educated in the USA, *Drosophila* might be the organism that comes to mind when we hear the phrase "model organism." Unlike yeast, the fly was (is) not of great economic and cultural value, so it is much easier to pinpoint its adoption as a model organism. When Thomas Hunt Morgan chose the pomace fly as suitable for his new experimental research program based on recommendations from colleagues and students, its chief attractions were practical (cheap to rear, adaptable to lab conditions, short generation time; Allen 1975). Morgan was an experimental embryologist swept, like others, into the study of heredity after the rediscovery of Mendel's work in 1900 (Allen 1969, 1975, 1980). He was skeptical of both Mendel's and Darwin's conclusions. Morgan favored de Vries' mutation theory of evolution and chose *Drosophila* as the organism with which to both refute Darwin and prove de Vries. Morgan eventually changed his views of Mendel's and, later, Darwin's work (Allen 1969, 1980).

In 1928, Morgan left Columbia for the California Institute of Technology, where he built a biology department focused on his perception of the future of the field: genetics, evolution, embryology, and biochemistry. He brought some of the fly lab researchers with him, including a scientist visiting from the Soviet Union, Theodosius Dobzhansky. Dobzhansky was interested in population genetics before he arrived in the USA (and may have coined the term). He pioneered population genetic studies of *Drosophila* in the western USA (Ayala 1976, 1985).

This is the nexus of ideas in which the study of cactophilic yeast is embedded: Morgan espoused a "new biology" that stressed the unity of life and, hence, unity in the aims and, to a degree, the methods of the science of biology (Allen 1969, 2005; Kay 1996). Kluyver visited California more than once and held a favorable view of the research being done there. Herman Phaff's choice of Berkeley brought him into contact with a host of new ideas and potential collaborators but in an environment compatible with his earliest training (Bennett and Phaff 1993; Phaff 1986). It was Dobzhansky who proposed to Phaff and Emil Mrak that it might be productive to look into the yeast in the crops of *Drosophila*. At the nexus, it is impossible to say if Phaff was responding to the "new biology" or continuing the program of research envisioned by Kluyver.

Morgan was not the only influential American biologist trained as an embryologist who felt the pull of genetics and evolution. John T. Patterson earned his doctorate from the University of Chicago in 1908, where he was mentored by such eminent embryologists as C.M. Child and C.O. Whitman (Painter 1965). But Patterson was also in contact with the "new biology." Patterson interacted with Morgan and Morgan's ideas on the importance of the gene (Painter 1965; Kenney and Borisy 2009) at the Marine Biological Laboratory in Woods Hole, MA, USA. When Patterson was hired to lead the University of Texas at Austin's Zoology Department, one of his hires was a member of Morgan's fly lab, H.J. Muller (Painter 1965).

At Austin, Patterson developed an interest in *Drosophila* and even taught undergraduate labs with flies gotten from Morgan's lab. He and Muller published together and Patterson developed an extensive research program around the evolution of *Drosophila* and it relatives based on his and his students' extensive field work (Painter 1965), including the pioneering work in the cactus-yeast-*Drosophila* system (Wagner 1944). His last graduate student, William B. Heed, made collecting trips to Mexico and Central America (Fogleman 1990).

After earning his doctorate, Heed was hired by the University of Arizona in 1958. He expanded his interest in the evolution, ecology, genetics, and behavior of Drosophilids with collections from the Caribbean and South America. Heed also participated in the Hawai'ian *Drosophila* Project begun in 1963 (Fogleman 1990; Etges 2007). As the difficulty of working with Hawai'ian Drosophilids from a laboratory in Arizona took its toll, he shifted more and more of his effort to the *Drosophila* native to the Sonoran Desert.

Heed and his collaborators worked on many aspects of desert fly ecology, establishing the importance of cactus chemistry on fly distribution and behavior (Etges 2007). As a natural extension of this work, Heed contacted Herman Phaff about investigating the microbiota found in the larval and adult habitats of Sonoran *Drosophila*. In 1972, Heed spent a sabbatical year in Herman Phaff's laboratory at UC Davis (Fogleman 1990). While there, he hired William T. Starmer to supervise his laboratory in Tucson (Fogleman 1990; Starmer, personal communication). Starmer was a population geneticist who had earned his doctorate at the University of Arizona. During Bill Heed's time in Davis, Starmer became interested in the relationships between the cacti, flies, and yeasts as well. He began collecting flies and yeasts from the Sonoran Desert, and after a trip to Baja California Sur, Mexico, in 1976, Phaff offered Starmer a postdoctoral position in his laboratory (Starmer, personal communication), where he met many students and visitors who subsequently contributed to the study of cactophilic yeasts, most notably Marc-André Lachance, who, in a career that has expanded our knowledge of almost all areas of yeast ecology, was to become Starmer's most frequent collaborator in studies of cactophilic yeast.

By 1980, Phaff, Heed, Starmer, and their collaborators had begun to reveal the diversity of yeasts from the cactus habitat. The new species included some found widely in the cactus habitat – *Cryptococcus cereanus* (*Sporop. cereana*) (Phaff et al. 1974), *Torulopsis sonorensis* (*C. sonorensis*) (Miller et al. 1976), *Candida* (*Myxozyma*) *mucilagina* (Phaff et al. 1980), and *P. cactophila* (Starmer et al. 1978b)—plus others confined to particular hosts and/or locations (these are often confounded factors): *Pichia* (*Phaffomyces*) *opuntiae* (Starmer et al. 1979), *P. heedii* (Phaff et al. 1978), and *Starm. amethionina* (Starmer et al. 1978a). Since that beginning, the study of cactophilic yeast and flies has expanded to new locations (the Caribbean, Hawaii, Australia, and South America) and attracted many new collaborators from labs in Italy, Australia, Spain, Argentina, and Brazil. An early workshop in Arizona (Ecological Genetics and Evolution: The Cactus-Yeast-*Drosophila* Model System, Oracle, AZ, USA) organized by Starmer and J. S. F. Barker, from the University of New England, Armidale, NSW, Australia, helped to establish the system as a model system and to attract new researchers.

From the beginning, the focus of the research into the cactus-yeast-*Drosophila* system has not been on description but on establishing associations between hosts, flies, and yeasts. Phaff and his associates developed an approach that Lachance describes as the "Phaff School of Yeast Ecology" (Lachance 2003). The underlying assumption of those who contributed to the study of the cactus-yeast-*Drosophila* system was that all components are part of a single system and that the physiology, ecology, and biogeography of any part of the system must be seen in the larger context if we are to understand it. This approach owes a debt to both Kluyver's "unity" and Morgan's "new biology."

8.4 The Cactophilic Yeast Community

Yeast species associated with cactus substrates can be generalists occurring in association with a wide variety of cactus species or specialists found only in certain species, subtribes, or tribes of the Cactaceae (Starmer et al. 1990; Freitas et al. 2015). *P. cactophila*, *C. sonorensis*, *Cl. opuntiae*, and members of the *Sporop. cereana* species complex occur in cactus rots in a wide range of species and regions (Starmer et al. 1990; Rosa et al. 1995; Lachance et al. 2001a; Ganter 2011). Several species are isolated in low frequencies from rotting stems of cacti or are restricted to a few host plants or to defined geographical areas (Starmer et al. 1990; Ganter et al. 2010; Cardinali et al. 2012; Freitas et al. 2015).

P. cactophila is the worldwide dominant yeast species, in terms of frequency of isolation, from cactus tissues (Ganter 2011). However, physiological and morphological similarities among many of the cactophilic *Pichia* species can make them difficult to distinguish. Ganter et al. (2010) described the cactophilic species *Pichia insulana* based on a collection of yeast isolates from necrotic tissue of cacti on Caribbean islands. This species is phylogenetically related to *P. cactophila* and *Pichia pseudocactophila*. The physiological profiles of these species are very similar. *P. insulana* represented 32% of the yeast community isolated from three cactus species in Curaçao (Ganter et al. 2010), where *P. cactophila* was not isolated from the same substrates. *P. insulana* is not confined to Curacao, but it is not the dominant species on the other islands with cactus communities similar to those on Curaçao (Ganter et al. 2010). In a similar study in Brazil, *P. cactophila* was not isolated from necrotic tissues of cacti in two different regions (Minas Gerais and Tocantins states), but it was the most frequent species isolated from necrotic stems of *Pilosocereus arrabidae* and *Cereus pernambucensis* in the restinga (coastal vegetation) ecosystems of Brazil (Rosa, unpublished data). One *Pichia* species related to *P. insulana* was isolated in low frequency in necroses of *Cereus saddianus* and *Micranthocereus dolichospermaticus* in the state of Tocantins in Brazil. This species (GenBank accession number KT373981) differs by nine nucleotide substitutions from *P. insulana* in the sequence of the D1/D2 region of the large rRNA gene and may represent a new cactophilic yeast species (not yet formally described) closely related to *P. cactophila* (Table 8.1).

Pichia barkeri was isolated from fruit and somatic tissues of *Opuntia stricta* on several islands of the Caribbean Sea, southern Mexico, and Australia. Rosa et al. (1992, 1994) reported also the isolation of *P. barkeri* from necrotic tissues of *Pil. arrabidae* and tunnels of the moth *Sigelgaita* sp. that attacks this cactus. However, recent collections of necrotic samples of the same species of cactus did not contain this yeast (Rosa, unpublished data). In the earlier studies of Rosa et al. (1992, 1994), the identification of yeast species was based on morphological and physiological characteristics, and strains of *P. cactophila* were probably misidentified as *P. barkeri*. If true, then *P. barkeri* has not been isolated from cacti sampled in Brazil (Rosa, unpublished data). Strains consistent with the *P. barkeri* physiological profile have been isolated from *Opuntia ficus-indica* pads and fruits near Zuñiga, Peru (Ganter, unpublished data).

Pichia cephalocereana, *Pichia deserticola*, *Pichia eremophila*, *P. heedii*, and *P. pseudocactophila* represent cactophilic species associated with necrotic tissues of columnar cacti or *Opuntia* collected in the USA and on islands in the Caribbean. These species occur in geographically limited areas or are restricted to specific cactus hosts, and they were not isolated from samples collected in South America. *P. deserticola* has been found in necroses of *Pachycereus lepidanthus* near the town of San Cristobal, Guatemala, and *Op. stricta (inermis)* necroses near the Angorichina Station in the Flinders Ranges, South Australia (Ganter, unpublished data). Consistent with the original description (Phaff et al. 1985), sporulating strains were found on the *Opuntia* in Australia but only non-sporulating strains from columnar cacti in Guatemala.

C. sonorensis is one of the most common cactophilic yeast species associated with necrotic cactus tissue, and it has been isolated from almost all cactus species studied. This yeast belongs to the clade *Ogataea* and has the ability to use methanol as a source of carbon (Starmer et al. 2003; Daniel et al. 2014). The species might use the methanol produced during cell-wall breakdown as the cactus tissue decomposes. *C. sonorensis* is able to benefit the insects by consuming 2-propanol present in cactus rots (Starmer et al. 1986; Ganter et al. 1989; Lachance et al. 2011).

Cl. opuntiae has been isolated from several different regions and is associated with prickly pear cactus (*Opuntia* spp.). Some studies suggest that this species is associated with the moth *Cact. cactorum* that attacks cacti (Rosa et al. 1992; Lachance et al. 2000). This moth was used for biological control of *Opuntia*, and it has been suggested that it is the primary vector of this yeast. Rosa et al. (1992) described the association of *Cl. opuntiae* and the moth *Sigelgaita* sp. (Lepidoptera: Phycitidae) in the columnar cactus *Pil. arrabidae* in restinga ecosystems in Brazil. This moth is a solitary internal cactus feeder, forming a tunnel with a single external exit, where the feces are deposited, forming a plug that obscures access to the tunnel (Fig. 8.3, Rosa et al. 1992). This moth larva is apparently one of the principal causes of mechanical damage with necrosis formation in *Pil. arrabidae*. *Cl. opuntiae* was the prevalent species associated with the larva, tunnels, and external fecal deposits

Fig. 8.3 External fecal deposit and tunnel of *Sigelgaita* sp. larva in the cactus *Pilosocereus arrabidae*

of the moth; however, the yeast was not isolated from the eggs of *Sigelgaita* sp. Probably, the microhabitat conditions in the tunnels of the moth favor the growth of *Cl. opuntiae*. In necroses not associated with the moth in the same species of cactus, *Cl. opuntiae* was a minor component of the yeast community (Rosa et al. 1994). The vectors of *Cl. opuntiae* among the *Sigelgaita* sp. tunnels are *Drosophila* flies and beetles that also feed and breed in the tunnels and external fecal deposits of this moth (Rosa et al. 1992).

Cactophilic *Starmera* species are represented by *Starmera amethionina*, *Starmera pachycereana*, *Starmera caribaea*, and *Starmera pilosocereana* (Kurtzman 2011a; Freitas et al. 2015). These species require an exogenous source of L-methionine or L-cysteine. *Starm. amethionina* is commonly recovered from rot pockets in cacti of the subtribe *Stenocereinae*, whereas *Starm. pachycereana* has its principal habitat in cacti of the subtribe *Pachycereinae* (Phaff et al. 1992). *Starm. caribaea* occurs associated with cereoid and *Opuntia* cacti in the Caribbean region. One cactophilic species of this genus, *Starm. pilosocereana*, was found associated with necrotic stems of the cactus *Pil. arrabidae* in the coastal regions of Brazil (Freitas et al. 2015). Four possible new species of *Starmera,* without a formal description, have been isolated from *Opuntia caracassana* on Curaçao, from *Pac. lepidanthus* in Guatemala, from *Opuntia quimilo* near Santiago del Estero in Argentina, and from *Opuntia streptacantha* near Blinman Pools in South Australia and *Opuntia tomentosa* at various sites in Queensland, Australia (Ganter, unpublished data). Each has a distinct physiological profile, and two, designed as *Starmera* sp. A and *Starmera* sp. B (from Australia and Guatemala, respectively), have partial LSU rDNA sequences sufficiently different from all other described species to justify their description as new species (the remaining two have not yet been sequenced).

Phaffomyces species associated with cacti are *Phaffomyces antillensis*, *Phaffomyces (Candida) coquimbonensis*, *Phaff. opuntiae*, *Phaffomyces (Candida) orba*, and *Phaffomyces thermotolerans*. The other species of this genus, *Phaffomyces usticensis*, was isolated from the cloaca of migratory birds (Francesca et al. 2014; see also Chap. 14 of this book). Cactophilic *Phaffomyces* species are restricted to a few host plants or to defined geographical areas. *Phaff. opuntiae* and *Phaff. orba* were isolated from *Opuntia* species only in Australia. *Phaff. coquimbonensis* was isolated in Chile and Australia. However, there is no overlap in the geographical distribution between *Phaff. coquimbonensis* and any other species in the *Phaffomyces* clade (Cardinali et al. 2012). *Phaff. thermotolerans* is associated with cardon and senita cacti of the subtribe *Pachycereinae* in the Sonoran Desert. In Brazil, no species of *Phaffomyces* has been isolated in collections of at least six cactus species in different regions (Rosa, unpublished data).

Lachance et al. (2001a) reported that *Sporop. cereana* species complex consists of three varieties related to the type species of *Sporop. cereana* (var. *cereana*, var. *oaxacaensis*, and var. *stenocereana*) and nine new species. These varieties of *Sporop. cereana sensu stricto* appear mutually exclusive in their distributions: the Sonoran region (var. *cereana*), southern Mexico (var. *oaxacaensis*), and Haiti (var. *stenocereana*). The other species isolated from cacti are *Sporopachydermia pachycereana* and some of the above new species informally reported in the literature as belonging to the *Sporop. cereana* species complex (Lachance et al. 2001a; Ganter 2011) but still lacking a formal description: *Sporopachydermia obscura*, *Sporopachydermia australis*, *Sporopachydermia opuntiana*, *Sporopachydermia trichocereana*, *Sporopachydermia centralis*, and *Sporopachydermia brasiliensis*. *Sporopachydermia agaves* and *Sporopachydermia paraquercuum* are not cactus-specific and were isolated from agave rots collected in the Mexican state of Jalisco and a red oak exudate in Ontario, Canada, respectively. *Sporop. obscura* has been sporadically isolated from various host plants and geographical locations in the tropical Americas, and *Sporop. australis* was isolated from three South American regions (Lachance et al. 2001a; Koch et al. 2015). The species *Sporop. opuntiana* comprises isolates that originated from Australia, the southeastern USA, and the Caribbean, mostly from prickly pears. *Sporop. trichocereana* and *Sporop. brasiliensis* were isolated from Argentinian cardon cactus and from columnar cacti in Brazil, respectively. *Sporop. centralis* was isolated from the Mexican state of Chiapas and two localities in Honduras (Lachance et al. 2001a). In Brazil, three species of the *Sporop. cereana* species complex are frequently isolated from cactus rots: *Sporop. australis*, *Sporop. brasiliensis*, and *Sporop. obscura* (Rosa, unpublished data).

Myx. mucilagina is a mucoid ascomycetous yeast and was described based on 40 strains, 11 from *Sten. gummosus* collected in Baja California, Mexico, and 29 from *Opuntia inermis* collected in New South Wales and Queensland, Australia (Phaff et al. 1980). It has been isolated from *Opuntia* in the northern Chihuahuan Desert (New Mexico, USA: Ganter, unpublished data). This species was also isolated from rotting tissues of *Pil. arrabidae* in Southeastern Brazil (Rosa et al. 1994) and *Op. quimilo* in Santiago del Estero, Argentina (Ganter, unpublished data). It probably has a widespread geographic distribution, particularly from *Opuntia* rots. *Myx. mucilagina* is the only species that forms exceptionally slimy colonies among the yeasts isolated from necrotic cactus tissue (Phaff et al. 1980).

Another interesting yeast genus that is predominately cactophilic is *Tortispora* (Lachance and Kurtzman 2013). This genus currently accommodates eight species and six are isolated from cacti. *Tortispora caseinolytica* has a strong extracellular proteolytic activity and was isolated from *Opuntia* species in the USA and Mexico. The type of species of the genus is represented by *Tortispora ganteri* and was described based on a single ascosporic strain producing helical ascospores isolated from sample of *Sten. gummosus* and another non-sporogenous strain from *Myrtillocactus cochal* collected in Baja California, Mexico. *Tortispora cuajiniquilana* was isolated form liquid rot of a dead, unidentified columnar cactus in Costa Rica; *Tortispora mauiana* from a rotting cladode of *Op. ficus-indica* in Hawaii; *Tortispora starmeri* from necrotic tissue of *Sten. thurberi* in Arizona, USA; and *Tortispora phaffii* from the necrotic cladode of *Opuntia bonaerensis* in Tucumán, Argentina. Additional strains of *Tort. phaffii* were isolated from lepidopteran larvae feeding in flowers of *Cer. saddianus* and from necrotic stems of *Pil. arrabidae* in Brazil (Rosa, unpublished data), suggesting that this species is endemic of

South America. However, according to Lachance and Kurtzman (2013), *Tortispora* species are only infrequently isolated, which makes it difficult to draw generalizations on habitat and biogeography.

Strains of *Magn. starmeri* were isolated from columnar cacti belonging to the subtribe Stenocereinae Baja California Mexico, Sonora in Mexico, and southern Arizona (Phaff et al. 1997). These cacti are characterized by high levels of triterpene glycosides, and this species is resistant to this toxic compound. Phaff et al. (1997) postulated that the original habitat of this species is *Stenocereus* spp. necroses and that *Magn. starmeri* may be occasionally transported to species of other cacti by various insects. Isolates of two possible new species associated with cacti and related to *Magn. starmeri* are reported by Phaff et al. (1997): five isolates designated as "species A" came from necroses in specimens of *Stenocereus hystrix* on the Caribbean islands, and nine strains isolated from cacti in Mexico are designated as "species C" and represent another genospecies related to *Magn. starmeri*. Unfortunately, the formal description of these two species was not done by the authors. Another species of this genus, *Magnusiomyces spicifer*, was isolated from cactus necrosis. This yeast was species isolated from *Op. phaeacantha* in Arizona, and it was described based on a single strain (de Hoog et al. 1986; Phaff et al. 1997). The arthroconidial species, *Dip. australiensis,* was also isolated from rotting cladodes of *Op. stricta* in Australia and South Africa. One strain was isolated from *Euph. ingens* in South Africa (de Hoog and Smith 2011). Other studies are necessary to confirm if these two species are cactophilic or occasional contaminants of cactus rotting tissues. In our studies in Brazil, the most frequent arthroconidial species associated with the cactus necroses was *Magnusiomyces capitatus* (Rosa, unpublished data). Ganter et al. (2010) reported that *Galactomyces geotrichum* represented 14% of the yeast community isolated from cacti in Curaçao. Three new species (not yet formally described) of *Dipodascus* from cactus necroses, *Dipodascus* sp. A, *Dipodascus* sp. B, and *Dipodascus* sp. C, respectively, were found in Australia, South America, and Caribbean islands, respectively (Table 8.2, Ganter, unpublished data). They were isolated from *Cereus repandus*, *Opuntia caracassana*, *Stenocereus griseus*, and *Cephalocereus urbanianus* in Curaçao and Saint Martin, from several cactus species in Peru and Brazil, and from *Op. stricta* in Australia. LSU rDNA and ITS sequencing is completed for these three species, and the sequences are consistent within strains from each species and distinct from other *Dipodascus* species (Roscini, unpublished data).

Yamadazyma mexicana (former *Pichia mexicana*) was described based on 13 isolates obtained from decayed tissues of ceroid cacti in Mexico and Arizona, USA (Miranda et al. 1982). This species accounted for 10% of the yeast communities from organ pipe cactus and 1% of the total yeast from other cactus species (Miranda et al. 1982; Starmer et al. 1990). Based on these data, it was suggested that this this species was specific to cactus tissue from *Stenocereus* spp. Subsequently, further collections showed that *Yam. mexicana* has a worldwide distribution and can be isolated from a variety of habitats: grape must, pressed olives, cow mastitis, insect frass, exudates, and cocoa fermentation (Daniel et al.

2009; Kurtzman 2011b). Strains with distinct physiological profiles have been isolated from *Op. tomentosa* in Queensland, Australia; a second variety from *Op. stricta (inermis)* in South Australia; and a third from *Opuntia* spp. in Argentina (Ganter, unpublished data).

Several other yeast species occur in necrotic tissues of cacti and could be considered as transitory component of this yeast community. *Kluyveromyces marxianus* occurs in low frequency associated with cactus rots in North America (Starmer et al. 1990) and Brazil (Rosa, unpublished data). However, in a recent study of yeasts associated with cacti in Brazil, a novel species of *Kluyveromyces* sp. not yet described was found (Table 8.2). Twenty-one strains of this species (GenBank accession number KT853036) were isolated from necrotic tissues, flowers, fruits, and associated insects of the columnar cacti *Cer. saddianus*, *Micranth. dolichospermaticus*, and *Pil. arrabidae* in two different ecosystems. This species is related to *K. marxianus* and could represent a cactophilic species in this genus as this yeast was not isolated from other substrates in these ecosystems.

Pichia norvegensis, *Pichia kluyveri*, *Meyerozyma guilliermondii*, *Pichia kudriavzevii*, *Wickerhamomyces rabaulensis*, *Hanseniaspora opuntiae*, *Hanseniaspora uvarum*, *Candida tropicalis*, *Rhodotorula mucilaginosa*, *Kwoniella* spp., *Naganishia* spp., *Papiliotrema* spp., *Vanrija* spp., *Sporidiobolus* spp., *Hannaella* spp., and the yeast-like algae *Prototheca* spp., among others, have been isolated from necrotic tissues of different species of cacti and could be considered as transitory components of these substrates. These species have a widespread distribution occurring in different substrates and ecosystems.

Rotting cactus flowers present a distinct yeast community, with *Kodamaea nitidulidarum*, *Candida (Kodamaea) restingae*, and *Wickerhamiella cacticola* as the prevalent species in several columnar cacti in Brazil (Lachance et al. 1998; Rosa et al. 1999). These species are vectored by beetles of an unidentified nitidulid species (Fig. 8.4). Another yeast associated with cactus flowers is *Kurtzmaniella cleridarum*, isolated from *Carpophilus* beetles, collected from flowers of different *Opuntia, Cylindropuntia, Echinocereus*, and *Carnegiea* species in the USA (Lachance et al. 2013). To this point, only a few studies of cactus flower yeasts have been done, and the biogeographic distribution of these species still needs to be elucidated.

Fig. 8.4 Flowers of *Pilosocereus arrabidae* with nitidulid beetles. (**a**) Open flower, (**b**) flower showing adults of nitidulid beetles, (**c**) rotting flowers after pollination and fruit formation

8.5 Cactophilic Yeast Phylogeny and Phylobiogeography

The first descriptions of cactophilic yeasts placed many species in *Pichia* (Phaff et al. 1978, 1985; Starmer et al. 1978a, b, 1979, 1984). The presence of so many congeners within the system seemed to suggest extensive speciation within the cactophilic habitat (Starmer et al. 1986). At that time, DNA data was limited to G + C content and reassociation studies, neither of which permitted the construction of phylogenies. As DNA techniques improved, it became possible to compare multiple species at the same time, and the earliest phylogenetic studies of cactophilic yeasts found considerable variation among cactophilic species (Shen and Lachance 1993; Yamada et al. 1999). They also suggested that the genus was polyphyletic. These phylogenies found strong support for subclades of phenotypically similar species that were removed from *Pichia* and described as *Phaffomyces* and *Starmera* (Yamada et al. 1998).

With the reconsideration of *Pichia*, was it apparent that no single clade dominated the cactophilic habitat (Table 8.2). All of the species designated as "cactophilic" are ascomycetes. Although species from some polyphyletic basidiomycetous genera (former *Cryptococcus*, *Rhodotorula*, and their related new genera) (Fell et al. 2000) are regularly isolated from stem rots, no exclusively cactophilic species have been described so far. A phylogeny of cactophilic Saccharomycotina would include members of most of the more speciose clades within the Saccharomycotina (Kurtzman and Robnett 2013; Shen et al. 2016). The inference is that the cactophilic habitat has been invaded multiple times by yeasts from many different Saccharomycotina lineages (Starmer et al. 2003).

However, there is evidence of within-habitat speciation in *Sporopachydermia*, *Dipodascus*, *Magnusiomyces*, *Starmera*, and *Phaffomyces*. The extent of within-habitat speciation is difficult to assess at this time as there are undescribed species in several of the genera. In addition, as much of the cactophilic habitat is unsampled or undersampled, more species are likely to be described. *Myx. mucilagina* strains vary widely in karyotype and overall genome size, so there may be cryptic species within the only species of cactophilic *Myxozyma* yet described (Ganter 2011).

Starmera and *Phaffomyces* once contained only cactophilic species. By 2011, several non-cactophilic species of *Starmera* had either been described or transferred into the genus based on sequence data (Kurtzman 2011a; Freitas et al. 2015). This leaves *Phaffomyces* as the only cactophilic genus, but its status has been put into doubt by recent isolations of a new *Phaffomyces* species, *Phaff. usticensis*, from the cloacae of birds migrating north across the Mediterranean Sea (Francesca et al. 2014). The new species tolerated the high body temperature of the birds and, being isolated from the cloaca, had successfully passed through the bird's gut. What the birds fed on before setting off is not known. *Opuntia* have been introduced on both shores of the Mediterranean Sea, so it is possible that *Phaff. usticensis* is cactophilic, but this is the first time a bird has acted as a vector for a (potentially) cactophilic yeast.

Starmer et al. (2003) mapped habitat data onto sequence-based phylogenies of cactophilic lineages and their relatives in order to reveal the ancestral habitats for cactophilic species and genera. The ancestral habitat for the *Pichia* clade, which has invaded the cactus habitat at least five times, was fruit necroses. Fruits are full of sugars and the ancestral species probably were able to ferment, but many of the *Pichia* lineages that invaded cactus stem necroses have lost the ability. *C. sonorensis* and cactophilic *Sporopachydermia* come from ancestors living in tree fluxes, another sugar-rich habitat. Only *C. sonorensis* retained the ability to ferment. The habitat of the immediate ancestor of the cactophilic *Sporopachydermia* was equivocal, but older ancestors lived in tree fluxes. *Starmera* and *Cl. opuntiae* followed the same pattern as *Sporopachydermia*: equivocal immediate ancestral habitat and tree fluxes for deeper ancestors. *Magn. starmeri* and *Myx. mucilagina* were like *C. sonorensis* in that their immediate ancestors came from tree habitats.

Anderson et al. (2004) investigated the relationship between phylogenetic relatedness and abundance (frequency of isolation, which positively correlates with within-rot population size) for cactophilic communities. They separated the *Opuntia* community from the rest of cacti that commonly host stem necroses, which they called the columnar community. In most native cactus habitat, the larger species of cacti are either *Opuntia*, with their flattened stems (called cladodes) or large tree- or bush-like cacti with cylindrical stems. These are the "columnar" cacti. As mentioned above, the chemistry of *Opuntia* and the columnar cacti differs in that *Opuntia* stems are less toxic. Anderson et al. (2004) point out other ecological differences (vectors, maximum temperatures, sugar abundance) and conclude that *Opuntia* stem necroses are less stressful for yeasts than are columnar rots. When they correlated genetic relatedness with abundance, the results differed for the two habitats. There was no relationship between relatedness and abundance for the *Opuntia* yeast community but a slightly negative relationship between the two for columnar yeasts. That is, abundant yeasts in the columnar community tend to be distantly related, and rare yeasts tend to be closely related. They suggest that the *Opuntia* system is dominated by competitive yeast-yeast interactions and that this overwhelms any relationship between phylogeny and abundance. For columnar cacti, limited resources may result in resource partitioning that results in common yeasts not being related to one another. The assumption is that related species are more likely to have similar resource utilization patterns. But the stressful columnar stems also favor mutualisms among the less abundant species. Low population size may lessen the probability of dispersal to the next resource patch (the next rot), and the growth mutualisms seen among columnar cactophilic species (Starmer and Fogleman 1986) may be important as a way to increase the likelihood of successful dispersal.

8.6 Fruit as Habitats for Yeast Communities

The pulp of fleshy fruits with soft nutritive tissues is a primary substrate for colonization for innumerable microorganisms, and those microbial communities are dominated by yeasts among all fungi. The fruit edible pulp is also a primary food resource for many frugivorous animals, notably mammals and birds, but also invertebrates such as insects (Jordano 2000). It has been hypothesized that the origin of fruiting plants (Angiosperms) brought microbial communities an abundant source of rich food based on simple sugars (Dashko et al. 2014). The pulp of fruits has repeatedly been considered of low nitrogen and protein content but presents an excess of highly digestible energy in the form of a variety of sugars and has a high water content (Jordano 2000). Water contributes by far the greatest weight to the bulk of fruit flesh, but a large number of other compounds also accumulate, many of which are water soluble (Hulme 1971). Those products include sugars (especially glucose, fructose, sucrose), acids (malate, tartrate, oxalate, citrate, succinate), ions (especially K^+, Cl^-, SO_4^{2-}, phosphate, Ca^{2+}, Mg^{2+}), alcohols (including inositol), esters, flavonoids and glycosides (including many pigments), phenolics, amino acids, amides, proteins, vitamins, alkaloids, terpenes, lipids, and many others. In certain cells, there are large bodies consisting of resins, tannins, or crystals (Thomas et al. 1973). Carbohydrate polymers, e.g., starch, predominate in some fruits but are often hydrolyzed by the time the fruit is ripe.

When fruits ripen, a fierce competition for the fruit sugars occurs within microbial communities (Dashko et al. 2014). Yeasts usually become the predominant group in niches with freely available mono- and oligosaccharides. The fast sugar consumption, ethanol production, high levels of ethanol tolerance, and the ability to propagate without oxygen are likely some of the "winning" traits responsible for the competition's outcome (Piskur et al. 2006). The Crabtree effect, the ability to ferment abundant glucose in the presence of sufficient oxygen for oxidative respiration, allows some yeast to dominate sugar-rich environments. The ability appeared in yeast contemporaneously with the origin of fruits and at the divergence of the *Kluyveromyces-Eremothecium* (Crabtree-negative) and *Lachancea-Saccharomyces* (Crabtree-positive) lineages (Hagman et al. 2013). Crabtree-positive yeast produce copious ethanol under aerobic conditions, and its toxicity, to which they are tolerant, gives them a competitive advantage. This metabolic pathway emerged in yeasts concomitantly with the shift by angiosperms from small wind-dispersed seeds to larger and fleshier vertebrate-dispersed fruits during the late Cretaceous into the Paleocene (Eriksson et al. 2000; Benner et al. 2002). The fermentative metabolism in yeasts appears to have specifically evolved to exclude bacterial competitors within ripe fruit (Ingram and Buttke 1984).

Ethanol plumes emanating from ripe fruit result in the attraction of insect vectors that disperse yeasts among available new fruits both locally and regionally. Those plumes might also have provided useful sensory information, both diurnally and nocturnally, for mammalian frugivores. The low molecular weight of ethanol and its substantial concentration within fruit pulp well suit this molecule for long-distance signaling of availability to appropriate consumers. Ripening involves production of a number of fruit volatiles (Nursten 1970), but ethanol is perhaps the only olfactory commonality to an otherwise bewildering taxonomic array of angiosperm fruits.

As resources, fruits are extremely aggregated both in space and time. Fruit production by trees is regulated by seasonality—a complex role of drought and rainfall seasons—and they are also spatially aggregated, usually in relatively isolated patches with high local abundance. The olfactory cues are thus important in the ability to attract dispersers and success in colonization of new habitats for fruit-producing plants. For example, female *Drosophila* flies are known to follow ethanol plumes to locate ripe fruit suitable for oviposition sites (Hoffmann and Parsons 1984). By visiting ripe fruits patches colonized by yeasts, drosophilids vector yeasts to new substrates. On the other hand, evolutionary exposure to ethanol has yielded flies adapted to high levels of ethanol (Geer et al. 1990; Merçot et al. 1994; Ashburner 1998; Fry 2001). Chronic environmental exposure to ethanol may thus result in physiological adaptation and overall fitness benefits to animal frugivores. For example, the life span of *Drosophila* species that naturally encounter fermenting nutritional substrates is increased at very low concentrations of ethanol but decreases at zero exposure and at higher concentrations (Starmer et al. 1977; Parsons 1980). Similarly, lifetime fecundity of *Drosophila* is enhanced by the presence of low-concentration ethanol vapor (Etges and Klassen 1989).

8.7 Fruit Yeast Diversity

Fruits are not a single habitat but a mosaic of patchy habitats for yeasts and other microorganisms. It is probable that niche partitioning and ecological succession occur in fruit yeasts. Fruit are composed of enlarged floral components including one or more carpels and, in some cases (depending on the species), tissues derived from the calyx, receptacle, bracts, or floral tube (the basal region of floral organ fusion) (Adams-Phillips et al. 2004). The first yeasts occupying green fruit may come from the flower. Flowers of various plants host extensive yeast communities that include members of the *Starmerella, Wickerhamiella, Kodamaea*, and *Metschnikowia* clades (Lachance 2013). Transient species in flowers are basidiomycetes, members of Tremellomycetes (Agaricomycotina) and Microbotryomycetes (Pucciniomycotina) (Mittelbach et al. 2016). They lack the ability to ferment sugars but effectively consume a broad spectrum of carbon sources (polysaccharides, polyols, acids, phenols) and produce extracellular polysaccharides (e.g., capsules), enzymes, and volatiles (Kurtzman et al. 2011). In a study from an orchard in Korea, Hyun et al. (2014) showed that only sixteen species among one hundred fifty-six yeast strains overlapped between fruits and flowers indicating clear separation between the two habitats on the same plants.

While fruits are attached to the plant, they begin the ripening process, a series of physical-chemical alterations that include pigment biosynthesis, production of volatiles (the main one being ethylene), and softening (Fisher and Bennett 1991), and become a different habitat for yeasts. Softening occurs through the synthesis of specific cell-wall hydrolases and enhances susceptibility to colonization by yeasts (Fisher and Bennett 1991) that may also contribute to the softening process with pectinases and polygalacturonases. Ripe fruit that has been damaged sufficiently to break the integument opens a new habitat for yeast, the soft tissues of the inner mass. Most fruit yeasts are ascomycetous species that assimilate only a very narrow set of carbon sources. Ascomycetous yeasts are described by Lachance (2013) as copiotrophs. They assimilate few carbon sources in a carbon-rich environment. They are often facultatively fermentative, may use some form of filamentous growth for local dispersal, and release odorous compounds that facilitate long-distance dispersal by insects.

Coexistence in yeast communities colonizing fruits is dynamic and depends on at least three fundamental factors: the ability to colonize and grow quickly, the ability to compete in an environment characterized by high colonization-extinction rates, and the ability to disperse to new substrates.

Fruits are usually sites for oviposition and feeding by insects that disperse yeasts to new substrates. Fruit yeast communities that are geographically close are more similar to each other, but community structure also depends on dispersal and also on a priority effect. In fact, high growth rates are expected to lead to stronger priority effects because species grow rapidly and modify the environment for the next to arrive (Peay et al. 2013). In fruits, colonization is strongly mediated by the high sugar concentrations, low pH, and constant presence of visiting insects. In masau fruits from Zimbabwe, *A. pullulans* was the dominant species on the unripe fruits but was not isolated from the fermented fruit pulp. *S. cerevisiae* and *P. kudriavzevii* (formerly *Issatchenkia orientalis*) were predominant in the fermented fruit pulp but were not detected on the unripe fruits (Niyanga et al. 2007). This is indicative of the strength of the constraints that promote different yeast communities on green and ripe fruits.

The first report on the microbiota of tropical fruits was Joly's monograph on the ripe fruits of Brazil in 1955 that pointed out the prevalence of ascomycetous yeasts in the communities of mango fruits. The monographs of Faparusi (1974), Leitão et al. (1976), Oliveira (1980), Ivo (1982), Robbs et al. (1989), and Santos et al. (1996) reported the isolation of *Hanseniaspora* and *Pichia* (*Issatchenkia*) predominantly from different Brazilian fruits. Joly (1955) and Leitão et al. (1976) studied the yeast microbiota of ripe fruits and isolated strains from three genera, *Hanseniaspora*, *Pichia*, and *Candida*, with the apiculates predominating (41%) but colonies of black yeasts were also observed. Oliveira (1980) reported the predominance of *H'spora uvarum* (anamorph: *Kloeckera apiculata*) in the external and internal parts of cashew and suggested that this yeast may be responsible for the rotting of these fruits. Robbs et al. (1989) verified that species of *Mey. guilliermondii*, *P. kudriavzevii*, and *Hanseniaspora guilliermondii* were associated with rotting fruit. Santos et al. (1996) observed that yeasts were not found in the internal tissues of unripe cashew, caja, and umbu fruits but that large numbers were present inside the very ripe cashew. Abranches et al. (1997) found 28 ascomycetous species associated with guava fruits. All species had similar functional profiles of carbon usage that could have converged in response to host chemistry and vectors. Trindade et al. (2002) found 42 ascomycetous and 28 basidiomycetous yeasts associated with ripe tropical fruits in northeastern Brazil. Among those yeasts, various pectinolytic and β-glucosidase-producing strains were isolated. Morais et al. (1995) found that about 50% of the yeasts associated with Amazonian fruits were ascomycetous. Physiological abilities of the yeast community associated with the fruits were mostly restricted to the use of glycerol, cellobiose, DL-lactate, salicin, and L-sorbose among the 36 carbon compounds tested. Fermentative species made up more than 82% of the isolates, as was expected for yeasts associated with deteriorating fruits. In a broad search for yeasts in tropical fruits from 20 different angiosperms in the Atlantic rain forest of Brazil, Prada and Pagnocca (1997) found that ascomycetous yeasts comprised 74% of the community. Most species produced a pseudomycelium, were strong fermenters (51%), and did not assimilate nitrate as a sole nitrogen source (58%). Spencer et al. (1992) showed that the range of yeasts found in decaying citrus fruits (oranges, mandarins, grapefruit, limes, and lemons) was rather narrow. Species isolated from the rotting fruit included *H'spora uvarum*, *Mey. guilliermondii*, *Candida stellata*, *P. kluyveri*, other *Pichia* species of the *Pichia membranifaciens* group, and *Galactomyces candidus*. The strains of the species *H'spora uvarum* isolated in this work appeared to have some pectinolytic activity.

Endophytic yeasts from the genera *Sporobolomyces*, *Rhodotorula*, *Debaryomyces*, and *Cryptococcus* (currently *Papiliotrema* and *Naganishia*) were isolated from the leaves, flowers, and fruit of healthy apple trees (*Malus domestica*, Borkh) in Brazil (Cannatti-Sartori et al. 2005). In the cerrado (a vast tropical savanna biome typical of Brazil), *A. pullulans* was the dominant species among fourteen ascomycetous and three basidiomycetous yeasts isolated from fruits of *Eugenia dysenterica* and *Byrsonima crassifolia*, plants endemic to this biome (Sperandio et al. 2015). Among the yeasts, only *A. pullulans* and *Pseudozyma hubeiensis* were found in both hosts indicating that different yeast communities may be associated with different host plants.

Several new yeast species have been isolated from ripe tropical fruits and associated insects. *Saccharomycopsis* (*Candida*) *amapae* was isolated from the Amazonian fruit, *Parahancornia amapa* (Morais et al. 1995). *Wickerhamomyces queroliae* and *Candida azymoides* (*Wickerhamiella* clade) were isolated from the larvae of *Anastrepha mucronata* (Diptera: Tephritidae) collected from ripe fruit of *Peritassa campestris* ("bacupari," Hippocrateaceae, Fig. 8.5a) in Brazil (Rosa et al. 2006, 2009). *Candida citri* (Metschnikowiaceae clade) was isolated from rotting lime fruits in Borneo (Sipiczki 2011). Three yeast strains, described as *Candida pruni,* were isolated from the surface of plums (cv Chinese Angelino) collected in the north of China. These strains reduced the brown rot caused by the fungus *Monilinia fructicola* incidence to 20% (Zhang et al. 2014). In the study about the yeast community associated with the fruit *Byrsonima* sp. (murici, Malpighiaceae, Fig. 8.5b), 11 novel yeast species were isolated (Rosa, unpublished data). In this fruit, the most frequent species were *A. pullulans*, *P. kudriavzevii*, and *Mey.*

Fig. 8.5 Ripe fruits. (**a**) Ripe fruit of *Peritassa campestris* (bacupari, Hippocrateaceae) with larvae of *Anastrepha mucronata* (Diptera: Tephritidae), (**b**) fruits of *Byrsonima* sp. (murici, Malpighiaceae)

caribbica. It is obvious that ripe fruits represent a rich source of new yeast species, and sampling of these substrates in regions where yeast diversity is understudied could produce many new species.

8.8 Yeast Succession and Ecological Interactions in Tropical Fruit

Successional patterns of yeast colonization of fruits have been associated with *Drosophila* vectors (Miller and Phaff 1962; Lachaise et al. 1979; Morais et al. 1995). Miller and Phaff (1962) studied the colonization of *Calimyrna* figs by yeasts vectored by the pollinating fig wasp *Blastophaga psenes* and other insects. This wasp introduces a specific microbiota which consists of *Mey. guilliermondii* and bacteria of the species *Serratia plymuthica*. These microorganisms persist and increase in number throughout the ripening period, attracting drosophilids that carry yeasts such as *Hanseniaspora valbyensis*, *H'spora uvarum*, and *C. stellata* that cause active fermentative spoilage of mature fruits.

On the African savanna, Lachaise et al. (1979) showed that the colonization of *Ficus* fruits proceeds with the aid of a successive series of drosophilid species influencing the microbial colonization of the fig receptacle. *Candida fructus* colonizes the early green receptacle, serving as food for *Lissocephala* flies, and persists after the inoculation of fermentative yeasts by the drosophilids *Zaprionus vittiger* and *Drosophila malerkotliana*. *P. membranifaciens*, *H'spora valbyensis*, *H'spora uvarum*, and *Candida sorboxylosa* cause rapid fermentative breakdown (souring) of the fig and stimulate oviposition by *Drosophila fima*, *Drosophila greeni*, and *Zaprionus ornatus*. Thus, the fig may be regarded as a microhabitat for

Drosophila-yeast interactions, and sequential fermentation of figs could be considered as a succession of microhabitat patches.

Morais et al. (1995) found a similar pattern of succession in the ripe fruits from an Amazonian tree, *Par. amapa*, that had fallen to the forest floor. The yeast populations and species frequencies fluctuated during the deterioration of the *Par. amapa* fruit. Yeast counts increased from 5.2×10^5 on the first day after the fruit fell to 1.7×10^6 cfu g^{-1} at day 14 and the odds measure of yeast species diversity went from 6.1 to 16.8. The first isolation of each species demonstrated a diagonal trend suggesting succession. The availability of simple sugars and favorable pH may have affected yeast colonization. Species frequently isolated during the first two days after the fruit fell disappeared after three or four days. After day 3, these initial species were replaced by others that were probably inoculated by visiting *Drosophila*. In the final stage of succession, another group of species replaced those that replaced the initial colonizers.

Colonization patterns were correlated with the nutritional profiles of the yeast species in the rotting amapa fruit (Morais et al. 1995). Yeasts colonizing the early phases of the fallen fruit had physiological profiles restricted mostly to assimilating a few simple sugars together with rapid fermentation of glucose and sucrose, whereas those from the latter successional stages presented a broader physiology. Among the factors affecting the succession patterns, pH, and nutrient content of the fruit, drosophilid visitors and killer toxins are listed as the main forces shaping the yeast communities in *Par. amapa* fruits. Janisiewicz and Buyer (2010) found that the composition of the fruit surface natural microbiota changes as the fruit season progresses. This may be due to the flux in the nutritional status of the fruit, such as differences in abundance of carbon and nitrogen compounds (sugar and amino acids) and tissue pH.

Biological interactions can alter the colonization rates and species' success in fruits, including production of mycocins and ethanol derived from fermentation of fruit sugars. Mycocinogenic yeasts are more abundant in habitats with relatively large yeast populations, while habitats such as soil, which contains relatively few yeasts, harbor a smaller percentage yeasts able to produce these antifungal compounds. Golubev and Golubeva (2004) isolated mycocinogenic yeast strains from plants with a total yeast count of 10^5–10^6 cfu g^{-1}, whereas none could be isolated from soil with a total yeast count of 10^4 cfu g^{-1}, indicating that the competitive ability of yeasts within a densely populated yeast community is enhanced by their ability to produce mycocins. Also, mycocinogenic yeasts isolated from particular habitats were found to have greater killing activity against yeasts from other habitats than against those in their own habitat (Golubev 2006). Starmer et al. (1987a) showed that *P. kluyveri* toxin killed 12% of the strains tested from fruit habitat, whereas the activity against yeasts from other habitats rose to 42% of the strains tested. Abranches et al. (2000) showed that mycocinogenic yeasts in guava (*Psidium guajava*) fruits in tropical Brazil presented narrow killing spectra against strains from the same habitat. Trindade et al. (2002) also found mycocin production among yeasts from ripe pitanga (*Eugenia uniflora* L.), mangaba (*Hancornia speciosa* Gom.), umbu (*Spondias tuberosa* Avr. Cam.), and acerola (*Malpighia glabra* L.) fruits in Northeast Brazil. Lima et al. (2012) showed that yeasts from tropical fruits produce killer toxins capable of controlling the germination of fungi of the species *Colletotrichum gloeosporioides* in vitro, and they suggest that antagonist yeasts isolated from tropical fruits can be used in the control of anthracnose caused by *Coll. gloeosporioides* in papaya. Antagonistic interactions played a role in yeast succession in *Par. amapa* fruits. Killer strains of *P. kluyveri* and *C. sorboxylosa* displaced colonizing species after arriving on the third day. Cessation of killing activity tied to a rise in the pH of the rot fruit led to a rise in yeast diversity during the third stage of succession (Morais et al. 1995). Golubev (2006) points that mycocinogeny has a role in defending an ecological niche against invading strains with the same nutritional requirements and excludes "foreign" competitor yeasts from the habitat (see also Klassen et al. 2017).

Another important factor shaping yeast communities from fruit is ethanol production by fermentation. Temperate zone fruits are unlikely, for reasons of low ambient temperatures alone, to be characterized by particularly high ethanol concentrations. Fermentation of fruit crops is more pronounced in warm and humid environments that promote both yeast growth and rapid decomposition. Yeasts in the Saccharomycetes subphylum ferment even in the presence of oxygen (the Crabtree effect). Fermentation is the least energetically efficient route of energy production from the substrate sugars, but it does have a greater rate of ATP production (Pfeiffer et al. 2001). This strategy produces ethanol, heat, and carbon dioxide, which in combination modifies the fruit niche and excludes other competing microbes from colonizing and utilizing the resource (Goddard 2008; Thomson et al. 2005; Piskur et al. 2006). Modification of the fruit resource to a form that is no longer suited to most other microbes has been described as an example of ecosystem engineering (Goddard 2008). Dudley (2002) presented ethanol data for three fruiting taxa in a Neotropical forest and found that pulp of ripe and very ripe palm fruits (*Astrocaryum standleyanum*) contained ethanol at concentrations of about 0.5% and 0.6%, respectively.

Another product of the manipulation of the fruit niche by yeast, which may or may not be linked to fermentation, is the array of odorous volatile chemicals (Swiegers and Pretorius 2005). Yeast will typically modify sugars, amino acids, and fatty acids to produce esters, higher alcohols, carbonyls, fatty acid derivatives, and sulfur compounds, or they may free volatiles from conjugated forms to release monoterpenes and thiols (Swiegers and Pretorius 2005; Styger et al. 2011; Romano et al. 2003). The probable function of volatile manipulation and production by yeasts is that these odors attract insects and thus enhance the dispersal and survival of otherwise nonmotile yeasts (Saerens et al. 2010).

Yeasts growing on fruit occupy a commonly overlooked trophic level between fruit and insects (Becher et al. 2012). Drosophilid flies benefit from accessing yeast-infested fruits. More attractive yeasts are dispersed more frequently, and flies associated with more attractive yeasts have higher fecundity. Although there may be multiple natural yeast and fly species combinations within fruit rots, there is evidence of a mutualistic interactions facilitated by the yeast's niche modification (Buser et al. 2014). Yeast odors represent the critical signal in the establishment of the fly-fruit-yeast relationship. Palanca et al. (2013) showed that yeasts manipulate fruit odors to mediate their interactions with their *Drosophila* vectors. *Drosophila* larval feeding may mediate yeast species densities in fruits (Piskur et al. 2006; Stamps et al. 2012), and *Drosophila* larvae may stabilize fruit yeast communities comprising mostly of *Candida*, *Pichia*, *Hanseniaspora*, and *Saccharomyces* species (Chen et al. 2004; Chandler et al. 2012). *Drosophila melanogaster* adults and their larval offspring together engage in "niche construction," facilitating the construction of a predictable microbial environment in which the larvae live and develop (Stamps et al. 2012). According to Becher et al. (2012), the traditional plant-herbivore niche concept needs to be updated, to accommodate for the role of microorganisms in insect-plant.

8.9 Concluding Remarks

Sixty-plus years of research have uncovered much about the diverse yeast communities of cactus rots, flowers, and fruits. All have a similar physical structure: discrete patches of rich habitat linked by animal activity. Each habitat differs in the abundance and types of sugars and plant secondary compounds present. Each is occupied by a distinct assemblage of yeast species, summarized above. However, there remains much work to do before a comprehensive picture of biodiversity in these habitats can be constructed.

More survey work is needed for all three habitats. Large geographical areas remain uncollected and, for tropical fruits, many hosts have not been investigated. Sample numbers are small for many hosts in all three habitats. The surveys done so far have not incorporated DNA technology, and we do not know if there is a significant uncultured component to the biodiversity of these communities. More data will be needed to make meaningful comparisons of the diversity within a flower, fruit, or rot versus local community diversity, larger-scale diversity, and global diversity.

In addition to more survey work, experiments are needed to understand the impact of ecological factors on biodiversity. The degree of habitat fragmentation, variation in host chemistry, yeast-yeast interactions, and the interaction between yeasts and vectors may all play important, intertwined roles in determining yeast biodiversity. It is significant that these microbial systems were experimental approaches that are feasible using standard techniques. Experimental microcosms can closely mimic natural rots, flowers, or fruits. Larger-scale microcosms can mimic local patches of the habitat and offer the ability to manipulate both yeast and vector over multiple generations. Field experiments involving all components of the system are also feasible, although the logistics may be daunting at times. These natural advantages offer opportunities for research into fundamental questions in microbial ecology.

Acknowledgments We would like to acknowledge William T. Starmer for contributing data for this chapter, for giving input to the planning of the chapter, and for commenting on an early draft. We would also like to acknowledge Kyria Boundy-Mills for providing a list of cactophilic strains available at the Phaff Yeast Culture Collection at the University of California, Davis, and to Luca Roscini of the University of Perugia for the sequence data for the new species of *Dipodascus*. We would like to thank the Conselho Nacional de Desenvolvimento Científico e Tecnológico (CNPq) and Fundação do Amparo a Pesquisa do Estado de Minas Gerais (FAPEMIG) for their financial support.

References

Abranches J, Mendonça-Hagler LC, Hagler AN, Morais PB, Rosa CA (1997) The incidence of killer activity and extracellular proteases in tropical yeast communities. Can J Microbiol 43: 328–336

Abranches J, Vital MJS, Starmer LC, Mendonça-Hagler LC, Hagler AN (2000) The yeast community and mycocin producers of guava fruit in Rio de Janeiro, Brazil. Mycologia 92:16–22

Adams-Phillips L, Barry C, Giovannoni J (2004) Signal transduction systems regulating fruit ripening. Trends Plant Sci 9:331–338

Allen GE (1969) T. H. Morgan and the emergence of a new American biology. Q Rev Biol 44: 168–188

Allen GE (1975) The introduction of *Drosophila* into the study of heredity and evolution: 1900–1910. Isis 66:322–333

Allen GE (1980) The evolutionary synthesis: Morgan and natural selection revisited. In: Mayr E, Provine WB (eds) The evolutionary synthesis. Harvard University Press, Cambridge, MA, pp 365–382

Allen GE (2005) Mechanism, vitalism and organicism in late nineteenth and twentieth-century biology: the importance of historical context. Stud Hist Phil Biol Biomed Sci 36:261–283

Anderson TM, Lachance MA, Starmer WT (2004) The relationship of phylogeny to community structure: the Cactus yeast community. Am Nat 164:709–721

Ashburner M (1998) Speculations on the subject of alcohol dehydrogenase and its properties in *Drosophila* and other flies. BioEssays 20:949–954

Ayala FJ (1976) Theodosius Dobzhansky: the man and the scientist. Annu Rev Genet 10:1–7

Ayala FJ (1985) Theodosius Dobzhansky. Biogr Mem Natl Acad Sci 55:163–213

Becher PG, Flick G, Rzpedowska E, Schmidt A, Hagman A, Lebreton S, Larsson MC, Hansson BS, Piskur J, Witzgall P, Bengtsson M (2012) Yeast, not fruit volatiles mediate *Drosophila melanogaster* attraction, oviposition and development. Funct Ecol 26:822–828

Benner SA, Caraco MD, Thomson JM, Gaucher EA (2002) Planetary biology—paleontological, geological, and molecular histories of life. Science 296:864–868

Bennett JW, Phaff HJ (1993) Early biotechnology: the Delft connection. ASM News 59:401–404

Botstein D, Fink GR (2011) Yeast: an experimental organism for 21st Century biology. Genetics 189:695–704

Botstein D, Chervitz SA, Cherry M (1997) Yeast as a model organism. Science 277:1259–1260

Buser CC, Newcomb RD, Gaskett AC, Goddard MR (2014) Niche construction initiates the evolution of mutualistic interactions. Ecol Lett 17:1257–1264

Cannatti-Sartori V, Silva-Ribeiro RT, Valdenito-Sanhueza RM, Pagnocca FC, Echeverrigaray S, Azevedo JL (2005) Endophytic yeasts and filamentous fungi associated with southern Brazilian apple *(Malus domestica)* orchards subjected to conventional, integrated or organic cultivation. J Basic Microbiol 45:397–402

Cardinali G, Antonielli L, Corte L, Roscini L, Ganter PF (2012) *Candida coquimbonensis* sp. nov., a link between Australian and Nearctic/Neotropical *Phaffomyces*. Int J Syst Evol Microbiol 62: 3067–3071

Castrezana S, Markow TA (2001) Arthropod diversity in necrotic tissue of three species of columnar cacti (Cactaceae). Can Entomol 133:301–309

Chandler JA, Eisen JA, Kopp A (2012) Yeast communities of diverse *Drosophila* species: comparison of two symbiont groups in the same hosts. Appl Environ Microbiol 78:7327–7336

Chen H, Fujita M, Feng Q, Clardy J, Fink GR (2004) Tyrosol is a quorum-sensing molecule in *Candida albicans*. Proc Natl Acad Sci U S A 101:5048–5052

Daniel HM, Vrancken G, Takrama JF, Camu N, de Vos P, de Vuyst L (2009) Yeast diversity of Ghanaian cocoa been heap fermentations. FEMS Yeast Res 9:774–783

Daniel HM, Lachance MA, Kurtzman CP (2014) On the reclassification of species assigned to *Candida* and other anamorphic ascomycetous yeast genera based on phylogenetic circumscription. A van Leeuwenhoek 106:67–84

Dashko S, Zhou N, Campagno C, Piskur J (2014) Why, when and how did yeast evolve alcoholic fermentation? FEMS Yeast Res 14:826–832

de Hoog GS, Smith MT (2011) *Magnusiomyces* Zender (1977). In: Kurtzman CP, Fell JW, Boekhout T (eds) The yeasts: a taxonomic study. Elsevier, London, pp 565–574

de Hoog GS, Smith MT, Guého E (1986) A revision of the genus *Geotrichum* and its telomorphs. Stud Mycol 29:1–131

Dodd AP, Published under the authority of the Commonwealth Prickly Pear Board (Australia) (1940) The biological campaign against the prickly pear. A.H. Tucker, Government Printer, Brisbane

Dudley R (2002) Fermenting fruit and the historical ecology of ethanol ingestion: is alcoholism in modern humans an evolutionary hangover? Addiction 97:381–388

Eriksson O, Friis EM, Löfgren P (2000) Seed size, fruit size, and dispersal systems in angiosperms from the Early Cretaceous to the Late Tertiary. Am Nat 156:47–58

Etges WJ (2007) In memoriam. William B. Heed 1926–2007. Evolution 61:2739–2742

Etges WJ, Klassen CS (1989) Influences of atmospheric ethanol on adult *Drosophila mojavensis*: altered metabolic rates and increases in fitnesses among populations. Physiol Zool 62:170–193

Faparusi SI (1974) Yeasts associated with cacao (*Theobroma cacao*) pods. Rev Microbiol 5:49–54

Fell JW, Boekhout T, Fonseca A, Scorzetti G, Statzell-Tallman A (2000) Biodiversity and systematics of basidiomycetous yeasts as determined by large-subunit rDNA D1/D2 domain sequence analysis. Int J Syst Evol Microbiol 50:1351–1371

Fisher RL, Bennett AB (1991) Role of cell wall hydrolases in fruit ripening. Annu Rev Plant Physiol Plant Mol Biol 42:675–703

Fogleman JC (1982) The role of volatiles in the ecology of cactophilic *Drosophila*. In: Barker JSF, Starmer WT (eds) *Ecological genetics and evolution: the cactus-yeast-*Drosophila *model system*. Academic, Sydney, pp 191–208

Fogleman JC (1990) William B. Heed, A Biography. In: Barker JSF, Starmer WT, MacIntyre RJ (eds) Ecological and evolutionary genetics of *Drosophila*, Armidale, Australia, 1989. Springer Science + Business Media, New York, pp 1–14

Fogleman JC, Abril R (1990) Ecological and evolutionary importance of host plant chemistry. In: Barker JSF, MacIntyre RJ, Starmer WT (eds) Ecological and evolutionary genetic of Drosophila. Plenum Press, New York, pp 121–143

Fogleman JC, Armstrong L (1989) Ecological aspects of cactus triterpene glycosides I. Their effect on fitness components of *Drosophila mogavensis*. J Chem Ecol 15:663–676

Fogleman JC, Danielson PB (2001) Chemical interactions in the Cactus-Microorganism-Drosophila model system of the Sonoran Desert. Am Zool 41:877–889

Fogleman JC, Foster JL (1989) Microbial colonization of injured cactus tissue (*Stenocereus gummosus*) and its relationship to the ecology of cactophilic *Drosophila mojavensis*. Appl Environ Microbiol 55:100–105

Fogleman JC, Heed WB (1989) Columnar cacti and desert *Drosophila*: the chemistry of host plant specificity. In: Schmidt JO (ed) Special biotic relationships in the arid Southwest. University of New Mexico Press, Albuquerque, NM
Fogleman JC, Starmer WT (1985) Analysis of the community structure of yeasts associated with the decaying stems of cactus. III *Stenocereus thurberi*. Microb Ecol 11:165–173
Fogleman JC, Starmer WT, Heed WB (1982) Comparisons of yeast florae from natural substrates and larval guts of southwestern *Drosophila*. Oecologia 52:187–191
Fogleman JC, Duperret SM, Kircher HW (1986) The role of phytosterols in host plant utilization by cactophilic Drosophila. Lipids 21:92–96
Foster JLM, Fogleman JC (1993) Identification and ecology of bacterial communities associated with necroses of three cactus species. Appl Environ Microbiol 59:1–6
Francesca N, Carvalho C, Sannino C, Guerreiro MA, Almeida PM, Settanni L, Massa B, Sampaio JP, Moschetti G (2014) Yeasts vectored by migratory birds collected in the Mediterranean island of Ustica and description of *Phaffomyces usticensis* f.a. sp. nov., a new species related to the cactus ecoclade. FEMS Yeast Res 14:910–921
Freitas LF, Barbosa R, Sampaio JP, Lachance MA, Rosa CA (2015) *Starmera pilosocereana* sp. nov., a yeast isolated from necrotic tissue of cacti in a sandy coastal dune ecosystem. Int J Syst Evol Microbiol 65:4474–4478
Fry JD (2001) Direct and correlated responses to selection for larval ethanol tolerance in *Drosophila melanogaster*. J Evol Biol 14:296–309
Ganter PF (2011) Everything is not everywhere: the distribution of cactophilic yeast. In: Fontaneto D (ed) The biogeography of small organisms: is everything small everywhere? The Systematics Association Special Volumes, Cambridge, pp 130–174
Ganter PF, Peris F, Starmer WT (1989) Adult life span of cactophilic *Drosophila*: interactions among volatiles and yeasts. Am Midl Nat 121:331–340
Ganter PF, Bustillo E, Pendola J (1993) Yeast interactions inferred from natural distribution patterns. Florida Sci 57:50–61
Ganter PF, Cardinali G, Boundy-Mills K (2010) *Pichia insulana* sp. nov., a novel cactophilic yeast from the Caribbean. Int J Syst Evol Microbiol 60(Pt 4):1001–1007
Geer BW, Heinstra PWH, Kapoun AM, van der Zel A (1990) Alcohol dehydrogenase and alcohol tolerance in *Drosophila melanogaster*. In: Barker JSF, Starmer WT, MacIntyre RJ (eds) Ecological and evolutionary genetics of *Drosophila*. Plenum Press, New York, pp 231–252
Goddard MR (2008) Quantifying the complexities of *Saccharomyces cerevisiae*'s ecosystem engineering via fermentation. Ecology 89:2077–2082
Golubev WI (2006) Antagonistic interactions among yeasts. In: Rosa CA, Peter G (eds) The yeast handbook. Biodiversity and ecophysiology of yeasts. Springer, Berlin, pp 197–219
Golubev WI, Golubeva EW (2004) Yeasts of the steppe and forest communities of PriokskoTerraced Biosphere Reserve. Mycol Phytophatol 6:20–27 (in Russian)
Gurevitch A (1984) Population biology of the yeasts associated with plant necroses: effects of growth, colonization, and dispersal in a patch-structured environment. Dissertation, Syracuse University
Hagman A, Sall T, Compagno C, Piskur J (2013) Yeast "make-accumulate-consume" life strategy evolved as a muti-step that predates the whole genome duplication. PLoS One 8:e68734
Heed WB, Kircher HW (1965) Unique sterol in the ecology and nutrition of *Drosophila pachea*. Science 149:758–761
Hoffmann AA, Parsons PA (1984) Olfactory response and resource utilization in *Drosophila*: interspecific comparisons. Biol J Linn Soc 22:45–53
Hulme AC (1971) The biochemistry of fruits and their products, vol 2. Academic, London
Hyun S-H, Lee J-G, Park W-J, Kim H-K, Lee JS (2014) Isolation and diversity of yeasts from fruits and flowers of orchard in Sinam-myeon of Yesan-gun, Chungcheongnam-do, Korea. Korean J Microbiol 42:21–27
Ingram LO, Buttke TM (1984) Effects of alcohols on microorganisms. Adv Microb Physiol 25: 253–300

Ivo MI (1982) Leveduras do abacaxi. M.Sc. Thesis, Departamento de Microbiologia, Escola Paulista de Medicina

Janisiewicz WJ, Buyer JS (2010) Culturable bacterial microflora associated with nectarine fruit and their potential for control of brown rot. Can J Microbiol 56:480–486

Joly S (1955) Contribuição ao estudo da sistemática de leveduras ocorridas em frutos maduros. Thesis, Escola Superior de Agricultura Luiz de Queiroz, Piracicaba, São Paulo

Jordano P (2000) Fruits and Frugivory. In: Fenner M (ed) Seeds: the ecology of regeneration in plant communities, 2nd edn. CABI Publishing, Wallingford, pp 125–166

Kay LE (1996) The molecular vision of life: Caltech, the Rockefeller Foundation, and the rise of the new biology. Oxford University Press on Demand

Kenney DE, Borisy GG (2009) Thomas Hunt Morgan at the marine biological laboratory: naturalist and experimentalist. Genetics 181:841–846

Kircher H (1980) Triterpenes in organ pipe cactus. Phytochemistry 19:2707–2712

Kircher H (1982) Chemical composition of cacti and its relationship to Sonoran desert *Drosophila*. In: Barker JSF, Starmer WT (eds) Ecological genetics and evolution: the cactus-yeast-*Drosophila* model-system. Academic, Sydney, pp 143–158

Klassen R, Schaffrath R, Buzzini P, Ganter PF (2017) Antagonistic interactions and killer yeasts. In: Buzzini P, Lachance MA, Yurkov AM (eds) Yeasts in natural ecosystems: ecology. Springer, Heidelberg, pp 229–275

Koch NM, Soto IM, Galvagno M, Hasson E, Iannone L (2015) Biodiversity of cactophilic microorganisms in western Argentina: community structure and species composition in the necroses of two sympatric cactus hosts. Fungal Ecol 13:167–180

Kurtzman CP (2011a) *Starmera* Y. Yamada, Higashi, Ando & Mikata (1997). In: Kurtzman CP, Fell JW, Boekhout T (eds) The yeasts: a taxonomic study. Elsevier, London, pp 805–810

Kurtzman CP (2011b) *Yamadazyma* Billon-Grand (1989). In: Kurtzman CP, Fell JW, Boekhout T (eds) The yeasts: a taxonomic study. Elsevier, London, pp 919–925

Kurtzman CP, Robnett CJ (2013) Relationships among genera of the Saccharomycotina (Ascomycota) from multigene phylogenetic analysis of type species. FEMS Yeast Res 13: 23–33

Kurtzman CP, Fell J, Boekhout T (eds) (2011) The yeasts, a taxonomic study, 5th edn. Elsevier, Amsterdam

Lachaise D, Pignal MC, Roualt J (1979) Yeast flora partitioning by drosophilid species inhabiting a tropical African savanna of the ivory-coast (*Diptera*). Ann Soc Entomol France 15:659–680

Lachance M-A (1990) Ribosomal DNA spacer variation in the cactophilic yeast *Clavispora opuntiae*. Mol Biol Evol 7:178–193

Lachance MA (2003) The Phaff school of yeast ecology. Int Microbiol 6:163–167

Lachance MA (2013) The Biodiversity, ecology, and biogeography of ascomycetous yeasts. In: Martin F (ed) The ecological genomics of fungi. Wiley, London, pp 355–370

Lachance MA, Kurtzman CP (2013) The yeast genus *Tortispora* gen. nov., description of *Tortispora ganteri* sp. nov., *Tortispora mauiana* f.a., sp., *Tortispora agaves* f.a., sp. nov., *Tortispora sangerardonensis* f.a., sp. nov., *Tortispora cuajiniquilana* f.a., sp. nov., *Tortispora starmeri* f.a., sp. nov. and *Tortispora phaffii* f.a., sp. nov., reassignment of *Candida caseinolytica* to *Tortispora caseinolytica* f.a., comb. nov., emendation of *Botryozyma*, and assignment of *Botryozyma*, *Tortispora* gen. nov. and *Trigonopsis* to the family Trigonopsidaceae fam. nov. Int J Syst Evol Microbiol 63:3104–3114

Lachance MA, Starmer WT (2008) *Kurtzmaniella* gen. nov. and description of the heterothallic, haplontic yeast species *Kurtzmaniella cleridarum* sp. nov., the teleomorph of *Candida cleridarum*. Int J Syst Evol Microbiol 58:520–524

Lachance MA, Rosa CA, Starmer WT, Schlag-Edler B, Barker JS, Bowles JM (1998) *Wickerhamiella australiensis, Wickerhamiella cacticola, Wickerhamiella occidentalis, Candida drosophilae* and *Candida lipophila*, five new related yeast species from flowers and associated insects. Int J Syst Bacteriol 48:1431–1443

Lachance MA, Starmer WT, Bowles JM, Phaff HJ, Rosa CA (2000) Ribosomal DNA, species structure, and biogeography of the cactophilic yeast *Clavispora opuntiae*. Can J Microbiol 46: 195–210

Lachance MA, Kaden JE, Phaff HJ, Starmer WT (2001a) Phylogenetic structure of the *Sporopachydermia cereana* species complex. Int J Syst Evol Microbiol 51:237–247

Lachance MA, Starmer WT, Rosa CA, Bowles JM, Barker JSF, Janzen DH (2001b) Biogeography of the yeasts of ephemeral flowers and their insects. FEMS Yeast Res 1:1–8

Lachance MA, Boekhout T, Scorzetti G, Fell JW, Kurtzman CP (2011) *Candida* Berkhout (1923). In: Kurtzman CP, Fell JW, Boekhout T (eds) The yeasts: a taxonomic study. Elsevier, London, pp 987–1278

Lachance MA, Perri AM, Farahbakhsh AS, Starmer WT (2013) Genetic structure of *Kurtzmaniella cleridarum*, a cactus flower beetle yeast of the Sonoran and Mojave Deserts: speciation by distance? FEMS Yeast Res 13:674–681

Latham BP (1998) Yeast community persistence in a spatially structured environment. Microb Ecol 36:60–65

Leitão MFF, Uboldi Eiroa MNR, Geraldine AM (1976) Microflora de leveduras em maracujá-amarelo (*Passiflora edulis*). Col ITAL 8:531–541

Lima JR, Gonçalves LRB, Brandão LR, Rosa CA, Viana FMP (2012) Isolation, identification, and activity *in vitro* of killer yeasts against *Colletotrichum gloeosporioides* isolated from tropical fruits. J Basic Microbiol 53:590–599

Mangan RL (1979) Reproductive behavior of the cactus fly, *Odontoloxozus longicornis*, male territoriality and female guarding as adaptive strategies. Behav Ecol Sociobiol 4:265–278

Merçot H, Defaye D, Capy P, Pla E, David JR (1994) Alcohol tolerance, ADH activity, and ecological niche of *Drosophila* species. Evolution 48:746–757

Meyer JM, Fogleman JC (1987) Significance of saguaro cactus alkaloids in ecology of *Drosophila mettleri*, a soil-breeding, cactophilic drosophlid. J Chem Ecol 13:2069–2081

Miller MW, Phaff HJ (1962) Successive microbial populations in *Calimyrna* figs. Appl Microbiol 10:394–400

Miller MW, Phaff HJ, Miranda M, Heed WB, Starmer WT (1976) *Torulopsis sonorensis*, a new species of the genus *Torulopsis*. Int J Syst Bacteriol 26:88–91

Miranda M, Holzschu DL, Phaff HJ, Starmer WT (1982) *Pichia mexicana*, a new heterothallic yeast from cereoid cacti in the North American Sonoran Desert. Int J Syst Bacteriol 32: 101–107

Mittelbach M, Yurkov AM, Stoll R, Begerow D (2016) Inoculation order of nectar-borne yeasts opens a door for transient species and changes nectar rewarded to pollinators. Fungal Ecol 22: 90–97

Morais PB, Martins MB, Klaczko LB, Mendonça-Hagler LC, Hagler AN (1995) Yeast succession in the Amazon fruit *Parahancornia amapa* as resource partitioning among *Drosophila* spp. Appl Environ Microbiol 61:4251–4257

Murray ND (1982) Ecology and evolution of the *Opuntia-Cactoblastis*-ecosystem in Australia. In: Barker JSF, Starmer WT (eds) Ecological genetics and evolution: the cactus-yeast-*Drosophila* model system. Academic, Sydney, pp 17–30

Myles TG (1986) Oviposition and development of *Volucella isabellina* (Diptera: Syrphidae) on saguaro cactus, *Cereus giganteus*. Entomol News 97:104–108

Niyanga LK, Nout MJR, Gadaga TH, Theelen B, Boekhout T, Zwietering MH (2007) Yeasts and lactic acid bacteria microbiota form masau (*Ziziphus mauritiana*) fruits and their fermented fruit pulp in Zimbabwe. Int J Food Microbiol 120:159–166

Nursten HE (1970) Volatile compounds: the aroma of fruits. In: Hulme AC (ed) The biochemistry of fruits and their products. Academic, London, pp 239–268

Nyffeler R, Eggli U (2010) Disintegrating Portulacaceae: a new familial classification of the suborder Portulacineae (Caryophyllales) based on molecular and morphological data. Taxon 59:227–240

Oliveira RB (1980) Taxonomia de leveduras associadas com cajus e mangas. M.Sc. Thesis. Instituto de Microbiologia, Universidade Federal do Rio de Janeiro, Rio de Janeiro

Painter TS (1965) John Thomas Patterson, November 3, 1878–December 4, 1960. Biogr Mem Natl Acad Sci 38:223–262

Palanca L, Gaskett AC, Gunther CS, Newcomb RD, Goddard MR (2013) Quantifying variation in the ability of yeasts to attract *Drosophila melanogaster*. PLoS One 8:e75332

Pareek OP, Singh RS, Vashishtha BB (2003) Performance of cactus pear [*Opuntia ficus-indica* (L.) Mill.] clones in hot arid region of India. J Prof Assoc Cactus Develop 7:121–130

Parsons PA (1980) Responses of *Drosophila* to environmental ethanol from ecologically optimal and extreme habitats. Experientia 36:1070–1071

Peay KG, Belisle M, Fukami T (2013) Phylogenetic relatedness predicts priority effects in nectar yeast communities. Proc R Soc B 279:749–758

Pemberton RW (1995) Cactoblastis cactorum (Lepidoptera: Pyralidae) in the United States:-an immigrant biological control agent or an introduction of the nursery industry? Am Entomol 41: 230–232

Pemberton RW, Cordo HA (2001a) *Nosema* (Microsporida: Nosematidae) species as potential biological control agents of *Cactoblastis Cactorum* (Lepidoptera: Pyralidae): surveys for the Microsporidia in Argentina and South Africa. Florida Entomol 84:527–530

Pemberton RW, Cordo HA (2001b) Potential and risks of biological control of *Cactoblastis cactorum* (Lepidoptera – Pyralidae) in North America. Florida Entomol 84:513–526

Pfeiffer T, Schuster S, Bonhoeffer S (2001) Cooperation and competition in the evolution of ATP-producing pathways. Science 292:504–507

Phaff HJ (1986) My life with yeasts. Ann Rev Microbiol 40:1–28

Phaff HJ, Miller MW, Miranda M, Heed WB, Starmer WT (1974) *Cryptococcus cereanus*, a new species of the genus *Cryptococcus*. Int J Systematic Bacteriol 24:486–490

Phaff HJ, Starmer WT, Miranda M, Miller MW (1978) *Pichia heedii*, a new species of yeast indigenous to necrotic cacti in the North American Sonoran Desert. Int J Syst Bact 28:326–331

Phaff HJ, Starmer WT, Miranda M, Miller MW (1980) *Candida mucilagina*, a new species of yeast found in decaying cladodes of *Opuntia inermis* and in necrotic tissue of cereoid cacti. Int J Syst Bacteriol 30:596–600

Phaff HJ, Starmer WT, Tredick J, Miranda M (1985) *Pichia deserticola* and *Candida deserticola*, two new species of yeasts associated with necrotic cactus stems. Int J Syst Bacteriol 35: 211–216

Phaff HJ, Starmer WT, Lachance MA, Aberdeen V, Tredick-kline J (1992) *Pichia caribaea*, a new species of yeast occurring in necrotic tissue of cacti in the Caribbean area. Int J Syst Evol Microbiol 42:459–462

Phaff HJ, Blue J, Hagler AN, Kurtzman CP (1997) *Dipodascus starmeri* sp. nov., a new species of yeast occurring in cactus necroses. Int J Syst Bacteriol 47:307–312

Piskur J, Rozpedowska E, Polakova S, Merico A, Compagno C (2006) How did *Saccharomyces cerevisiae* evolve to become a good brewer? Trends Genet 22:183–186

Prada GMM, Pagnocca FC (1997) Ascomycetous yeasts associated with naturally occurring fruits in a tropical rain forest. Folia Microbiol 42:39–46

Robbs PG, Hagler AN, Mendonça-Hagler LC (1989) Yeasts associated with a pinneapple plantation. Yeast 5:485–489

Romano P, Fiore C, Paraggio M, Caruso M, Capece A (2003) Function of yeast species and strains in wine flavour. Int J Food Microbiol 86:169–180

Rosa CA, Hagler AN, Mendonça-Hagler LC, Morais PB, Gomes NCM, Monteiro RF (1992) *Clavispora opuntiae* and other yeasts associated with the moth *Sigelgaita* sp. in the cactus *Pilosocereus arrabidae* of Rio de Janeiro, Brazil. A van Leeuwenhoek 62:267–272

Rosa CA, Morais PB, Hagler AN, Mendonça-Hagler LC, Monteiro RF (1994) Yeast communities of the cactus *Pilosocereus arrabidae* and associated insects in the Sandy coastal plains of southeastern Brazil. A van Leeuwenhoek 65:55–62

Rosa CA, Morais PB, Santos SR, Peres Neto PR, Mendoça-Hagler LC, Hagler NA (1995) Yeast communities associated with different plant resources in Sandy coastal plains of southeastern Brazil. Mycol Res 99:1047–1054

Rosa CA, Lachance MA, Starmer WT, Barker JS, Bowles JM, Schlag-Edler B (1999) *Kodamaea nitidulidarum, Candida restingae* and *Kodamaea anthophila*, three new related yeast species from ephemeral flowers. Int J Syst Bacteriol 49:309–318

Rosa CA, Morais PB, Lachance MA, Pimenta RS, Santos RO, Trindade RC, Figueroa DL, Resende MA, Bragança MAL (2006) *Candida azymoides* sp. nov., a yeast species from tropical fruits and larva of *Anastrepha mucronota* (Diptera:Tephritidae). Lundiana 7:82–86

Rosa CA, Morais PB, Lachance MA, Santos RO, Melo WGP, Viana RHO, Bragança MAL, Pimenta RS (2009) *Wickerhamomyces queroliae* sp. nov. and *Candida jalapaonensis* sp. nov., two yeast species isolated from Cerrado ecosystem in North Brazil. Int J Syst Evol Microbiol 59:1232–1236

Saerens SMG, Delvaux FR, Verstrepen KJ, Thevelein JM (2010) Production and biological function of volatile esters in *Saccharomyces cerevisiae*. J Microbiol Biotechnol 3:165–177

Santos EA, Oliveira RB, Mendonça-Hagler LC, Hagler AN (1996) Yeasts associated with flowers and fruits from a semi-arid region of northeastern Brazil. Rev Microbiol 27:33–40

Shen R, Lachance M-A (1993) Phylogenetic study of ribosomal DNA of cactophilic *Pichia* species by restriction mapping. Yeast 9:315–330

Shen XX, Zhou X, Kominek J, Kurtzman CP, Hittinger CT, Rokas A (2016) Reconstructing the backbone of the Saccharomycotina yeast phylogeny using genome-scale data. G3: Genes| Genomes| Genetic 6:3927–3939

Simmonds FJ, Bennett FD (1966) Biological control of *Opuntia* spp. by *Cactoblastis cactorum* in the Leeward Islands (West Indies). Entomophaga 1:183–189

Sipiczki M (2011) Dimorphic cycle in *Candida citri* sp. nov., a novel yeast species isolated from rotting fruit in Borneo. FEMS Yeast Res 11:202–208

Soberón J (2002) The routes of invasion of *Cactoblastis cactorum*. Comisión Nacional para el Conocimiento y uso de la Biodiversidad (CONABIO), Puerto Vallarta, Mexico

Soberon J, Golubov J, Sarukhán J (2001) The importance of *Opuntia* in Mexico and routes of invasion and impact of *Cactoblastis cactorum* (Lepidoptera: Pyralidae). Fla Entomol 84: 486–492

Spencer DM, Spencer JFT, de Figueroa L, Heluane H (1992) Yeasts associated with rotting citrus fruits in Tucumán, Argentina. Mycol Res 96:891–892

Sperandio EM, Martins do Vale HM, Moreira GAM (2015) Yeasts from native Brazilian Cerrado plants: occurrence, diversity and use in the biocontrol of citrus green mould. Fungal Biol 119: 984–993

Stamps JA, Yang LH, Morales VM, Boundy-Mills KL (2012) *Drosophila* regulate yeast density and increase yeast community similarity in a natural substrate. PLoS One 7:e42238

Starmer WT (1982) Analysis of the community structure of yeasts associated with the decaying stems of cactus. I. *Stenocereus gummosus*. Microb Ecol 8:71–81

Starmer WT, Fogleman JC (1986) Coadaptation of *Drosophila* and yeasts in their natural habitat. J Chem Ecol 12:1035–1053

Starmer WT, Phaff HJ (1983) Analysis of the community structure of yeasts associated with the decaying stems of cactus. II. *Opuntia* species. Microb Ecol 9:247–259

Starmer WT, Heed WB, Rockwood-Sluss ES (1977) Extension of longevity in *Drosophila mojavensis* by environmental ethanol: differences between subraces. Proc Natl Acad Sci U S A 74: 387–391

Starmer WT, Phaff HJ, Miranda M, Miller MW (1978a) *Pichia amethionina*, a new heterothallic yeast associated with the decaying stems of cereoid cacti. Int J Syst Bacteriol 28:433–441

Starmer WT, Phaff HJ, Miranda M, Miller MW (1978b) *Pichia cactophila*, a new species of yeast found in decaying tissue of cacti. Int J Syst Bacteriol 28:318–325

Starmer WT, Phaff HJ, Miranda M, Miller MW, Barker JSF (1979) *Pichia opuntiae*, a new heterothallic yeast found in decaying cladodes of *Opuntiae inermis* and in necrotic tissue of cereoid cacti. Int J Syst Bacteriol 29:159–167

Starmer WT, Kircher HW, Phaff HJ (1980) Evolution and speciation of host plant specific yeasts. Evolution 34:137–146

Starmer WT, Phaff HJ, Miranda M, Miller MW, Heed WB (1982) The yeast flora associated with the decaying stems of columnar cacti and *Drosophila* in North America. In: Hecht MK, Wallace B, Prance CT (eds) Evolutionary biology, vol 14. Plenum Press, New York, pp 269–295

Starmer WT, Phaff HJ, Tredick J, Miranda M, Aberdeen V (1984) *Pichia antillensis*, a new species of yeast associated with necrotic stems of cactus in the Lesser Antilles. Int J Syst Bacteriol 34: 350–354

Starmer WT, Ganter PF, Phaff HJ (1986) Quantum and continuous evolution of DNA base composition in the yeast genus *Pichia*. Evolution 40:1263–1274

Starmer WT, Ganter PF, Aberdeen V, Lachance M-A, Phaff HJ (1987a) The ecological role of killer yeasts in natural communities of yeasts. Can J Microbiol 33:783–793

Starmer WT, Lachance M-A, Phaff HJ (1987b) A comparison of yeast communities found in necrotic tissue of cladodes and fruits of *Opuntia stricta*-on islands in the Caribbean Sea and where introduced into Australia. Microb Ecol 14:179–192

Starmer WT, Aberdeen V, Lachance M-A (1988a) The yeast community associated with *Opuntia stricta* (Haworth) in Florida, with regard to the moth *Cactoblastis cactorum* (Berg). Florida Sci 51:7–11

Starmer WT, Phaff HJ, Bowles JM, Lachance M-A (1988b) Yeasts vectored by insects feeding on decaying saguaro cactus. Southwest Nat 33:362–363

Starmer WT, Phaff HJ, Miranda M, Miller MW (1990) The yeast flora associated with the decaying stems of columnar cacti and *Drosophila* in North America. Evol Biol 14:269–294

Starmer WT, Fogleman JC, Lachance M-A (1991) The yeast community of cacti. In: Andrews JH, Hirano SS (eds) Microbial ecology of leaves. Springer, New York, pp 158–178

Starmer WT, Schmedicke RA, Lachance M-A (2003) The origin of the cactus-yeast community. FEMS Yeast Res 3:441–448

Starmer WT, Aberdeen V, Lachance MA (2006) The biogeographic diversity of cactophilic yeasts. In: Rosa CA, Peter G (eds) The yeast handbook. Biodiversity and ecophysiology of yeasts. Springer, Berlin, pp 485–499

Stiling P, Moon D, Gordon D (2004) Endangered cactus restoration: mitigating the non-target effects of a biological control agent (*Cactoblastis cactorum*) in Florida. Restor Ecol 12: 605–610

Styger G, Jacobson D, Bauer FF (2011) Identifying genes that impact on aroma profiles produced by *Saccharomyces cerevisiae* and the production of higher alcohols. Appl Microbiol Biotechnol 91:713–730

Swiegers JH, Pretorius IS (2005) Yeast modulation of wine flavor. Adv Appl Microbiol 57: 131–175

Thomas SM, Ranson SL, Richardson SA (1973) Plant physyiology. Longman, London

Thomson JM, Gaucher EA, Burgan MF, De Kee DW, Li T et al (2005) Resurrecting ancestral alcohol dehydrogenases from yeast. Nat Gen 37:630–635

Trindade RC, Resende MA, Silva CM, Rosa CA (2002) Yeasts associated with fresh and frozen pulps of Brazilian tropical fruits. Syst Appl Microbiol 25:294–300

Vigueras AL, Portillo L (2001) Uses of *Opuntia* species and the potential impact of *Cactoblastis cactorum* (Lepidoptera: Pyralidae) in Mexico. Florida Entomol 84:493–498

Wagner RP (1944) The nutrition of *Drosophila mulleri* and *D. aldrichi*. Growth of the larve on cactus extract and the microorganisms found in cactus. Stud Gen *Drosophila* 4445:104–128

Woods DD (1957) Albert Jan Kluyver. 1888–1956. Biogr Mem Fell R Soc 3:109–128

Yamada Y, Higashi T, Ando S, Mikata K (1998) The phylogeny of strains of species of the genus *Pichia* Hansen (Saccharomycetaceae) based on the partial sequences of 18S ribosomal RNA:

the proposals of *Phaffomyces* and *Starmera*, the new genera. Bull Fac Agr-Shizuoka Univ 47: 23–35

Yamada Y, Kawasaki H, Nagatsuka Y, Mikata K, Seki T (1999) The phylogeny of the cactophilic yeasts based on the 18S ribosomal RNA gene sequences: the proposals of *Phaffomyces antillensis* and *Starmera caribaea*, new combinations. Biosci Biotechnol Biochem 63:827–832

Yang Y, Moore MJ, Brockington SF, Soltis DE, Wong GK, Carpenter EJ, Zhang Y, Chen L, Yan Z, Xie Y, Sage RF, Covshoff S, Hibberd JM, Nelson MN, Smith SA (2015) Dissecting molecular evolution in the highly diverse plant clade caryophyllales using transcriptome sequencing. Mol Biol Evol 32:2001–2014

Zhang DP, Lu CG, Zhang TT, Spadaro D, Liu DW, Liu WC (2014) *Candida pruni* sp. nov. is a new yeast species with antagonistic potential against brown rot of peaches. Arch Microbiol 196:525–530

Chapter 9
Yeasts Associated with Decomposing Plant Material and Rotting Wood

Raquel M. Cadete, Mariana R. Lopes, and Carlos A. Rosa

Abstract Yeasts play different roles in the decomposition of plant materials, actively participating as producers of hydrolytic enzymes or as transient fungi that use products released during decomposition by other organisms such as fungi and bacteria. Yeasts able to assimilate lignocellulose-related sugars are usually found in tree bark, leaf litter, and rotting wood, which are suitable habitats for these organisms. Many yeast species that are found in live or decaying plant parts are associated with insects that also use these habitats as feeding or breeding sites. Among the major clades of yeasts associated with rotting wood habitats, four are worth mentioning for their ability to assimilate/ferment lignocellulose-related sugars and/or to produce enzymes that act on this substrate: *Scheffersomyces*, *Spathaspora*, *Spencermartinsiella*, and *Sugiyamaella*. In addition to these clades, *Candida tropicalis* (*Candida albicans*/*Lodderomyces* clade) is a species that deserves attention because it is frequently found in decaying plant materials. *Aureobasidium pullulans* and basidiomycetous species of the genera *Apiotrichum*, *Cutaneotrichosporon*, *Cystofilobasidium*, *Naganishia*, *Papiliotrema*, *Pseudotremella*, *Saitozyma*, *Solicoccozyma*, *Tausonia*, *Trichosporon*, *Vanrija*, and *Vishniacozyma* are also frequently reported occurring in these substrates. The physiological characteristics that allowed these yeasts to adapt to decaying plant substrates make these organisms interesting for biotechnological applications.

Keywords Yeast diversity • Plant materials • Decomposition • Hydrolytic enzymes

Contents

R.M. Cadete • M.R. Lopes • C.A. Rosa (✉)
Departamento de Microbiologia, Instituto de Ciências Biológicas, C.P. 486, Universidade Federal de Minas Gerais, CEP 31270-901, Belo Horizonte, MG, Brazil
e-mail: carlrosa@icb.ufmg.br

P. Buzzini et al. (eds.), *Yeasts in Natural Ecosystems: Diversity*,
DOI 10.1007/978-3-319-62683-3_9

9.1 Introduction

Plant biomass is primarily composed of lignocellulose, a complex matrix consisting of cellulose, hemicelluloses, and lignin, which are intimately associated to form the structural framework of the plant cell wall. Cellulose is the main polymeric component of lignocellulose, which is considered the most abundant polysaccharide on Earth. This polymer has a simple chemical composition, consisting of D-glucose residues linked by β-1,4-glycosidic bonds, and it is highly crystalline with less-ordered amorphous regions (Hon 1994). Hemicelluloses, the second most common group of polysaccharides in nature, are heterogeneous polymers of pentoses (xylose and arabinose), hexoses (mannose, glucose, and galactose), and sugar acids. Xylans, the main hemicelluloses found in hardwoods, are heteropolysaccharides with homopolymeric backbone chains of 1,4-linked β-D-xylopyranose units. Basidiomycetous fungi are the most potent degraders of plant biomass because many species grow on dead wood or leaf litter, environments rich in cellulose and hemicelluloses. To degrade cellulose and xylans, these organisms utilize a set of hydrolytic enzymes, typically comprising cellobiohydrolases, endoglucanases, β-glucosidases, endoxylanases, and β-xylosidases (Baldrian 2008). Wood-associated yeasts also exhibit cellulolytic and xylanolytic enzyme activities (Jiménez et al. 1991; Lara et al. 2014). The efficiency and regulation of plant material degradation differs among wood-rotting, litter-decomposing, mycorrhizal, and plant-pathogenic fungi and yeasts because of the different effects of cellulose and hemicellulose degradation in the physiology and ecology of individual groups (Baldrian and Valášková 2008).

Yeasts form an important association with basidiomycetes during the wood decay process, in which bacteria are also closely associated. These organisms are present in both the initial and advanced stages of wood decay, where they establish mutualistic associations with other microorganisms (Blanchette and Shaw 1978; Maksimova and Chernov 2004). In fact, basidiomycetous yeasts and xylose-assimilating strains are found during different stages of degradation of wood. The presence of yeasts able to assimilate xylose is related to the increased availability of this pentose sugar in the decayed wood (González et al. 1989). On the other hand, basidiomycetous yeasts can produce hydrolytic enzymes that act on plant polymers (Štursová et al. 2012; Lara et al. 2014), and some species have the ability to degrade aromatic substances released during the decay process (Sampaio 1999; Botha 2011). Moreover, methanol-assimilating yeasts are frequently found in decaying wood, plant leaves, and other plant residues. This is due to the emission of methanol by metabolizing leaves and methoxy groups during the breakdown of lignin, thus explaining the presence of yeasts that can utilize this alcohol in such substrates (Kurtzman 2011).

Filamentous fungi are the key players in plant material decomposition because of their ability to produce a wide range of extracellular enzymes that efficiently attack the recalcitrant lignocellulose matrix (van der Wal et al. 2013; Voříšková and Baldrian 2013; Purahong et al. 2016). However, the presence of yeasts during the different stages of wood and leaf litter breakdown highlights the ecological role of these microorganisms. The occurrence of yeast primarily in the early and middle stages of plant material decomposition can be directly related to the processes inherent to each of these steps: the initial stage of decomposition is generally dominated by the colonization of organisms expressing enzymes that can attack easily accessible polysaccharides such as extractable hemicelluloses and pectins, whereas the middle stage is characterized by an attack on non-lignified tissues such as parenchyma and collenchyma, in which cellulose and hemicellulose are not protected (Berg and McClaugherty 2008). The production of enzymes acting on cellulose, hemicelluloses, and pectin is detected in yeasts (Table 9.1). However, xylan-degrading activities are more widespread than cellulose-degrading abilities in these organisms (Jiménez et al. 1991). Few studies on cellulolytic activities in yeasts have been conducted, which is probably because, yeasts, with the exception of the species *Cutaneotrichosporon cutaneum* and *Tausonia pullulans* (Dennis 1972; Štursová et al. 2012), are unable to hydrolyze insoluble cellulose (Leclerc et al. 1987; Biely and Kremnický 1998). Nevertheless, yeasts able to develop detectable halos when grown on carboxymethylcellulose (CMC) medium, a hydrosoluble polymer derived from cellulose, are recognized (Brandão et al. 2011; Vaz et al. 2011; Thongekkaew et al. 2012; Carvalho et al. 2013; Gomes et al. 2015), including abundant species or those commonly isolated from decaying wood and leaf litter. Additionally, many yeast species show β-glucosidase activity; this enzyme is responsible for the conversion of cellobiose to glucose and acts synergistically with cellobiohydrolases and endoglucanases for the enzymatic saccharification of cellulose (Leclerc et al. 1987; Otero et al. 2003; Bedriñana et al. 2012; Carvalho et al. 2013; Molnárová et al. 2014). Xylanolytic activities are mainly assigned to species from the genera *Aureobasidium*, *Pseudozyma*, *Scheffersomyces*, *Sugiyamaella*, and the former genera *Cryptococcus* and *Trichosporon* (Biely et al. 1980; Hrmová et al. 1984; Leathers 1986; Özcan et al. 1991; Bhadra et al. 2008; Adsul et al. 2009; Carvalho et al. 2013; Morais et al. 2013a, b; Borges et al. 2014; Gomes et al. 2015; Watanabe et al. 2015; Lara et al. 2014; Sena et al. 2017). Yeasts showing pectinolytic activities are common and mainly include *Saccharomyces cerevisiae* strains and *Kluyveromyces* species (Blanco et al. 1999; da Silva et al. 2005; Alimardani-Theuil et al. 2011).

Along with the decomposition of plant material, depolymerization of sugar-rich polymers occurs, releasing carbon constituents (de Vries and Visser 2001), which mainly include cellooligosaccharides and xylooligosaccharides, among other compounds (Pettersen 1984). Even though traditional fermentative metabolism of

Table 9.1 Yeast species widely associated with rotting wood and/or leaf litter able of producing cellulase, xylanase, or pectinase enzymes and/or metabolizing the plant cell wall primarily sugars

Species	Enzyme activity[a]			Carbon source[b]				References
	CelA	XylA	PecA	Xyl	Cel	Ara	Gal	
Ascomycota[c]								
Aureobasidium pullulans	+	+	+					Leathers (1986), Sampaio et al. (2004, 2007), Wuczkowski et al. (2005), Bhadra et al. (2008), Brandão et al. (2011), Thongekkaew et al. (2012), Molnárová et al. (2014), Gomes et al. (2015)
Barnettozyma californica	+			+	+	–	–	Morais et al. (2013b), Molnárová et al. (2014), Guamán-Burneo et al. (2015)
Candida boidinii (*Ogataea* clade)		+	+	+	–	v	–	Blanco et al. (1999), Alimardani-Theuil et al. (2011), Cadete et al. (2012b), Morais et al. (2013b)
Candida tropicalis (*Lodderomyces*/*Candida albicans* clade)	+	+		+	v	–	+	Randhawa et al. (2001), Cadete et al. (2012b), Thongekkaew et al. (2012), Lara et al. (2014), Molnárová et al. (2014), Guamán-Burneo et al. (2015)
Cyberlindnera saturnus		+		+	+	–	–	Sampaio et al. (2004, 2007), Cadete et al. (2012b), Morais et al. (2013b)
Debaryomyces hansenii	+		+	+	+	+/w	+	Jiménez et al. (1991), Sampaio et al. (2004, 2007), da Silva et al. (2005), Brandão et al. (2011), Vaz et al. (2011), Cadete et al. (2012b), Molnárová et al. (2014)

(continued)

Table 9.1 (continued)

Species	Enzyme activity[a]			Carbon source[b]				References
	CelA	XylA	PecA	Xyl	Cel	Ara	Gal	
Meyerozyma guilliermondii	+	+	+	+	+	+	+	Randhawa et al. (2001), Buzzini and Martini (2002), Otero et al. (2003), da Silva et al. (2005), Vaz et al. (2011), Bedriñana et al. (2012), Cadete et al. (2012b), Thongekkaew et al. (2012), Lara et al. (2014), Molnárová et al. (2014), Gomes et al. (2015)
Scheffersomyces shehatae		+		+	+	v	+	Morais et al. (2013b), Lara et al. (2014), Ren et al. (2014)
Scheffersomyces stipitis		+		+	+	v	+	Özcan et al. (1991), Cadete et al. (2012b), Morais et al. (2013b), Ren et al. (2014)
Schwanniomyces polymorphus	+		+	+	+	v	+	Otero et al. (2003), da Silva et al. (2005), Cadete et al. (2012b), Morais et al. (2013b)
Sugiyamaella smithiae		+	+	+	+	+	+	Jiménez et al. (1991), da Silva et al. (2005), Lara et al. (2014)
Basidiomycota[d]								
Cutaneotrichosporon cutaneum	+	+	+	+	+	+	+	Dennis (1972), Hrmová et al. (1984), Randhawa et al. (2001), Sampaio et al. (2007)
Cystofilobasidium capitatum	+		+	+	+	+	+	Maksimova and Chernov (2004), Wuczkowski et al. (2005), Brizzio et al. (2007), Sampaio et al. (2007), Alimardani-Theuil et al. (2011), Brandão et al. (2011)

(continued)

Table 9.1 (continued)

Species	Enzyme activity[a]			Carbon source[b]				References
	CelA	XylA	PecA	Xyl	Cel	Ara	Gal	
Leucosporidium scottii	+		+	+	+	s	+/s	Maksimova and Chernov (2004), Sampaio et al. (2004, 2007), Vaz et al. (2011)
Naganishia albida	+	+	+	+	+	+	−/s	Biely et al. (1980), González et al. (1989), Jiménez et al. (1991), Blanco et al. (1999), Randhawa et al. (2001), Maksimova and Chernov (2004), Sampaio et al. (2004, 2007), Molnárová et al. (2014)
Papiliotrema laurentii	+	+		+	+	+	+	Jiménez et al. (1991), Randhawa et al. (2001), Maksimova and Chernov (2004), Sampaio et al. (2004, 2007), Wuczkowski et al. (2005), Carvalho et al. (2013), Morais et al. (2013b), Lara et al. (2014), Molnárová et al. (2014), Gomes et al. (2015)
Rhodotorula glutinis	+			s	+	+	+	Randhawa et al. (2001), Maksimova and Chernov (2004), Sampaio et al. (2004, 2007), Molnárová et al. (2014)
Rhodotorula mucilaginosa	+		+	+	+/s/w	+	+/s	Jiménez et al. (1991), Randhawa et al. (2001), Sampaio et al. (2004), Vaz et al. (2011), Brandão et al. (2011), Molnárová et al. (2014)

(continued)

Table 9.1 (continued)

Species	Enzyme activity[a]			Carbon source[b]				References
	CelA	XylA	PecA	Xyl	Cel	Ara	Gal	
Saitozyma podzolica		+	+	+	+	+	+	Maksimova and Chernov (2004), Morais et al. (2013b), Gomes et al. (2015)
Tausonia pullulans	+		+	+	+	+	+	Dennis (1972), Jiménez et al. (1991), Maksimova and Chernov (2004), Brandão et al. (2011), Štursová et al. (2012)
Vanrija humicola	+	+		+	+	+	+	Cadete et al. (2012b), Thongekkaew et al. (2012), Carvalho et al. (2013), Morais et al. (2013b)

[a]Qualitative and/or quantitative assays detecting CelA (carboxymethylcellulose and/or β-glucosidase), XylA (xylanase and/or β-xylosidase), and PecA (pectinase) activities in solid or liquid media

[b]Comsuption of Xyl (D-xylose), Cel (D-cellobiose), Man (D-mannose), Ara (L-arabinose), and Gal (D-galactose) as sole carbon sources according to Kurtzman et al. (2011)

[c]Species reclassified according to Kurtzman et al. (2011) and Daniel et al. (2014)

[d]Species reclassified according to Kurtzman et al. (2011) and Liu et al. (2016a)

+ (positive), − (negative), s (slow), v (variable), w (weak)

sugars is most commonly attributed to yeasts, these microorganisms exhibit oxidative utilization of a much broader range of organic carbon compounds such as pentose sugars, alcohols, amino sugars, hydrocarbons, and aromatic compounds (Kurtzman et al. 2011). This feature allows decaying plant materials to serve as excellent habitats for yeasts. Particularly, in recent years, many efforts to isolate from rotting plant material yeasts that are able to assimilate and/or ferment the wood-sugar xylose have been made, with the aim of biotechnological exploitation of these organisms (Cadete et al. 2012b, 2014; Morais et al. 2013b; Sena et al. 2017).

9.2 Fungi and Decomposition of Plant Material

Fungi play a predominant role in the decomposition of lignocellulose complexes, mainly in recalcitrant fractions (van der Wal et al. 2013). In natural ecosystems, decomposer fungi form communities of several or many species, with microbial succession occurring by sequential selection and development of microbial populations, depending on the decomposition stage and thus substrate quality (Wei et al. 2009). The decomposition process starts with the degradation of soluble and low-molecular-weight compounds. Next, the easily accessible polysaccharides, pectins and hemicelluloses, mainly those that are arabinan-based, are hydrolyzed, followed by degradation of cellulose and xylan-based hemicelluloses. The final stage is characterized by decomposition of the more recalcitrant components, the lignified tissues. Although individual processes may dominate a particular stage of decomposition, any or all processes may occur, to some extent, throughout the decay continuum (Berg and McClaugherty 2008).

Functional concepts have been used to define plant degradation. White rot, brown rot, and soft rot refer to different types of lignin, cellulose, and hemicellulose degradation. These terminologies refer to the type of rot and were adopted to commonly refer to groups of microorganisms, specifically fungi (Worrall et al. 1997). White rot fungi are those basidiomycetes with the ability to completely degrade lignin, and the cellulolytic ascomycetes (van der Wal et al. 2013). Brown rot fungi decompose mainly cellulose and hemicellulose, but they can also contribute to partial lignin degradation (van der Wal et al. 2013). The soft rot fungi are traditionally thought to be active in soften and moist wood degradation. They comprise the cellulolytic ascomycetes and some other fungi (Berg and McClaugherty 2008). In addition, the soft rots have the ability to degrade lignin to some extent (Crawford 1981; Nilsson and Daniel 1989).

Cellulose can be degraded by numerous species of bacteria and fungi that produce extracellular hydrolytic enzymes, but few cellulolytic organisms possess the complete set of enzymes necessary to degrade this structure (Eriksson et al. 2012). Three main hydrolytic enzymes form the cellulolytic complex: endo-1,4-β-glucanase breaks β-glucosidic links randomly, producing oligosaccharide units of different lengths; exo-1,4-β-glucanase acts at the ends of the cellulose chain to release cellobiose or glucose; and, finally, 1,4-β-glucosidase hydrolyzes cellobiose and other water-soluble oligosaccharides such as triose and tetraose to glucose. These enzymes with different specificities exert synergistic actions and are able to degrade both crystalline and amorphous cellulose (Zhang et al. 2012). The cellulolytic ascomycetes are critically responsible for the decomposition of lignin-rich organic matter, once the thin perforation hyphae of these fungi can reach cellulose-rich layers in woody cell walls (Schmidt 2006). They encompass both the soft rot and white rot fungi, which have cellulose-degrading systems. The brown rot fungi appear not to have exoglucanases. Wolter et al. (1980) suggested that another mechanism is involved, with a multifunctional enzyme (nonspecific) able to degrade different polysaccharides, even cellulose, and that this mechanism can be found in some basidiomycetous species. In addition, an oxidative mechanism has also been proposed for depolymerization of amorphous cellulose by brown rot fungi (Green and Highley 1997).

The enzyme system that degrades hemicellulose is more complex than the cellulolytic one. The variety of heteropolysaccharide structures requires a complex set of hydrolytic enzymes for degradation (Patel and Savanth 2015). The main enzyme acting in xylan-based hemicellulose degradation is an endo-1,4-β-xylanase that randomly attacks unbranched chains to produce oligomers of different lengths. However, other enzymes with diverse specificities and modes of action, such as xylan-1,4-β-xylosidase, α-glucosiduronase, α-L-arabinofuranosidase, and acetylxylan esterase, are also necessary (Patel and Savanth 2015; Kumar and Shukla 2016).

Pectinases are enzymes that contribute to the breakdown of pectins, acting on their homogalacturonan (HG) and rhamnogalacturonan (RG) regions. Pectinase classification is based on the enzymatic modes of action against galacturonan backbones. The esterases catalyze the de-esterification of pectins and are represented by pectinesterases (PEs) and pectin-acetyl esterases (PAE). The depolymerases break α-(1,4)-glycosidic bonds by hydrolysis, polygalacturonase (PG) enzymes, or by β-elimination, pectinlyases (PL) and pectate lyases (PAL). A third type of pectolytic enzyme comprises the protopectinases (PPase). Arabinases and galactanases also contribute to pectin degradation by acting on side chains (Alimardani-Theuil et al. 2011). Pectinases are naturally found in higher plants because they are involved in the maturation of some fruits (Ward et al. 1989). However, pectinolytic enzymes are also produced by bacteria, fungi, and yeasts (Jayani et al. 2005).

Lignin decomposers can be found in the three groups: white rot, soft rot, and brown rot fungi. The basidiomycete white rot fungi are the most studied group and are considered the main lignin degraders (Baldrian 2008; Floudas et al. 2012). Despite this, lignin breakdown by the Xylariales in the Ascomycota has been reported (Osono et al. 2011). Brown rot fungi also contribute to the degradation of lignin, because they are able to modify lignin and provide access to cellulose (Yelle et al. 2008; Eastwood et al. 2011). Different enzymatic mechanisms may contribute to lignin degradation. Manganese peroxidase enzyme (MnP) has been found in many basidiomycetes white rot fungi and it is considered the most widely spread lignin-degrading enzyme (Hofrichter 2002). Other known lignolytic enzymes are laccase and lignin peroxidase (Berg and McClaugherty 2008).

Yeasts are also found in decayed organic matter. In addition to their capacity to produce hydrolytic enzymes and consume a variety of sugars, these organisms can also degrade some small aromatic compounds (Middelhoven 1993; Botha 2011; Kurtzman et al. 2011). Examples of yeast species that are associated with rotting plant materials and able to produce hydrolytic enzymes that can contribute to lignocellulose degradation and metabolize plant cell wall sugars are shown in Table 9.1.

9.3 Yeasts Associated with Decomposing Plant Material

9.3.1 Leaf Litter

Decomposition of plant litter refers to the physical and chemical processes that reduce litter to its elemental chemical constituents. Decomposition is triggered and mediated to a large extent by microorganisms that colonize the litter material (Romaní et al. 2006; Purahong et al. 2016). In fact, leaf decomposition is a major source of carbon and energy for microbial growth in freshwater mountain streams (Fisher and Likens 1973). The microbial community in leaf detritus consists mainly of filamentous fungi, yeast, and bacteria that are involved in the degradation of macromolecules such as cellulose, hemicelluloses, pectin, proteins, and lignin, thus regenerating metabolites (Sampaio et al. 2007). In general, fungi (including yeasts) are more significant leaf-decaying agents than are bacteria, but the relevance of these two groups depends mainly on the duration of the decomposition process and nutrient levels (Baldy et al. 1995; Sampaio et al. 2007). In general, bacteria colonize rapidly during the initial stages of decomposition, acting upon easily assimilated molecules (Tanaka 1991; Gonçalves et al. 2006), whereas fungi have the capacity to metabolize molecules that are not easy to decompose such as cellulose and lignin (Romaní et al. 2006).

Microorganisms play a crucial role in the biological decomposition of plant litter in terrestrial ecosystems. Purahong et al. (2016) investigated microbial community succession in the decomposing leaf litter of a temperate beech forest by sequencing the bacterial 16S and ITS region of the rRNA gene. The results revealed that both communities underwent rapid changes. Proteobacteria, Actinobacteria, and Bacteroidetes dominated throughout the study period, but their taxonomic composition and abundances changed markedly with each sampling. The fungal community also changed dynamically as decomposition progressed, with ascomycetes increasingly replaced by basidiomycetes. During this process, macronutrients, micronutrients, C:N ratio, and pH were significantly correlated with fungal and bacterial communities (Purahong et al. 2016).

In terms of yeast composition in leaf litter, the number and occurrence of species can be influenced by climate, season, type of vegetation, locality, and degree of substrate decomposition, as well as whether the breakdown process occurs in a terrestrial or aquatic environment. Nevertheless, these organisms can be isolated during all decomposition stages and are predominantly basidiomycetous. According to Štursová et al. (2012), the yeast species *Apiotrichum porosum* (formerly *Trichosporon porosum*) and *Solicoccozyma terricola* (formerly *Cryptococcus terricola*) were the most abundant representatives of the cellulolytic Basidiomycota fungi associated with litter and soil in a *Picea abies* forest in Central Europe. During long-term studies of yeasts inhabiting forest habitats in subboreal forests of the European part of Russia, Maksimova and Chernov (2004) observed that the yeast populations in forest leaf litter were characterized by cohabiting epiphytes, pedobionts, and leaf litter species. Forty-seven species belonging to 20 genera were identified based on morphological and physiological characteristics, and the highest diversity was shown in basidiomycetous yeasts (35 species). The cell number and species diversity of yeast communities in the array of substrates corresponding to succession stages in decomposing plant residues were shown to steadily decline, with each stage characterized by its specific set of dominant species. The mean total yeast counts in the deeper forest litter layers were low (10^3 CFU g^{-1}), and the taxonomic structure of these communities differed substantially from that found on the aboveground parts of plants. In decaying plant residues, typical species were *Ta. pullulans* and *Cystofilobasidium capitatum*. The dominant *Naganishia albida* and *Papiliotrema laurentii* (formerly *Cryptococcus albidus* and *Cryptococcus laurentii*, respectively) and subdominant *Cystofil. capitatum*, *Leucosporidium scottii*, *Rhodotorula fujisanensis*, *Saitozyma podzolica*, and *Ta. pullulans* species in leaf litter were basidiomycetes (Maksimova and Chernov 2004).

In a study of the development of a fungal community over 24 months in decaying litter in a temperate forest dominated by oak trees (*Quercus petraea*) using sequencing techniques, Voříšková and Baldrian (2013) observed a rapid succession of fungi with dramatic changes in the composition of the fungal community, with most of the abundant taxa only temporarily dominating in the substrate. A total of 387 fungal genera were identified as the closest hits of individual operational taxonomical units (OTUs). The filamentous fungi of the genera *Mycosphaerella*, *Naevala*, and *Troposporella* and the yeast genera *Apiotrichum* and *Cutaneotrichosporon* were the most abundant fungi in the amplicon pool. In particular, the latest stages of litter decomposition, *Ap. porosum* and *Cutaneotrichosporon moniliiforme*, were prevalent. In addition to these species, representatives of *Aureobasidium* and the basidiomycetous genera *Mrakia*, *Rhodotorula*, *Sporobolomyces*, and *Tremella* and the former *Cryptococcus* genus were identified among the 50 most abundant fungal genera in the leaves. The majority of fungal sequences were assigned to the phyla Ascomycota (71%) and Basidiomycota (26%). Sequences assigned to the Ascomycota showed the highest relative abundances in live leaves and during the early stages of decomposition. In contrast, the relative abundance of sequences assigned to the Basidiomycota phylum, particularly basidiomycetous yeasts, increased with time. In fact, the basidiomycetous former yeast genus *Trichosporon* comprised 50% of all sequences at month 24 (Voříšková and Baldrian 2013). This finding is consistent with reports that identify species from this genus as cellulose decomposers (Dennis 1972; Štursová et al. 2012).

The diversity of yeasts in soil and litter samples taken from three Austrian natural forest reserves was investigated by Wuczkowski et al. (2005). A total of 82 yeast strains were isolated and identified through molecular methods, leading to the identification of 21 different species belonging to 11 genera. The vast majority of the isolated strains were basidiomycetous yeasts. Specially, the species associated with litter were the black yeast *Aureobasidium pullulans*, the ascomycetes *Taphrina wiesneri*, and the basidiomycetes *Ap. porosum*, *Cystofil. capitatum*, *Dioszegia hungarica*, *Goffeauzyma gastricus*, *Leucosporidium fellii*, *Pa. laurentii*, *Pseudotremella moriformis*, *Rhodotorula bacarum*, *Sporidiobolus pararoseus*, *Sporobolomyces salicinus*, *Vishniacozyma carnescens*, and *Vishniacozyma victoriae* (Wuczkowski et al. 2005).

Many studies investigating leaf breakdown in aquatic environments take place in streams. Sampaio et al. (2001) examined the decomposition of three leaf species (*Alnus glutinosa*, *Eucalyptus globulus*, and *Quercus robur*) in a headstream over two months, during which the decomposing leaves were periodically analyzed for nutrient content, soluble sugars, phenols, protein precipitation capacity, total fiber, weight loss, and microbial and macroinvertebrate colonization. Interestingly, the results indicated that the most important parameters to describe litter quality were related to the abundance of microorganisms, namely, the number of yeast and total heterotrophic colonies. Additionally, the findings revealed an increase in the number of yeasts and yeast-like fungi that appeared later in the decay process, suggesting these transient organisms are dependent on substances released during degradation by other fungi and/or bacteria that are easier to assimilate. However, this group was strongly correlated with soluble sugars (positively) and with phenols (negatively). The authors proposed that the unstable pattern in yeast colonization reflected the variation in the availability of substrate type, to which these organisms are very sensitive during incubation. Similar results were found by Gonçalves et al. (2006) during an investigation of leaf breakdown in two streams of different magnitudes (third and fourth orders) by applying litter bags to a tropical stream and evaluating colonization by invertebrates and microorganisms during the processing of detritus over 270 days of immersion. The authors observed that the chemical composition of detritus was important to biological colonization, with the density of bacteria greatest in the initial stages, whereas, in the intermediate and final stages, fungi appeared at high densities. In particular, yeasts were most common on the 2nd, 3rd, 10th, and 90th days, and the highest density was observed in the fourth order reach on the 90th day. However, from this study, it was not possible to identify the effective role of yeasts in leaf breakdown. The results showed fluctuation in the density of yeasts, suggesting that this group of fungi consists of transient organisms that assimilate substances released during the degradation of leaves by other fungi or bacteria (Gonçalves et al. 2006).

Investigations regarding the composition of yeast and macroinvertebrate communities on black alder, blue gum eucalyptus, and English oak leaves decaying in a stream over a 6-month period revealed rapid yeast colonization and significant differences in yeast densities between the three leaf litters (Sampaio et al. 2004). Yeast dynamics fluctuated considerably during the different stages of decomposition. Both litter type and the length of immersion affected yeast counts; generally after rapid initial colonization, however, yeast counts stabilized until the final stages of decay. The yeasts were positively correlated with the total number of invertebrates and collectors-gatherers. The species were identified using morphological, physiological, and biochemical techniques. In total, 36 yeast species were isolated, represented by typical inhabitants of phylloplane and soils. Basidiomycetous yeasts accounted for 50% or more of the total number of species found, with some yeast species such as *Na. albida*, *Pa. laurentii*, *Rhodotorula glutinis*, *Rhodotorula colostri*, and the ascomycetous *Debaryomyces hansenii* present in all litter types, whereas other yeasts were restricted to a specific type of litter. The authors suggested that such different distribution of yeast populations among litter types was associated with distinct foliar chemical compositions. For instance, eucalyptus leaves have poor nutrient (nitrogen) content but are rich in polyphenols, condensed tannins, and essential oils, substances that delay microbial attack. The high percentages of basidiomycetous yeasts on eucalyptus and oak leaf litters might be related to the fact that some of these yeasts—*Pa. laurentii*, *Rh. glutinis*, *Rhodotorula graminis*, *Rhodotorula mucilaginosa*, and *Solicoccozyma terreus*—are able to degrade aromatic substances (Sampaio 1999; Sampaio et al. 2004).

Saprophytic yeasts colonizing autochthonous and exotic leaf litters submerged in a natural mountain stream at different phases of decomposition were investigated in studies carried out over two consecutive years (Sampaio et al. 2007). Leaf litter mass loss, total yeast counts, yeast community structure, and physiologic abilities were analyzed to determine the dynamics of yeast communities during decay. Seventy-two yeast species belonging to 26 genera were identified through phenotypic or sequencing methods. Three groups of yeasts were isolated: those common to all litters, those shared by two types of litter, and those that appeared only in one particular leaf litter. However, in all litter types, basidiomycetous species predominated over ascomycetous species by up to 85%. Still, despite being less abundant, ascomycetous species showed greater diversity than basidiomycetous yeasts and were primarily isolated in the first and later decomposition stages. During an analysis of the colonization patterns of yeast species with regard to their order of appearance during leaf decomposition, only a few species were observed to be persistently present (at three or more sampling times) in litter, namely, *Cut. cutaneum*, *Cystofil. capitatum*, *Leuc. scottii*, *Na. albida*, *Pa. laurentii*, *Rh. colostri*, *Rh. glutinis*, and *Sporid. pararoseus*, whereas most were rare and only recorded once or twice during decomposition. All litter types shared generalist species that are frequently isolated worldwide or from aquatic environments, decomposing organic matter, phylloplane, and soils, such as *A. pullulans*, *Cut. cutaneum*, *Cut. moniliiforme*, *Cyberlindnera saturnus*, *Cystofil. capitatum*, *Deb. hansenii*, *Leuc. scottii*, *Na. albida*, *Rh. colostri*, and *Rh. glutinis*. An analysis of the presence/absence of yeast species showed significant differences by substrate types and throughout the decay period. Likewise, carbon and nitrogen source utilization by yeast strains varied with substrate and decomposition time, suggesting that yeast communities respond to habitat modifications, which, in this case, may be equivalent to the distinct chemical foliar compositions among litters and throughout decomposition. This study also concluded that the order of yeast appearance and their substrate assimilation patterns strongly suggest a succession phenomenon (Sampaio et al. 2007), which can be related to differences in metabolic abilities of the yeast species or even different physiological capacities (Sampaio et al. 2004). In fact, in this study, species richness was higher at the beginning of the experiment than at the end, reflecting the fact that the mass of leaves initially rich in soluble sugars was rapidly colonized by yeasts, resulting in high-diversity values (Sampaio et al. 2001, 2007). Moreover, the appearance of a new yeast population at the end of the experiment/later colonization provides evidence of a new niche and supports tolerance and/or competition behavior among the species (Sampaio et al. 2007).

9.3.2 Rotting Wood

The chemical composition of wood varies by tree part, type of wood, geographic location, climate, and soil conditions, thus influencing the associated microbial communities. The carbohydrate portion of wood comprises cellulose and hemicelluloses, including the most abundant of the wood hemicelluloses, xylans, and glucomannans. Thus, the wood carbon content mainly corresponds to the hexoses and pentoses glucose, xylose, arabinose, mannose, and galactose (Pettersen 1984; Martínez et al. 2005), sugars widely assimilated by yeasts found in this substrate (Table 9.1).

The number and occurrence of yeasts in rotting wood can be influenced by the type of wood (tree), vegetation, climate, season, locality, degree of the substrate decomposition, and insects participating in the decomposition process. However, basidiomycetous yeasts, as observed for leaf litter, or ascomycetous yeasts, which can assimilate or ferment xylose and other wood-related sugars, are predominantly present in such material (Table 9.1). In addition, these organisms are present in both the initial and advanced stages of wood decay, when they establish mutualistic associations with other microorganisms (Blanchette and Shaw 1978).

González et al. (1989) studied the presence and distribution of yeasts during the fungal transformation of wood at different stages of degradation (namely, initial, mid, advanced, and final) in a Chilean rain forest. A total of 327 different strains were isolated and phenotypically identified as belonging to 68 yeast species of 14 genera. Thirty-seven species were most representative (total relative abundance $\geq$ 1%). Variation in the numbers of yeasts identified in the different types of decaying wood corresponded to the natural abundance and distribution of basidiomycetes fungi in the study areas. Most of the samples studied corresponded to wood decayed by the fungus *Ganoderma applanatum*, and yeast succession during decay was clearly observed when the distribution of this fungus during the decomposition process was considered. During wood decay by *Gan. applanatum*, *Candida railenensis* (*Kurtzmaniella* clade) was very abundant at the initial stages of decay, whereas *Sugiyamaella chiloensis* (formerly *Candida bertae*) and *Candida sake* were predominant at the final stages. *Na. albida* was the only species present at all four stages of degradation, showing an increasing abundance through the decay process. The relative abundance of basidiomycetous yeasts was low at the initial stage of decay but was significantly higher at more advanced stages when compared with other natural substrates. Basidiomycetous yeast species, particularly the genus *Apiotrichum*, were abundant (81% of the total) in samples of wood decayed by *Armillaria limonea*. A high percentage (88%) of xylose-assimilating strains among the yeast isolated was observed. The diversity of yeasts in this study was found to be very high, with species present during the advanced stages of decay that were different than those frequently reported at the initial stages of wood decay (González et al. 1989).

Jiménez et al. (1991) tested 51 of 68 yeast species isolated from decaying wood and identified by González et al. (1989) for growth in carboxymethylcellulose, straw hemicelluloses, and wood xylan and β-glucosidase and β-xylosidase enzyme activities. Twenty-three species showed growth in at least one substrate or enzymatic activity. The black yeast *Aureobasidium microstictum* was the sole yeast able to grow in the three substrates tested and that displayed activities of both enzymes. *Sugiyamaella castrensis*, *Sugiyamaella santjacobensis*, *Deb. hansenii*, *Middelhovenomyces petrohuensis* (formerly *Candida ancudensis*), *Middelhovenomyces tepae*, *Pichia membranifaciens*, *Rh. mucilaginosa* (*Rhodotorula grinbergsii*), and *Ta. pullulans* grew in carboxymethylcellulose; *Deb. hansenii*, *Filobasidium uniguttulatum*, *Midd. petrohuensis*, *Na. albida*, *Pa. laurentii*, and *Ta. pullulans* showed growth in straw hemicelluloses and wood xylan; *Nakazawaea ishiwadae*, *Peterozyma toletana*, *Sug. chiloensis*, and *Sugiyamaella smithiae* (*Candida edax*) grew solely on wood xylan. Besides *A. microstictum*, *Candida oleophila* (*Kurtzmaniella* clade) showed β-glucosidase activity, and *F. uniguttulatum*, *Kluyveromyces marxianus*, and *Sug. chiloensis* displayed β-xylosidase activity (Jiménez et al. 1991).

Yeasts able to assimilate lignocellulose-related sugars are usually found in tree bark and rotting wood, which are suitable habitats for these organisms. In past years, many studies have focused on the isolation of xylose-assimilating yeast species from rotting plant materials and related substrates, including mainly wood and wood-boring insects, to obtain strains and/or species capable of converting this pentose into products of high-added value or great economic interest, such as xylitol and biofuel ethanol. In the same context, efforts to isolate yeasts able to ferment cellobiose—the disaccharide resulting from the activities of endoglucanases and cellobiohydrolases on cellulose—have also been conducted (Cadete et al. 2014). In this scenario, Brazilian, Chinese, the USA, and Thai ecosystems stand out for their important contributions as sources of new D-xylose- and/or cellobiose-assimilating and fermenting yeasts.

The first study to investigate the Brazilian Amazonian Forest aiming to isolate and identify new D-xylose-fermenting yeasts associated with rotting wood was conducted by Cadete et al. (2012b). A total of 224 yeast strains were isolated from 40 rotting wood samples collected in Amazonian Forest areas in the north of Brazil and cultured on xylose or xylan media as sole carbon sources. Of the 33 yeast species identified, 26 species were previously known and seven were new. *Candida tropicalis* was the most frequently isolated yeast, followed by *Vanrija humicola* and *Candida boidinii* (*Ogataea* clade). *Cyb. saturnus*, *Deb. hansenii*, *Schwanniomyces polymorphus*, and *Meyerozyma guilliermondii* were also abundant. Six strains of *Spathaspora passalidarum*, as well as four new species belonging to the clade *Spathaspora*—recently described as *Spathaspora brasiliensis*, *Spathaspora suhii*, *Spathaspora roraimanensis*, and *Spathaspora xylofermentans* (Cadete et al. 2013)—were obtained. The novel species *Scheffersomyces amazonensis* (Cadete et al. 2012a) and *Cyberlindnera xylosilytica* (Cadete et al. 2015) were also identified in this study.

Based on the successful isolation of yeasts associated with rotting wood in the Brazilian Amazonian Forest, an investigation of yeast species on rotting wood in Brazilian Atlantic Rainforest ecosystems that focused on the identification of

D-xylose-fermenting and/or xylanase-producing species was conducted by Morais et al. (2013b). A total of 321 yeast strains belonging to 69 different species were isolated from 100 rotting wood samples collected in two areas of the Atlantic Rainforest in southeastern and northeastern Brazil. *Barnettozyma californica*, *Ogataea boidinii*, *Schw. polymorphus*, and *Scheffersomyces queiroziae* were the most frequently isolated yeasts. *Candida melibiosica* (*Metschnikowia* clade), *Galactomyces geotrichum*, *Spencermartinsiella cellulosicola*, *Sugiyamaella boreocaroliniensis*, and *Va. humicola* were also abundant. Fifteen possibly novel species were obtained, among which eight were described as *Sugiyamaella xylanicola* (Morais et al. 2013a), *Sugiyamaella bahiana*, *Sugiyamaella xylolytica*, and *Sugiyamaella ligni* (Sena et al. 2017); *Spathaspora girioi* and *Spathaspora gorwiae* (Lopes et al. 2016); *Spencermartinsiella silvicola* (Morais et al. 2016a); and *Saturnispora bothae* (Morais et al. 2016b).

Other new species that were isolated from rotting wood in Brazil and able to ferment xylose/cellobiose and/or producing xylanase were also described: three species belonging to the *Spathaspora* clade, *Candida materiae* (Barbosa et al. 2009), *Spathaspora arborariae* (Cadete et al. 2009), and *Spathaspora hagerdaliae* (Lopes et al. 2016); one species of the *Scheffersomyces* clade, *Scheff. queiroziae* (Santos et al. 2011); and four species harbored by the *Sugiyamaella* clade, *Sugiyamaella ayubii*, *Sugiyamaella bonitensis*, *Sugiyamaella carassensis*, and *Sugiyamaella valenteae* (Sena et al. 2017). New strains of the traditionally xylose-fermenting species *Scheffersomyces stipitis* and *Scheffersomyces shehatae* were also isolated from rotting wood in Brazilian ecosystems (Cadete et al. 2012b; Morais et al. 2013b).

Studies aiming to isolate xylanase-producing yeasts from rotting wood and other decaying plant materials were also conducted in Brazil. Morais et al. (2013b) identified strains from 20 species (of the 69 species isolated from rotting wood in the Atlantic Rainforest) that showed xylanolytic activities on xylan-agar medium, including *Pa. laurentii*, *Spenc. cellulosicola*, and *Sug. xylanicola*. Lara et al. (2014) evaluated a total of 358 yeast isolates obtained from decaying wood and decomposing sugarcane bagasse sampled in Brazilian biomes and sugar-ethanol plants for xylanase production. Seventy-five isolates showed xylanase activity in solid medium and were identified as belonging to nine species: *Apiotrichum mycotoxinivorans*, *C. tropicalis*, *C. intermedia* (*Clavispora* clade), *Kwoniella heveanensis*, *Mey. guilliermondii*, *Naganishia diffluens*, *Pa. laurentii*, *Scheff. shehatae*, and *Sug. smithiae* (Lara et al. 2014).

Investigations of the yeast community associated with rotten wood in the Baotianman Nature Reserve of Henan Province, central China, led to the discovery and description of a novel yeast species belonging to the clade *Scheffersomyces*—*Scheffersomyces henanensis* (Ren et al. 2014). In this study, 105 yeast strains were isolated from 23 rotten wood samples. Seventeen strains showed the ability to ferment D-xylose and were identified as belonging to clades *Spathaspora* and *Scheffersomyces*, namely, *Spath. passalidarum*, *Scheffersomyces insectosa*, *Scheffersomyces lignosus*, *Scheffersomyces segobiensis*, *Scheff. stipitis*, *Scheff. shehatae*, and *Scheff. henanensis* (Ren et al. 2014). Recently, a search for

D-xylose-assimilating yeasts associated with rotting wood in the Galápagos Archipelago, Ecuador, was undertaken (Guamán-Burneo et al. 2015). A total of 140 yeast strains belonging to 15 genera and 35 species were obtained from 35 rotting wood samples. The species *Barn. californica*, *Candida sinolaborantium*, *C. tropicalis*, *Kazachstania unispora*, and *Candida natalensis* (*Kurtzmaniella* clade) were the most frequently isolated species. A new species, *Cyberlindnera galapagoensis*, has been described in this study (Guamán-Burneo et al. 2015).

Among the major clades of yeasts associated with rotting wood habitats, four are worth mentioning for their ability to assimilate/ferment lignocellulose-related sugars and/or to produce enzymes that act on this substrate: *Scheffersomyces*, *Spathaspora*, *Spencermartinsiella*, and *Sugyiamaella*. In addition to these clades, *C. tropicalis* is a species that deserves attention because it is frequently found in decaying plant materials, including mainly rotting wood and even decaying sugarcane bagasse (Rosa, unpublished data), suggesting that this is the natural habitat for this yeast. Members of clades *Scheffersomyces* and *Spathaspora* are capable of fermenting xylose to produce ethanol (Cadete et al. 2014). Clades *Spencermartinsiella* and *Sugiyamaella* belong to the family Trichomonascaceae and have been recently found to include xylose-fermenting and xylanase-producing species (Morais et al. 2013b, 2016b; Handel et al. 2016; Sena et al. 2017). *C. tropicalis* belongs to the *Lodderomyces*/*Candida albicans* clade and is extensively studied for its potential for bioproduction of xylitol from xylose (Ur-Rehman et al. 2015; Rao et al. 2016).

Currently, the genus *Scheffersomyces* comprises 20 recognized yeast species, although results from a multilocus DNA sequence comparison with extended taxon sampling indicate that the genus is not monophyletic (Kurtzman and Robnett 2013; Suh et al. 2013; Liu et al. 2016b). All of the species assigned to this clade are associated with rotting wood and/or wood-boring insects (Kurtzman et al. 2011; Chang et al. 2011; Santos et al. 2011; Cadete et al. 2012a; Urbina and Blackwell, 2012; Urbina et al. 2013a; Suh et al. 2013; Ren et al. 2014; Liu et al. 2016b), except for *Scheffersomyces spartinae*, which was isolated from oyster grass and water (Kurtzman et al. 2011). In fact, *Scheff. spartinae* was recently placed in a clade with *Spath. passalidarum*, distinct from the type species of *Scheffersomyces*, *Scheff. stipitis*, although the clade was weakly supported statistically (Kurtzman and Robnett 2013). Results from a combined sequence comparison strongly suggested that this genus should be circumscribed to a monophyletic group of xylose-fermenting yeasts closely related to the type species, i.e., *Scheff. stipitis*, *Scheff. segobiensis*, *Scheffersomyces illinoinensis*, *Scheffersomyces xylosifermentans*, *Scheff. insectosa*, *Scheff. lignosus*, *Scheff. shehatae*, *Scheffersomyces cryptocercus*, *Scheffersomyces quercinus*, *Scheffersomyces virginianus*, *Scheffersomyces parashehatae*, *Scheff. henanensis*, and *Scheffersomyces titanus*. The remaining species, which have been treated previously as members of *Scheffersomyces*, may become representatives of several novel genera (Suh et al. 2013; Liu et al. 2016b).

The *Spathaspora* clade was first described as including the species *Spath. passalidarum*, the first teleomorphic species of the genus, and the anamorphic species *Candida jeffriesii* (Nguyen et al. 2006). Currently, the genus is composed of a core group that is well supported by phylogenetic analyses of D1/D2 LSU sequences and includes *Spath. passalidarum*, *Spath. arborariae*, *Spath. brasiliensis*, *Spath. suhii*, *Spath. girioi*, *C. jeffriesii*, and *C. materiae* (Lopes et al. 2016). The species *Spath. roraimanensis* exhibits the unique ascospore morphology of the genus and is closely related to *Spath. xylofermentans*, in which ascospore production has not been observed (Cadete et al. 2013). The recently described species *Spath. gorwiae*, *Spath. hagerdaliae*, and *Spathaspora allomyrinae* are closely related to *Candida lyxosophila* (Lopes et al. 2016; Wang et al. 2016). *Candida alai*, *Candida insectamans*, *C. sake*, *Candida xylanilytica*, and *Candida subhashii* are weakly associated with this clade (Daniel et al. 2014). As observed for the genus *Scheffersomyces*, the *Spathaspora* clade harbors yeasts associated with rotting wood and/or wood-boring insects (Barbosa et al. 2009; Cadete et al. 2009, 2013; Kurtzman et al. 2011; Wang et al. 2016; Lopes et al. 2016).

The genus *Spencermartinsiella* was proposed by Péter et al. (2011) to accommodate a novel yeast species, *Spencermartinsiella europaea*, isolated from rotten wood sampled in different locations in Hungary. Later, a second species belonging to the same genus, *Spencermartinsiella ligniputridi*, found in association with decaying wood in Hungary, was described by Dlauchy et al. (2012). Recently, a third ascosporic species in the genus, *Spenc. silvicola*, was isolated from rotting wood samples collected in the Brazilian Atlantic Rainforest (Morais et al. 2016b). *Spenc. cellulosicola* is the only recognized anamorphic species of the clade *Spencermartinsiella*. This species was also isolated from rotten wood collected in a rainforest across different mountains in southern China (Guo et al. 2012). Xylanolytic activities for the species *Spenc. cellulosicola* and *Spenc. silvicola* have been reported by Morais et al. (2013b, 2016b).

The *Sugiyamaella* clade was proposed by Kurtzman and Robnett (2007) to accommodate the former species *Stephanoascus smithiae* and related species in a clade distinct from that including *Stephanoascus ciferrii*, the type species of *Stephanoascus*, as well as from *Stephanoascus farinosus*. Currently, this genus comprises six ascosporic members and 21 species whose sexual stage is not known. These species are associated with rotting plant materials (mainly wood), soil, and guts of insects (Kurtzman et al. 2011; Morais et al. 2013a; Urbina et al. 2013a; Handel et al. 2016; Sena et al. 2017). Although recognized as able to assimilate xylose, *Sugiyamaella* species have received attention recently because of their ability to produce xylanolytic enzymes or to ferment xylose to ethanol. Indeed, in the last few years, at least nine new species have been described, with eight associated with rotting wood in Brazil (Morais et al. 2013a; Sena et al. 2017) and one (*Sugiyamaella mastotermitis*) isolated from the hindgut and nest material of a termite species native to Australia. All of these novel species are able to produce xylanase.

Many yeast species that are found in live or decaying plant parts are associated with insects that also use these habitats as feeding or breeding sites. In general, these three-part associations (insect-yeast-plant) are dependent on reciprocal benefits exchanged between the insect-yeast partners. Often, the yeast supplies essential nutrients or beneficial supplements to the insect, while the insect provides transportation of the yeast to new habitats (Kurtzman et al. 2011). Particularly, the guts of wood-feeding insects in several families are especially rich in yeasts that carry out the fermentation of the wood-related sugars cellobiose and xylose (Urbina et al. 2013b) or are able to produce xylanases. Aiming to expand knowledge on the diversity of ascomycete and basidiomycete yeasts in the guts of Guatemalan passalids, beetles that live in dead wood which is ingested as their primary food source, Urbina et al. (2013b) isolated and identified 771 yeasts that corresponded to approximately 78 species. The most abundant Saccharomycotina clades were *Scheffersomyces* and *Spathaspora* (560 isolates, approximately 76.5%). The xylose-fermenting yeasts *Scheff. shehatae* (314 isolates, 42.9%) and *Scheff. stipitis* (109 isolates, 14.9%) were the most common ascomycete yeasts identified. The guts of the beetles also contained undescribed cellobiose-fermenting and xylose-fermenting species in the genera *Lodderomyces*, *Scheffersomyces*, and *Spathaspora* and undescribed species in the *Sugiyamaella* clade, as well as rare yeast species in the clades *Phaffomyces* and *Spencermartinsiella*. Basidiomycete yeasts (60 isolates, 0.8%) were also recovered from the guts of the passalids, including mainly *Vanrija musci*, *Apiotrichum laibachii*, *Ap. porosum*, *Cutaneotrichosporon dermatis*, and *Cut. moniliiforme* (Urbina et al. 2013b). The fact that yeasts belonging to clades *Scheffersomyces*, *Spathaspora*, and *Sugiyamaella*, as well as basidiomycetous yeasts primarily in the Tremellomycetes, are reported in wood-boring insects and in decaying wood strongly suggests a close association between this habitat and these microorganisms. In fact, according to Suh et al. (2003), beetle invasions into woody substrates may be facilitated by their association with yeasts. This may be because xylose is not usually found as a soluble sugar in nature, unlike sucrose and fructose, so the insect gut offers a place in which hemicellulose can be broken down for assimilation (Jeffries and Jin 2000).

In addition to the strong association of yeasts that assimilate xylose and other sugars found in wood and yeasts that produce xylanases and cellulases from this habitat, methanol-assimilating species are frequently found in decaying wood, plant leaves, and other residues (Péter et al. 2003; Santos et al. 2015). Methanol is common in nature and is emitted by metabolizing leaves as well as from methoxy groups during the breakdown of lignin, thus explaining the presence of yeasts able of consume this alcohol in plant sources (Kurtzman 2011). Yeasts capable of growing with methanol as the sole carbon source belong to the genera *Komagataella*, *Kuraishia*, and *Ogataea* (Kurtzman 2011; Kurtzman et al. 2011). Studies on the diversity of yeasts in rotting wood in Brazil have frequently isolated the methanol-assimilating species *O. boidinii* and other species belonging to the same clade. This yeast is the third most frequently isolated species from rotting wood sampled in the Brazilian Amazonian Forest and Atlantic Rainforest (Cadete et al. 2012b; Morais et al. 2013b). In addition, Morais et al. (2013b) reported *Candida* (*Ogataea*) *nanaspora* and *Candida cylindracea* (basal to the *Ogataea* clade) and two novel *Ogataea* species on decaying wood.

Yeasts of clinical interest are also found in rotting wood, bark, and other plant materials. *Cryptococcus neoformans* has been reported in the final stage of wood decay (González et al. 1989) and in 3 of 45 (6.6%) decaying trunk hollows of two tree species and one of 390 *Eucalyptus* bark samples (Randhawa et al. 2001). In agreement with their essentially saprobic character, many yeast species were sporadically isolated in this last study. These yeasts, identified by morphological and physiological methods, include *C. tropicalis*, *Clavispora lusitaniae*, *Cut. cutaneum*, *Geotrichum klebahnii*, *Candida zeylanoides* (*Kurtzmaniella* clade), *Magnusiomyces capitatus*, *Mey. guilliermondii*, *Na. albida*, *Pa. laurentii*, *Pichia kudriavzevii*, *Rh. mucilaginosa*, *Rh. glutinis*, and *Sporobolomyces salmonicolor* (Randhawa et al. 2001). *Cryptococcus gattii* has also been isolated from decayed wood inside trunk hollows of many tree species (Lazéra et al. 1998; Grover et al. 2007). These findings suggest that decaying wood inside trunk hollows of some trees may be a natural habitat of pathogenic *Cryptococcus* yeasts. In our studies in Brazil, *C. albicans* has been isolated in low frequency from rotting wood samples (Rosa, unpublished data), and this substrate could also represent a natural habitat for this opportunistic yeast species.

9.4 Concluding Remarks

Yeasts play different roles in the decomposition of plant materials, actively participating as producers of hydrolytic enzymes or passively participating as opportunists that use products released during decomposition by other organisms such as fungi and bacteria. The order that yeasts appear during the different stage of plant material decomposition strongly suggests a succession phenomenon that might be related to differences in metabolic abilities or physiological behaviors of the yeast species. In particular, the most abundant yeast taxa associated with decayed wood and leaf litter are basidiomycetous (Agaricomycotina) and xylose-assimilating species, reflecting the environment where these organisms live.

The same characteristics that allowed these yeasts to adapt to decaying plant substrates make these organisms interesting for biotechnological applications. Hydrolytic enzymes—cellulases, xylanases, and pectinases—can be employed in several industrial processes. Cellulases are applied in textile, pulp and paper, laundry and detergent, animal feed, food processing, biorefinery, and pharmaceutical industries (Sharma et al. 2016). Although few yeast species are reported able to produce enzymes from the cellulolytic complex, many species exhibit β-glucosidase activities. β-glucosidases are crucial for the production of bioethanol from cellulose as a raw material—so-called second-generation (2G) ethanol. Once this enzyme converts cellobiose in glucose, it allows the fermentation of glucose to ethanol by *S. cerevisiae* strains—the most commonly applied species in alcoholic fermentation industries, which, however, lack the ability to ferment cellobiose (Lynd et al. 2002). In addition, β-glucosidases-producing yeasts are sources of genes that could be expressed in *S. cerevisiae* to construct strains able to ferment cellobiose (Hu et al. 2016). Lastly, native cellobiose-fermenting species can be used directly in cellulose saccharification and fermentation steps. Several species in the genus *Scheffersomyces* have potential to be applied in this way (Santos et al. 2011;

Urbina and Blackwell 2012; Cadete et al. 2014). Xylanases are used by pulp and paper industries, as well as in the textile and food sectors (Ahmed et al. 2009; Knob et al. 2010; Kumar and Shukla 2016). Like β-glucosidases, xylanases can be applied in the production of 2G bioethanol, but using hemicelluloses (xylan) as a feedstock, and hydrolyzing this polymer to release xylose (Gírio et al. 2010). Several yeast species feature the ability to produce xylanase, including species able to ferment xylose, like *Scheff. stipitis* and *Scheff. shehatae* (Cadete et al. 2012b; Morais et al. 2013b). Pectinases are applied in several industrial processes, including juice and wine processing and other food applications, paper manufacturing, textile and plant fiber processing, oil extraction, treatment of industrial wastewater, and degradation of pectinacious materials (Alimardani-Theuil et al. 2011; Irshad et al. 2014). Yeasts are also sources of pectinolytic enzymes, similar to ascomycetes and basidiomycetes species. Pectinolytic yeasts can be directly applied for simultaneous fermentation and beverage clarification (Alimardani-Theuil et al. 2011).

Xylose-assimilating yeasts are promising microbial sources of 2G bioethanol and xylitol, a high-value added product resulting from the reduction of D-xylose by the enzyme xylose reductase (Cadete et al. 2014, 2015). To date, *Spath. passalidarum* is considered the best D-xylose-fermenting yeast producing ethanol ever reported. This species is able to produce ethanol from xylose at high rates under anaerobic conditions, an ability unprecedented in other xylose-fermenting species (Hou 2012). Recently, a *S. cerevisiae* strain able to ferment xylose anaerobically has been constructed using *Spath. passalidarum* xylose-metabolism-related genes. *S. cerevisiae* lacks the ability to metabolize xylose, and the construction of strains to produce ethanol from xylose at high yields is of substantial interest for the 2G bioethanol industry (Cadete et al. 2016a). Xylitol, a five-carbon sugar alcohol, presents anticariogenic properties and participates in insulin-independent metabolism; these features make xylitol an important sweetener used in food and pharmaceutical industries. Beyond this, xylitol has been applied industrially as an antioxidant, moisturizer, stabilizer, and cryoprotectant, because it reduces the freezing point of some products (Prakasham et al. 2009; Mohamad et al. 2015). Today, *C. tropicalis*, *Cyb. xylosilytica*, *Mey. guilliermondii*, and *Scheff. amazonensis* are considered efficient xylitol bioproducers, and efforts are being made to employ these species in the production of this polyol from plant hydrolysates in industrial conditions (Cadete et al. 2015, 2016b; Guamán-Burneo et al. 2015; Wang et al. 2015; de Arruda et al. 2016). Lastly, methylotrophic yeasts have been extensively exploited for high-level production of native and heterologous proteins, because the secretion of proteins is more efficient in these yeasts than in *S. cerevisiae* (Fernández et al. 2016).

In summary, this chapter detailed the association of yeasts with decaying plant materials. These substrates are promising environments for the isolation and description of novel yeast species. The frequent association of yeasts with leaf litter and rotten wood provides great support for research on yeast ecology, biodiversity, physiology, and biotechnology.

Acknowledgments We would like to thanks the Conselho Nacional de Desenvolvimento Científico e Tecnológico (CNPq) and Fundação do Amparo à Pesquisa do Estado de Minas Gerais (FAPEMIG) for their financial support.

References

Adsul MG, Bastawde KB, Gokhale DV (2009) Biochemical characterization of two xylanases from yeast *Pseudozyma hubeiensis* producing only xylooligosaccharides. Biores Technol 100: 6488–6495

Ahmed S, Riaz S, Jamil A (2009) Molecular cloning of fungal xylanases: an overview. Appl Microbiol Biotechnol 84:19–35

Alimardani-Theuil P, Gainvors-Claisse A, Duchiron F (2011) Yeasts: an attractive source of pectinases—from gene expression to potential applications: a review. Process Biochem 46: 1525–1537

Baldrian P (2008) Enzymes of saprotrophic basidiomycetes. In: Boddy L, Frankland JC, van West P (eds) British mycological society symposia series, vol 28. Elsevier, Amsterdam, pp 19–41

Baldrian P, Valášková V (2008) Degradation of cellulose by basidiomycetous fungi. FEMS Microbiol Rev 32:501–521

Baldy V, Gessner MO, Chauvet E (1995) Bacteria, fungi and the breakdown of leaf litter in a large river. Oikos 74:93–102

Barbosa AC, Cadete RM, Gomes FC, Lachance M-A, Rosa CA (2009) *Candida materiae* sp. nov., a yeast species isolated from rotting wood in the Atlantic Rain Forest. Int J Syst Evol Microbiol 59:2104–2106

Bedriñana RP, Queipo AL, Valles BS (2012) Screening of enzymatic activities in non-*Saccharomyces* cider yeasts. J Food Biochem 36:683–689

Berg B, McClaugherty C (2008) Plant litter. Decomposition, humus formation, carbon sequestration, 2nd edn. Springer, Heidelberg

Bhadra B, Rao RS, Singh PK, Sarkar PK, Shivaji S (2008) Yeasts and yeast-like fungi associated with tree bark: diversity and identification of yeasts producing extracellular endoxylanases. Curr Microbiol 56:489–494

Biely P, Kremnicky L (1998) Yeasts and their enzyme systems degrading cellulose, hemicelluloses and pectin. Food Technol Biotechnol 36:305–312

Biely P, Vršanská M, Krátký Z (1980) Xylan-degrading enzymes of the yeast Cryptococcus albidus. Eur J Biochem 108:313–321

Blanchette RA, Shaw CG (1978) Associations among bacteria, yeasts, and basidiomycetes during wood decay. Phytopathology 68:631–637

Blanco P, Sieiro C, Villa TG (1999) Production of pectic enzymes in yeasts. FEMS Microbiol Lett 175:1–9

Borges TA, de Souza AT, Squina FM, Riaño-Pachón DM, dos Santos RAC, Machado E, Oliveira JVC, Damásio ARL, Goldman GH (2014) Biochemical characterization of an endoxylanase from *Pseudozyma brasiliensis* sp. nov. strain GHG001 isolated from the intestinal tract of *Chrysomelidae* larvae associated to sugarcane roots. Process Biochem 49:77–83

Botha A (2011) The importance and ecology of yeasts in soil. Soil Biol Biochem 43:1–8

Brandão LR, Libkind D, Vaz ABM, Santo LCE, Moliné M, de García V, van Broock M, Rosa CA (2011) Yeasts from an oligotrophic lake in Patagonia (Argentina): diversity, distribution and synthesis of photoprotective compounds and extracellular enzymes. FEMS Microbiol Ecol 76:1–13

Brizzio S, Turchetti B, de García V, Libkind D, Buzzini P, van Broock M (2007) Extracellular enzymatic activities of basidiomycetous yeasts isolated from glacial and subglacial waters of northwest Patagonia (Argentina). Can J Microbiol 53:519–525

Buzzini P, Martini A (2002) Extracellular enzymatic activity profiles in yeast and yeast-like strains isolated from tropical environments. J Appl Microbiol 93:1020–1025

Cadete RM, Santos RO, Melo MA, Mouro A, Gonçalves DL, Stambuk BU, Gomes FCO, Lachance M-A, Rosa CA (2009) *Spathaspora arborariae* sp. nov., a D-xylose-fermenting yeast species isolated from rotting wood in Brazil. FEMS Yeast Res 9:1338–1342

Cadete RM, Melo MA, Lopes MR, Pereira GMD, Zilli JE, Vital MJS, Gomes FCO, Lachance M-A, Rosa CA (2012a) *Candida amazonensis* sp. nov., an ascomycetous yeast isolated from rotting wood in the Amazonian forest. Int J Syst Evol Microbiol 62:1438–1440

Cadete RM, Melo MA, Dussan KJ, Rodrigues RC, da Silva SS, Zilli JE, Vital MJS, Gomes FCO, Lachance M-A, Rosa CA (2012b) Diversity and physiological characterization of D-xylose-fermenting yeasts isolated from the Brazilian Amazonian forest. PLoS One 7:e43135

Cadete RM, Melo MA, Zilli JE, Vital MJ, Mouro A, Prompt AH, Gomes FCO, Stambuk BU, Lachance M-A, Rosa CA (2013) *Spathaspora brasiliensis* sp. nov., *Spathaspora suhii* sp. nov., *Spathaspora roraimanensis* sp. nov. and *Spathaspora xylofermentans* sp. nov., four novel D-xylose-fermenting yeast species from Brazilian Amazonian forest. A van Leeuwenhoek 103:421–431

Cadete RM, Fonseca C, Rosa CA (2014) Novel yeast strains from Brazilian biodiversity: biotechnological applications in lignocellulose conversion into biofuels. In: da Silva SS, Chandel AK (eds) Biofuels in Brazil. Springer, Cham, pp 255–279

Cadete RM, Cheab MAM, Santos RO, Safar SVB, Zilli JE, Vital MJS, Basso LC, Lee C-F, Kurtzman CP, Lachance M-A, Rosa CA (2015) *Cyberlindnera xylosilytica* sp. nov., a xylitol-producing yeast species isolated from lignocellulosic materials. Int J Syst Evol Microbiol 65: 2968–2974

Cadete RM, Alejandro M, Sandström AG, Ferreira C, Gírio F, Gorwa-Grauslund MF, Rosa CA, Fonseca C (2016a) Exploring xylose metabolism in *Spathaspora* species: *XYL1. 2* from *Spathaspora passalidarum* as the key for efficient anaerobic xylose fermentation in metabolic engineered *Saccharomyces cerevisiae*. Biotechnol Biofuels 9:1–14

Cadete RM, Melo-Cheab MA, Viana AL, Oliveira ES, Fonseca C, Rosa CA (2016b) The yeast *Scheffersomyces amazonensis* is an efficient xylitol producer. World J Microbiol Biotechnol 32:207

Carvalho P, de Souza AC, Magalhatilde KT, Dias DR, Silva CF, Schwan RF (2013) Yeasts diversity in Brazilian Cerrado soils: study of the enzymatic activities. Afr J Microbiol Res 7:4176–4190

Chang CF, Yao CH, Young SS, Limtong S, Kaewwichian R, Srisuk N, Lee CF (2011) *Candida gosingica* sp. nov., an anamorphic ascomycetous yeast closely related to *Scheffersomyces spartinae*. Int J Syst Evol Microbiol 61:690–694

Crawford RL (1981) Lignin biodegradation and transformation. Wiley, New York

da Silva EG, Borges MF, Medina C, Piccoli RH, Schwan RF (2005) Pectinolytic enzymes secreted by yeasts from tropical fruits. FEMS Yeast Res 5:859–865

Daniel HM, Lachance MA, Kurtzman CP (2014) On the reclassification of species assigned to *Candida* and other anamorphic ascomycetous yeast genera based on phylogenetic circumscription. A van Leeuwenhoek 106:67–84

de Arruda PV, dos Santos JC, Rodrigues RCLB, da Silva DDV, Yamakawa CK, Rocha GJM, Nolasco Júnior J, Pradella JGC, Rossell CEV, Felipe MGA (2016) Scale up of xylitol production from sugarcane bagasse hemicellulosic hydrolysate by *Candida guilliermondii* FTI 20037. J Ind Eng Chem. doi:10.1016/j.jiec.2016.11.046

de Vries RP, Visser J (2001) *Aspergillus* enzymes involved in degradation of plant cell wall polysaccharides. Microbiol Mol Biol Rev 65:497–522

Dennis C (1972) Breakdown of cellulose by yeast species. Microbiology 71:409–411

Dlauchy D, Lee CF, Péter G (2012) *Spencermartinsiella ligniputridi* sp. nov., a yeast species isolated from rotten wood. Int J Syst Evol Microbiol 62:2799–2804

Eastwood DC, Floudas D, Binder M, Majcherczyk A, Schneider P, Aerts A, Asiegbu FO, Baker SE, Barry K, Bendiksby M, Blumentritt M, Coutinho PM, Cullen D, de Vries RP, Gathman A,

Goodell B, Henrissat B, Ihrmark K, Kauserud H, Kohler A, LaButti K, Lapidus A, Lavin JL, Lee YH, Lindquist E, Lilly W, Lucas S, Morin E, Murat C, Oguiza JA, Park J, Pisabarro AG, Riley R, Rosling A, Salamov A, Schmidt O, Schmutz J, Skrede I, Stenlid J, Wiebenga A, Xie X, Kües U, Hibbett DS, Hoffmeister D, Högberg N, Martin F, Grigoriev IV, Watkinson SC (2011) The plant cell wall–decomposing machinery underlies the functional diversity of forest fungi. Science 333:762–765

Eriksson KEL, Blanchette RA, Ander P (2012) Microbial and enzymatic degradation of wood and wood components. Springer Science & Business Media, New York

Fernández FJ, López-Estepa M, Querol-García J, Vega MC (2016) Production of protein complexes in non-methylotrophic and methylotrophic yeasts. In: Vega MC (ed) Advanced technologies for protein complex production and characterization. Springer, Cham, pp 137–153

Fisher SG, Likens GE (1973) Energy flow in Bear Brook, New Hampshire: an integrative approach to stream ecosystem metabolism. Ecol Monogr 43:421–439

Floudas D, Binder M, Riley R, Barry K, Blanchette RA, Henrissat B, Martínez AT, Otillar R, Spatafora JW, Yadav JS, Aerts A, Benoit I, Boyd A, Carlson A, Copeland A, Coutinho PM, de Vries RP, Ferreira P, Findley K, Foster B, Gaskell J, Glotzer D, Górecki P, Heitman J, Hesse C, Hori C, Igarashi K, Jurgens JA, Kallen N, Kersten P, Kohler A, Kües U, Kumar TKA, Kuo A, LaButti K, Larrondo LF, Lindquist E, Ling A, Lombard V, Lucas S, Lundell T, Martin R, McLaughlin DJ, Morgenstern I, Morin E, Murat C, Nagy LG, Nolan M, Ohm RA, Patyshakuliyeva A, Rokas A, Ruiz-Dueñas FJ, Sabat G, Salamov A, Samejima M, Schmutz J, Slot JC, St. John F, Stenlid J, Sun H, Sun S, Syed K, Tsang A, Wiebenga A, Young D, Pisabarro A, Eastwood DC, Martin F, Cullen D, Grigoriev IV, Hibbett DS (2012) The Paleozoic origin of enzymatic lignin decomposition reconstructed from 31 fungal genomes. Science 336:1715–1719

Gírio FM, Fonseca C, Carvalheiro F, Duarte LC, Marques S, Bogel-Łukasik R (2010) Hemicelluloses for fuel ethanol: a review. Bioresour Technol 101:4775–4800

Gomes FCO, Safar SVB, Marques AR, Medeiros AO, Santos ARO, Carvalho C, Lachance MA, Sampaio JP, Rosa CA (2015) The diversity and extracellular enzymatic activities of yeasts isolated from water tanks of *Vriesea minarum*, an endangered bromeliad species in Brazil, and the description of *Occultifur brasiliensis* f.a., sp. nov. A van Leeuwenhoek 107:597–611

Gonçalves JF, França JS, Medeiros AO, Rosa CA, Callisto M (2006) Leaf breakdown in a tropical stream. Int Rev Hydrobiol 91:164–177

González AE, Martinez AT, Almendros G, Grinbergs J (1989) A study of yeasts during the delignification and fungal transformation of wood into cattle feed in Chilean rain forest. A van Leeuwenhoek 55:221–236

Green F, Highley TL (1997) Mechanism of brown-rot decay: paradigm or paradox. Int Biodeterior Biodegrad 39:113–124

Grover N, Nawange SR, Naidu J, Singh SM, Sharma A (2007) Ecological niche of *Cryptococcus neoformans* var. *grubii* and *Cryptococcus gattii* in decaying wood of trunk hollows of living trees in Jabalpur City of Central India. Mycopathologia 164:159–170

Guamán-Burneo MC, Dussán KJ, Cadete RM, Cheab MAM, Portero P, Carvajal-Barriga EJ, da Silva SS, Rosa CA (2015) Xylitol production by yeasts isolated from rotting wood in the Galápagos Islands, Ecuador, and description of *Cyberlindnera galapagoensis* fa, sp. nov. A van Leeuwenhoek 108:919–931

Guo X, Zhu H, Bai FY (2012) *Candida cellulosicola* sp. nov., a xylose-utilizing anamorphic yeast from rotten wood. Int J Syst Evol Microbiol 62:242–245

Handel S, Wang T, Yurkov AM, König H (2016) *Sugiyamaella mastotermitis* sp. nov. and *Papiliotrema odontotermitis* fa, sp. nov. from the gut of the termites *Mastotermes darwiniensis* and *Odontotermes obesus*. Int J Syst Evol Microbiol 66:4600–4608

Hofrichter M (2002) Review: lignin conversion by manganese peroxidase (MnP). Enzym Microb Technol 30:454–466

Hon DNS (1994) Cellulose: a random walk along its historical path. Cellulose 1:1–25

Hou X (2012) Anaerobic xylose fermentation by *Spathaspora passalidarum*. Appl Microbiol Biotechnol 94:205–214

Hrmová M, Biely P, Vršanská M, Petráková E (1984) Induction of cellulose-and xylan-degrading enzyme complex in the yeast *Trichosporon cutaneum*. Arch Microbiol 138:371–376

Hu ML, Zha J, He LW, Lv YJ, Shen MH, Zhong C, Li BZ, Yuan YJ (2016) Enhanced bioconversion of cellobiose by industrial *Saccharomyces cerevisiae* used for cellulose utilization. Front Microbiol 7:241

Irshad M, Asgher M, Anwar Z, Ahmad A (2014) Biotechnological valorization of pectinolytics and their industrial applications: a review. Nat Prod Commun 9:1649–1654

Jayani RS, Saxena S, Gupta R (2005) Microbial pectinolytic enzymes: a review. Process Biochem 40:2931–2944

Jeffries TW, Jin YS (2000) Ethanol and thermotolerance in the bioconversion of xylose by yeasts. Adv Appl Microbiol 47:221–268

Jiménez M, Gonzalez AE, Martinez MJ, Martinez AT, Dale BE (1991) Screening of yeasts isolated from decayed wood for lignocellulose-degrading enzyme activities. Mycol Res 95: 1299–1302

Knob A, Terrasan CRF, Carmona EC (2010) β-Xylosidases from filamentous fungi: an overview. World J Microbiol Biotechnol 26:389–407

Kumar V, Shukla P (2016) Functional aspects of xylanases toward industrial applications. In: Shukla P (ed) Frontier discoveries and innovations in interdisciplinary microbiology. Springer, India, pp 157–165

Kurtzman CP (2011) A new methanol assimilating yeast, *Ogataea parapolymorpha*, the ascosporic state of *Candida parapolymorpha*. A van Leeuwenhoek 100:455–462

Kurtzman CP, Robnett CJ (2007) Multigene phylogenetic analysis of the *Trichomonascus*, *Wickerhamiella* and *Zygoascus* yeast clades, and the proposal of *Sugiyamaella* gen. nov. and 14 new species combinations. FEMS Yeast Res 7:141–151

Kurtzman CP, Robnett CJ (2013) Relationships among genera of the Saccharomycotina (Ascomycota) from multigene phylogenetic analysis of type species. FEMS Yeast Res 13: 23–33

Kurtzman C, Fell JW, Boekhout T (eds) (2011) The yeasts: a taxonomic study, 5th edn. Amsterdam, Elsevier

Lara CA, Santos RO, Cadete RM, Ferreira C, Marques S, Gírio F, Oliveira ES, Rosa CA, Fonseca C (2014) Identification and characterization of xylanolytic yeasts isolated from decaying wood and sugarcane bagasse in Brazil. A van Leeuwenhoek 105:1107–1119

Lazéra MS, Cavalcanti MAS, Trilles L, Nishikawa MM, Wanke B (1998) *Cryptococcus neoformans* var. *gattii*—evidence for a natural habitat related to decaying wood in a pottery tree hollow. Med Mycol 36:119–122

Leathers TD (1986) Color variants of *Aureobasidium pullulans* overproduce xylanase with extremely high specific activity. Appl Environ Microbiol 52:1026–1030

Leclerc M, Arnaud A, Ratomahenina R et al (1987) Yeast β-glucosidases. Biotechnol Genet Eng Rev 5:269–296

Liu XJ, Cao WN, Ren YC, Xu LL, Yi ZH, Liu Z, Hui FL (2016a) Taxonomy and physiological characterisation of *Scheffersomyces titanus* sp. nov., a new D-xylose-fermenting yeast species from China. Sci Rep 6:32181

Liu XZ, Wang QM, Göker M, Groenewald M, Kachalkin AV, Lumbsch HT, Millanes AM, Wedin M, Yurkov AM, Boekhout T, Bai FY (2016b) Towards an integrated phylogenetic classification of the Tremellomycetes. Stud Mycol 81:85–147

Lopes MR, Morais CG, Kominek J, Cadete RM, Soares MA, Uetanabaro APT, Fonseca C, Lachance MA, Hittinger CT, Rosa CA (2016) Genomic analysis and D-xylose fermentation of three novel *Spathaspora* species: *Spathaspora girioi* sp. nov., *Spathaspora hagerdaliae* f.a., sp. nov. and *Spathaspora gorwiae* f.a., sp. nov. FEMS Yeast Res 16:fow044

Lynd LR, Weimer PJ, van Zyl WH, Pretorius IS (2002) Microbial cellulose utilization: fundamentals and biotechnology. Microbiol Mol Biol Rev 66:506–577

Maksimova IA, Chernov IY (2004) Community structure of yeast fungi in forest biogeocenoses. Microbiology 73:474–481

Martínez AT, Speranza M, Ruiz-Duenãs FJ, Ferreira P, Camarero S, Guillén F, Martínez MJ, Gutiérrez A, del Río JC (2005) Biodegradation of lignocellulosics: microbial, chemical, and enzymatic aspects of the fungal attack of lignin. Int Microbiol 8:195–204

Middelhoven WJ (1993) Catabolism of benzene compounds by ascomycetous and basidiomycetous yeasts and yeastlike fungi. A van Leeuwenhoek 63:125–144

Mohamad NL, Mustapa Kamal SM, Mokhtar MN (2015) Xylitol biological production: a review of recent studies. Food Rev Intl 31:74–89

Molnárová J, Vadkertiová R, Stratilová E (2014) Extracellular enzymatic activities and physiological profiles of yeasts colonizing fruit trees. J Basic Microbiol 54:S74–S84

Morais CG, Lara CA, Marques S, Fonseca C, Lachance MA, Rosa CA (2013a) *Sugiyamaella xylanicola* sp. nov., a xylan-degrading yeast species isolated from rotting wood. Int J Syst Evol Microbiol 63:2356–2360

Morais CG, Cadete RM, Uetanabaro APT, Rosa LH, Lachance MA, Rosa CA (2013b) D-xylose-fermenting and xylanase-producing yeast species from rotting wood of two Atlantic Rainforest habitats in Brazil. Fungal Genet Biol 60:19–28

Morais CG, Lara CA, Borelli BM, Cadete RM, Moreira JD, Lachance MA, Rosa CA (2016a) *Saturnispora bothae* sp. nov., isolated from rotting wood. Int J Syst Evol Microbiol 66: 3810–3813

Morais CG, Lara CA, Oliveira ES, Péter G, Dlauchy D, Rosa CA (2016b) *Spencermartinsiella silvicola* sp. nov., a yeast species isolated from rotting wood. Int J Syst Evol Microbiol 66: 604–608

Nguyen NH, Suh SO, Marshall CJ et al (2006) Morphological and ecological similarities: wood-boring beetles associated with novel xylose-fermenting yeasts, *Spathaspora passalidarum* gen. sp. nov. and *Candida jeffriesii* sp. nov. Mycol Res 110:1232–1241

Nilsson T, Daniel G (1989) Chemistry and microscopy of wood decay by some higher ascomycetes. Holzforschung [ZDB] 43:11–18

Osono T, To-Anun C, Hagiwara Y, Hirose D (2011) Decomposition of wood, petiole and leaf litter by *Xylaria* species from northern Thailand. Fungal Ecol 4:210–218

Otero RRC, Iranzo JFU, Briones-Perez AI, Potgieter N, Villena MA, Pretorius IS, Rensburg PV (2003) Characterization of the β-glucosidase activity produced by enological strains of non-*Saccharomyces* yeasts. J Food Sci 68:2564–2569

Özcan S, Kötter P, Ciciary M (1991) Xylan-hydrolysing enzymes of the yeast *Pichia stipitis*. Appl Microbiol Biotechnol 36:190–195

Patel SJ, Savanth VD (2015) Review on fungal xylanases and their applications. Int J Adv Res 3: 311–315

Péter G, Tornai-Lehoczki J, Fülöp L, Dlauchy D (2003) Six new methanol assimilating yeast species from wood material. A van Leeuwenhoek 84:147–159

Péter G, Dlauchy D, Tornai-Lehoczki J, Suzuki M, Kurtzman CP (2011) *Spencermartinsiella europaea* gen. nov., sp. nov., a new member of the family Trichomonascaceae. Int J Syst Evol Microbiol 61:993–1000

Pettersen RC (1984) The chemical composition of wood. Chem Solid Wood 207:57–126

Prakasham RS, Rao RS, Hobbs PJ (2009) Current trends in biotechnological production of xylitol and future prospects. Curr Trends Biotechnol Pharm 3:8–36

Purahong W, Wubet T, Lentendu G (2016) Life in leaf litter: novel insights into community dynamics of bacteria and fungi during litter decomposition. Mol Ecol 25:4059–4074

Randhawa HS, Mussa AY, Khan ZU (2001) Decaying wood in tree trunk hollows as a natural substrate for *Cryptococcus neoformans* and other yeast-like fungi of clinical interest. Mycopathologia 151:63–69

Rao LV, Goli JK, Gentela J, Koti S (2016) Bioconversion of lignocellulosic biomass to xylitol: an overview. Biores Technol 213:299–310

Ren Y, Chen L, Niu Q, Hui F (2014) Description of *Scheffersomyces henanensis* sp. nov., a new D-xylose-fermenting yeast species isolated from rotten wood. PLoS One 9:e92315

Romaní AM, Fischer H, Mille-Lindblom C, Tranvik LJ (2006) Interactions of bacteria and fungi on decomposing litter: differential extracellular enzyme activities. Ecology 87:2559–2569

Sampaio JP (1999) Utilization of low molecular weight aromatic compounds by heterobasidiomycetous yeasts: taxonomic implications. Can J Microbiol 45:491–512

Sampaio A, Cortes R, Leão C (2001) Invertebrate and microbial colonisation in native and exotic leaf litter species in a mountain stream. Int Rev Hydrobiol 86:527–540

Sampaio A, Cortes R, Leão C (2004) Yeast and macroinvertebrate communities associated with leaf litter decomposition in a second order stream. Int Rev Hydrobiol 89:453–466

Sampaio A, Sampaio JP, Leão C (2007) Dynamics of yeast populations recovered from decaying leaves in a nonpolluted stream: a 2-year study on the effects of leaf litter type and decomposition time. FEMS Yeast Res 7:595–603

Santos RO, Cadete RM, Badotti F, Mouro A, Wallheim DO, Gomes FCO, Stambuk BU, Lachance MA, Rosa CA (2011) *Candida queiroziae* sp. nov., a cellobiose-fermenting yeast species isolated from rotting wood in Atlantic Rain Forest. A van Leeuwenhoek 99:635–642

Santos ARO, Faria ES, Lachance MA, Rosa CA (2015) *Ogataea mangiferae* sp. nov., a methylotrophic yeast isolated from mango leaves. Int J Syst Evol Microbiol 65:1855–1859

Schmidt O (2006) Wood cell wall degradation. In: Schmidt O (ed) Wood and tree fungi: biology, damage, protection, and use. Springer, Heidelberg, pp 87–107

Sena LM, Morais CG, Lopes MR, Santos RO, Uetanabaro AP, Morais PB, Vital MJS, de Morais Jr MA, Lachance MA, Rosa CA (2017) D-Xylose fermentation, xylitol production and xylanase activities by seven new species of *Sugiyamaella*. A van Leeuwenhoek. doi:10.1007/s10482-016-0775-5

Sharma A, Tewari R, Rana SS, Soni R, Soni SK (2016) Cellulases: classification, methods of determination and industrial applications. Appl Biochem Biotechnol 199:1346–1380

Štursová M, Žifčáková L, Leigh MB, Burgess R, Baldrian P (2012) Cellulose utilization in forest litter and soil: identification of bacterial and fungal decomposers. FEMS Microbiol Ecol 80: 735–746

Suh SO, Marshall CJ, Mchugh JV, Blackwell M (2003) Wood ingestion by passalid beetles in the presence of xylose-fermenting gut yeasts. Mol Ecol 12:3137–3145

Suh SO, Houseknecht JL, Gujjari P, Zhou JJ (2013) *Scheffersomyces parashehatae* fa, sp. nov., *Scheffersomyces xylosifermentans* fa, sp. nov., *Candida broadrunensis* sp. nov. and *Candida manassasensis* sp. nov., novel yeasts associated with wood-ingesting insects, and their ecological and biofuel implications. Int J Syst Evol Microbiol 63:4330–4339

Tanaka Y (1991) Microbial decomposition of reed (*Phragmites communis*) leaves in a saline lake. Hydrobiologia 220:119–129

Thongekkaew J, Khumsap A, Chatsa-nga P (2012) Yeasts in mixed deciduous forest areas of Phujong Nayoy National Park and their ability to produce xylanase and carboxymethyl cellulose. Songklanakarin J Sci Techonol 34:157–163

Urbina H, Blackwell M (2012) Multilocus phylogenetic study of the *Scheffersomyces* yeast clade and characterization of the N-terminal region of xylose reductase gene. PLoS One 7:e39128

Urbina H, Frank R, Blackwell M (2013a) *Scheffersomyces cryptocercus*: a new xylose-fermenting yeast associated with the gut of wood roaches and new combinations in the *Sugiyamaella* yeast clade. Mycologia 105:650–660

Urbina H, Schuster J, Blackwell M (2013b) The gut of Guatemalan passalid beetles: a habitat colonized by cellobiose-and xylose-fermenting yeasts. Fungal Ecol 6:339–355

Ur-Rehman S, Mushtaq Z, Zahoor T, Jamil A, Murtaza MA (2015) Xylitol: a review on bioproduction, application, health benefits, and related safety issues. Crit Rev Food Sci Nutr 55: 1514–1528

van der Wal A, Geydan TD, Kuyper TW, de Boer W (2013) A thready affair: linking fungal diversity and community dynamics to terrestrial decomposition processes. FEMS Microbiol Rev 37:477–494

Vaz ABM, Rosa LH, Vieira MLA, de Garcia V, Brandão LR, Teixeira LCRS, Moliné M, Libkind D, van Broock M, Rosa CA (2011) The diversity, extracellular enzymatic activities and photoprotective compounds of yeasts isolated in Antarctica. Braz J Microbiol 42:937–947

Voříšková J, Baldrian P (2013) Fungal community on decomposing leaf litter undergoes rapid successional changes. ISME J 7:477–486

Wang W, Ling H, Zhao H (2015) Steam explosion pretreatment of corn straw on xylose recovery and xylitol production using hydrolysate without detoxification. Process Biochem 50: 1623–1628

Wang Y, Ren YC, Zhang ZT, Ke T, Hui FL (2016) *Spathaspora allomyrinae* sp. nov., a D-xylose-fermenting yeast species isolated from a scarabeid beetle *Allomyrina dichotoma*. Int J Syst Evol Microbiol. doi:10.1099/ijsem.0.000979

Ward OP, Moo-Young M, Venkat K (1989) Enzymatic degradation of cell wall and related plant polysaccharides. Crit Rev Biotechnol 8:237–274

Watanabe T, Suzuki K, Sato I, Morita T, Koike H, Shinozaki Y, Ueda H, Koitabashi M, Kitamoto HK (2015) Simultaneous bioethanol distillery wastewater treatment and xylanase production by the phyllosphere yeast *Pseudozyma antarctica* GB-4 (0). AMB Express 5:36

Wei H, Xu Q, Taylor LE, Baker JO, Tucker MP, Ding SY (2009) Natural paradigms of plant cell wall degradation. Curr Opin Biotechnol 20:330–338

Wolter KE, Highley TL, Evans FJ (1980) A unique polysaccharide-and glycoside-degrading enzyme complex from the wood-decay fungus *Poria placenta*. Biochem Biophys Res Commun 97:1499–1504

Worrall JJ, Anagnost SE, Zabel RA (1997) Comparison of wood decay among diverse lignicolous fungi. Mycologia 89:199–219

Wuczkowski M, Metzger E, Sterflinger K, Prillinger H (2005) Diversity of yeasts isolated from litter and soil of different natural forest sites in Austria. Bodenkultur-Wie and Munchen 56:201

Yelle DJ, Ralph J, Lu F, Hammel KE (2008) Evidence for cleavage of lignin by a brown rot basidiomycete. Environ Microbiol 10:1844–1849

Zhang Z, Donaldson AA, Ma X (2012) Advancements and future directions in enzyme technology for biomass conversion. Biotechnol Adv 30:913–919

Chapter 10
Yeasts in Hypersaline Habitats

Janja Zajc, Polona Zalar, and Nina Gunde-Cimerman

Abstract For a long time, halotolerant yeasts were known exclusively as contaminants of food preserved with high concentrations of salt or sugar. Their presence in natural thalassohaline hypersaline environments was unknown until 2000, when they were first reported to be active inhabitants of man-made solar salterns in Slovenia. Since then, they have been described on the surface of halophytic plants, in salt mines, in cold and temperate saline lakes, in brine and bittern of different solar salterns on three continents and in $MgCl_2$-dominated waters of the Dead Sea. Yeasts in these environments can be described as halotolerant, extremely halotolerant and even extremely chaotolerant. The dominant representatives are different ascomycetous black yeast species, mainly of the genera *Hortaea* and *Phaeotheca*; non-melanised ascomycetous yeasts from the genera *Candida*, *Debaryomyces*, *Meyerozyma*, *Metschnikowia*, *Pichia* and *Yarrowia*; and basidiomycetous yeasts from the genera *Bulleromyces*, *Cryptococcus*, *Cutaneotrichosporon*, *Papiliotrema*, *Rhodosporidium*, *Rhodotorula*, *Solicoccozyma*, *Sterigmatomyces* and *Vishniacozyma*. Until the discovery and description of indigenous saltern mycobiota, the physiological and molecular mechanisms relating to salt tolerance in eukaryotic microorganisms were studied using salt-sensitive *Saccharomyces cerevisiae* as the model organism. Nowadays, most studies focus on halotolerant yeast species like *Debaryomyces hansenii*, *Aureobasidium pullulans* and *Hortaea werneckii*, which have been isolated globally from natural hypersaline environments and can tolerate up to 10%, 15% and 30% NaCl, respectively. Studies of halotolerant yeasts at the molecular level continue to unravel the complexity of the adaptations needed for yeasts to cope with the problems of ion toxicity and low water activity that are characteristic of hypersaline environments.

J. Zajc
National Institute of Biology, Večna pot 111, 1000 Ljubljana, Slovenia

Department of Biology, Biotechnical Faculty, University of Ljubljana, Večna pot 111, 1000 Ljubljana, Slovenia

P. Zalar • N. Gunde-Cimerman (✉)
Department of Biology, Biotechnical Faculty, University of Ljubljana, Večna pot 111, 1000 Ljubljana, Slovenia
e-mail: nina.gunde-cimerman@bf.uni-lj.si

P. Buzzini et al. (eds.), *Yeasts in Natural Ecosystems: Diversity*,
DOI 10.1007/978-3-319-62683-3_10

Keywords Salterns • NaCl • $MgCl_2$ • *Hortaea werneckii* • *Debaryomyces hansenii*

Contents

10.1 Introduction

By definition, hypersaline environments are generally considered to be those with halite (e.g. NaCl) concentrations >10% (w/v) (Oren 2002a). Many such environments represent some of the most extreme habitats in the world (Trüper and Galinski 1986). Hypersaline environments have long been considered to be populated almost exclusively by prokaryotic organisms (Pierce 1914) and the alga *Dunaliella salina* (Oren 2005). For more than 80 years after the discovery of halophilic Bacteria and Archaea, salt-tolerant fungi were known only as contaminants of food preserved with high concentrations of either NaCl or sugar (Pitt and Hocking 1977; Hocking and Pitt 1979; Filtenborg et al. 2000); thus natural hypersaline environments were not investigated. Only in 2000 were fungi, including yeasts, first described as an integral part of hypersaline microbial communities in Slovenian 700-year-old seasonal salterns in Sečovlje (Schneider and Herrmann 1979; Gunde-Cimerman et al. 2000). Yeasts were isolated by filtration of the brine, microbial baiting and spreading of pellicules covering the surface of the brine. Their population dynamics were then followed using filtration and plating on selective media, with lowered water activity (a_w) provided by high concentrations of NaCl (17%–32%) or sugars (50%–70%).

Before the discovery of xerophilic fungi in natural hypersaline environments, it was assumed that their growth at low a_w (Pitt and Hocking 1977; Hocking and Pitt 1979) reflected a general xerophilic/osmophilic phenotype, as these fungi can grow at $a_w < 0.85$. This corresponds to a medium supplemented with 17% NaCl or 50% glucose (Gunde-Cimerman et al. 2000, 2005). After their isolation from salterns (Gunde-Cimerman et al. 2000), it became clear that the growth of some fungi depends on the chemical nature of the solutes that lower the a_w (de Hoog et al. 2005; Gunde-Cimerman et al. 2005, 2009). Hence, osmotolerant/osmophilic yeasts grow better in environments that are rich in sugar, while those that are halotolerant/halophilic can grow in environments that are dominated by salts.

The initial studies in the Slovenian Sečovlje salterns were later supplemented by isolation of mycobiota from the crystallisers of salterns that operated throughout the year on three continents, including the salterns along the Red Sea coast in Israel (Eilat), the Mediterranean coast in Spain (Santa Pola, Ebro River Delta) and France (Camargue), the Atlantic Coast in Namibia (Skeleton Coast) and the coasts of the Dominican Republic (Monte Cristi), Puerto Rico (Fraternidad) and Portugal (Samouco). Yeasts have also been isolated from the Dead Sea (Ein Bokek, Ein Gedi), Great Salt Lake (Utah, USA) and the Enriquillo Salt Lake (Dominican Republic) (Gunde-Cimerman et al. 2005). The presence of yeasts has not only been documented in brine but also on wood immersed in brine (Zalar et al. 2005a), on the surface of halophytic plants (EI-Morsy 2000; Finkel et al. 2011; Ma et al. 2011; Crous et al. 2016), in saltern microbial mats (Cantrell et al. 2006; Tkavc 2012) and even in brine ponds in deep seas (Burgaud et al. 2015).

In summary, yeasts in hypersaline waters around the world include meristematic melanised black yeast-like fungi (here referred to as 'black yeasts', for simplicity) (Gunde-Cimerman et al. 2000; Butinar et al. 2005a; Zalar et al. 2007) and different species of non-melanised yeasts (Butinar et al. 2005b). The dominant black yeasts have almost exclusively been isolated on selective media with high concentrations of NaCl; however, they did not require NaCl for viability. These black yeasts can grow and adjust across the whole salinity range, from freshwater to almost saturated NaCl solutions (Plemenitaš et al. 2008). Their growth in vitro was optimal across a broad salinity range, from 5% to 17% NaCl, and they have been regularly isolated from global environments at salinities >10% NaCl (Gunde-Cimerman et al. 2005). To accommodate this particular type of adaptation, a new term of 'extremely halotolerant' yeast was introduced (Gunde-Cimerman and Plemenitaš 2006; Plemenitaš et al. 2008). The term halophilic is thus now used only for the few species of filamentous fungi that have an obligate requirement for NaCl in their growth medium, such as two species of the genus *Wallemia* (Zalar et al. 2005b; Jančič et al. 2015), *Basipetospora halophila* (Wheeler et al. 1988), now *Aspergillus baarnensis* and the *Phialosimplex*-like taxa *Aspergillus salisburgensis* and *Aspergillus atacamensis* (Martinelli et al. 2017). No yeast species described so far has shown such a requirement.

Halotolerance and halophily are usually tested only on NaCl, and tolerance to other salts has only become known more recently, such as for the kosmotropic KCl, $MgSO_4$ and others and the chaotropic $MgCl_2$ and $CaCl_2$. Consequently, the term 'chaophile' has been introduced (Williams and Hallsworth 2009). When ionic chaotropes like $MgCl_2$ were tested in vitro for active microbial growth, the upper limit was set at 1.26 M, with 2.3 M $MgCl_2$ determined as the upper limit for active life, based on in situ measurements of microbial activity (Hallsworth et al. 2007). The screening of over 130 strains of fungi for their growth in the presence of a variety of salts revealed many chaophilic, or at least extremely chaotolerant, fungi. As a consequence, the upper concentration of $MgCl_2$ that can support fungal life in vitro in the absence of counteracting kosmotropes has been set at 2.1 M $MgCl_2$, which is much higher than reported previously for bacteria (1.26 M) (Zajc et al. 2014a). The best growing yeasts among those that have been tested is *Hortaea werneckii*, which can tolerate 2.1 M $MgCl_2$, while the other tested yeasts, i.e. *Naganishia albida* and *Vishniacozyma victoriae* (formerly *Cryptococcus albidus* and *Cryptococcus victoriae*), *Rhodotorula babjevae* and *Rhodotorula glutinis* can tolerate up to 1.5 M $MgCl_2$. Despite this extreme tolerance, none have been shown to be obligately chaophilic.

In conclusion, fungi/yeasts can thrive at the lowest a_w ever recorded as supporting active life (Williams and Hallsworth 2009). They are also widespread globally in hypersaline environments, and many can be defined as extremely halotolerant and chaotolerant (Gunde-Cimerman et al. 2009; Zajc et al. 2012).

10.2 Hypersaline Thalassohaline Environments Sampled for Yeasts

Two types of hypersaline brines have been defined with respect to the origin of their formation: thalassohaline and athalassohaline (Oren 2002a). Thalassohaline waters either originated by evaporation of seawater (e.g. marine ponds, salt marshes, solar salterns) or they have similar salt composition to seawater (e.g. Great Salt Lake). These brines are dominated by sodium and chloride ions. During evaporation, their ionic composition changes due to the consecutive precipitation of calcite ($CaCO_3$), gypsum ($CaSO_4{\cdot}2H_2O$), halite (NaCl), sylvite (KCl) and finally carnallite ($KCl{\cdot}MgCl_2{\cdot}6H_2O$), after their maximum solubilities have been surpassed (Oren 2002a, 2013). The major change in the ratio of divalent to monovalent cations occurs when the total salt concentration exceeds 300–350 g l^{-1}, and most of the sodium precipitates as halite. In the remaining brine, the so-called bittern, the dominant ion becomes Mg^{2+} (Oren 2013).

10.2.1 Food Preserved with High Concentrations of NaCl

Low a_w is generally considered to be hostile to most forms of life, and thus crystalline salt (NaCl) and sugar have been used for centuries as food preservatives.

Until 2000 (Gunde-Cimerman et al. 2000), it was believed that yeasts do not show any particular preference for the chemical nature of the solute that lowered the a_w of their substrate or the medium used for their isolation (Hocking 1993; Pitt and Hockering 1997), and thus they were considered as xerophiles. The decisive criterion for defining xerophilic properties of food-borne yeasts was their growth on media that included 17% NaCl or 50% glucose, which correspond to $a_w \leq 0.85$ (Northolt et al. 1995). Dakal et al. (2014) proposed a new classification of halotolerance and osmotolerance that was based on combined behaviours:

(i) Moderately osmotolerant and moderately halotolerant, which has lack of growth at >50% (w/v) D-glucose and at >2.0 M NaCl (e.g. *Saccharomyces cerevisiae*; *Schizosaccharomyces pombe*; *Zygotorulaspora florentina*, formerly *Zygosaccharomyces florentinus*; *Candida glabrata*)
(ii) Osmotolerant and moderately halotolerant, which has growth at >50% (w/v) D-glucose and lack of growth at >2.0 M NaCl (e.g. *Candida tropicalis*, *Zygosaccharomyces mellis*, *Zygosaccharomyces sapae*, *Zygosaccharomyces bailii*, *Zygosaccharomyces bisporus*)
(iii) Moderately osmotolerant and halotolerant, which has growth at >50% (w/v) D-glucose and lack of growth at >2.0 M NaCl, or lack of growth at 50% (w/v) D-glucose and growth at >2.0 M NaCl (e.g. *Candida parapsilosis*; *Pichia membranifaciens*; *Pichia kudriavzevii*, formerly *Issatchenkia orientalis*)
(iv) Osmotolerant and halotolerant, which has growth at >50% (w/v) D-glucose and at >2.0 M NaCl (e.g. *Millerozyma farinosa*, formerly *Pichia sorbitophila*; *Zygosaccharomyces rouxii*; *Candida magnoliae*; *Meyerozyma guilliermondii*, formerly *Pichia guilliermondii*; *Ho. werneckii*; *Debaryomyces hansenii*; *Candida versatilis*, formerly *Candida halophila*)

As an example, the halotolerant yeast *Citeromyces siamensis* isolated from dry salted squid and fermented soybeans in Thailand was tolerant to high concentrations of cations (i.e. 3 M NaCl, 0.8 M LiCl) as well as to 60% glucose (Nagatsuka et al. 2002).

Most of the isolates of classes (i) and (ii) occur almost exclusively in sugary food products, such as juices, wine and dairy food, whereas classes (iii) and (iv) are generally isolated from salty food (e.g. cheese brine, olives, pickles, sauces, salty fish and meat).

Yeasts are more tolerant to low a_w than bacteria but less so than filamentous fungi. Most yeast species grow in a range of a_w 0.90–0.95, but some species can also grow in a_w 0.65–0.85. In food, the highest concentrations of NaCl (5%–15% NaCl) occur in cured meat products (e.g. bacon, ham, corned beef), vegetables in fermented brine, dried and salted meat and fish and fermented soy products, such as soy sauce and miso. Also, the aqueous phases of many cheeses, margarines and mayonnaises represent a saline habitat with 10–12% NaCl. Spoilage of such foods is usually dominated by *Deb. hansenii*, *Zygosacch. bailii*, *Zygosacch. rouxii*, *Yarrowia lipolytica* and, less frequently, some species of *Pichia* and *Candida* (Fleet 2011). Halotolerant yeast species, like *Citeromyces matritensis* (syn. *Candida globosa*), *Candida boidinii* and *Candida bombi*, have been isolated from olive brine wastewater, which contains phenolic compounds and high salt concentrations and has a very acidic pH (Crognale et al. 2012), which selects for polyextremophilic species.

10.2.2 Halophytic Plants

Halophytes are a small group of higher plants that can grow in saline soils. According to Flowers and Colmer (2015), who summarised salt tolerance of halophytes and their underlying mechanisms of adaptation, plants cope with NaCl in different ways. Such plants can tolerate high concentrations of chloride in their cell sap through increased succulence (e.g. *Salicornia herbacea*); they can resist salts by desalinisation of their tissues and excretion of the excess salts through salt glands (e.g. *Spartina alterniflora*) or they can accumulate salts in their tissues until their death (e.g. *Juncus gerardii*). Studies of cultivation of yeasts on halophytes are rare, as most research has been focussed either on endophytic fungi (Okane and Nakagiri 2015) or on nonculturable techniques. The globally distributed desert tree *Tamarix* (family Tamaricaceae) excretes salt through glands on the surface of its leaves (Ma et al. 2011). Finkel et al. (2011) studied the total phyllosphere microbial community from leaves of different *Tamarix* species and from different regions. The majority of their sequences belonged to unnamed Dothideomycetes, although these also included traces of the Saccharomycotina and basidiomycetous genera *Bensingtonia* and *Sporobolomyces*. The black yeasts *Aureobasidium pullulans* and *Exophiala salmonis* have been reported with nonculturable techniques on the halophyte *Inula crithmoides* in sea dunes close to Alicante, Spain (Maciá-Vicente et al. 2012). Yeasts were recorded but not identified on *Zilla spinosa* from the Red Sea coast in Egypt (El-Morsy 2000). A single collection from a *Tamarix* tree close to the Dead Sea revealed the presence of an unidentified black yeast *Phaeotheca* sp. (Oren and Zalar, unpublished data), and a new species named *Phaeotheca salicorniae* was recently described from South African *Salicornia* sp. (Crous et al. 2016).

Black yeasts not only inhabit the surface of halophytes but also their wood when immersed in hypersaline waters. *Ho. werneckii* and *Trimmatostroma salinum* were isolated from black patches on wood boards, and these showed complementary xylanolytic and lignolytic activities under both hypersaline and nonsaline conditions (Zalar et al. 2005a).

10.2.3 Evaporitic Solar Salterns

Solar salterns located in tropical and subtropical areas worldwide are composed of multiple artificial shallow ponds for production of halite from seawater. Seawater is pumped into the first set of ponds to evaporate, and when the required salinity is reached, it is then transferred to the next series of ponds, with the salinity increasing at each stage. In the early stages, when the salinity is two to three times greater than that of seawater, $CaCO_3$ precipitates in the form of aragonite and/or calcite, followed by gypsum ($CaSO_4{\cdot}2H_2O$) when the salt concentration has reached four times the seawater concentration. NaCl (halite) precipitates when the total salt concentration is >300 g l^{-1}. The concentrated brine that remains (the bittern) contains mainly Mg^{2+}, K^+, Cl^- and SO_4^{2-}. Generally, these are returned to the sea or processed further to harvest potash (KCl) or other salts. These conditions can also be enriched in fungi, as well as halotolerant and halophilic Bacteria and Archaea (Gunde-Cimerman et al. 2000, 2009; Oren 2002a, 2008; Butinar et al. 2005a).

The presence of yeasts in salterns has to date been most thoroughly investigated in the seasonal Sečovlje solar salterns in Slovenia (Gulf of Trieste, northern Adriatic Sea). Salt is produced traditionally by the daily gathering of the saturated brine on the cultivated microbial mat, called 'petola' that covers the crystallisers. The petola has a dual role: it prevents the mixing of the halite with the mud at the bottom of the ponds, and it prevents the incorporation of undesired ions of iron and manganese into the crystals (Pahor and Poberaj 1963). The yeasts *C. tropicalis*, *Rhodotorula mucilaginosa* and *Cryptococcus diffluens* (now *Naganishia diffluens*) were isolated from the upper layers of the petola. Only *Candida* has also been confirmed by nonculturable methods (Tkavc 2012).

The fungi that have been isolated from the hypersaline waters of the Slovenian Sečovlje salterns have included many that are halotolerant and extremely halotolerant (Zalar et al. 1999a, b, c, 2005b, 2007, 2008a; Gunde-Cimerman et al. 2000; Butinar et al. 2005a, b, c, 2011a) and a few that are halophilic (Zalar et al. 2005a). Most of the represented halotolerant species have included the meristematic black yeasts *Ho. werneckii*, *Phaeotheca triangularis*, *Trimm. salinum* and *A. pullulans* (Gunde-Cimerman et al. 2000) and the non-melanised yeasts *Mey. guilliermondii*, *Deb. hansenii* and *C. parapsilosis* (Butinar et al. 2005a).

The halotolerant and halophilic yeasts that were initially isolated from the Slovenian Sečovlje salterns (Gunde-Cimerman et al. 2000) were later isolated from different salterns around the world: La Trinitat in the Ebro River Delta and Santa Pola on the Mediterranean coast of Spain, Camargue in France and the salterns on the Atlantic Coast in Portugal and in Namibia and the Dominican Republic (Butinar et al. 2005a, b, c; Gunde-Cimerman and Zalar 2014).

10.2.4 Great Salt Lake

Great Salt Lake is a hypersaline shallow desert lake in the Great Basin of North America. The present-day lake has a surface area of 3900 km^2, and it represents a remnant of the much larger Bonneville glacial lake. The air and water temperature in the area of Great Salt Lake is known to vary from −30 °C in the winter to 48 °C in the summer. Although the lake has no connection with any of the world oceans, the composition of its salts is similar to that of seawater. NaCl represents 86% of the total salts, while nitrogen is the limiting nutrient, and the levels of phosphate are relatively high. Dramatic fluctuations in the water level and salinity have taken place throughout the history of Great Salt Lake, mainly as a result of changes in the amount of snowfall. Additionally, in the 1950s, a rock-fill railroad causeway across the middle part of the lake divided it into two independent water bodies. As about 90% of the freshwater inflow enters the southern arm of the lake, its salinity has decreased rapidly, while the salt concentration in the northern arm of the lake has increased to up to 33% to 35% NaCl. *Deb. hansenii* and *Metschnikowia bicuspidata* were isolated from the less saline southern arm (Butinar et al. 2005b), while only filamentous fungi, and no yeasts, have been isolated from the more saline northern arm (Cronin and Post 1977; Zalar et al. unpublished data).

10.2.5 Salt Mines

Hallstatt is a salt mine in Upper Austria, and it is one of the most important archaeological sites in Europe. The salt mining in the galleries is documented as far back as 1500 years B.C. Hallstatt contains a wooden staircase that was built in the year 1108 B.C. which was in use for approximately 100 years, until it was buried by a landslide in 1000 B.C. This important artefact was recently contaminated by fungi (Piñar et al. 2016). Only two yeast species were isolated, such as *Sterigmatomyces halophilus* and the black yeast *Exophiala* sp., although the filamentous isolates prevailed. There are no data on the presence of either yeasts or fungi from other salt mines.

10.2.6 Hypersaline Soil

Changes in the physical and chemical parameters of soil can be drastic over time and space. Dry soil has a very low level of available water that is further reduced by the presence of salts. Measurement of soil salinity is based on the transmission of an electric current through a soil extract solution; i.e. its electric conductivity. For instance, seawater has an electric conductivity of 55 dS m^{-1}, whereas in the Dead Sea this increases to 110 dS m^{-1}. An electric conductivity of around 8 dS m^{-1} is generally too saline for any crop to grow (Hardie and Doyle 2012).

In contrast to the extensively studied ecology of aquatic hypersaline environments, in terrestrial environments, only a few studies have investigated yeasts in soda-rich soil, layers of dried lakes, saline sediments or arid desert soils (Lisichkina et al. 2003; Al-Musallam et al. 2010; Evans et al. 2013; Yuan et al. 2016, respectively). Alkalitolerant yeasts were isolated from samples of soda-rich soils of Armenia (Arazdayan) and the Transbaikal region (the Kungur Steppe). Representatives of capsulated yeasts belonging to the species *Papiliotrema laurentii*, *Na.*

albida, *Rh. glutinis*, *Rh. mucilaginosa* and *Sporobolomyces roseus* were isolated and identified phenotypically. Among them, *Pa. laurentii* clearly dominated (Lisichkina et al. 2003). Another study focussed on the Urmia Lake National Park (northwest part of Iran), which is a unique place that is protected by the UNESCO biosphere reserve. This National Park comprises two ecosystems: Urmia Lake basin and the hypersaline land surrounding the lake. Urmia Lake is the largest lake in the Middle East, and it is also the second saltiest lake in the world, after the Dead Sea (Mokhtarnejad et al. 2016). Its water salinity varies from 12% to 34% NaCl, and the surrounding soils have a generally high, but variable, salinity (Mokhtarnejad et al. 2016). Recent studies have revealed different ascomycetous and basidiomycetous yeasts in the Urmia Lake soil. These generally tolerate 10% (w/v) NaCl in vitro, while a smaller proportion of strains within species can also tolerate 15% NaCl, such as the ascomycetous *Candida baotianensis*, *Debaryomyces subglobosus*, *Metschnikowia sinensis* and *Torulaspora delbrueckii* and representatives of basidiomycetous genera *Cystobasidium*, *Holtermanniella*, *Naganishia*, *Rhodotorula*, *Saitozyma*, *Solicoccozyma*, *Tausonia*, *Vanrija* and *Vishniacozyma* (Mokhtarnejad et al. 2016). The dominant species was *Solicoccozyma aeria* (Mokhtarnejad et al. 2015, 2016).

The Great Salt Plains of Oklahoma are an inland terrestrial area of flats that are covered by evaporate salt crusts and are fed by capillary flow of subterranean saturated brine. Here, as well as filamentous fungi, *Deb. hansenii* was also isolated (Evans et al. 2013). So far it appears that the terrestrial hypersaline environments harbour less characteristic mycobiota that observed for hypersaline waters (Ranzoni 1968; Guiraud et al. 1995; Grishkan et al. 2003; Evans et al. 2013). The recently published study on fungal diversity of soda-rich soils around soda lakes in Asia and Africa did not retrieve any yeast but numerous filamentous fungi (Grum-Grzhimaylo et al. 2016).

10.2.7 Deep-Sea Sediment and Brine

Some deep-sea sedimentary environments contain bottom brine pools, such as in the Mediterranean, the Red Sea, the Gulf of Mexico and others. Deep-sea sites can harbour novel sequences that are relevant to the early evolution of eukaryotes (Lai et al. 2007), as was shown by the unexpected diversity of the micro-eukaryotic communities dominated by fungi (reviewed in Lopez-Garcia et al. 2003; Singh et al. 2013; Zhang et al. 2014; Burgaud et al. 2015). Targeted environmental sequencing of deep-sea sediments from the East Indian Ocean defined 54% yeasts clones mostly related to the *Cryptococcus*: *Cryptococcus curvatus* (now *Cutaneotrichosporon curvatus*), *Cryptococcus fragicola* (now *Vanrija fragicola*), *Cryptococcus podzolicus* (now *Saitozyma podzolica*) and *Galactomyces candidus* (now *Dipodascus geotrichum*) (Zhang et al. 2014). Phylotypes related to *Cryptococcus*, *Rhodotorula slooffiae* (now *Cystobasidium slooffiae*), *Sterig. halophilus* and *Trichosporon moniliiforme* (now *Cutaneotrichosporon moniliiforme*) have been reported for the majority of deep subseafloor samples, while phylotypes related to *Dip. geotrichum* and *Candida* spp. (*Candida etchellsii*, *Candida inconspicua*, etc.) were only rarely recovered. Phylotypes related to the *Cut. moniliiforme* and *Sterig. halophilus* suggest the potential presence of opportunistic pathogens or parasites of deep-sea animals. The remaining yeasts (i.e. *Sporobolomyces lactosus*, *Guehomyces pullulans* (now *Tausonia pullulans*), *Ho. werneckii*) were singletons or doubletons and indicated low abundance (Zhang et al. 2014). The culturing of yeasts from these sediments revealed some additional taxa, including *A. pullulans* and *Exophiala dermatitidis*.

When Alexander et al. (2009) analysed hypersaline (up to 365 g l^{-1} NaCl) anoxic water of the L'Atalante deep-sea basin in the eastern Mediterranean Sea at a depth of 3500 m below sea level, the estimate from the 18S ribosomal RNA of total DNA from the upper halocline showed high diversity of fungal phylotypes (17%), which was drastically reduced to 2% in the lower hypersaline halocline. Most of the fungal taxa were closely related to the described ascomycetes and basidiomycetes. A single clone of a sole yeasts representative in the hypersaline community was 99% similar to the 18S rDNA of the basidiomycetous *Meira argovae*, which has otherwise been isolated from coastal plains of Israel and from a spider mite.

The two deep-sea super-haline anoxic basins of Bannock and Discovery at the Mediterranean Basin brine and brine/seawater interface were also enriched with fungi, which represented 17% of total 18S rRNA gene sequences identified. No fungi were detected in the NaCl-rich Bannock brine and at the Discovery interface, while there were numerous fungi at the Bannock interface: 24 operational taxonomic units (OTUs) related to *Malassezia* spp. and 10 OTUs related to *Schizosaccharomyces* spp. (Edgcomb et al. 2009). The presence of yeasts in such extreme anoxic hypersaline environments needs further evaluation.

10.3 Athalassohaline Hypersaline Environments Sampled for Yeasts

In contrast to the thalassohaline waters, the ionic composition of athalassohaline environments is primarily influenced by the geology of the surroundings (Oren 2002a), either through leaching of ions or dissolving of salt deposits from former evaporation (Grant 2004). Many such deposits are dominated by $MgCl_2$ (Hallsworth et al. 2007) and $CaCl_2$ (Dickson et al. 2013). The effects of ions in hypersaline environments are either chaotropic or kosmotropic; chaotropes weaken electrostatic interactions and destabilise biological macromolecules, whereas the contrary is true for kosmotropes (Zhang and Cremer 2006; reviewed in Oren 2013). The differences between the kosmotropic effects of NaCl on the one hand and the chaotropic effects of $MgCl_2$ and $CaCl_2$ on the other might explain why high concentrations of Mg^{2+} and Ca^{2+} are toxic even to the most halophilic microorganisms (McGenity and Oren 2012). However, to some extent, the chaotropic effects of

Mg^{2+} and Ca^{2+} can be counteracted by kosmotropic ions (Williams and Hallsworth 2009).

10.3.1 The Dead Sea

The Dead Sea is the deepest terrestrial position in the world (Gavrieli et al. 1999). It is a hypersaline desert lake on the border between Israel and Jordan, and it has a surface area of about 760 km^2. The water of the Dead Sea is characterised by its unique ionic composition, due to the prevalence of divalent cations (1.9 M Mg^{2+}, 0.4 M Ca^{2+}) over monovalent cations (1.6 M Na^+, 0.14 M K^+). Its extreme salinity (in 2007: 347 g l^{-1}), low a_w (<0.683) (Hallsworth, personal communication) and near-toxic magnesium levels (~2.0 M Mg^{2+}) make it one of the most hostile water bodies in the world (Oren 2013). This extreme ratio between divalent and monovalent cations is steadily increasing as a result of massive precipitation of halite and due to the return of brine containing mainly $MgCl_2$ and $CaCl_2$ after commercial exploitation of the Dead Sea water for the production of potash and bromide (Oren 2013). Life in the Dead Sea in its present state primarily depends on abundant rain that can sufficiently dilute the epilimnion and enable propagation of dormant halophilic microorganisms.

The first fungus isolated from Dead Sea water was an osmophilic yeast. Unfortunately, no culture was preserved (Kritzman 1973). Later, a variety of fungi were isolated (Kis-Papo et al. 2001; Zajc et al. 2014a), including the yeasts *C. parapsilosis* and *Candida* sp. *Mey. guilliermondii*, *Cutaneotrichosporon mucoides* and *Cystobasidium laryngis* (formerly *Trichosporon mucoides*, and *Rhodotorula laryngis*, respectively).

10.3.2 Non-halite Hypersaline Environments

10.3.2.1 $MgCl_2$-Rich Waters

Seawater in salterns evaporates through a series of sequential ponds to arrive in the final crystalliser pond, where the halite precipitation occurs. The remaining brine, or bittern, has a high content of highly soluble salts, like $MgCl_2$, and consequently it has an extremely low a_w. Until recently, it was believed that this bittern is aseptic (Javor 1989). However, the level of the toxic ion Mg^{2+} of bittern brine is not completely inhibitory for microbial growth, as it can be compensated for by the relatively higher concentration of Na^+ (Zajc et al. 2014a). Surprisingly, fungi, including the black yeasts *Ho. werneckii* and *Phaeoth. triangularis* and basidiomycetous *Bullera alba* (*Bulleromyces albus*), have been isolated from bittern of Slovenian Sečovlje salterns (Zalar, unpublished data).

10.3.2.2 $CaCl_2$-Rich Waters

Lake Vanda is a hypersaline lake in the McMurdo Dry Valley in Antarctica, and it contains high concentrations of dissolved $CaCl_2$ (0.6 M) and $MgCl_2$ (0.3 M) but low content of NaCl (0.25 M). However, the total concentration of salts exceeds 35%. The epilimnion of the ice-covered lake is oxic, oligotrophic and cold (4–6 °C), while the layer below is warm (23 °C) and hypersaline (Tregoning et al. 2015). Microbiological studies have identified various bacteria (Tregoning et al. 2015) and fungi, (Sugiyama et al. 1967; Kriss et al. 1976) including yeasts from the genera *Cryptococcus*, *Rhodotorula*, *Sporobolomyces*, *Trichosporon* and *Candida* (Goto et al. 1969; Nagashima et al. 1990). Since these references are rather old, no modern taxonomic names could be applied to these taxa.

10.4 Hypersaline Environments with a Reported Prokaryotic Biodiversity But with No Yeasts

10.4.1 Permanently Cold Saline and Hypersaline Lakes

Antarctica contains a surprisingly large number of saline and hypersaline lakes. The high salt concentrations in the brine depress the freezing point of the water, thereby preventing it from freezing. The most intensively investigated hypersaline lakes of Antarctica are in the Vestfold Hills. These lakes are relics of seawater catchments that were isolated by uplifts about 6000 years ago. They include Organic Lake, Ekho Lake and Deep Lake and the $CaCl_2$-dominated Don Juan Pond close to Lake Vanda. With over 400 g l^{-1} dissolved salts, Don Juan Pond is the saltiest water body in the world. Its water is concentrated $CaCl_2$ brine (up to 3.2 M), and it has very low concentrations of Mg^{2+}, K^+ and Na^+ (Matsubaya et al. 1979). Due to the extremely high salinity, it is also the only Antarctic lake that almost never freezes, even at temperatures below -50 °C (Marion 1997). Despite some attempts (Siegel et al. 1979, 1983), it has remained unanswered whether life is possible under these hostile conditions. No yeasts have been detected to date. When growth of fungi was tested in vitro, $CaCl_2$ concentrations higher than 1.9 M completely inhibited fungal growth (Zajc et al. 2014a).

10.4.2 Hypersaline Alkaline Lakes

Another example of hostile environments with so far no detected yeasts are the hypersaline alkaline lakes that are located primarily in continental interiors at tropical or subtropical latitudes. Alkaline saline brines form due to the geology of the drainage waters, restricted outflow of the surface water and evaporative concentration. Examples of such lakes are Lake Magadi (Kenya); a number of lakes in the East African Rift valley; the shallow lakes of Wadi Natrun (Egypt); the soda lakes in China and India; Mono Lake, California, USA; and Big Soda Lake, Nevada, USA. The major ions in these brines are Na^+, Cl^-, HCO_3^- and CO_3^{2-}. The salinity can reach values close to the halite concentration, carbonate can be very high, and the pH might reach 10 to 11, or even higher. Although considerable research efforts have been dedicated to investigations of some of these hypersaline alkaline lakes, and in particular Wadi Natrun in Egypt and Lake Magadi in Kenya (Oren 2002b), the available information is limited almost exclusively to prokaryotic microbiota.

Hot and hypersaline solar lakes represent less extreme environments, such as Sinai and other warm brines, although the level of fungal/yeast diversity in these remains unknown.

10.4.3 Salt Efflorescence on the Walls of Buildings and Wall Paintings

There is some superficial resemblance between the properties of dry saline soils and historic wall paintings in old buildings, especially under noncontrolled climatic conditions. The core groups of microorganisms include the pigmented halophilic Archaea and Bacteria (Oren 2009), which produce typical rosy stains that affect the optical appearance of wall paintings and historical plaster (Sterflinger and Piñar 2013). Practically all walls contain soluble salts, either dispersed within the porous materials or concentrated locally. The main salts known to occur in walls are carbonates, sulphates, chlorides and nitrates and oxalates of sodium, potassium, calcium, magnesium and ammonia, depending on the materials present. Salts can thus be available as a result of biological processes (e.g. ammonium salts) or due to comigration with infiltrating water. Due to changes in physical parameters of temperature and humidity, these salts can precipitate on the exposed wall surfaces. The resulting salt efflorescence mimics the natural hypersaline conditions. There are hardly any data on the ecology of yeasts in such environments, except for a clone of *Candida* spp. on an ancient wall painting in China (Ma et al. 2014), an isolate of *Sterig. halophilus* from a brick from a historical building in Poland (Adamiak et al. 2016) and the isolates of *A. pullulans* and *Sporobolomyces ruberrimus* from a wall painting from a sixteenth-century Lutheran cellar (Slovenia) (Zalar, unpublished data).

10.5 Halotolerant and Halophilic Yeasts

The data on isolation sources of halotolerant/halophilic yeasts from different saline environments, as well as the important reference literature, are given in Table 10.1.

Table 10.1 Yeast species reported for different saline environments

Species	Other known names	Sample								Location				Reference
		Sw	Dsw	Hsal	Hbit	Hs	Hal	Sf	Mm	Med	Str	Tr	Ar	
Black yeast-like organisms														
Aureobasidium pullulans	*Dematium pullulans*	■	■	■					■	■	■		■	Butinar et al. 2005a; Li et al. 2007; Tkavc 2012
Hortaea werneckii	*Cladosporium werneckii*	■	■	■	■					■	■	■		Gunde Cimerman et al. 2000; Cantrel et al. 2006; Diaz-Munos and Montalvo-Rodriguez 2005
Exophiala sp.									■					Tkavc 2012
Phaeotheca triangularis			■	■	■		■			■	■			Zalar et al. 1999; Butinar et al. 2005b
Trimmatostroma salinum				■						■				Zalar et al. 1999
Ascomycetous yeasts														
Candida spp.									■	■				Tkavc 2012
Candida baotianensis						■					■			Mokhtarnejad et al. 2016
Candida parapsilosis		■		■						■	■			Gadanho et al. 2003; Butinar et al. 2005b;
Candida etchellsii	*Candida nodaensis*		■					■						Kurtzman et al. 2011; Zhang et al. 2014
Candida inconspicua			■											Zhang et al. 2014
Candida tropicalis									■		■			Tkavc 2012
Candida versatilis	*Candida halophila*							■						Kurtzman et al. 2011
Citeromyces matritensis								■						Crognale et al. 2012
Debaryomyces hansenii	*Candida famata*	■	■	■	■	■		■		■	■		■	Butinar et al. 2005b; Burgaud et al. 2010; Suzuki et al. 2011; Duarte et al. 2013; Evans et al. 2013; Jacques et al. 2015
Debaryomyces subglobosus						■					■			Mokhtarnejad et al. 2015
Dipodascus geotrichum	*Galactomyces candidus*		■								■			Zhang et al. 2014
Metschnikowia bicuspidata		■		■							■			van Uden et al. 1968; Butinar et al. 2005b
Metschnikowia sinensis						■					■			Mokhtarnejad et al. 2015
Meyerozyma guilliermondii	*Pichia guillermondii, Candida guillermondii*		■	■				■	■	■				Butinar et al. 2005b; Cantrell et al. 2006; Burgaud et al. 2010
Millerozyma farinosa	*Pichia farinosa, Pichia sorbitophila*							■						Kutrzman et al. 2011
Schwanniomyces etchellsii	*Pichia etchellsii*							■						Kutrzman et al. 2011
Yarrowia lipolytica		■		■				■		■	■			Butinar et al. 2005b; Kurtzman et al. 2011
Torulaspora delbrueckii						■					■			Mokhtarnejad et al. 2015
Wickerhamomyces anomalus	*Pichia anomala*							■						Kurtzman et al. 2011
Zygosaccharomyces rouxii								■						Kurtzman et al. 2011

Basidiomycetous yeasts														
Bullera alba	*Bulleromyces albus*				■									Zalar, unpublished data
Cryptococcus sp.								■						Tkavc 2012
Cutaneotrichosporon curvatus	*Cryptococcus curvatus*		■											Zhang et al. 2014
Cutaneotrichosporon moniliiforme	*Trichosporon moniliiforme*		■											Zhang et al. 2014
Cutaneotrichosporon mucoides	*Trichosporon mucoides*	■		■		■								Butinar et al. 2005b
Cystobasidium laryngis	*Rhodotorula laryngis*			■						■				Butinar et al. 2005b
Cystobasidium slooffiae		■				■		■		■	■		■	
Naganishia albida	*Cryptococcus albidus*	■				■		■		■	■		■	Tkavc 2012; Mokhtarnejad et al. 2016; Butinar et al. 2011b
Papiliotrema laurentii	*Cryptococcus laurentii*					■								Lisichkina et al. 2003
Papiliotrema taeanensis	*Cryptococcus taeanensis*					■								Shin et al. 2005
Rhodosporidium sp.								■				■		Tkavc 2012
Rhodotorula babjevae	*Rhodosporidium babjevae*	■												Gadanho et al. 2003; Kurtzman et al. 2011
Rhodotorula glutinis						■								Lisichkina et al. 2003
Rhodotorula mucilaginosa	*Rhodotorula rubra*	■	■	■	■			■						Lisichkina et al. 2003; Gadanho and Sampaio 2005; Butinar et al. 2005b; Tkavc 2012
Rhodotorula sphaerocarpa	*Rhodosporidium sphaerocarpum*	■	■											van Uden et al. 1968; Butinar et al. 2005
Sporobolomyces roseus						■								Lisichkina et al. 2003
Solicoccozyma aeria	*Cryptococcus aerius*					■								Mokhtarnejad et al. 2016
Sterigmatomyces halophilus		■	■							■		■		Al.i et al. 2013; Zhang et al. 2014; Piñar et al. 2016; Zalar, unpublished data
Vanrija fragicola	*Cryptococcus fragicola*		■											Zhang et al. 2014
Vishniacozyma victoriae	*Cryptococcus victoriae*	■												Butinar et al. 2007

Sw seawater, *Dsw* deep seawater, *Hsal* hypersaline saltern water, *Hbit* hypersaline bittern, *Hs* hypersaline soil, *Hal* halophytic plants, *Sf* salty food, *Mm* microbial mats in hypersaline water, *Med* Mediterranean, *Str* subtropical, *Tr* tropical, *Ar* Arctic

10.5.1 Ascomycetous Black Yeasts

Black yeasts are a polyphyletic group of melanised polymorphic fungi that can grow in filamentous and yeast-like forms. They have an 'extremophilic ecotype' (Gunde-Cimerman et al. 2000) that is characterised by thick, melanised cell walls; slow, often meristematic growth; and proliferation with endoconidiation. Globally, black yeasts populate different extreme environments, from hypersaline coastal salterns worldwide (Butinar et al. 2005a) to surfaces and subsurfaces of rocks at high or low temperatures (Wollenzien et al. 1995; Selbmann et al. 2005) and Arctic glacial ice (Gunde-Cimerman et al. 2003).

The representatives of black yeasts that inhabit hypersaline environments include *Ho. werneckii* (Zalar et al. 1999c) and *Phaeoth. triangularis* (Zalar et al. 1999b), both from the order Capnodiales; *A. pullulans* (Zalar et al. 1999c), from the order Dothideales; and *Trimm. salinum*, from the order Helotiales (Zalar et al. 1999a). Both Capnodiales and Dothideales are known for their xerotolerant representatives.

The black yeasts listed in Table 10.1 that were initially isolated from the Slovenian Sečovlje salterns, have since also been detected in hypersaline waters of salterns on three continents, with the exception of *Trimm. salinum*.

10.5.1.1 *Hortaea werneckii*

Ho. werneckii is a cosmopolitan black yeast (Capnodiales, Dothideomycetidae, Dothideomycetes, Pezizomycotina) without any known teleomorph. Historically, *Ho. werneckii* has been placed in numerous genera (e.g. *Cladosporium*, *Exophiala*, *Dematium*, *Pullularia*, *Aureobasidium*, *Sarcinomyces*, *Phaeoannellomyces*), before being ascribed to a new genus due to its special conidiogenesis (Nishimura and Miyaji, 1983). The genus now contains three species: as well as *Ho. werneckii*, there is *Hortaea thailandica*, a rare halotolerant species that was originally described from plant material in Thailand (Schoch et al. 2009) and subglacial ice in Svalbard (Zalar and Gunde-Cimerman, unpublished data), and the phylogenetically unrelated *Hortaea acidophila*, which was isolated from saline and acidic soils in the Czech Republic (Hujslová et al. 2010).

Ho. werneckii was for a long time only known as the primary aetiological agent of a skin disorder known as *Tinea nigra*, which is a surface infection of human hands and feet that can occur in the warmer areas of the world (Bonifaz et al. 2008). The ecology of *Ho. werneckii* has been linked to the presence of NaCl, as it has been isolated from seawater (Iwatsu and Udagawa 1988), marine fish (Todaro et al. 1983), salted freshwater fish (Mok et al. 1981), sea sponges (Liu et al. 2010), mangrove water and sediments (Nayak et al. 2012), beach soil (de Hoog and Guého 2010), scuba diving equipment (Cabañes et al. 2012) and recently deep-sea environments (Singh et al. 2013; Zhang et al. 2014). In spite of its wide distribution in marine environments, brine in eutrophic solar salterns is its primary ecological niche. *Ho. werneckii* was dominant in Slovenian Sečovlje salterns at salinities above 20% NaCl (Gunde-Cimerman et al. 2000; Butinar et al. 2005a), with it representing from 70% to 80% of all of the isolates, with a frequency of occurrence of up to 1400 CFU l^{-1} at high salinities, and it is present throughout the year at low CFU also at lower salinities. As well as these Slovenian salterns, *Ho. werneckii* has been isolated from saltern brines from temperate, subtropical and tropical salterns all over the world (Gunde-Cimerman et al. 2000; Diaz-Munos and Montalvo-Rodriguez 2005; Gunde-Cimerman and Zalar 2014) and also from saltern microbial mats (Cantrell et al. 2006).

Ho. werneckii is the only black yeast that can grow across the whole range of NaCl concentrations, from 0% to NaCl saturation at 32%, with a broad optimum from 6% to 14% NaCl (Butinar et al. 2005a). As well as NaCl, *Ho. werneckii* also tolerates high concentrations of the chaotropic salts $MgCl_2$ and $CaCl_2$ (Zajc et al. 2014a, b), as evident also from its isolation from bitterns (Zalar, unpublished data). At present, *Ho. werneckii* is the most extremely halotolerant fungus, and therefore it is an important model organism for the study of halotolerance in Eukarya (Plemenitaš et al. 2008).

Genome sequencing (Lenassi et al. 2013) revealed recent genome duplication, and several studies have shown differential expression of most isogenes, depending on the salinity (Vaupotič and Plemenitaš 2007a, b).

10.5.1.2 *Phaeotheca triangularis*

Phaeoth. triangularis (Capnodiales, Dothideomycetidae, Dothideomycetes, Pezizomycotina) resembles yeast colonies due to its excessive production of extra-cellular polymeric substances; however, *Phaeoth. triangularis* cells proliferate as endoconidia from hyphae or multicellular clumps (Zalar et al. 1999b). This rare black yeast was originally described from a humidifier of an air-conditioning system in Belgium (de Hoog et al. 1997) but was later reported in the brine of Slovenian Sečovlje salterns throughout the year (Butinar et al. 2005a). It appeared most abundantly (up to 340 CFU l^{-1}) at higher salinities (18%–25% NaCl), while at lower salinities, it decreased to <50 CFU l^{-1}. In comparison with *Ho. werneckii*, *Phaeoth. triangularis* has a lower maximum and narrower salinity amplitude. In vitro, *Phaeoth. triangularis* can grow in up to 26% NaCl, with its optimal concentration at ~5% NaCl (Zalar et al. 1999c; Burgaud et al. 2010). In salterns, *Phaeoth. triangularis* has been isolated most frequently from the storage ponds, which have a relatively stable salinity. *Phaeoth. triangularis* forms pellicules in brine and biofilms on solid saline substrates (Gunde-Cimerman et al. 2000; Butinar et al. 2005a) and on rocks immersed in seawater in Sardinia (Egidi et al. 2014) and from deep-sea hydrothermal vents (Burgaud et al. 2010). *Phaeoth. triangularis* has been recorded on halophytic plants, such as *Salicornia salina* in Slovenia and *Tamarix* leaves in Israel (Oren and Zalar, unpublished data). Recently, a new species of *Phaeotheca* was described, from the surface of *Salicornia meyeriana*, as *Phaeoth. salicorniae* (Crous et al. 2016).

10.5.1.3 *Aureobasidium pullulans*

A. pullulans (Dothideales, Dothideomycetes, Pezizomycotina) is a cosmopolitan black yeast. Until recently, the species was comprised of four well-defined varieties that differed in their ecology and physiology (Zalar et al. 2008b). However, based on comparisons of the genomic sequences of representative strains of each variety, three new species were described. As for *A. pullulans*, these can all tolerate NaCl: *Aureobasidium melanogenum*, *Aureobasidium subglaciale* and *Aureobasidium namibiae* (Gostinčar et al. 2014). *A. pullulans* has been most frequently found among these, under various extremophilic environmental conditions, from hypersaline (Gunde-Cimerman et al. 2000) to acidic, basic (Ranta 1990; Shiomi et al. 2004), cold and oligotrophic (Onofri 1999). Due to its exceptional extremotolerance, *A. pullulans* is considered to be a polyextremotolerant black yeast (Gostinčar et al. 2010, 2011). As these four species were until recently still grouped as *A. pullulans sensu lato*, most of the known ecology and physiology remain ambiguous. The occurrence of *A. pullulans sensu stricto* is cosmopolitan under mostly mildly osmotic environments, such as hypersaline water, where it occurred with the highest frequency (800 CFU l^{-1}) at 5% NaCl; at higher salinities, its levels did not exceed 50 CFU l^{-1}. *A. pullulans* has also been isolated from polar glacial ice; cold-, salt- and desiccation-preserved foods; bathroom surfaces; household appliances; polluted and tap water; coconut water and the surface of the human skin (reviewed in Gostinčar et al. 2014), across different climatic zones, and from tropical, temperate to polar areas. *A. pullulans* has also been frequently associated with the phyllosphere (Andrews et al. 2002; Grube et al. 2011), as an epiphyte or endophyte (reviewed in Gostinčar et al. 2014). It grows optimally on medium without NaCl (Butinar et al. 2005a) but can tolerate up to 18% NaCl in vitro; thus it is regarded as a halotolerant species (Gostinčar et al. 2014).

A. pullulans is a well-known producer of pullulan (Leathers 2003; Cheng et al. 2011), a large spectrum of extracellular enzymes (Buzzini and Martini 2002; Molnarova et al. 2013; for review, see Chi et al. 2009), the antifungal cyclic peptide aureobasidin A (Takesako et al. 1991) and the antibacterial compound exophilin A (Price et al. 2013). *A. pullulans* is also a plant probiotic, or biocontrol agent, against detrimental diseases in fruit (Schena et al. 1999).

10.5.1.4 *Trimmatostroma salinum*

Trimm. salinum (Helotiales, Leotiomycetidae, Leotiomycetes, Pezizomycotina) is the only halotolerant black yeast that has so far only been detected in the Adriatic salterns (Zalar et al. 1999a), in the brine and on a wooden fence immersed in the brine of a single pond (Zalar et al. 1999a, 2005a). It appeared at salinities from 8% to 25% NaCl, with a peak (700 CFU l^{-1}) at 25% NaCl (Butinar et al. 2005a). In vitro, it can grow in 0% to 24% NaCl, with the optimal at 2% to 8% NaCl.

10.5.2 Non-melanised Yeasts

Butinar et al. (2005b) first reported on the presence of non-melanised yeasts in hypersaline waters of salterns in 2005. Yeasts were reported from eight different salterns and three salt lakes worldwide (the Dead Sea, Enriquillo Lake in the Dominican Republic and Great Salt Lake in Utah, USA). Their diversity was shown by *Mey. guilliermondii*, *Deb. hansenii*, *Y. lipolytica*, *M. bicuspidata* and *C. parapsilosis* and the basidiomycetous *Rhodotorula sphaerocarpa* and *Rh. babjevae* (both formerly *Rhodosporidium*) and *Cut. mucoides* (Liu et al. 2015). The most frequently occurring species in Slovenian Sečovlje salterns were *Mey. guilliermondii* and *C. parapsilosis*, which were sporadically isolated also in other salterns (Butinar et al. 2005b). Among these listed yeasts, *Mey. guilliermondii*, *Deb. hansenii*, *Y. lipolytica* and *C. parapsilosis* were previously known as contaminants of food preserved with low a_w, while *Rh. sphaerocarpa*, *Rh. babjevae* and *Cut. mucoides* were identified for the first time in hypersaline habitats. The ascomycetous *M. bicuspidate* is a known parasite of the brine shrimp, and it was isolated as the free-living yeasts in Great Salt Lake brine. In the bittern of the La Trinitat salterns in Spain, two new species were discovered that are closely related to *Candida atmosphaerica* and *Yamadazyma philogaea*, formerly *Pichia philogaea* (Butinar et al. 2005b).

In contrast to the black yeasts that increase in occurrence at salinities above 20% NaCl (Gunde-Cimerman et al. 2000; Butinar et al. 2005a), non-melanised yeasts are isolated primarily in brine at lower salinity, outside the peak of the salt production (Butinar et al. 2005b). In addition, these yeasts were never isolated on 32% NaCl medium, occasionally isolated on media with 17% to 25% NaCl and isolated most frequently on media with 10% NaCl added (Butinar et al. 2005b).

10.5.2.1 Ascomycetous Yeasts

Blastobotrys (Trichomonascaceae)

Blastobotrys adeninivorans, formerly *Arxula adeninivorans* (Kurtzman et al. 2011), is a thermotolerant and halotolerant yeast (Yang et al. 2000) that has so far been isolated from soil, silage, wood hydrolysate and an infected lizard (Kurtzman et al. 2011), and not yet from natural hypersaline environments. Its genome was published recently (Kunze et al. 2014).

Candida (*Incertae sedis*)

The halotolerant species *C. baotianensis* has so far been isolated only from nonsaline soil in China (Hui et al. 2012) and hypersaline soil in Urmia Lake National Park, Iran (Mokhtarnejad et al. 2015).

The occurrence of *C. parapsilosis* in saline habitats was reported for seawater in the south of Portugal (Gadanho et al. 2003) and, surprisingly, in Dead Sea brine (Butinar et al. 2005b). After *C. albicans* and *C. tropicalis*, the opportunistic *C. parapsilosis* is the third most common cause of candidemias in patients (Silva et al. 2012). It is present in a variety of man-made habitats, such as dishwashers, washing machines (Novak Babič et al. 2015; Zupančič et al. 2016) and tap water (Pires-Gonçalves et al. 2008; Novak Babič et al. 2016); however, its natural ecology remains poorly understood (Kurtzman et al. 2011). *C. parapsilosis* is considered to be primarily associated with mammals and wild birds, either as an opportunist or as an intestinal coloniser (Lord et al. 2010).

The halotolerant *Candida nodaensis*, now renamed to *C. etchellsii*, has been sporadically isolated from a broad variety of sources. Although it is regarded as a spoilage organism of salty foods and can tolerate up to 10% NaCl (Kurtzman et al. 2011), *C. etchellsii* has never been isolated from natural hypersaline environments. An association with floricolous insects is suspected (Kurtzman et al. 2011). *C. etchellsii* synthesises the resilient killer toxin CnKT under saline conditions. This zymocin is active across a broad range of pH and temperatures, and it tolerates freezing, displays increased stability under very high ionic strength and stimulates activity in the presence of sodium ions (da Silva et al. 2008). The antagonistic interactions and killer yeasts have been thoroughly reviewed by Klassen et al. (2017). *C. versatilis* (formerly *C. halophila*) has so far only been isolated from fermented cucumber brine (Kurtzman et al. 2011).

Citeromyces (Saccharomycetaceae)

Cit. matritens has only been isolated from food products with high osmotic potential, such as fruit syrup, sugar, condensed milk and olive brine wastewater (Crognale et al. 2012), and not from natural saline environments. *Cit. matritens* can tolerate up to 5% NaCl.

Debaryomyces (Saccharomycetaceae)

Deb. hansenii is a ubiquitous generalistic species that inhabits natural cold saline environments, like seawater and brine in salterns (Butinar et al. 2005b), as well as various manufactured salty foodstuffs, like fermented soy products, cheeses, meat products and similar (Suzuki et al. 2011). *Deb. hansenii* was also isolated from Namibian salterns (Butinar et al. 2005b), Great Salt Lake (Utah), Slovenian Sečovlje saltern bittern (Zalar, unpublished data), Arctic soils, overcooled brine cryopegs in permafrost, polar seawaters, Arctic glacial ice (Butinar et al. 2011b), marine sediments, lichens, ornithogenic soils and various marine invertebrates in Antarctica (Duarte et al. 2013). The putative hybrids isolated from Arctic glacier samples (Jacques et al. 2015) were differentiated from non-hybrid isolates and other *Debaryomyces* species on the basis of their carbon assimilation patterns. As well as cold and saline environments, this cosmopolitan yeast has also been found in soil, fruit, plants, insects and the air (Suzuki et al. 2011). The maximal reported NaCl concentrations to support its growth are 12% to 24% NaCl. *Deb. hansenii* is one of the model organisms for understanding fungal mechanisms of adaptation to hypersaline conditions, particularly on the level of ion transporters (Prista et al. 1997). The *Deb. hansenii* genome has been sequenced recently (Kumar et al. 2012).

The ecology of the halotolerant *Deb. subglobosus* is almost unknown. It has been isolated from hypersaline soil (Mokhtarnejad et al. 2015) and humans (Kurtzman et al. 2011). The strains tested showed growth at 10% NaCl.

Metschnikowia (Metschnikowiaceae)

The ecology of *M. bicuspidata* is usually associated with diseased brine shrimps (*Artemia salina*), as reported for salt ponds in Southern California (Lachance et al. 1976), Great Salt Lake (Butinar et al. 2005b), Sečovlje salterns (Slovenia) and Eilat salterns (Israel) (Tkavc 2012). It was also isolated from seawater and from kelp off the coast and even sporadically from freshwater ponds. For growth, it requires the addition of 2% NaCl to the medium.

M. sinensis has so far been isolated from fruit surfaces and hypersaline soils (Mokhtarnejad et al. 2015).

Meyerozyma (Debaryomycetaceae)

Mey. guilliermondii is globally distributed in nature. It was isolated from insect frass, flowers and fruit and from other food products. This species is often described as an opportunistic pathogen of humans and animals, and it was once isolated from hypersaline water of the Dead Sea (Butinar et al. 2005b).

Millerozyma (Debaryomycetaceae)

Pichia farinosa (currently *Mill. farinosa*) is widely distributed. Strains have been isolated from such diverse substrates as agricultural products, food (e.g. miso, soy mash), animal dung, petroleum and human infections (Kurtzman et al. 2011).

Schwanniomyces (Saccharomycetaceae)

Pichia etchellsii, now renamed *Schwanniomyces etchellsii*, has so far been isolated only from cucumber brine (Kurtzman et al. 2011).

Yarrowia (Dipodascaceae)

Y. lipolytica is widespread in nature. It grows best on substrates high in lipids in marine environments, including hypersaline waters of the Dead Sea and Great Salt Lake (Butinar et al. 2005b; Kurtzman et al. 2011). Its maximal reported NaCl concentration was in the 10% to 12 % NaCl range (Kurtzman et al. 2011).

Torulaspora (Saccharomycetaceae)

T'spora delbrueckii was initially found in hypersaline soils (Mokhtarnejad et al. 2015) and has not been isolated from any other hypersaline source. Instead, it is widely distributed in nature in soil, fermenting grapes and other berry juices, agave juice, tea beer and tree bark. This primarily food spoilage yeast has some biotechnological importance (Kurtzman et al. 2011).

Wickerhamomyces (Wickerhamomycetaceae)

Wickerhamomyces anomalus is widespread in nature, as it occurs in soil, on plant material and as an opportunistic pathogen of humans and animals. It has been reported to be a killer yeast (Aguiar and Lucas 2000). The primary habitat of this species is believed to be plants (Kurtzman et al. 2011), and it has also been reported from sugar, dry salted beans and sauerkraut and cucumber brines (Westerdijk Institute database, http://www.westerdijkinstitute.nl/). Its maximal reported NaCl concentrations were up to 10% to 15% NaCl (Lages et al. 1999; Kurtzman et al. 2011).

Zygosaccharomyces (Saccharomycetaceae)

Zygosacch. rouxii, like its close relative *Zygosacch. mellis*, is typically found in highly osmotic habitats. Strains of *Zygosacch. rouxii* have been isolated from cane sugar, chocolate syrup, concentrated black grape must, honey, jam, maple syrup, marmalade, marzipan, miso, red wine, salted beans, soft drinks and soy sauce (Kurtzman et al. 2011). Its maximal reported NaCl concentrations were up to 7% to 20% NaCl (Kurtzman et al. 2011).

10.5.2.2 Basidiomycetous Yeasts

Bullera

B. alba is considered an important phyllosphere yeast. Additionally, this species has been isolated from air and insect frass in Japan, New Zealand, Canada, Portugal,

South Africa and the USA (Liu et al. 2015). The only isolate from hypersaline environments is from bittern of Sečovlje salterns (Slovenia) (Zalar, unpublished data).

Cutaneotrichosporon (Trichosporonaceae, Trichosporonales, *Incertae sedis*, Tremellomycetes, Agaricomycotina)

Detailed ecological information on the ecology of *Tr. mucoides* is not available. But this species has been placed in the genus *Cutaneotrichosporon* together with many strains originating from humans (Liu et al. 2015). The only record from hypersaline environments is hypersaline waters of salterns in Eilat (Israel) and the Dead Sea (Butinar et al. 2005b).

Naganishia (Tremellaceae, Tremellales, *Incertae sedis*, Tremellomycetes, Agaricomycotina)

The species *Cr. albidus* has been transferred to the genus *Naganishia* as *Na. albida*. This species has been isolated from different sources, including soils (Vishniac 2006; Yurkov et al. 2015), humans (Kurtzman et al. 2011) and the Dead Sea (Turk, unpublished data). *Na. albida* is also known as marine yeast, and it has also been detected in the Arctic sea (Butinar et al. 2007).

Papiliotrema (Tremellaceae, Tremellales, *Incertae sedis*, Tremellomycetes, Agaricomycotina)

Cryptococcus laurentii, now *Pa. laurentii*, was isolated from soda-rich saline soils of Armenia and the Transbaikal region (Lisichkina et al. 2003). *Cryptococcus taeanensis*, now *Papiliotrema taeanensis*, was isolated from hypersaline soil in Korea (Shin et al. 2005).

Rhodotorula (Incertae sedis, Sporidiobolales, *Incertae sedis*, Microbotryomycetes, Pucciniomycotina)

The genus *Rhodotorula* was recently amended to include *Rhodotorula* species and their sexual counterpart *Rhodosporidium* in the *Rhodosporidium* clade (Wang et al. 2015a), which is a well-supported clade within Sporidiobolaceae (Sporidiobolales) (Wang et al. 2015b).

Rh. mucilaginosa is a ubiquitous yeast isolated from a variety of sources, including humans, marsh and marine waters in Portugal (Kurtzman et al. 2011), a Mid-Atlantic Ridge hydrothermal vent near the Azores archipelago (Gadanho and

Sampaio 2005) and hypersaline water of Sečovlje salterns (Slovenia) (Butinar et al. 2005b).

Rh. sphaerocarpa is a commonly occurring species in marine habitats, from the Antarctic Peninsula to Palmer Archipelago and the Caribbean Sea (van Uden and Fell 1968). It was also isolated from hypersaline water of Sečovlje salterns (Slovenia), with frequencies of occurrence up to 1000 CFU l^{-1} hypersaline water (Butinar et al. 2005b).

Rh. babjevae is a ubiquitous yeast species that was originally isolated in Europe (Russia, Portugal, UK, Germany), the Americas (USA, Argentina) and Asia (Japan). It is usually associated with plants, soil, air, freshwater and seawater (Gadanho et al. 2003; Kurtzman et al. 2011; Yurkov et al. 2015).

Sterigmatomyces (Agaricostilbaceae, Agaricostilbales, *Incertae sedis*, Agaricostilbomycetes, Pucciniomycotina)

Sterig. halophilus strains were primarily isolated from seawater or habitats adjacent to it. Isolates have originated from the equatorial and upper deep water mass of the Indian Ocean and intermediate and central water mass of the Antarctic waters and the Atlantic Ocean. The Indian Ocean isolates originated from water depths of 110 m–1977 m, with temperatures as low as 3 °C. *Sterig. halophilus* was also recorded on seawater sponges (Liu et al. 2010) and from brine in Thailand salterns (Ali et al. 2013) and Slovenian Sečovlje salterns (Zalar, unpublished data). Additional hypersaline habitats were dampness-induced salt efflorescence on bricks, plaster and paint coatings from the former Auschwitz II-Birkenau concentration camp and a nineteenth-century chateau in Poland. The reported optimal salinity range for *Sterig. halophilus* was 15% to 30% NaCl (Adamiak et al. 2016). This species also has pathogenic potential, as indicated by a clinical isolate from Brazil (Kurtzman et al. 2011).

Vishniacozyma (Tremellaceae, Tremellales, *Incertae sedis*, Tremellomycetes, Agaricomycotina)

Vishn. victoriae is a cosmopolitan species, known mainly from soil, sea and glacial water in the polar areas (Antarctica, Arctic), and also from the phyllosphere in Europe (reviewed in Yurkov et al. 2015). It appears to be a pervasive psychrotolerant species that is not restricted to cold climates. The yeasts diversity in polar and subpolar regions and in non-polar worldwide cold habitats is reported in Chaps. 11 and 12 of this book. The only record from hypersaline environments was from bittern of Sečovlje salterns (Zalar, unpublished data).

10.6 Adaptations of Yeasts to Hypersaline Conditions

The responses of yeasts to high concentrations of toxic ions involve rigorous changes in gene expression that lead to subsequent synthesis of compatible solutes, regulation of intracellular alkali-metal cations, changed composition of the cell membrane and changed cell-wall ultrastructure and morphology. Extremely saline conditions can induce an extremophilic ecotype, which is characterised by enhanced melanisation and meristematic growth and changes in size and appearance of the colonies (Kogej et al. 2006a; Kralj Kunčič et al. 2010).

The mechanisms underlying the salt tolerance of fungi were initially extensively studied in the salt-sensitive baker yeast *S. cerevisiae* (Blomberg and Adler 1992; Blomberg 2000; Hohmann 2002) and in the halotolerant yeast *Deb. hansenii* (Larsson and Gustafsson 1987, 1993; Andre et al. 1988; Larsson et al. 1990; Prista et al. 1997; Almagro et al. 2000) and only later on in the halotolerant *A. pullulans* (Kogej et al. 2005; Gostinčar et al. 2008) and the extremely halotolerant black yeast *Ho. werneckii* (Turk and Plemenitaš 2002; Turk et al. 2004; Kogej et al. 2005, 2006b, 2007; Plemenitaš et al. 2008, 2014; Lenassi et al. 2011). The physiological and molecular adaptations of the obligate halophilic filamentous fungus, the basidiomycete *Wallemia ichthyophaga* (Zajc et al. 2013, 2014b), were used for comparison.

Sensing an increased osmolarity change is of major importance for survival in hypersaline environments. The pathway required for cellular adaptation to hypersaline stress, as well as to oxidative, heavy metal and temperature stress, is the branched mitogen-activated protein kinases (MAPK) signal transduction system and high-osmolarity glycerol (HOG) signalling pathway (Hohmann 2002). This pathway consists of two signalling modules: the MAPK module of MAPK, MAPK kinase (MAPKK) and MAPKK kinase (MAPKKK), and a two-component phosphorelay module that is composed of hybrid sensor kinases, a histidine-containing phosphotransfer protein and response regulators (Bahn 2008; Hohmann 2009). Homologues of the HOG pathway components have been extensively studied in *Ho. werneckii* (Turk and Plemenitaš 2002; Plemenitaš et al. 2008, 2014) and in *A. pullulans* (Gostinčar et al. 2014) and *Ex. dermatitidis* (Chen et al. 2014).

Most halotolerant and halophilic yeasts counterbalance the osmotic pressure and prevent loss of intracellular water by accumulation and synthesis of small compatible organic solutes, and at the same time, they maintain low concentrations of alkali-metal cations within the cytoplasm (Kogej et al. 2005; Zajc et al. 2014b). The most common compatible solute among eukaryotic microorganisms is glycerol, as also among halotolerant and halophilic yeasts, which is energetically the cheapest such compound to produce at high concentrations (Oren 1999). The highest glycerol levels have been measured in the extremely halotolerant black yeast *Ho. werneckii* (Petrovič et al. 2002; Kogej et al. 2007) and *A. pullulans* (Kogej, unpublished data). The genes that encode the main enzyme involved in glycerol biosynthesis, nicotinamide adenine dinucleotide (NAD)-dependent glycerol-3-phosphate dehydrogenase (Gpd), have shown salt-responsive expression under saline conditions (Lenassi et al. 2011). *Ho. werneckii* also accumulates other complementary compatible solutes, such as erythritol, arabitol and mannitol (Kogej et al. 2007), and interestingly, a mycosporine-glutaminol-glucoside that is primarily known as being involved in fungal sporulation and UV protection (Kogej et al. 2006b). In combination with complementary solutes like trehalose and polyols and at sufficient concentrations, glycerol helps to maintain positive turgor pressure at high salinities in many salt-tolerant yeast species, like *Deb. hansenii*, *C. versatilis* (formerly *C. halophila*), *Rh. mucilaginosa* and *Mey. guilliermondii* (Andre et al. 1988; Almagro et al. 2000; Prista et al. 2005).

Including for the halotolerant *Ho. werneckii* and *A. pullulans*, yeasts can maintain low intracellular concentrations of sodium and potassium cations, and they are therefore considered as Na^+ excluders (Kogej et al. 2005). This suggests an effective transport system for extrusion of Na^+ cations, accompanied by prevention of their influx. One of the most important transporters for the extrusion of Na^+ ions is the P-type (ENA-like) ATPase, a sodium–potassium pump (for review, see Arino et al. 2010). Genome analysis of *Ho. werneckii* uncovered an enrichment in genes coding for alkali-metal cation transporters (Lenassi et al. 2013). Two salt-responsive ENA pumps and their salt-dependent activities are in line with this concept (Gorjan and Plemenitaš 2006). Likewise, in the genome of *A. pullulans*, a plethora of plasma-membrane transporters have been found, including rare types that have arisen through several duplications (Gostinčar et al. 2014).

When under saline stress, relatively high internal concentrations of sodium together with the glycerol accumulation have only been detected in rare cases. This can be interpreted as a mixed strategy for the halotolerant yeast *Deb. hansenii* (Prista et al. 1997) and *Mey. guilliermondii* (Lahav et al. 2002).

Eukaryotes that use glycerol accumulation as their osmoadaptation strategy need to prevent its leaking through the lipid bilayer of the cell membrane. This is achieved through structural changes to the cell membrane, like increased sterol content or reduced membrane fluidity, and/or recuperation of the loss of glycerol by energetically costly transport systems (Oren 1999). Surprisingly, reduction in membrane fluidity is not the case for the halotolerant black yeasts, as the plasma membranes of *Ho. werneckii* and *Phaeoth. triangularis* were more fluid over a wide range of salinities than the membranes of salt-sensitive fungi, due to an increase in the content and length of the unsaturated fatty acids (Turk et al. 2004, 2007). Indeed, in *Ho. werneckii* and *A. pullulans*, changes in the expression of fatty-acid-modifying enzymes (e.g. desaturases, elongases) can allow for precise regulation of membrane fluidity (Gostinčar et al. 2008, 2009). Furthermore, the melanised cell wall of *Ho. werneckii* appears to contribute to the glycerol retention and high membrane fluidity (Kogej et al. 2006b). A continuous layer of melanin granules in the outer part of the cell wall creates a mechanical permeability barrier for glycerol, by reducing the size of the pores in the cell wall (Jacobson and Ikeda 2005), and thereby minimising glycerol loss from the cells (Kogej et al. 2006b).

10.7 Concluding Remarks

This overview of halotolerant and extremely halotolerant yeasts (Table 10.1) has revealed surprising diversity of both the yeast species and the habitats they populate. Over the years, many new species have been described, and natural ecological niches have been found for many species that were previously known primarily as food contaminants. The ecology of some of these yeasts remains enigmatic, and their further isolation is required. In particular, future isolation should be aimed at extreme hypersaline environments that have so far only been investigated for prokaryotic diversity; e.g. tropical and subtropical hypersaline alkaline lakes in continental interiors and saline and hypersaline lakes in polar areas. Due to their general xerophilic nature, quite a lot of halotolerant yeasts can populate cold environments and can be exemplified by the rich abundance of *Deb. hansenii* in polar glaciers. On the basis that soil salinisation is becoming an extremely important agricultural challenge worldwide, more research into yeast biodiversity should also focus on the hypersaline/saline microbiome of halophytes and the soil and the symbiotic associations with plants, as this appears to provide a highly perspective approach to improve the salt tolerance of plants.

The addition of yeasts to other microbial communities in saline environments has improved our understanding of complex microbial processes, which are often closely connected to evaporation and mineral precipitation. One such intimately linked biological system is man-made salterns, where the presence of different microorganisms can affect both the quality and quantity of the salt produced. Only recently has it been acknowledged that certain yeasts/fungi together with prokaryotes can form microbial mats and can aid or harm salt production.

In comparison to their prokaryotic counterparts, most yeast species that inhabit saline and hypersaline environments show either halotolerant or extremely halotolerant behaviour. Thus, they do not require NaCl for viability, as they can grow and adjust across the whole salinity range, from freshwater to 10% NaCl, and some also to near-saturated NaCl solutions. This type of adaptation shows promise for targeted searches for genes that have biotechnological importance, with the aim to improve halotolerance of industrially important yeasts and plants, thus providing an important asset for our increasingly salinised world.

Acknowledgements The scientific studies integral to this report were financed partly through the 'Centre of Excellence for Integrated Approaches in Chemistry and Biology of Proteins' (N° OP13.1.1.2.02.0005) of the European Regional Development (30%); partly by the Slovenian Ministry of Higher Education, Science and Technology (35%); and partly by the Slovenian Research Agency (35%).

References

Adamiak J, Otlewska A, Gutarowska B, Pietrzak A (2016) Halophilic microorganisms in deteriorated historic buildings: insights into their characteristics. Acta Biochim Pol 63:335–341

Aguiar C, Lucas C (2000) Yeasts killer/sensitivity phenotypes and halotolerance. Food Technol Biotechnol 38:39–46

Alexander E, Stock A, Breiner HW, Behnke A, Bunge J, Yakimov MM, Stoeck T (2009) Microbial eukaryotes in the hypersaline anoxic L'Atalante deep-sea basin. Environ Microbiol 11:360–381

Ali I, Kanhayuwa L, Rachdawong S, Rakshit SK (2013) Identification, phylogenetic analysis and characterization of obligate halophilic fungi isolated from a man-made solar saltern in Phetchaburi province, Thailand. Ann Microbiol 63:887–895

Almagro A, Prista C, Castro S, Quintas C, Madeira-Lopes A, Ramos J, Loureiro-Dias MC (2000) Effects of salts on *Debaryomyces hansenii* and *Saccharomyces cerevisiae* under stress conditions. Int J Food Microbiol 56:191–197

Al-Musallam AAS, Al-Sammar HF, Al-Sané NA (2010) Diversity and dominance of fungi inhabiting the sabkha area in Kuwait. Bot Mar 54:83–94

Andre L, Nilsson A, Adler L (1988) The role of glycerol in osmotolerance of the yeast *Debaryomyces hansenii*. J Gen Microbiol 134:669–677

Andrews JH, Spear RN, Nordheim EV (2002) Population biology of *Aureobasidium pullulans* on apple leaf surfaces. Can J Microbiol 48:500–513

Arino J, Ramos J, Sychrova H (2010) Alkali metal cation transport and homeostasis in yeasts. Microbiol Mol Biol Rev 74:95–120

Bahn YS (2008) Master and commander in fungal pathogens: the two-component system and the HOG signaling pathway. Eukaryot Cell 7:2017–2036

Blomberg A (2000) Metabolic surprises in *Saccharomyces cerevisiae* during adaptation to saline conditions: questions, some answers and a model. FEMS Microbiol Lett 182:1–8

Blomberg A, Adler L (1992) Physiology of osmotolerance in fungi. Adv Microb Physiol 33:145–212

Bonifaz A, Badali H, de Hoog GS, Cruz M, Araiza J, Cruz MA, Fierro L, Ponce RM (2008) Tinea nigra by *Hortaea werneckii*, a report of 22 cases from Mexico. Stud Mycol 61:77–82

Burgaud G, Arzur D, Durand L, Cambon-Bonavita MA, Barbier G (2010) Marine culturable yeasts in deep-sea hydrothermal vents: diversity and association with fauna. FEMS Microbiol Ecol 73:121–133

Burgaud G, Huéb NTM, Arzur D, Coton M, Perrier-Cornet JM, Jebbar M, Barbier G (2015) Effects of hydrostatic pressure on yeasts isolated from deep-sea hydrothermal vents. Res Microbiol 166:700–709

Butinar L, Sonjak S, Zalar P, Plemenitaš A, Gunde-Cimerman N (2005a) Melanized halophilic fungi are eukaryotic members of microbial communities in hypersaline waters of solar salterns. Bot Mar 48:73–79

Butinar L, Santos S, Spencer-Martins I, Oren A, Gunde-Cimerman N (2005b) Yeast diversity in hypersaline habitats. FEMS Microbiol Lett 244:229–234

Butinar L, Zalar P, Frisvad JC, Gunde-Cimerman N (2005c) The genus *Eurotium* – members of indigenous fungal community in hypersaline waters of salterns. FEMS Microbiol Ecol 51:155–166

Butinar L, Spencer-Martins I, Gunde-Cimerman N (2007) Yeasts in high Arctic glaciers: the discovery of a new habitat for eukaryotic microorganisms. A van Leeuwenhoek 91:277–289

Butinar L, Frisvad JC, Gunde-Cimerman N (2011a) Hypersaline waters – a potential source of foodborne toxigenic *Aspergilli* and *Penicillia*. FEMS Microbiol Ecol 77:186–199

Butinar L, Strmole T, Gunde-Cimerman N (2011b) Relative incidence of ascomycetous yeasts in arctic coastal environments. Microb Ecol 61:832–843

Buzzini P, Martini A (2002) Extracellular enzymatic activity profiles in yeast and yeast-like strains isolated from tropical environments. J Appl Microbiol 93:1020–1025

Cabañes FJ, Bragulat MR, Castellá G (2012) *Hortaea werneckii* isolated from silicone scuba diving equipment in Spain. Med Mycol 50:852–857

Cantrell SA, Casillas-MartiNez L, Molina M (2006) Characterization of fungi from hypersaline environments of solar salterns using morphological and molecular techniques. Mycol Res 110:962–970

Chen Z, Martinez DA, Gujja S, Sykes SM, Zeng Q, Szaniszlo PJ, Wang Z, Cuomo CA (2014) Comparative genomic and transcriptomic analysis of *Wangiella dermatitidis*, a major cause of phaeohyphomycosis and a model black yeast human pathogen. Genes Gen Genet 4:561–578

Cheng KC, Demirci A, Catchmark JM (2011) Pullulan: biosynthesis, production, and applications. Appl Microbiol Biotechnol 92:29–44

Chi ZM, Wang F, Chi Z, Yue L, Liu G, Zhang T (2009) Bioproducts from *Aureobasidium pullulans*, a biotechnologically important yeast. Appl Microbiol Biotechnol 82:793–804

Crognale S, Pesciaroli L, Petruccioli M, D'Annibale A (2012) Phenoloxidase-producing halotolerant fungi from olive brine wastewater. Process Biochem 47:1433–1437

Cronin AE, Post FJ (1977) Report of a dematiaceous Hyphomycetes from the Great Salt Lake, Utah. Mycologia 69:846–847

Crous PW, Wingfield MJ, Richardson DM, Leroux JJ, Strasberg D, Edwards J, Roets F, Hubka V, Taylor PWJ, Heykoop M, Martín MP, Moreno G, Sutton DA, Wiederhold NP, Barnes CW, Carlavilla JR, Gené J, Giraldo A, Guarnaccia V, Guarro J, Hernández-Restrepo M, Kolařík M, Manjón JL, Pascoe IG, Popov ES, Sandoval-Denis M, Woudenberg JHC, Acharya K, Alexandrova AV, Alvarado P, Barbosa RN, Baseia IG, Blanchette RA, Boekhout T, Burgess TI, Cano-Lira JF, Čmoková A, Dimitrov RA, Dyakov MY, Dueñas M, Dutta AK, Esteve-Raventós F, Fedosova AG, Fournier J, Gamboa P, Gouliamova DE, Grebenc T, Groenewald M, Hanse B, Hardy GESTJ, Held BW, Jurjević Ž, Kaewgrajang T, Latha KPD, Lombard L, Luangsa-ard JJ, Lysková P, Mallátová N, Manimohan P, Miller AN, Mirabolfathy M, Morozova OV, Obodai M, Oliveira NT, Ordóñez ME, Otto EC, Paloi S, Peterson SW, Phosri C, Roux J, Salazar WA, Sánchez A, Sarria GA, Shin H-D, Silva BDB, Silva GA, Smith MTH, Souza-Motta CM, Stchigel AM, Stoilova-Disheva MM, Sulzbacher MA, Telleria MT, Toapanta C, Traba JM, Valenzuela-Lopez N, Watling R, Groenewald JZ (2016) Fungal Planet description sheets: 400–468. Persoonia 36:316–458

da Silva S, Calado S, Lucas C, Aguiar C (2008) Unusual properties of the halotolerant yeast *Candida nodaensis* Killer toxin, CnKT. Microbiol Res 163:243–251

Dakal TC, Solieri L, Giudici P (2014) Adaptive response and tolerance to sugar and salt stress in the food yeast *Zygosaccharomyces rouxii*. Int J Food Microbiol 185:140–157

de Hoog GS, Guého E (2010) White Piedra, Black Piedra, and Tinea Nigra. Topley & Wilson's Microbiology and Microbial Infections. Wiley

de Hoog GS, Beguin H, Batenburg-van de Vegte WH (1997) *Phaeotheca triangularis*, a new meristematic black yeast from a humidifier. A van Leeuwenhoek 71:289–295

de Hoog GS, Zalar P, van den Ende BG, Gunde-Cimerman N (2005) Relation of halotolerance to human pathogenicity in the fungal tree of life: an overview of ecology and evolution under stress. In: Gunde-Cimerman N, Oren A, Plemenitas A (eds) Adaptation to life at high salt-concentration in Archaea, Bacteria and Eukarya, vol 9. Springer, Dordrecht, The Netherlands, pp 373–395

Diaz-Munos G, Montalvo-Rodriguez R (2005) Halophilic black yeast *Hortaea werneckii* in the Cabo Rojo solar salterns: it's first record for this extreme environment in Puerto Rico. Caribb J Sci 41:360–365

Dickson JL, Head JW, Levy JS, Marchant DR (2013) Don Juan Pond, Antarctica: near-surface $CaCl_2$-brine feeding Earth's most saline lake and implications for Mars. Sci Rep 3:1166

Duarte AW, Dayo-Owoyemi I, Nobre FS, Pagnocca FC, Chaud LC, Pessoa A, Felipe MG, Sette LD (2013) Taxonomic assessment and enzymes production by yeasts isolated from marine and terrestrial Antarctic samples. Extremophiles 17:1023–1035

Edgcomb VP, Orsi W, Leslin C, Epstein SS, Bunge J, Jeon S, Yakimov MM, Behnke A, Stoeck T (2009) Protistan community patterns within the brine and halocline of deep hypersaline anoxic basins. Extremophiles 13:151–167

Egidi E, de Hoog GS, Isola D, Onofri S, Quaedvlieg W, de Vries M, Verkley GJM, Stielow JB, Zucconi L, Selbmann L (2014) Phylogeny and taxonomy of meristematic rock-inhabiting black fungi in the Dothideomycetes based on multi-locus phylogenies. Fungal Divers 65:127–165

El-Morsy EM (2000) Fungi isolated from the endorhizosphere of halophytic plants from the Red Sea Coast of Egypt. Fungal Divers 5:43–54

Evans S, Hansen RW, Schneegurt MA (2013) Isolation and characterization of halotolerant soil fungi from the Great Salt Plains of Oklahoma. Cryptogam Mycol 34:329–341

Filtenborg O, Frisvad JC, Samson RA (2000) Specific association of fungi to foods and influence of physical environmental factors. In: Samson RA, Hoekstra ES, Frisvad JC, Filtenborg O (eds) Introduction of food- and airborne fungi. Centraalbureau voor Schimmelcultures, Utrecht, pp 306–320

Finkel OM, Burch AY, Lindow SE, Post AF, Belkin S (2011) Geographical location determines the population structure in phyllosphere microbial communities of a salt-excreting desert tree. Appl Environ Microbiol 77:7647–7655

Fleet GH (2011) Yeast spoilage of foods and beverages. In: Kurtzman CP, Fell JW, Boekhout T (eds) The yeasts, a taxonomic study, 5th edn. Elsevier, Amsterdam, pp 53–63

Flowers TJ, Colmer TD (2015) Plant salt tolerance: adaptations in halophytes. Ann Bot 115:327–331

Gadanho M, Sampaio JP (2005) Occurrence and diversity of yeasts in the mid-atlantic ridge hydrothermal fields near the Azores Archipelago. Microb Ecol 50:408–417

Gadanho M, Almeida JM, Sampaio JP (2003) Assessment of yeast diversity in a marine environment in the south of Portugal by microsatellite-primed PCR. A van Leeuwenhoek 84:217–227

Gavrieli I, Beyth M, Yechieli Y (1999) The Dead Sea – a terminal lake in the Dead Sea rift: a short overview. In: Oren A (ed) Microbiology and biogeochemistry of hypersaline environments. CRC Press, Boca Raton, FL, pp 121–127

Gorjan A, Plemenitaš A (2006) Identification and characterization of ENA ATPases HwENA1 and HwENA2 from the halophilic black yeast *Hortaea werneckii*. FEMS Microbiol Lett 265:41–50

Gostinčar C, Turk M, Trbuha T, Vaupotič T, Plemenitaš A, Gunde-Cimerman N (2008) Expression of fatty-acid-modifying enzymes in the halotolerant black yeast *Aureobasidium pullulans* (de Bary) G. Arnaud under salt stress. Stud Mycol 61:51–59

Gostinčar C, Turk M, Plemenitaš A, Gunde-Cimerman N (2009) The expressions of Δ9-, Δ12-desaturases and an elongase by the extremely halotolerant black yeast *Hortaea werneckii* are salt dependent. FEMS Yeast Res 9:247–256

Gostinčar C, Grube M, de Hoog S, Zalar P, Gunde-Cimerman N (2010) Extremotolerance in fungi: evolution on the edge. FEMS Microbiol Ecol 71:2–11

Gostinčar C, Grube M, Gunde-Cimerman N (2011) Evolution of fungal pathogens in domestic environments? Fungal Biol 115:1008–1018

Gostinčar C, Ohm R, Kogej T, Sonjak S, Turk M, Zajc J, Zalar P, Grube M, Sun H, Han J, Sharma A, Chiniquy J, Ngan CY, Lipzen A, Barry K, Grigoriev IV, Gunde-Cimerman N (2014) Genome sequencing of four *Aureobasidium pullulans* varieties: biotechnological potential, stress tolerance, and description of new species. BMC Genomics 15:549

Goto S, Sugiyama J, Iizuka H (1969) A taxonomic study of Antarctic yeasts. Mycologia 61:748–774

Grant WD (2004) Life at low water activity. Philos Trans R Soc Lond B 359:1249–1266

Grishkan I, Nevo E, Wasser SP (2003) Soil micromycete diversity in the hypersaline Dead Sea coastal area, Israel. Mycol Prog 2:19–28

Grube M, Schmid F, Berg G (2011) Black fungi and associated bacterial communities in the phyllosphere of grapevine. Fungal Biol 115:978–986

Grum-Grzhimaylo AA, Georgieva ML, Bondarenko SA, Debets AJM, Bilanenko EN (2016) On the diversity of fungi from soda soils. Fungal Divers 76:27–74

Guiraud P, Steiman R, Seigle-Murandi F, Sage L (1995) Mycoflora of soil around the Dead Sea II – Deuteromycetes (except *Aspergillus* and *Penicillium*). Syst Appl Microbiol 18:318–322

Gunde-Cimerman N, Plemenitaš A (2006) Ecology and molecular adaptations of the halophilic black yeast *Hortaea werneckii*. Rev Environ Sci Biotechnol 5:323–331

Gunde-Cimerman N, Zalar P (2014) Extremely halotolerant and halophilic fungi inhabit brine in solar salterns around the globe. Food Technol Biotechnol 52:170–179

Gunde-Cimerman N, Zalar P, de Hoog S, Plemenitaš A (2000) Hypersaline waters in salterns – natural ecological niches for halophilic black yeasts. FEMS Microbiol Ecol 32:235–240
Gunde-Cimerman N, Sonjak S, Zalar P, Frisvad JC, Diderichsen B, Plemenitaš A (2003) Extremophilic fungi in arctic ice: a relationship between adaptation to low temperature and water activity. Phys Chem Earth 28:1273–1278
Gunde-Cimerman N, Butinar L, Sonjak S, Turk M, Uršič V, Zalar P, Plemenitaš A (2005) Halotolerant and halophilic fungi from coastal environments in the Arctics. In: Seckbach J (ed) Adaptation to life at high salt concentrations in Archaea, Bacteria, and Eukarya, vol 9. Springer, Netherlands, pp 397–423
Gunde-Cimerman N, Ramos J, Plemenitaš A (2009) Halotolerant and halophilic fungi. Mycol Res 113:1231–1241
Hallsworth JE, Yakimov MM, Golyshin PN, Gillion JL, D'Auria G, de Lima AF, La Cono V, Genovese M, McKew BA, Hayes SL, Harris G, Giuliano L, Timmis KN, McGenity TJ (2007) Limits of life in $MgCl_2$-containing environments: chaotropicity defines the window. Environ Microbiol 9:801–813
Hardie M, Doyle R (2012) Measuring soil salinity. In: Shabala S, Cuin TA (eds) Plant salt tolerance: methods and protocols. Humana Press, Totowa, NJ, pp 415–425
Hocking AD (1993) Responses in xerophilic fungi to changes in water activity. In: Jennings DH (ed) Stress tolerance of fungi. Marcel Dekker, New York, pp 233–243
Hocking AD, Pitt JI (1979) Water relations of some *Penicillium* species at 25 °C. Trans Br Mycol Soc 73:141–145
Hohmann S (2002) Osmotic stress signalling and osmoadaptation in yeasts. Microbiol Mol Biol Rev 66:300–372
Hohmann S (2009) Control of high osmolarity signalling in the yeast *Saccharomyces cerevisiae*. FEBS Lett 583:4025–4029
Hui FL, Ke T, Niu Q-H, Du PC (2012) *Candida baotianensis* sp. nov. an ascomycetous yeast species from forest soil in China. J Gen Appl Microbiol 58:59–63
Hujslová M, Kubátová A, Chudíčková M, Kolařík M (2010) Diversity of fungal communities in saline and acidic soils in the Soos National Natural Reserve, Czech Republic. Mycol Prog 9:1–15
Iwatsu TU, Udagawa SI (1988) *Hortaea werneckii* isolated from sea-water. Jpn J Med Mycol 29:142–145
Jacobson ES, Ikeda R (2005) Effect of melanization upon porosity of the cryptococcal cell wall. Med Mycol 43:327–333
Jacques N, Zenouche A, Gunde-Cimerman N, Casaregola S (2015) Increased diversity in the genus *Debaryomyces* from Arctic glacier samples. A van Leeuwenhoek 107:487–501
Jančič S, Nguyen HDT, Frisvad JC, Zalar P, Schroers H-J, Seifert KA, Gunde-Cimerman N (2015) A taxonomic revision of the *Wallemia sebi* species complex. PLoS One 10(5):e0125933
Javor BJ (1989) Hypersaline environments In: Schiewer U (ed) Microbiology and biogeochemistry vol 76. vol 2. Springer/Heidelberg GmbH & Co. K, Berlin, pp 287–287
Kis-Papo T, Grishkan I, Oren A, Wasser P, Nevo E (2001) Spatiotemporal diversity of filamentous fungi in the hypersaline Dead Sea. Mycol Res 105:749–756
Klassen R, Schaffrath R, Buzzini P, Ganter PF (2017) Antagonistic interactions and killer yeasts. In: Buzzini P, Lachance MA, Yurkov AM (eds) Yeasts in natural ecosystems: ecology. Springer, Heidelberg, pp 229–275
Kogej T, Ramos J, Plemenitaš A, Gunde-Cimerman N (2005) The halophilic fungus *Hortaea werneckii* and the halotolerant fungus *Aureobasidium pullulans* maintain low intracellular cation concentrations in hypersaline environments. Appl Environ Microbiol 71:6600–6605
Kogej T, Gorbushina AA, Gunde-Cimerman N (2006a) Hypersaline conditions induce changes in cell-wall melanization and colony structure in a halophilic and a xerophilic black yeast species of the genus *Trimmatostroma*. Mycol Res 110:713–724
Kogej T, Gostinčar C, Volkmann M, Gorbushina AA, Gunde-Cimerman N (2006b) Mycosporines in extremophilic fungi – novel complementary osmolytes? Environ Chem 3:105–110

Kogej T, Stein M, Volkmann M, Gorbushina AA, Galinski EA, Gunde-Cimerman N (2007) Osmotic adaptation of the halophilic fungus *Hortaea werneckii*: role of osmolytes and melanization. Microbiology 153:4261–4273

Kralj Kunčič M, Kogej T, Drobne D, Gunde-Cimerman N (2010) Morphological response of the halophilic fungal genus *Wallemia* to high salinity. Appl Environ Microbiol 76:329–337

Kriss AE, Mitskevich IN, Rozanova EP, Osnitskaya LK (1976) Microbiological investigations of Lake Vanda (Antarctica). Microbiology (Russian Academy of Sciences) 45:917–922

Kritzman G (1973) Observations on the microorganisms in the Dead Sea. M.Sc. thesis, The Hebrew University of Jerusalem (in Hebrew)

Kumar S, Randhawa A, Ganesan K, Singh Raghava GP, Mondal AK (2012) Draft genome sequence of salt-tolerant yeast *Debaryomyces hansenii* var. *hansenii* MTCC 234. Eukaryot Cell 11:961–962

Kunze G, Gaillardin C, Czernicka M, Durrens P, Martin T, Böer E, Gabaldón T, Cruz JA, Talla E, Marck C, Goffeau A, Barbe V, Baret P, Baronian K, Beier S, Bleykasten C, Bode R, Casaregola S, Despons L, Fairhead C, Giersberg M, Gierski PP, Hähnel U, Hartmann A, Jankowska D, Jubin C, Jung P, Lafontaine I, Leh-Louis V, Lemaire M, Marcet-Houben M, Mascher M, Morel G, Richard GF, Riechen J, Sacerdot C, Sarkar A, Savel G, Schacherer J, Sherman DJ, Stein N, Straub ML, Thierry A, Trautwein-Schult A, Vacherie B, Westhof E, Worch S, Dujon B, Souciet JL, Wincker P, Scholz U, Neuvéglise C (2014) The complete genome of *Blastobotrys* (*Arxula*) *adeninivorans* LS3 – a yeast of biotechnological interest. Biotechnol Biofuels 66

Kurtzman CP, Fell JW, Boekhout T (eds) (2011) The yeasts: a taxonomic study, 5th edn. Amsterdam, Elsevier

Lachance MA, Miranda M, Miller MW, Phaff HJ (1976) Dehiscence and active spore release in pathogenic strains of the yeast *Metschnikowia bicuspidata* var. *australis*: possible predatory implication. Can J Microbiol 22:1756–1761

Lages F, Silva-Graça M, Lucas C (1999) Active glycerol uptake is a mechanism underlying halotolerance in yeasts: a study of 42 species. Microbiology 145:2577–2585

Lahav R, Fareleira P, Nejidat A, Abeliovich A (2002) The identification and characterization of osmotolerant yeast isolates from chemical wastewater evaporation ponds. Microb Ecol 43:388–396

Lai X, Cao L, Tan H, Fang S, Huang Y, Zhou S (2007) Fungal communities from methane hydrate-bearing deep-sea marine sediments in South China Sea. ISME J 1:756–762

Larsson C, Gustafsson L (1987) Glycerol production in relation to the ATP pool and heat-production rate of the yeasts *Debaryomyces hansenii* and *Saccharomyces cerevisiae* during salt stress. Arch Microbiol 147:358–363

Larsson C, Gustafsson L (1993) The role of physiological-state in osmotolerance of the salt-tolerant yeast *Debaryomyces hansenii*. Can J Microbiol 39:603–609

Larsson C, Morales C, Gustafsson L, Adler L (1990) Osmoregulation of the salt-tolerant yeast *Debaryomyces hansenii* grown in a chemostat at different salinities. J Bacteriol 172:1769–1774

Leathers TD (2003) Biotechnological production and applications of pullulan. Appl Microbiol Biotechnol 62:468–473

Lenassi M, Zajc J, Gostinčar C, Gorjan A, Gunde-Cimerman N, Plemenitaš A (2011) Adaptation of the glycerol-3-phosphate dehydrogenase Gpd1 to high salinities in the extremely halotolerant *Hortaea werneckii* and halophilic *Wallemia ichthyophaga*. Fungal Biol 115:959–970

Lenassi M, Gostinčar C, Jackman S, Turk M, Sadowski I, Nislow C, Jones S, Birol I, Gunde Cimerman N, Plemenitaš A (2013) Whole genome duplication and enrichment of metal cation transporters revealed by de novo genome sequencing of extremely halotolerant black yeast *Hortaea werneckii*. PLoS One 8(8):e71328

Li H, Chi Z, Wang X, Ma C (2007) Amylase production by the marine yeast *Aureobasidium pullulans* N13d. J Ocean Univ China 6:60

Lisichkina GA, Bab'eva IP, Sorokin DY (2003) Alkalitolerant yeasts from natural biotopes. Microbiology 72:618–620

Liu WC, Li CQ, Zhu P, Yang JL, Cheng KD (2010) Phylogenetic diversity of culturable fungi associated with two marine sponges: *Haliclona simulans* and *Gelliodes carnosa*, collected from the Hainan Island coastal waters of the South China Sea. Fungal Divers 42:1–15

Liu X-Z, Wang Q-M, Göker M, Groenewald M, Kachalkin AV, Lumbsch HT, Millanes AM, Wedin M, Yurkov AM, Boekhout T, Bai F-Y (2015) Towards an integrated phylogenetic classification of the Tremellomycetes. Stud Mycol 81:85–147

Lopez-Garcia P, Philippe H, Gail F, Moreira D (2003) Autochthonous eukaryotic diversity in hydrothermal sediment and experimental microcolonizers at the Mid-Atlantic Ridge. Proc Natl Acad Sci U S A 100:697–702

Lord AT, Mohandas K, Somanath S, Ambu S (2010) Multidrug resistant yeasts in synanthropic wild birds. Ann Clin Microbiol Antimicrob. doi:10.1186/1476-0711-9-11

Ma H, Tian C, Feng G, Yuan JF (2011) Ability of multicellular salt glands in *Tamarix* species to secrete Na^+ and K^+ selectively. Sci China Life Sci 54:282–289

Ma Y, Zhang H, Du Y, Tian T, Xiang T, Liu X, Wu F, An L, Wang W, Gu J-D, Feng H (2014) The community distribution of bacteria and fungi on ancient wall paintings of the Mogao Grottoes. Sci Rep 5:7752

Maciá-Vicente JG, Ferraro V, Burruano S, Lopez-Llorca LV (2012) Fungal assemblages associated with roots of halophytic and non-halophytic plant species vary differentially along a salinity gradient. Microb Ecol 64:668–679

Marion GM (1997) A theoretical evaluation of mineral stability in Don Juan Pond, Wright Valley, Victoria Land. Antarct Sci 9:92–99

Martinelli L, Zalar P, Gunde-Cimerman N, Azua-Bustos A, Sterflinger K, Piñar G (2017) *Aspergillus atacamensis* and *A. salisburgensis*: two new halophilic species from hypersaline/arid habitats with a phialosimplex-like morphology. Extremophiles 21(4):755–773

Matsubaya O, Sakai H, Torii T, Burton H, Kerry K (1979) Antarctic saline lake: stable isotopic ratios, chemical composition and evolution. Geochim Cosmochim Acta 43:7–25

McGenity TJ, Oren A (2012) Life in saline environments. In: Bell EM (ed) Life at extremes. Environments, organisms and strategies for survival. CABI International, Wallingford, pp 402–437

Mok WY, Castelo FP, Dasilva MSB (1981) Occurrence of *Exophiala werneckii* on salted freshwater fish *Osteoglossum bicirrhosum*. J Food Technol 16:505–512

Mokhtarnejad L, Arzanlou M, Babai-Ahari A (2015) Molecular and phenotypic characterization of ascomycetous yeasts in hypersaline soils of Urmia Lake basin (NW Iran). Rostaniha 16:174–185

Mokhtarnejad L, Arzanlou M, Babai-Ahari A, Di Mauro S, Onofri A, Buzzini P, Turchetti B (2016) Characterization of basidiomycetous yeasts in hypersaline soils of the Urmia Lake National Park, Iran. Extremophiles 20:915–928

Molnarova J, Vadkertiova R, Stratilova E (2013) Extracellular enzymatic activities and physiological profiles of yeasts colonizing fruit trees. J Basic Microbiol 53

Nagashima H, Nishikawa J, Matsumoto GI, Tizuka H (1990) Characterization and habitats of bacteria and yeasts isolated from Lake Vanda in Antarctica. Proc NIPR Symp Polar Biol 3:190–200

Nagatsuka Y, Kawasaki H, Limtong S, Mikata K, Seki T (2002) *Citeromyces siamensis* sp. nov. a novel halotolerant yeast isolated in Thailand. Int J Syst 52:2315–2319

Nayak S, Gonsalves V, Nazareth S (2012) Isolation and salt tolerance of halophilic fungi from mangroves and solar salterns in Goa India. Indian J Geo-Mar Sci 41:164–172

Nishimura K, Miyaji M (1983) Studies on the phylogenesis of pathogenic black yeasts. Mycopathologia 81:135–144

Northolt MD, Frisvad JC, Samson RA (1995) Occurrence of foodborne fungi and factors for growth. In: Samson RA, Hoekstra ES, Frisvad JC, Filtenborg O (eds) Introduction to food borne fungi. CBS, Delft, pp 243–250

Novak Babič M, Zalar P, Ženko B, Schroers H-J, Džeroski S, Gunde-Cimerman N (2015) *Candida* and *Fusarium* species known as opportunistic human pathogens from customer-accessible parts of residential washing machines. Fungal Biol 119:95–113

Novak Babič M, Zalar P, Ženko B, Džeroski S, Gunde-Cimerman N (2016) Yeasts and yeast-like fungi in tap water and groundwater, and their transmission to household appliances. Fungal Ecol 20:30–39

Okane I, Nakagiri A (2015) Assemblages of endophytic fungi on *Salicornia europaea* disjunctively distributed in Japan: towards clarification of the ubiquity of fungal endophytes on halophytes and their ecological roles. Curr Sci 109(1):62–71

Onofri S (1999) Antarctic microfungi. In: Seckbach J (ed) Enigmatic microorganisms and life in extreme environments, Cellular origin and life in extreme habitats. Kluwer Academic, Dordrecht, London, pp 323–336

Oren A (1999) Bioenergetic aspects of halophilism. Microbiol Mol Biol Rev 63:334–348

Oren A (2002a) Hypersaline environment and their biota. In: Oren A (ed) Halophilic microorganisms and their environments, Cellular origin, life in extreme habitats and astrobiology, vol 5. Kluwer Academic, Dordrecht/Boston/London, p 575

Oren A (2002b) Diversity of halophilic microorganisms: environments, phylogeny, physiology and applications. J Ind Microbiol Biotechnol 28:56–63

Oren A (2005) A hundred years of *Dunaliella* research: 1905–2005. Saline Syst 1:2

Oren A (2008) Microbial life at high salt concentrations: phylogenetic and metabolic diversity. Saline Syst 4:2

Oren A (2009) Microbial diversity and microbial abundance in salt-saturated brines: why are the waters of hypersaline lakes red? Nat Resour Environ Issues 15:247–255

Oren A (2013) Life in magnesium- and calcium-rich hypersaline environments: salt stress by chaotropic ions. In: Seckbach J, Oren A, Stan-Lotter H (eds) Polyextremophiles, vol 27. Springer Netherlands, Dordrecht, pp 215–232

Pahor M, Poberaj T (1963) Stare Piranske soline. Mladinska knjiga, Ljubljana

Petrovič U, Gunde-Cimerman N, Plemenitaš A (2002) Cellular responses to environmental salinity in the halophilic black yeast *Hortaea werneckii*. Mol Microbiol 45:665–672

Pierce GJ (1914) The behavior of certain micro-organisms in brine: the Salton Sea. Carnegie Inst Wash Publ 193:49–69

Piñar G, Dalnodar D, Voitl C, Reschreiter H, Sterflinger K (2016) Biodeterioration risk threatens the 3100 year old staircase of Hallstatt (Austria): possible involvement of halophilic microorganisms. PLoS One 11:e0148279

Pires-Gonçalves RH, Sartori FG, Montanari LB, Zaia JE, Melhem MS, Mendes-Giannini MJ, Martins CH (2008) Occurrence of fungi in water used at a haemodialysis centre. Lett Appl Microbiol 46:542–547

Pitt JI, Hockering AD (1997) Fungi and food spoilage, 2nd edn. Blackie Academic & Professional, London

Pitt JI, Hocking AD (1977) Influence of solute and hydrogen-ion concentration on water relations of some xerophilic fungi. J Gen Microbiol 101:35–40

Plemenitaš A, Vaupotič T, Lenassi M, Kogej T, Gunde-Cimerman N (2008) Adaptation of extremely halotolerant black yeast *Hortaea werneckii* to increased osmolarity: a molecular perspective at a glance. Stud Mycol 61:67–75

Plemenitaš A, Lenassi M, Konte T, Kejžar A, Zajc J, Gostinčar C, Gunde-Cimerman N (2014) Adaptation to high salt concentrations in halotolerant/halophilic fungi: a molecular perspective. Front Microbiol 5:199

Price NPJ, Manitchotpisit P, Vermillion KE, Bowman MJ, Leathers TD (2013) Structural characterization of novel extracellular liamocins (mannitol oils) produced by *Aureobasidium pullulans* strain NRRL 50380. Carbohydr Res 370:24–32

Prista C, Almagro A, Loureiro-Dias MC, Ramos J (1997) Physiological basis for the high salt tolerance of *Debaryomyces hansenii*. Appl Environ Microbiol 63:4005–4009

Prista C, Loureiro-Dias MC, Montiel V, García R, Ramos J (2005) Mechanisms underlying the halotolerant way of *Debaryomyces hansenii*. FEMS Yeast Res 5:693–701

Ranta HM (1990) Effect of simulated acid rain on quantity of epiphytic microfungi on Scots pine (Pinus sylvestris L.) needles. Environ Pollut 67:349–359

Ranzoni FV (1968) Fungi isolated in culture from soils of the Sonoran desert. Mycologia 60:356–371

Schena L, Ippolito A, Zahavi T, Cohen L, Nigro F, Droby S (1999) Genetic diversity and biocontrol activity of *Aureobasidium pullulans* isolates against postharvest rots. Postharvest Biol Technol 17:189–199

Schneider J, Herrmann GA (1979) Saltworks – natural laboratories for microbiological and geochemical investigations during the evaporation of seawater. In: Coogan AH, Hauder L (eds), Proc Fifth Int Symp on Salt, Northern Ohio Geological Society, Cleveland, pp 371–381

Schoch CL, Crous PW, Groenewald JZ, Boehm EWA, Burgess TI, de Gruyter J, de Hoog GS, Dixon LJ, Grube M, Gueidan C, Harada Y, Hatakeyama S, Hirayama K, Hosoya T, Huhndorf SM, Hyde KD, Jones EBG, Kohlmeyer J, Kruys Å, Li YM, Lücking R, Lumbsch HT, Marvanová L, Mbatchou JS, McVay AH, Miller AN, Mugambi GK, Muggia L, Nelsen MP, Nelson P, Owensby CA, Phillips AJL, Phongpaichit S, Pointing SB, Pujade-Renaud V, Raja HA, Rivas Plata E, Robbertse B, Ruibal C, Sakayaroj J, Sano T, Selbmann L, Shearer CA, Shirouzu T, Slippers B, Suetrong S, Tanaka K, Volkmann-Kohlmeyer B, Wingfield MJ, Wood AR, Woudenberg JHC, Yonezawa H, Zhang Y, Spatafora JW (2009) A class-wide phylogenetic assessment of Dothideomycetes. Stud Mycol 64:1–15

Selbmann L, de Hoog GS, Mazzaglia A, Friedmann EI, Onofri S (2005) Fungi at the edge of life: cryptoendolithic black fungi from Antarctic desert. Stud Mycol 51:1–32

Shin KS, Park YH, Park DJ, Kim CJ (2005) *Cryptococcus taeanensis* sp. nov. a new anamorphic basidiomycetous yeast isolated from a salt farm. Int J Syst Evol Microbiol 55:1365–1368

Shiomi N, Yasuda T, Inoue Y, Kusumotoa N, Iwasaki S, Katsuda T, Katoh S (2004) Characteristics of neutralization of acids by newly isolated fungal cells. J Biosci Bioeng 97:54–58

Siegel BZ, McMurty G, Siegel SM, Chen J, Larock P (1979) Life in the calcium chloride environment of Don Juan Pond, Antarctica. Nature 280:828–829

Siegel B, Siegel S, Chen J, LaRock P (1983) The algal mat of Don Juan Pond, an extraterrestrial habitat on Earth. Adv Space Res 3:39–42

Silva S, Negri M, Henriques M, Oliveira R, Williams DW, Azeredo J (2012) *Candida glabrata, Candida parapsilosis* and *Candida tropicalis*: biology, epidemiology, pathogenicity and antifungal resistance. FEMS Microbiol Rev 36:288–305

Singh P, Raghukumar C, Verma P, Shouche Y (2013) Assessment of fungal diversity in deep-sea sediments by multiple primer approach. World J Microbiol Biotechnol 28:659–667

Sterflinger K, Piñar G (2013) Microbial deterioration of cultural heritage and works of art — tilting at windmills? Appl Microbiol Biotechnol 97:9637–9646

Sugiyama J, Sugiyama Y, Iizuka H, Torii T (1967) Report of the Japanese summer parties in dry valleys, Victoria Land, 1963–1965. IV. Mycological studies of the Antarctic fungi. Part. 2. Mycoflora of Lake Vanda, an ice-free lake. Antarctic Rec 28:23–32

Suzuki M, Prasad GS, Kurtzman CP (2011) *Debaryomyces* Lodder & Kreger-van Rij. In: Kurtzman CP, Kurtzman CP, Fell JW, Boekhout T (eds) The yeasts, a taxonomic study, 5th edn, vol 2. Elsevier, Amsterdam, pp 361–372

Takesako K, Ikai K, Haruna F, Endo M, Shimanaka K, Sono E, Nakamura T, Kato I, Yamaguchi H (1991) Aureobasidins, new antifungal antibiotics taxonomy, fermentation, isolation, and properties. J Antibiot 44:919–924

Tkavc R (2012) Microbial communities of the brine shrimp *Artemia* sp. and selected hypersaline microbial mats. Doctoral Dissertation, University of Ljubljana

Todaro F, Berdar A, Cavaliere A, Criseo G, Pernice L (1983) Gasophthalmus in Black-Sea bream (*Spondyliosoma cantharus*) caused by *Sarcinomyces crustaceus* Lindner. Mycopathologia 81:95–97

Tregoning GS, Kempher ML, Jung DO, Samarkin VA, Joye SB, Madigan MT (2015) A halophilic bacterium inhabiting the warm, $CaCl_2$-rich brine of the perennially ice-covered Lake Vanda, McMurdo Dry Valleys, Antarctica. Appl Environ Microbiol 81:1988–1995

Trüper HG, Galinski EA (1986) Concentrated brines as habitats for microorganisms. Experientia 42:1182–1187

Turk M, Plemenitaš A (2002) The HOG pathway in the halophilic black yeast *Hortaea werneckii*: isolation of the HOG1 homolog gene and activation of HwHog1p. FEMS Microbiol Lett 216:193–199

Turk M, Mejanelle L, Šentjurc M, Grimalt JO, Gunde-Cimerman N, Plemenitaš A (2004) Salt-induced changes in lipid composition and membrane fluidity of halophilic yeast-like melanized fungi. Extremophiles 8:53–61

Turk M, Abramović Z, Plemenitaš A, Gunde-Cimerman N (2007) Salt stress and plasma-membrane fluidity in selected extremophilic yeasts and yeast-like fungi. FEMS Yeast Res 7:550–557

van Uden N, Fell JW (1968) Marine yeasts. Adv Microbiol Sea 1:167–201

Vaupotič T, Plemenitaš A (2007a) Differential gene expression and HogI interaction with osmoresponsive genes in the extremely halotolerant black yeast *Hortaea werneckii*. BMC Genomics 8:280

Vaupotič T, Plemenitaš A (2007b) Differential gene expression and Hog1 interaction with osmoresponsive genes in the extremely halotolerant black yeast *Hortaea werneckii*. BMC Genomics 8:1–34

Vishniac HS (2006) A multivariate analysis of soil yeasts isolated from a latitudinal gradient. Microb Ecol 52:90–10

Wang QM, Groenewald M, Takashima M (2015a) Phylogeny of yeasts and related taxa within Pucciniomycotina determined from multigene sequence analyses. Stud Mycol 81:27–53

Wang QM, Begerow D, Groenewald M (2015b) Multigene phylogeny and taxonomic revision of yeasts and related fungi in the Ustilaginomycotina. Stud Mycol 81:54–80

Wheeler KA, Hocking AD, Pitt JI (1988) Influence of temperature on the water relations of *Polypaecilum pisce* and *Basipetospora halophila*, two halophilic fungi. J Gen Microbiol 134:2255–2260

Williams JP, Hallsworth JE (2009) Limits of life in hostile environments: no barriers to biosphere function? Environ Microbiol 11:3292–3308

Wollenzien U, de Hoog GS, Krumbein WE, Urzí C (1995) On the isolation of microcolonial fungi occurring on and in marble and other calcareous rocks. Sci Total Environ 167:287–294

Yang XX, Wartmann T, Stoltenburg R, Kunze G (2000) Halotolerance of the yeast *Arxula adeninivorans* LS3. A van Leeuwenhoek 77:303–311

Yuan Z, Druzhinina IS, Labbé J, Redman R, Qin Y, Rodriguez R, Zhang C, Tuskan GA, Lin F (2016) Specialized microbiome of a halophyte and its role in helping non-host plants to withstand salinity. Sci Rep 6:32467

Yurkov A, Inácio J, Chernov IY, Fonseca A (2015) Yeast biogeography and the effects of species recognition approaches: the case study of widespread basidiomycetous species from birch forests in Russia. Curr Microbiol 70:587–601

Zajc J, Zalar P, Plemenitaš A, Gunde-Cimerman N (2012) The mycobiota of the salterns. In: Raghukumar C (ed) Biology of marine fungi, Progress in molecular and subcellular biology, vol 53. Springer, Berlin, Heidelberg, pp 133–158

Zajc J, Liu Y, Dai W, Yang Z, Hu J, Gostinčar C, Gunde-Cimerman N (2013) Genome and transcriptome sequencing of the halophilic fungus *Wallemia ichthyophaga*: haloadaptations present and absent. BMC Genomics 14:617

Zajc J, Dzeroski S, Kocev D, Oren A, Sonjak S, Tkavc R, Gunde-Cimerman N (2014a) Chaophilic or chaotolerant fungi: a new category of extremophiles? Front Microbiol 5:708

Zajc J, Kogej T, Galinski EA, Ramos J, Gunde-Cimerman N (2014b) The osmoadaptation strategy of the most halophilic fungus *Wallemia ichthyophaga*, growing optimally at salinities above 15% NaCl. Appl Environ Microbiol 80:247–256

Zalar P, de Hoog GS, Gunde-Cimerman N (1999a) *Trimmatostroma salinum*, a new species from hypersaline water. Stud Mycol 43:57–62

Zalar P, de Hoog GS, Gunde-Cimerman N (1999b) Taxonomy of the endoconidial black yeast genera *Phaeotheca* and *Hyphospora*. Stud Mycol 43:49–56

Zalar P, de Hoog GS, Gunde-Cimerman N (1999c) Ecology of halotolerant dothideaceous black yeasts. Stud Mycol 43:38–48

Zalar P, Kocuvan MA, Plemenitaš A, Gunde-Cimerman N (2005a) Halophilic black yeasts colonize wood immersed in hypersaline water. Bot Mar 48:323–326

Zalar P, de Hoog GS, Schroers H-J, Frank JM, Gunde-Cimerman N (2005b) Taxonomy and phylogeny of the xerophilic genus *Wallemia* (Wallemiomycetes and Wallemiales, cl. et ord. nov.) A van Leeuwenhoek 87:311–328

Zalar P, de Hoog GS, Schroers H-J, Crous PW, Groenewald JZ, Gunde-Cimerman N (2007) Phylogeny and ecology of the ubiquitous saprobe *Cladosporium sphaerospermum*, with descriptions of seven new species from hypersaline environments. Stud Mycol 58:157–183

Zalar P, Frisvad JC, Gunde-Cimerman N, Varga J, Samson RA (2008a) Four new species of *Emericella* from the Mediterranean region of Europe. Mycologia 100:779–795

Zalar P, Gostinčar C, de Hoog GS, Uršič V, Sudhadham M, Gunde-Cimerman N (2008b) Redefinition of *Aureobasidium pullulans* and its varieties. Stud Mycol 61:21–38

Zhang Y, Cremer PS (2006) Interactions between macromolecules and ions: the Hofmeister series. Curr Opin Chem Biol 10:658–663

Zhang X-Y, Tang G-l, Xu X-Y, Nong X-H, Qi S-H (2014) Insights into deep-sea sediment fungal communities from the East Indian Ocean using targeted environmental sequencing combined with traditional cultivation. PLoS One 9:e109118

Zupančič J, Novak M, Zalar P, Gunde-Cimerman N (2016) The black yeast *Exophiala dermatitidis* and other selected opportunistic human fungal pathogens spread from dishwashers to kitchens. PLoS One 11:1–29

Chapter 11
Yeasts in Polar and Subpolar Habitats

Pietro Buzzini, Martina Turk, Laura Perini, Benedetta Turchetti, and Nina Gunde-Cimerman

Abstract The yeasts that thrive in polar and subpolar areas have to be adapted to extreme environments with low temperatures and the consequential desiccation due to freezing of water into ice crystals, with relatively high concentrations of ions, generally low levels of nutrients and, sometimes, high UV irradiation and hypoxia. Yeast communities in polar areas include circumpolar, endemic and cosmopolitan species. Although some endemic yeast species show psychrophilic behaviour, the majority of them are psychrotolerant yeasts that can adapt to growth across a wide range of temperatures. Most investigations on yeasts in polar and subpolar areas have remained limited to their biodiversity and the quantification of rare or new species. Comparative taxonomic studies of polar and subpolar habitats from Antarctic, Arctic and sub-Arctic have shown that the yeast communities belong prevalently to Basidiomycota, in contrast to the general fungal community distribution, which shows Ascomycota as the dominant phylum. The reviews on yeast diversity in cold habitats worldwide that have been published in recent years have reported the unambiguous prevalence of the former basidiomycetous genera *Cryptococcus* and *Rhodotorula*. But the recent taxonomic revision of the Pucciniomycotina and Tremellomycetes taxa positioned these polyphyletic genera into many new taxa, thus modifying the known taxonomical picture and ecological significance of the yeast distribution in polar and subpolar ecosystems. To overcome the problems associated with the quantification of unculturable microbial communities, the new high-throughput sequencing of both DNA and RNA is proving to be a valuable tool in deciphering the microbial diversity in cold environments.

Keywords *Aureobasidium* • *Candida* • *Cryptococcus* • *Rhodotorula*

P. Buzzini • B. Turchetti
Department of Agricultural, Food and Environmental Sciences, University of Perugia, Borgo XX Giugno 74, 06121 Perugia, Italy

M. Turk • L. Perini • N. Gunde-Cimerman (✉)
Department of Biology, Biotechnical Faculty, University of Ljubljana, Večna pot 111, 1000 Ljubljana, Slovenia
e-mail: nina.gunde-cimerman@bf.uni-lj.si

P. Buzzini et al. (eds.), *Yeasts in Natural Ecosystems: Diversity*,
DOI 10.1007/978-3-319-62683-3_11

Contents

11.1 Introduction

The terrestrial cryosphere includes more than 198,000 glaciers and 3 ice sheets (Pfeffer et al. 2014). Cryospheric environments were considered abiotic in the past (Byrd 1938), while they are now recognised as ecosystems with a major reservoir of microbial diversity. It is this reservoir that drives the fundamental dynamics of the cryosphere (Edwards et al. 2014), although it can be discharged annually in the ice meltwaters (Irvine-Fynn and Edwards 2014) at accelerating rates.

The polar areas include Antarctica (~14 million km^2) and the Arctic (~4 million km^2), which represent together over 95% of the Earth cryosphere and where almost all of the surface is covered by ice and snow (Holdgate 1977). The sub-Arctic polar region falls approximately between 50°N and 70°N in latitude and includes both the continental lands and islands that surround the Arctic Circle: Siberia, Scandinavia, northern Canada, Alaska and some large islands, namely, Novaya Zemlya (Russia), Spitsbergen (Norway), Iceland and Southern Greenland (Denmark) (Benn and Evans 1998; Nesje and Dahl 2000).

The yeasts that survive, and even thrive, in polar and subpolar areas have to be adapted to low temperatures and freezing of water into ice crystals and the consequential desiccation and low water availability (low water activity: a_w). Additionally they have to be adapted to relatively high concentrations of ions, generally low levels of nutrients and, sometimes, high UV irradiation and hypoxia. Studies conducted to date have shown that some yeast genera are particularly well adapted to such conditions. These yeasts use different strategies to survive and thrive in these cold environments, such as production of cold-active enzymes, 'anti-freeze' compounds and extracellular polymers. They also need to maintain the fluidity and plasticity of their cellular membranes down to subzero temperatures, which can be achieved by an increased proportion of unsaturated fatty acids and by a decreased mean fatty acid chain length (Siddiqui and Cavicchioli 2005; Turk et al. 2011; Buzzini et al. 2012; Gunde-Cimerman et al. 2014). Due to these properties, cold-adapted yeasts represent novel biotechnological targets and promising model organisms for the study of adaptations to life at low temperatures (Robinson 2001; Ruisi et al. 2007; Gostinčar et al. 2008; Simon et al. 2009; Pathan et al. 2010; Xiao et al. 2010; Turk et al. 2011; Buzzini and Margesin 2014; Kim et al. 2015; Alcaíno et al. 2015).

Early investigations into yeast diversity in polar areas date back to the early 1870s and were carried out in Antarctica (Bridge and Newsham 2009). The initial diversity studies were exclusively culture based and focused on the scarce vegetation, wooden remains and surface layers of the permafrost. These were followed some decades later by studies on Antarctic rock-inhabiting black yeast-like fungi (here referred as black yeasts, for simplicity), air-borne yeasts and yeasts in the polar soil and polar offshore sea waters (Jones 1976; Abyzov 1993; Abyzov et al. 2004; Onofri et al. 2004; Vishniac 2006a, b). However, it was not until the late 1960s and onwards that studies of yeasts were initiated for the water-based environments, such as sea water, glacial meltwater, snow and glacial, subglacial and lake ice (di Menna 1966a, b; Goto et al. 1969; Vishniac and Hempfling 1979; Baharaeen and Vishniac 1982; Vishniac 1985a, b; Baublis et al. 1991; Vishniac and Kurtzman 1992; Montes et al. 1999; Butinar et al. 2007; Connell et al. 2008, 2010). The yeast diversity in non-polar worldwide cold habitats is reported in Chap. 12 of this book.

Yeast species in polar areas include circumpolar, endemic and cosmopolitan species (Yurkov 2017). Although some endemic yeast species show psychrophilic behaviour (i.e. they only grow at low temperatures) (Margesin 2008; Buzzini et al. 2012; de Garcia et al. 2012), the majority are psychrotolerant (i.e. they grow at low temperatures but also at temperatures slightly higher than 20 °C) and can be found in both of the polar areas and even globally. Interestingly, it appears that certain yeasts from polar environments have discovered new indoor habitats, as inhabitants of freezers, cold-storage rooms and refrigerated or even-frozen food (Altunatmaz et al. 2012; NSF International 2013; Catellani et al. 2014; Maraz and Kovacs 2014).

The early studies were primarily based on cultivation and microscopy, and over the years, these were supplemented with studies of sequence marker genes (i.e. DNA barcodes) (Kurtzman et al. 2015) and more recently with metagenomic studies (Simon et al. 2009; Gittel et al. 2014; Lutz et al. 2016). Although next-generation sequencing approaches have become much cheaper, and have therefore been more frequently used over the last few years, high-throughput DNA sequencing datasets exist from <30 of the 198,000 global ice-covered areas (Edwards et al. 2013; Pfeffer et al. 2014). The majority of such datasets is still limited to prokaryotic diversity, and only a few also take into account the eukaryotic contributions to the microbial communities studied (Shtarkman et al. 2013; Maccario et al. 2014; Boetius et al. 2015; Antony et al. 2016; Lutz et al. 2016).

Thus, most investigations on yeasts in polar and subpolar areas have remained limited to their biodiversity and the quantification of rare or new species. The majority of these data can be retrieved as scattered reports (Gilichinsky et al. 2007; Frisvad 2008; Ludley and Robinson 2008; Libkind et al. 2009; Ozerskaya et al. 2008; Buzzini et al. 2012; Buzzini and Margesin 2014) or as information in public fungal databases, such as the Westerdijk Fungal Biodiversity Institute (formerly CBS Fungal Biodiversity Centre) and the American Type Culture Collection. These collections also serve as valuable sources of data on yeasts from cold regions, and they were investigated for information relevant for the present report.

Comparative taxonomic studies of both polar areas have shown that the dominant yeasts across both of these areas belong to Basidiomycota (Branda et al. 2010; Buzzini and Margesin 2014) and include a few dominant genera, which will be presented in more detail in the following paragraphs.

11.2 Yeasts in Polar and Subpolar Habitats

The yeast communities of the polar and subpolar habitats belong prevalently to Basidiomycota, in contrast to the general fungal community distribution, which shows Ascomycota as the dominant phylum (Timling et al. 2014). The diversity of the ascomycetous and basidiomycetous yeasts in the Antarctic, Arctic and sub-Arctic and across their different habitats and locations is reported in Tables 11.1 and 11.2.

11.2.1 Dominant Ascomycetous Yeast Genera in Polar and Subpolar Habitats

11.2.1.1 *Candida* spp. and Former *Candida* spp.

Different species of *Candida* have been described for the Antarctic and Arctic regions (Tables 11.1 and 11.2, respectively). *Candida parapsilosis*, *Diutina rugosa* (former *Candida rugosa*) and other not well-characterised *Candida* spp. have been described for both poles, while *Candida albicans*, *Candida intermedia* and *Candida mycetangii* were described only using non-culturable approaches (i.e. next-generation sequencing; Bridge and Newsham 2009; Kochkina et al. 2012; Bellemain et al. 2013; Davey et al. 2015; Rämä et al. 2016). *C. parapsilosis* is the most frequently found *Candida* sp. (which is sometimes regarded as cosmopolitan) in both the Antarctic and Arctic areas and in different habitats (Jacobs and Shafer 1964; Arenz et al. 2006; Butinar et al. 2011; Farrell et al. 2011; Bellemain et al. 2013; Shtarkman et al. 2013; Santiago et al. 2015; Kachalkin 2014).

11.2.1.2 *Debaryomyces* spp.

A few species of the genus *Debaryomyces* are known to be psychrotolerant and xero-/halotolerant (Maraz and Kovacs 2014; Zalar and Gunde-Cimerman 2014). The particular physiological features of *Debaryomyces hansenii* allow its colonisation of a wide range of niches and also of several environments with low a_w, including Antarctic and Arctic glacial habitats (Tables 11.1 and 11.2, respectively). As a consequence *Deb. hansenii* has been described for both poles in sea water, subglacial ice, permafrost, organic material (e.g. root-associated microbial

Table 11.1 Yeast species described for Antarctic habitats

Species	Original taxonomic designation	Samples									Locality										References
		S	Sn	Ic	Om	Cl	R	Pf	Sw	Hh	SV	NV	RD	AS	VI	EA	LH	QM	Ap	Mi	
Ascomycetous yeasts																					
Candida sp.		■			■											■			■		Arenz and Blanchette 2011; Gesheva et al. 2012; Duarte et al. 2013; Rovati et al. 2013; Santiago et al. 2015; Martinez et al. 2016
Candida albicans #								■			■										Kochkina et al. 2012
Candida glaebosa		■				■			■										■		Arenz and Blanchette 2011; Vaz et al. 2011; Duarte et al. 2013; Martinez et al. 2016
Candida intermedia #		■																	■		Bridge and Newsham 2009
Candida mesenterica		■																	■		Arenz and Blanchette 2011
Candida parapsilosis §			■	■	■					■	■		■	■	■				■		Jacobs and Shafer 1964; Arenz et al. 2006; Farrell et al. 2011; Shtarkman et al. 2013; Santiago et al. 2015
Candida psychrophila	*Torulopsis psychrophila*				■						■		■								Goto et al. 1969
Candida saitoana	*Torulopsis candida*	■																			di Menna 1966b
Candida sakè	*Candida australis*	■			■				■		■					■			■		Goto et al. 1969; Fell and Hunter 1974; Vaz et al. 2011; Carrasco et al. 2012; Duarte et al. 2013; Godinho et al. 2013; Duarte et al. 2016; Martinez et al. 2016
Candida spencermartinsiae									■										■		Vaz et al. 2011
Candida zeylanoides		■							■				■						■		Vaz et al. 2011; Connell and Staudigel 2013; Martinez et al. 2016
Clavispora lusitaniae	*Candida lusitaniae*	■									■		■								Baublis et al. 1991; Connell et al. 2008; Connell and Staudigel 2013
Debaryomyces sp.		■			■														■		Arenz et al. 2011; Arenz and Blanchette 2011
Debaryomyces hansenii		■			■		■		■		■	■	■				■	■	■		Soneda 1961; di Menna 1966b; Arenz et al. 2006; Connell et al. 2008; Pavlova et al. 2009; Arenz et al. 2011; Arenz and Blanchette 2011; Vaz et al. 2011; Duarte et al. 2013; Godinho et al. 2013; Rovati et al. 2013; Selbmann et al. 2014a; Santiago et al. 2015; Martinez et al. 2016
Debaryomyces macquariensis					■				■										■		Duarte et al. 2013; Martinez et al. 2016

(continued)

Table 11.1 (continued)

Debaryomyces subglobosus		■											■							■	di Menna 1966b; Ferrari et al. 2011
Dipodascus australiensis #					■								■								Arenz et al. 2006
Diutina rugosa	*Candida rugosa*	■											■								di Menna 1966b
Galactomyces candidus	*Geotrichum candidum*	■									■										Baublis et al. 1991
(*Kluyveromyces* sp.) #				■											■						Shtarkman et al. 2013
Kodamaea ohmeri						■													■		Vaz et al. 2011
Komagataella pastoris #	*Pichia pastoris*			■											■						Shtarkman et al. 2013
Metschnikowia sp.		■																	■		Carrasco et al. 2012
Metschnikowia australis		■			■				■		■								■		Vaz et al. 2011; Duarte et al. 2013; Godinho et al. 2013; Vaca et al. 2013; Duarte et al. 2016; Martinez et al. 2016
Meyerozyma caribbica					■														■		Godinho et al. 2013
Meyerozyma guilliermondii		■			■	■													■		Vaz et al. 2011; Duarte et al. 2013; Godinho et al. 2013
Pichia kluyveri #								■			■										Kochkina et al. 2012
Pichia kudriavzevii	*Candida krusei, Issatchenkia orientalis*	■				■					■								■		Baublis et al. 1991; Vaz et al. 2011
(*Saccharomyces* sp.) #				■											■						Shtarkman et al. 2013
Saccharomyces cerevisiae §		■		■									■		■						Shtarkman et al. 2013; Connell and Staudigel 2013
Taphrina sp. #		■									■										Dreesens et al. 2014
Taphrina antarctica	*Taphrina* sp.						■					■									Selbmann et al. 2014a, b
Trichomonascus ciferrii	*Candida ciferrii*	■									■										Baublis et al. 1991
Wickerhamomyces anomalus		■			■				■										■		Carrasco et al. 2012 ; Duarte et al. 2013
Yamadazyma mexicana					■														■		Godinho et al. 2013
(*Yarrowia* sp.) #				■	■										■						Shtarkman et al. 2013
Yarrowia lipolytica					■														■		Santiago et al. 2015
Basidiomycetous yeasts																					
Bannozyma arctica	*Rhodotorula arctica*				■														■		Duarte et al. 2016
Bannozyma yamatoana	*Bensingtonia yamatoana*	■			■														■		Vaz et al. 2011; Santiago et al. 2015
Bulleromyces albus	*Bullera alba*	■			■								■					■			Arenz et al. 2006

Colacogloea diffluens	*Candida diffluens*, *Rhodotorula diffluens*					■					■										Goto et al. 1969
Cryptococcus sp.		■			■		■					■	■						■		Arenz et al. 2011; Carrasco et al. 2012; Rovati et al. 2013; Selbmann et al. 2014a; Duarte et al. 2016
Cryptococcus neoformans #				■											■						Shtarkman et al. 2013
Cutaneotrichosporon moniliiforme	*Trichosporon moniliiforme*					■					■										Goto et al. 1969
(*Cystobasidium* sp.)	*Rhodotorula* sp.						■				■										Selbmann et al. 2014a
Cystobasidium laryngis	*Rhodotorula laryngis*	■			■		■		■		■		■						■		Arenz et al. 2006; Connell et al. 2008; Arenz et al. 2011; Arenz and Blanchette 2011; Carrasco et al. 2012; Duarte et al. 2013; Rovati et al. 2013; Santiago et al. 2015; Duarte et al. 2016; Martinez et al. 2016
Cystobasidium minutum	*Rhodotorula texensis*, *Rhodotorula minuta*	■			■	■					■		■						■		di Menna 1966b; Goto et al. 1969; Cameron 1971; Atlas et al. 1978; Pavlova et al. 2009; Martinez et al. 2016
Cystobasidium pinicola	*Rhodotorula pinicola*				■														■		Vaca et al. 2013
Cystobasidium pallidum	*Rhodotorula pallida*	■					■						■						■		di Menna 1966b; Rovati et al. 2013
Cystobasidium portillonensis	*Rhodotorula portillonensis*	■																	■		Laich et al. 2013
Cystobasidium slooffiae	*Rhodotorula slooffiae*	■			■								■								Arenz et al. 2011; Arenz and Blanchette 2011
Cystofilobasidium capitatum	Cystofilobasidium *lari-marini*				■														■		Duarte et al. 2013
Cystofilobasidium infirmominiatum				■	■				■						■				■		D'Elia et al. 2009; Duarte et al. 2013; Vaca et al. 2013; Martinez et al. 2016
Cystofilobasidium macerans	*Rhodotorula macerans*, *Cryptococcus macerans*	■											■								di Menna 1966b; Connell and Staudigel 2013
Dioszegia sp.		■																	■		Vaz et al. 2011; Carrasco et l. 2012; Rovati et al. 2013
Dioszegia xingshanensis						■													■		Duarte et al. 2016

(continued)

Table 11.1 (continued)

Dioszegia antarctica		■									■										Connell et al. 2008
Dioszegia athyri						■													■		Duarte et al. 2016
Dioszegia aurantiaca		■																	■		Vaz et al. 2011
Dioszegia crocea		■																	■		Vaz et al. 2011
Dioszegia cryoxerica		■									■										Connell et al. 2008
Dioszegia fristingensis		■																	■		Carrasco et al. 2012
Dioszegia hungarica §	*Cryptococcus hungaricus*, *Bullera armeniaca*	■			■								■						■		Arenz et al. 2006; Vaz et al. 2011; Rovati et al. 2013
Dioszegia rishiriensis #				■											■						Shtarkman et al. 2013
Dioszegia statzelliae	*Cryptococcus statzelliae*	■														■					Thomas-Hall et al. 2002
Fibulobasidium murrhardtense		■																	■		Arenz and Blanchette 2011
Filobasidium sp.		■																	■		Vaz et al. 2011
Filobasidium magnum	*Cryptococcus magnus*				■														■		Duarte et al. 2016
Filobasidium stepposum	*Cryptococcus stepposus*							■			■										Zucconi et al. 2012
Filobasidium wieringae #	*Cryptococcus wieringae*	■			■								■								Arenz et al. 2006; Connell and Staudigel 2013
(*Genolevuria* sp.)	*Cryptococcus* sp.				■														■		Duarte et al. 2016
Glaciozyma sp.		■									■					■					Turchetti et al. 2011
Glaciozyma antarctica	*Leucosporidium antarcticum*	■							■		■								■		Turkiewicz et al. 2003, 2005; Turchetti et al. 2011; Carrasco et al. 2012; Martinez et al. 2016
Glaciozyma litoralis	*Glaciozyma litorale*				■														■		Duarte et al. 2016
Glaciozyma martinii		■			■				■		■					■			■		Turchetti et al. 2011; Duarte et al. 2016
Glaciozyma watsonii		■							■		■		■			■					Turchetti et al. 2011; Connell and Staudigel 2013
Goffeauzyma gastrica	*Cryptococcus gastricus*	■																	■	■	Arenz and Blanchette 2011; Ferrari et al. 2011; Carrasco et al. 2012; Martinez et al. 2016
Goffeauzyma gilvescens	*Cryptococcus gilvescens*	■			■														■		Carrasco et al. 2012; Rovati et al. 2013; Santiago et al. 2015; Martinez et al. 2016
Hannaella luteola	*Cryptococcus luteolus*	■											■								di Menna 1966b; Atlas et al. 1978
Holtermanniella festucosa					■														■		Duarte et al. 2016
Holtermanniella nyarrowii	*Cryptococcus nyarrowii*	■	■								■					■					Thomas-Hall and Watson 2002; Connell et al. 2008

Holtermanniella nyarrowii					■														■		Duarte et al. 2016
Holtermanniella wattica	*Cryptococcus watticus*	■	■		■											■			■		Guffogg et al. 2004; Carrasco et l. 2012; Santiago et al. 2015; Duarte et al. 2016
Leucosporidium sp. §				■		■									■				■		Shtarkman et al. 2013; de Garcia et al. 2015
Leucosporidium sp.	*Leucosporidiella* sp.				■														■		Vaca et al. 2013
Leucosporidium creatinivorum	*Rhodotorula creatinivora*, *Leucosporidiella creatinivora*	■			■	■	■												■		Vaz et al., 2011; Rovati et al., 2013; Carrasco et al. 2012; Vaca et al., 2013; Santiago et al., 2015
Leucosporidium escuderoi					■														■		Laich et al., 2014
Leucosporidium fragarium	*Leucosporidiella fragaria*	■			■														■		Vaz et al. 2011; Carrasco et l. 2012; Duarte et al. 2016; Martinez et al. 2016
Leucosporidium muscorum	*Rhodotorula muscorum*, *Leucosporidiella muscorum*	■			■														■		Vaz et al. 2011; Duarte et al. 2016; Martinez et al. 2016
Leucosporidium yakuticum	*Rhodotorula yakutica*, *Leucosporidiella yakutica*	■									■										Arenz and Blanchette 2011; de Garcia et al. 2015
Leucosporidum scottii	*Candida scottii*	■			■	■			■		■		■						■		di Menna 1966b; Goto et al. 1969; Atlas et al. 1978; Vaz et al. 2011; Duarte et al. 2013; de Gracia et al. 2015
Malassezia sp.		■											■								Connell and Staudigel 2013
Malassezia globosa		■											■								Connell and Staudigel 2013
Malassezia restricta §		■			■								■								Arenz et al. 2006; Bridge and Newsham 2009; Connell and Staudigel 2013
Moesziomyces antarcticus	*Candida antarctica*, *Pseudozyma antarctica*	■				■					■										Goto et al. 1969; Atlas et al. 1978
Mrakia sp.		■			■														■		Carrasco et al. 2012; Rovati et al. 2013; Duarte et al. 2016;
Mrakia aquatica	*Cryptococcus aquaticus*, *Mrakiella aquatica*				■														■		Santiago et al. 2015; Duarte et al. 2016
Mrakia blollopis		■														■			■		Thomas-Hall et al. 2010; Carrasco et l. 2012; Tsuji et al. 2015; Tsuji 2016

(continued)

Table 11.1 (continued)

Mrakia frigida #		■			■				■				■						■		di Menna 1966a,b; Krishnan et al. 2011; Santiago et al. 2015; Martinez et al. 2016
Mrakia gelida	*Mrakia stokesii*	■				■					■		■			■			■		di Menna 1966a, b; Connell et al. 2008; Tsuji et al. 2015
Mrakia niccombsii	*Mrakiella niccombsii*, *Cryptococcus niccombsii*	■														■					Thomas-Hall et al. 2010
Mrakia psychrophila		■																	■		Xin and Zhou 2007; Carrasco et al. 2012
Mrakia robertii		■														■			■		Thomas-Hall et al. 2010; Carrasco et l. 2012; Tsuji et al. 2015
Naganishia adeliensis	*Cryptococcus adeliensis*	■			■							■							■		Petrescu et al. 2000; Scorzetti et al. 2000; Duarte et al. 2013
Naganishia albida	*Cryptococcus albidus*	■	■		■	■					■		■		■		■		■		Soneda 1961; di Menna 1966b; Goto et al. 1969; Cameron 1971; Atlas et al. 1978; Baublis et al. 1991; Abyzov 1993; Pavlova et al. 2002
Naganishia albidosimilis §	*Cryptococcus albidosimilis*	■			■						■		■						■		Vishniac and Kurtzmann 1992; Arenz et al. 2006; Connell et al. 2008; Duarte et al. 2013; Martinez et al. 2016
Naganishia antarctica	*Cryptococcus antarcticus*	■			■					■	■		■						■		Vishniac and Kurtzmann 1992; Vishniac and Onofri 2002; Arenz et al. 2006; Farrell et al. 2011; Vaz et al. 2011; Duarte et al. 2016
Naganishia diffluens	*Cryptococcus diffluens*	■											■								di Menna 1966b; Arenz and Blanchette 2011
Naganishia friedmannii	*Cryptococcus friedmannii*				■		■			■	■	■	■						■		Vishniac 1985a; Arenz et al. 2006; Farrell et al. 2011; Selbmann et al. 2014a; Santiago et al. 2015
Naganishia liquefaciens	*Cryptococcus liquefaciens*	■											■								Arenz and Blanchette 2011
Naganishia vaughanmartiniae	*Cryptococcus vaughanmartiniae*	■										■							■		Turchetti et al. 2015
Naganishia vishniacii §	*Cryptococcus socialis*, *Cryptococcus vishniacii*	■			■		■			■	■	■	■								Vishniac and Hempfling 1979; Baharaeen and Vishniac 1982; Vishniac 1985b; Arenz et al. 2006; Connell et al. 2008; Farrell et al. 2011; Dreesens et al. 2014; Selbmann et al. 2014a
Papiliotrema laurentii §	*Cryptococcus laurentii*	■			■						■		■				■		■		Soneda 1961; di Menna 1966b; Vishniac and Hempfling 1979; Arenz et al. 2006; Pavlova et al. 2009; Duarte et al. 2013; Godinho et al. 2013; Santiago et al. 2015

Papiliotrema pseudoalba	*Bullera pseudoalba*				■														■		Duarte et al. 2013
Phaeotremella skinneri #	*Cryptococcus skinneri*				■								■								Arenz et al. 2006
Phenoliferia glacialis	*Rhodotorula glacialis*	■																	■		Vaz et al. 2011; Carrasco et al. 2012; Duarte et al. 2013; Tsuji et al. 2013; Duarte et al. 2016
Phenoliferia psychrophenolica	*Rhodotorula psychrophenolica*								■										■		Martinez et al. 2016
Piskurozyma fildesensis	*Cryptococcus fildesensis*				■														■		Zhang et al. 2014; Duarte et al. 2016
Pseudozyma sp.					■														■		Duarte et al. 2016
Pseudozyma tsukubaensis					■														■		Duarte et al. 2016
Rhodotorula sp. §		■		■	■			■			■				■				■		Arenz et al. 2011; Arenz and Blanchette 2011; Rovati et al. 2013; Shtarkman et al. 2013; Duarte et al. 2016; Goordial et al. 2016
Rhodotorula glutinis		■		■		■					■		■						■		di Menna 1966b; Goto et al. 1969; Abyzov 1993; Rovati et al. 2013
Rhodotorula graminis		■									■		■								di Menna 1966b; Cameron 1971
Rhodotorula kratochvilovae	*Rhodosporidium kratochvilovae*	■									■										Connell et al. 2008
Rhodotorula mucillaginosa §	*Rhodotorula rubra*	■		■	■	■			■		■		■		■		■	■	■		di Menna 1966b; Goto et al. 1969; Cameron 1971; Atlas et al. 1978; Baublis et al. 1991; Arenz et al. 2006; Connell et al. 2008; D'Elia et al. 2009; Vaz et al. 2011; Connell and Staudigel 2013; Duarte et al. 2013; Godinho et al. 2013; Knowlton et al. 2013; Rovati et al. 2013; Santiago et al. 2015; Martinez et al. 2016
Rhodotorula sphaerocarpa	*Rhodosporidium sphaerocarpum*								■											■	Newell and Fell 1970
Sakaguchia dacryoidea #				■											■						Shtarkman et al. 2013
Sakaguchia lamellibrachiae #	*Rhodotorula lamellibrachiae*			■											■						Shtarkman et al. 2013
Saitozyma flava	*Cryptococcus flavus*				■														■		Pavlova et al. 2009
Sampaiozyma ingeniosa	*Candida ingeniosa*, *Rhodotorula ingeniosa*	■																■			
Solicoccozyma aeria	*Cryptococcus aerius*	■																	■		Martinez et al. 2016

(continued)

Table 11.1 (continued)

Solicoccozyma keelungensis	*Cryptococcus keelungensis*	■																	■		Martinez et al. 2016
Solicoccozyma terricola	*Cryptococcus terricola*, *Cryptococcus terricolus*	■			■														■	■	Ferrari et al. 2011; Santiago et al. 2015
Sporidiobolus metaroseus	*Sporobolomyces roseus*	■					■	■			■								■		Vishniac and Hempfling 1979; Zucconi et al. 2012; Barahona et al. 2016
Sporidiobolus pararoseus					■														■		Duarte et al. 2016
Sporidiobolus salmonicolor	*Sporobolomyces salmonicolor*	■			■	■	■						■						■		Atlas et al. 1978; Arenz et al. 2006; Poli et al. 2010; Vaz et al. 2011; Carrasco et al. 2012; Barahona et al. 2016
Sporobolomyces sp.		■																	■		Rovati et al. 2013
Sporobolomyces johnsonii	*Sporobolomyces holsaticus*, *Sporidiobolus johnsonii*	■									■										Vishniac and Hempfling 1979
Sporobolomyces lactosus		■																	■		Arenz and Blanchette 2011
Sporobolomyces ruberrimus		■																		■	Ferrari et al. 2011
Symmetrospora marina	*Rhodotorula marina*	■			■								■						■		di Menna 1966b; Duarte et al. 2016
Symmetrospora symmetrica	*Sporobolomyces symmetricus*				■								■								Arenz et al. 2006
Tausonia pullulans	*Trichosporon pullulans*, *Guehomyces pullulans*	■			■	■							■						■		di Menna 1966b; Loperena et al. 2012; Duarte et al. 2013; Martinez et al. 2016
Tremella indecorata		■			■														■		Arenz and Blanchette 2011; Duarte et al. 2013
Trichosporon asteroides		■																	■		Goncalves et al. 2015
Vanrija humicola	*Cryptococcus humicola*, *Candida humicola*	■				■					■							■			Goto et al. 1969; Ray et al. 1992
Vanrija pseudolonga #	*Cryptococcus pseudolongus*							■			■										Kochkina et al. 2012
(*Vishniacozyma* sp.)					■														■		Godinho et al. 2013
Vishniacozyma carnescens §	*Cryptococcus carnescens*	■			■						■		■						■		Arenz et al. 2006; Connell et al. 2008; Duarte et al. 2016

Vishniacozyma foliicola	*Cryptococcus foliicola*				■														■		Arenz et al. 2011; Santiago et al. 2015
Vishniacozyma tephrensis §	*Cryptococcus tephrensis*				■				■				■						■		Arenz et al. 2006; Martinez et al. 2016
Vishniacozyma victoriae §	*Cryptococcus victoriae*	■			■	■			■		■		■			■			■		Montes et al. 1999; Thomas-Hall et al. 2002; Arenz et al. 2006; Arenz et al. 2011; Arenz and Blanchette 2011; Vaz et al. 2011; Carrasco et al. 2012; Loperena et al. 2012; Duarte et al. 2013; Rovati et al. 2013; Santiago et al. 2015; Duarte et al. 2016; Martinez et al. 2016
Yunzhangia auriculariae	*Candida auriculariae, Rhodotorula auriculariae*	■																■			
Yeast-like organisms																					
Aureobasidium sp.		■											■						■		Krishnan et al. 2011; Connell and Staudigel 2013
Aureobasidium pullulans		■			■	■		■			■		■		■				■		Atlas et al. 1978; Baublis et al. 1991; D'Elia et al. 2009; Vaz et al. 2011; Kochkina et al. 2012; Zucconi et al. 2012; Connell and Staudigel 2013; Henriquez et al. 2014
Exophiala sp.		■						■			■		■						■	■	Arenz and Blanchette 2011; Ferrari et al. 2011; Kochkina et al. 2012; Rovati et al. 2013
Exophiala equina #		■									■										Dreesens et al. 2014
Exophiala xenobiotica						■													■		Vaz et al. 2011

Genera in brackets, arbitrary new genus/species nomenclature designation based on recent literature for that strains/operational taxonomic units for which a clear phylogenetic position was defined in the original report

S soil; *Sn* snow; *Ic* ice cores; *Om* organic material; *Cl* cold lakes; *R* rock; *Pf* permafrost; *Sw* sea water; *Hh* historic huts; *SV* South Victoria Land; *NV* North Victoria Land; *RD* Ross Dependency; *AS* Amundsen-Scott Station; *Vl* Vostok lake; *EA* East Antarctica; *LH* Lutzow-Holm Bay; *QM* Queen Maud Land; *Ap* Antarctic peninsula; *Mi* Macquarie Island

[a]Species described using culture-independent approaches

[b]Species described using both culture-dependent and culture-independent approaches

Table 11.2 Yeast species described for Arctic and sub-Arctic habitats

Species	Original taxonomic designation	Sample								Locality						References
		Mw	S	Sw	Ic	Pf	Ch	R	Om	Sib	Sa	Sc	NA	Il	Gl	
Ascomycetous yeasts																
Candida sp. #									■		■	■				Davey et al. 2015; Rämä et al. 2016
Candida mycetangii #						■				■						Bellemain et al. 2013
Candida parapsilosis §		■				■			■	■	■	■				Butinar et al. 2011; Bellemain et al. 2013 ; Kachalkin 2014
Candida sakè	*Candida australis*		■									■				Kachalkin 2014
Debaryomyces sp.		■		■	■				■		■	■				Jacques et al. 2015; Rämä et al. 2016
Debaryomyces hansenii §	*Candida famata*; *Debaryomyces kloeckerii*; *Torulopsis famata*			■	■	■				■	■					Gilichinsky et al. 2005; Butinar et al. 2011; Bellemain et al. 2013; Jacques et al. 2015
Debaryomyces maramus				■							■					Butinar et al., 2011
Diutina rugosa	*Candida rugosa*		■												■	di Menna 1966b
Kluyveromyces sp. #									■			■				Rämä et al. 2016
Komagataella pastoris #									■		■					Zhang et al. 2015a, Zhang and Yao 2015
Metschnikowia sp. #									■		■					Zhang et al. 2015a, Zhang and Yao 2015
Metschnikowia bicuspidata				■					■		■	■				Butinar et al. 2011; Kachalkin 2014
Metschnikowia zobellii		■							■		■	■				Butinar et al. 2011; Kachalkin 2014
Meyerozyma guilliermondii §				■	■	■			■	■	■	■				Gilichinsky et al. 2005; Butinar et al. 2011; Kachalkin 2014; Davey et al. 2015
Pichia sp. #									■			■				Rämä et al. 2016
(*Pichia* sp.) #				■					■		■					Zhang et al. 2015a, b; Zhang and Yao 2015
Protomyces inouyei					■						■					Butinar et al. 2011
Torulaspora delbrueckii #						■				■						Bellemain et al. 2013
Basidiomycetous yeasts																
(*Apiotrichum* sp.) #									■		■					Zhang et al. 2015a
Apiotrichum porosum #				■							■					Zhang et al. 2015b
Bannozyma arctica	*Rhodotorula arctica*		■							■						Vishniac 1996; Vishniac and Takashima 2010
Bensingtonia sp. #			■										■			Timling et al. 2014

Bullera sp.			■										■			Geml et al. 2015
Bulleribasidium sp. #			■										■			Timling et al. 2014
Bulleromyces albus	*Bullera alba*				■						■					Butinar et al. 2007
Cryptococcus sp. §			■						■		■	■	■			Blaalid et al. 2014; Kachalkin 2014; Timling et al. 2014; Zhang et al. 2015a; Rämä et al. 2016
Curvibasidium cygneicollum #				■							■					Zhang et al. 2015b
Cutaneotrichosporon mucoides	*Trichosporon mucoides*				■						■					Butinar et al. 2007
Cystobasidium sp.			■						■			■				Kachalkin 2014
Cystobasidium laryngis	*Rhodotorula laryngis*				■						■					Butinar et al. 2007
Cystobasidium minutum §	*Rhodotorula texensis*, *Rhodotorula minuta*		■		■	■				■	■				■	di Menna 1966b; Butinar et al. 2007; Bellemain et al. 2013
Cystofilobasidium capitatum	*Cystofilobasidium lari-marini*		■						■			■		■		Birgisson et al. 2003; Kachalkin 2014
Cystofilobasidium macerans	*Rhodotorula macerans*, *Cryptococcus macerans*		■		■						■			■		Birgisson et al. 2003; Butinar et al. 2007
Dioszegia sp. #			■						■		■		■			Timling et al. 2014; Zhang et al. 2015a
Dioszegia fristingensis #									■		■					Davey et al. 2015; Zhang et al. 2015a; Zhang and Yao 2015
Dioszegia hungarica	*Cryptococcus hungaricus*, *Bullera armeniaca*				■								■			Uetake et al. 2012
(*Fibulobasidium* sp.)	*Cryptococcus* sp.							■							■	Selbmann et al. 2014a
Filobasidium magnum	*Cryptococcus magnus*				■				■		■	■			■	Butinar et al. 2007; Knowlton et al. 2013; Kachalkin 2014
Filobasidium oeirense	*Cryptococcus oeirensis*			■	■						■	■				Butinar et al. 2007; Kachalkin 2014
Filobasidium uniguttulatum §	*Cryptococcus uniguttulatus*				■						■				■	Ma et al. 1999; Butinar et al. 2007
Glaciozyma litoralis	*Glaciozyma litorale*		■									■				Kachalkin 2014
Glaciozyma watsonii							■				■					Edwards et al. 2013
Goffeauzyma gastrica	*Cryptococcus gastricus*	■									■					Pathan et al. 2010
Goffeauzyma gilvescens	*Cryptococcus gilvescens*		■		■			■			■	■			■	Butinar et al. 2007; Selbmann et al. 2014a; Coleine et al. 2015; Tsuji et al. 2016
Holtermanniella sp.#									■			■				Rämä et al. 2016
Holtermanniella festucosa			■						■			■				Kachalkin 2014

(continued)

Table 11.2 (continued)

		1	2	3	4	5	6	7	8	9	10	11	12	13	14	
Kurtzmanomyces sp.#			■										■			Timling et al. 2014
(*Leucosporidium* sp.) #									■		■					Blaalid et al. 2014
Leucosporidium creatinivorum	*Rhodotorula creatinivora*, *Leucosporidiella creatinivora*		■			■				■						Golubev 1998; de Garcia et al. 2015
Leucosporidium drummii			■									■				Kachalkin 2014
Leucosporidium fragarium	*Leucosporidiella fragaria*				■				■			■				Butinar et al. 2007; Kachalkin 2014
Leucosporidium muscorum §	*Rhodotorula muscorum*, *Leucosporidiella muscorum*	■							■		■					Pathan et al. 2010; Zhang et al. 2015a, b
Leucosporidium yakuticum	*Rhodotorula yakutica*, *Leucosporidiella yakutica*					■				■						Golubev 1998
Leucosporidum scottii	*Candida scottii*		■	■					■			■			■	di Menna 1966b; de Garcia et al. 2015; Kachalkin 2014
Lipomyces sp. #									■			■				Rämä et al. 2016
Malassezia sp. #			■	■					■		■		■			Timling et al. 2014; Zhang et al. 2015a, b; Rämä et al. 2016
Malassezia globosa						■				■						Bellemain et al. 2013
Malassezia restricta #			■			■			■	■	■		■			Bellemain et al. 2013; Blaalid et al. 2014; Geml et al. 2015; Zhang et al. 2015a, b
Mrakia sp. §			■		■				■		■	■	■			Uetake et al. 2012; Blaalid et al. 2014; Botnen et al. 2014; Timling et al. 2014; Zhang et al. 2015a, b; Rämä et al. 2016
Mrakia aquatica §	*Cryptococcus aquaticus*, *Mrakiella aquatica*		■			■			■	■	■			■		Birgisson et al. 2003; Bellemain et al. 2013; Blaalid et al. 2014
Mrakia blollopis §				■			■		■		■					Edwards et al. 2013; Zhang et al. 2015a, b; Singh et al. 2016
Mrakia cryoconiti	*Mrakiella cryoconiti*		■							■						Gounot 1986; Margesin and Fell 2008
Mrakia frigida									■			■				Kachalkin 2014
Mrakia gelida	*Candida gelida*; *Mrakia stokesii*	■	■								■				■	di Menna 1966b; Pathan et al. 2010; Tsuji et al. 2016
Mrakia niccombsii #	*Mrakiella niccombsii*, *Cryptococcus niccombsii*								■		■					Zhang et al. 2015a
Mrakia psychrophila		■	■								■					Pathan et al. 2010; Tsuji et al. 2016
Mrakia robertii			■				■				■					Edwards et al. 2013; Tsuji et al. 2016
Mrakiella sp.#			■										■			Timling et al. 2014

Naganishia adeliensis	*Cryptococcus adeliensis*, *Cryptococcus adeliae*		■		■						■	■				Butinar et al. 2007; Kachalkin 2014
Naganishia albida §	*Cryptococcus albidus*		■		■	■				■	■				■	di Menna 1966b; Ma et al. 1999; Butinar et al. 2007; Bellemain et al. 2013
Naganishia albidosimilis	*Cryptococcus albidosimilis*				■						■					Butinar et al. 2007
Naganishia diffluens	*Cryptococcus diffluens*		■												■	di Menna 1966b
Naganishia globosa	*Cryptococcus saitoi*				■						■					Butinar et al. 2007
Naganishia liquefaciens	*Cryptococcus liquefaciens*				■						■					Butinar et al. 2007, Butinar et al. 2011
(*Papiliotrema* sp.)	*Crypococcus* sp.				■						■					de Garcia et al. 2012
Papiliotrema fonsecae	*Cryptococcus fonsecae*				■							■				de Garcia et al. 2012
Papiliotrema laurentii	*Cryptococcus laurentii*		■	■	■	■				■	■	■			■	di Menna 1966b; Gilichinsky et al. 2005; Butinar et al. 2007; Kachalkin 2014
Phenoliferia glacialis	*Rhodotorula glacialis*	■	■		■						■		■			Pathan et al. 2010; Uetake et al. 2012
Phenoliferia psychrophenolica	*Rhodotorula psychrophenolica*		■		■		■				■		■			Uetake et al. 2012; Edwards et al. 2013
Rhodotorula sp. §			■				■		■		■	■	■			Edwards et al. 2013; Timling et al. 2014; Coleine et al. 2015; Geml et al. 2015; Rämä et al. 2016
Rhodotorula diobovata	*Rhodosporidium diobovatum*				■						■					Butinar et al. 2007
Rhodotorula glutinis§			■	■					■			■			■	di Menna 1966b; Kachalkin 2014; Zhang et al. 2015a, b
Rhodotorula mucillaginosa §	*Rhodotorula rubra*				■				■		■				■	Ma et al. 1999; Butinar et al. 2007, Butinar et al. 2011; Knowlton et al. 2013; Singh et al. 2016
Rhodotorula svalbardensis							■				■					Singh et al. 2014
(*Sakaguchia* sp.) #									■		■					Blaalid et al. 2014; Davey et al. 2015; Zhang et al. 2015a
Solicoccozyma terricola §	*Cryptococcus terricola*, *Cryptococcus terricolus*	■							■		■					Pathan et al. 2010; Davey et al. 2015
Sporidiobolus metaroseus	*Sporobolomyces roseus*								■			■				Kachalkin 2014
Sporobolomyces sp.#			■										■			Timling et al. 2014
Tausonia pullulans §	*Trichosporon pullulans*, *Guehomyces pullulans*		■										■		■	di Menna 1966b; Semenova et al. 2016
Tremella sp. #			■						■				■			Timling et al. 2014; Rämä et al. 2016

(continued)

Table 11.2 (continued)

Vanrija pseudolonga #	*Cryptococcus pseudolongus*					■				■						Bellemain et al. 2013
(*Vishniacozyma* sp.) #									■		■					Zhang et al. 2015a, b
Vishniacozyma carnescens	*Cryptococcus carnescens*				■						■					Butinar et al. 2007; de Garcia et al. 2012
Vishniacozyma tephrensis		■							■		■	■				de Garcia et al. 2012; Kachalkin 2014
Vishniacozyma heimaeyensis	*Cryptococcus heimaeyensis*		■											■		Vishniac 2002
Vishniacozyma psychrotolerans	*Cryptococcus psychrotolerans*				■						■					de Garcia et al. 2012
Vishniacozyma tephrensis §	*Cryptococcus tephrensis*		■						■		■			■		Vishniac 2002; Zhang et al. 2015a, b
Vishniacozyma victoriae §	*Cryptococcus victoriae*	■	■	■	■	■			■	■	■	■	■			Gilichinsky et al. 2005; Butinar et al. 2007; de Garcia et al. 2012; Blaalid et al. 2014; Kachalkin 2014; Davey et al. 2015; Zhang et al. 2015a, b; Tsuji et al. 2016
Yeast-like organisms																
Aureobasidium pullulans §			■	■	■	■			■	■			■		■	Gilichinsky et al. 2005; Osono et al. 2012; Knowlton et al. 2013; Kachalkin 2014; Geml et al. 2015
Exophiala sp. #			■						■		■		■			Blaalid et al. 2014; Timling et al. 2014; Zhang et al. 2015a; Rämä et al. 2016

Genera in brackets, arbitrary new genus/species nomenclature designation based on recent literature for that strains/operational taxonomic units for which a clear phylogenetic position was defined in the original report

Mw meltwater; *S* soil; *Sw* sea water; *Ic* ice cores; *Pf* permafrost; *Ch* cryoconite holes; *R* rock; *Om* organic material; *Sib* Siberia; *Sa* Svalbard archipelago; *Sc* Scandinavia; *NA* North America; *Il* Iceland; *Gl* Greenland

[a]Species described using culture-independent approaches

[b]Species described using both culture-dependent and culture-independent approaches

communities, lichens) and rock. The genome of *Deb. hansenii* has been fully sequenced and was annotated by the French Consortium Genolevures (Dujon et al. 2004), and it is available at http://cbi.labri.fr/Genolevures.

Other *Debaryomyces* species have also been described for these regions but with lower frequencies. These have included *Debaryomyces maramus* (Svalbard Islands only; Butinar et al. 2011), *Debaryomyces macquariensis* (King George Island, Antarctica; Duarte et al. 2013; Martinez et al. 2016) and *Debaryomyces subglobosus* (Ross Dependency and Macquarie Island, Antarctica; di Menna 1966b; Ferrari et al. 2011) (Tables 11.1 and 11.2).

11.2.1.3 *Pichia* spp. and Former *Pichia* spp.

The genus *Pichia* is widely distributed in nature, and it has been isolated across different natural habitats (Kurtzman 2011). Focusing only on the polar regions, and taking into consideration the taxonomic rearrangement of this genus over the last 5 years (Kurtzman 2011), species of the genus *Pichia* have only been described for Antarctica. These include, in particular, *Pichia kluyveri* in permafrost (Kochkina et al. 2012) and *Pichia kudriavzevii* in soil (Baublis et al. 1991) and cold water (Vaz et al. 2011). In contrast, both culturable strains and non-culturable operational taxonomic units associated to *Meyerozyma guilliermondii* and *Komagataella pastoris* (former *Pichia guilliermondii* and *Pichia pastoris*, respectively) have been described for Antarctic and Arctic habitats (Tables 11.1 and 11.2, respectively). These species are widely distributed in nature, and isolates have been obtained from many different substrates, which illustrate its worldwide distribution (Kurtzman 2011). In 2005, through the work of the Fungal Genome Initiative Nucleotide, the genome of *Mey. guilliermondii* became publicly available and can be found at http://www.broad.mit.edu.

11.2.1.4 *Metschnikowia* spp.

The species *Metschnikowia bicuspidata* and *Metschnikowia zobellii* have been described only for Svalbard sea water and puddles on snow/ice (Butinar et al. 2011), and they have been characterised according to their psychrotolerant and xero-/halotolerant aptitudes. In contrast, *Metschnikowia australis* was only described for Antarctica, as a result of different research campaigns (Tables 11.1 and 11.2). Thus, *M. australis* was described for sea water (Vaz et al. 2011), lake water and soil (Martinez et al. 2016), marine sediments (Vaz et al. 2011; Duarte et al. 2013) and in association with algae (Godinho et al. 2013; Duarte et al. 2016), marine sponges (Vaca et al. 2013) and other marine organisms (Duarte et al. 2013). Again, as this is only related to Antarctica, *M. australis* can at present be considered endemic. *M. australis* is also a psychrotolerant species, with optimal growth at 18 °C and extracellular enzymatic activities at low temperatures (Vaz et al. 2011; Duarte et al. 2013; Martinez et al. 2016).

11.2.1.5 *Taphrina* spp.

The genus *Taphrina* includes ascomycetous species known for their ability to cause specific infections in different vascular plants. Members of this genus show a dimorphic lifestyle: while the sexual filamentous state lives exclusively as a biotrophic plant pathogen, the asexual state is capable of saprobic growth, including laboratory culture media (Fonseca and Rodrigues 2011; Mittelbach and Vannette 2017).

Selbmann et al. (2014a, b) were the first to isolate two strains of *Taphrina* from Antarctica (where there are no vascular plants), and they also described them as new obligate psychrophilic species *Taphrina antarctica*. Likewise, Dreesens et al. (2014) reported operational taxonomic units classified as *Taphrina* spp. in the McMurdo Dry Valleys of Antarctica (Table 11.2). The presence of this genus, which is commonly considered an obligate plant parasite, suggests that *Taphrina* can successfully invade and exploit diverse environments and survive in its anamorphic state while acting as a commensal host inside microbial consortia (Fonseca and Rodrigues 2011).

11.2.2 Dominant Basidiomycetous Yeast Genera in Polar and Subpolar Habitats

The data reported in Tables 11.1 and 11.2 show the prevalence of yeasts in polar and subpolar habitats that belong to the Basidiomycota phylum. The reviews on yeast diversity in cold habitats worldwide that have been published in recent years have reported the unambiguous prevalence of the polyphyletic phenotypic basidiomycetous genera *Cryptococcus* and *Rhodotorula* (Buzzini et al. 2012; Connell et al. 2014; Zalar and Gunde-Cimerman 2014). Speculation thus indicated more efficient adaptation of these two genera to the selective pressures that are typical of glacial ecosystems (e.g. formation of a polysaccharide capsule; Vishniac 2006b; Shivaji and Prasad 2009). However, the recent taxonomic revision of the Pucciniomycotina and Tremellomycetes taxa (Liu et al. 2015; Wang et al. 2015) positioned these two polyphyletic genera into many new taxa, thus modifying the known taxonomical picture and ecological significance of the yeast distribution in polar and subpolar ecosystems.

11.2.2.1 *Cryptococcus* spp. and Former *Cryptococcus* spp.

In the last decades of the twentieth century, the former genus *Cryptococcus* was shown to be one of the most represented yeasts in cold habitats, and a few novel species of *Cryptococcus* with psychrotolerant aptitudes have been described in the last few years, including *Naganishia vaughanmartiniae* (*Cryptococcus vaughanmartiniae*) (Turchetti et al. 2015) and *Piskurozyma fildesensis* (*Cryptococcus fildesensis*) (Zhang et al. 2014). The taxonomic reorganisation of the genus implicated the decline of *Cryptococcus* as one of the dominant genera in many environments, whereby the only currently recognised *Cryptococcus* species reported for polar habitats is *Cryptococcus neoformans* (Shtarkman et al. 2013), which was detected using non-culturable approaches. Furthermore, taxonomic reorganisation resulted in the simultaneous rising importance and prevalence of some new genera, which included a number of former *Cryptococcus* species that are phylogenetically related and have similar physiological features (Tables 11.1 and 11.2). In particular, the following new genera were the most prominent: *Filobasidium* (principally as *Filobasidium magnum*, the former *Cryptococcus magnus*), *Goffeauzyma* (*Goffeauzyma gastrica* and *Goffeauzyma gilvescens*, the former *Cryptococcus gastricus* and *Cryptococcus gilvescens*), *Naganishia* (*Naganishia antarctica*, *Naganishia albida*, *Naganishia albidosimilis* and *Naganishia liquefaciens*, the former *Cryptococcus antarcticus*, *Cryptococcus albidus*, *Cryptococcus albidosimilis* and *Cryptococcus liquefaciens*), *Papiliotrema* (*Papiliotrema laurentii*, the former *Cryptococcus laurentii*), *Solicoccozyma* (*Solicoccozyma terricola*, the former *Cryptococcus terricola*) and *Vishniacozyma* (mainly *Vishniacozyma carnescens*, *Vishniacozyma tephrensis* and *Vishniacozyma victoriae*, the former *Cryptococcus carnescens*, *Cryptococcus tephrensis* and *Cryptococcus victoriae*) (Tables 11.1 and 11.2). *Vishn. victoriae* was one of the most frequently isolated species in both polar and subpolar environments, as it was described for many of the Antarctic localities investigated and all of the Arctic areas explored (Tables 11.1 and 11.2). The isolation of *Vishn. victoriae* also from temperate habitats demonstrates its psychrotolerant nature (i.e. flowers in Germany, surface of young apples in the Netherlands, rhizosphere soil in Korea and other isolation substrates around the world) (Fonseca et al. 2011; Yurkov et al. 2015a; see also Yurkov 2017 and Chap. 6 of this book).

11.2.2.2 *Rhodotorula* spp. and Former *Rhodotorula* spp.

As for the genus *Cryptococcus*, in the last decades of the twentieth century, the genus *Rhodotorula* was shown to represent the core of the basidiomycetous yeast communities of glacial environments (Butinar et al. 2005; Buzzini et al. 2012). After the taxonomic reorganisation of the species falling within the Pucciniomycotina taxa, a number of former *Rhodotorula* species that are phylogenetically related and have similar physiological features were rearranged into several new genera (Wang et al. 2015).

In parallel with *Deb. hansenii* and *Vishn. victoriae*, *Rhodotorula mucilaginosa* is apparently cosmopolitan, and its description for different habitats around the world, which include both polar and subpolar habitats, confirms its ubiquity (Sampaio 2011). The temperature range of growth for *Rh. mucilaginosa* is relatively wide (de Garcia et al. 2007; Singh et al. 2014). Together with the stability of its membrane fluidity (Turk et al. 2011), this temperature range demonstrates its ability to adapt its physiology to different unfavourable environmental conditions.

On the other hand, the species *Rhodotorula svalbardensis* was described for glacier cryoconite holes of Svalbard (Singh et al. 2014) and *Cystobasidium portillonensis* (i.e. the former *Rhodotorula portillonensis*) for a rock surface on King George Island, Antarctica (Laich et al. 2013). As these two species have only been described for these respective regions, they can be defined as apparently endemic (Tables 11.1 and 11.2). Wang et al. (2015) and Yurkov et al. (2015b) included a large group of *Rhodotorula* belonging to the class Cystobasidiomycetes, which included a few *Cystobasidium* species. Many species of this genus have been described as common inhabitants in polar and subpolar habitats, among which the most representative are *Cystobasidium laryngis* and *Cystobasidium minutum* (i.e. the former *Rhodotorula laryngis* and *Rhodotorula minuta*) (Tables 11.1 and 11.2). *Rhodotorula arctica* (*Bannozyma arctica*) was described by Vishniac and Takashima (2010) as a new psychrotolerant species that is endemic of the Arctic region. Although it was subsequently described for Antarctica a few years later (Duarte et al. 2016), it remains a typical yeast that colonises cold environments (Tables 11.1 and 11.2). Also, *Phenoliferia glacialis* and *Phenoliferia psychrophenolica* (i.e. the former *Rhodotorula glacialis* and *Rhodotorula psychrophenolica*; Margesin et al. 2007) were considered typical of cold environments although *Phen. psychrophenolica* was also isolated from *Sphagnum* mosses in the central regions of Russia (Kachalkin 2010). Both of these species were described, in particular, for soil (Carrasco et al. 2012; Duarte et al. 2013; Tsuji et al. 2013), snow and ice (Uetake et al. 2012), cryoconite holes or puddles on glaciers (Pathan et al. 2010; Edwards et al. 2013) and sea-water habitats (Vaz et al. 2011; Martinez et al. 2016).

11.2.2.3 *Dioszegia* spp., *Glaciozyma* spp., *Leucosporidium* spp. and *Mrakia* spp.

These four genera include a number of both psychrophilic and psychrotolerant species that have been described for cold habitats. Some of these were recently described as new species as a consequence of the increasing number of yeast strains isolated from sampling campaigns conducted in cold habitats and in particular in the Antarctic and Arctic regions (Buzzini et al. 2012; Buzzini and Margesin 2014). Examples here include (1) *Dioszegia antarctica* and *Dioszegia cryoxerica*, described for soil in Taylor Valley, South Victoria Land, Antarctica (Connell et al. 2010); (2) the recently described new genus *Glaciozyma* (in particular, the new species *Glaciozyma martinii* and *Glaciozyma watsonii*) which was also established to include the former species *Leucosporidium antarcticum* (*Glaciozyma antarctica*) (Turchetti et al. 2011); (3) *Leucosporidium escuderoi*, a new species associated with an Antarctic marine sponge (Laich et al. 2014); and (4) *Mrakia blollopis* and *Mrakia robertii* (Thomas-Hall et al. 2010), which, together with the other *Mrakia* species, make this genus one of the most representative in both polar and non-polar cold environments (Tables 11.1 and 11.2).

11.3 Black Yeasts in Polar and Subpolar Habitats

A special taxonomically heterogeneous group of yeast-like pleomorphic fungi are particularly adapted to different types of extreme environments (Selbmann et al. 2008; Zalar et al. 2008), and these are otherwise known as the 'black yeasts' (de Hoog et al. 1999). These sturdy organisms like *Friedmanniomyces endolithicus* and *Friedmanniomyces simplex, Elasticomyces elasticus, Recurvomyces mirabilis* and genus *Cryomyces* can inhabit habitats as different and extreme as Antarctic rock (Onofri et al. 2007; Selbmann et al. 2014c) and Arctic glacial ice with high salt concentrations, like representatives of the genera *Aureobasidium* and *Exophiala* (Branda et al. 2010; Butinar et al. 2011). Some black yeasts show high levels of viability after freezing and thawing, and also after UV exposure, and their tolerance to osmotic imbalance has shown their uncommon ability to survive under harsh external pressures. These black yeasts can colonise Earth habitats that have very extreme conditions, which will be akin to those that presumably occur extra-terrestrially. Accordingly, they have been used as the closest eukaryotic model for exobiological speculation (Selbmann et al. 2005; Onofri et al. 2008).

11.3.1 Aureobasidium *spp.*

The genus *Aureobasidium* is a member of the order Dothideales (Ascomycota, Dothideomycetes), and it includes over 30 species, with *Aureobasidium pullulans* being by far the most studied of these. *A. pullulans* is a common, ubiquitous black yeast that shows exceptional stress tolerance. It can inhabit various environments, from tropical to temperate and polar areas, as it populates the phyllosphere (Andrews et al. 2002), preserved foods (Pitt and Hocking 1999), various indoor habitats (e.g. dishwashers; Zalar et al. 2011; see also Chap. 5 of this book) and extreme environments, like hypersaline waters (Gunde-Cimerman et al. 2000), glacial ice and meltwaters from Arctic and Alpine glaciers (Gunde-Cimerman et al. 2003; Turchetti et al. 2008), accretion ice (D'Elia et al. 2009), freshwater lakes (Vaz et al. 2011) and Antarctic soil (Baublis et al. 1991) and permafrost (Gilichinsky et al. 2005). Due to its polyextremotolerant nature, *A. pullulans* is of considerable biotechnological importance (Takesako et al. 1991; Castoria et al. 2001; Chi et al., 2009; Cheng et al. 2011; Price et al. 2013). *A. pullulans* can tolerate oligotrophic conditions of up to 15% NaCl, and it can grow from 4 °C to 30 °C, with optimal growth at 25 °C (Zalar et al. 2008) and at pHs from acidic to basic (Zalar et al. 2008; Gostinčar et al. 2014). As a reflection of its global distribution, *A. pullulans* can keep its plasma membrane stable and flexible under diverse extreme conditions, which is one of the necessities for survival in extreme habitats that undergo changes (Turk et al. 2011).

Aureobasidium subglaciale has been described from subglacial ice of Svalbard glaciers and the immediate surroundings (Zalar et al. 2008), but it was also isolated from *Sphagnum* mosses in the central regions of Russia (Kachalkin 2010). It shows psychrotolerance, as it can grow between 4 °C and 25 °C, and halotolerance, as it can grow with up to 10% NaCl in the medium (Zalar et al. 2008). The third *Aureobasidium* species that has been described for cold environments is *Aureobasidium melanogenum*. This was isolated mainly from oligotrophic watery habitats, such as melted glacial water from the Italian Apennines (Branda et al. 2010) and Arctic Svalbard glaciers (Zalar et al. 2008). *A. melanogenum* is also considered an opportunistic human pathogen (Chan et al. 2012; Gostinčar et al. 2014), as it can grow between 10 °C and 35 °C and it can tolerate up to 10% NaCl (Salkin et al. 1986; Bolignano and Criseo 2003; Zalar et al. 2008).

The genomes of these three *Aureobasidium* species are relatively small (from 25.8 to 29.62 Mb) compared to those of other Dothideomycetous fungi, and they contain unusually high numbers of genes coding for extracellular carbohydrate-degrading enzymes (CAZy) and proteases, major facilitator superfamily membrane sugar transporters and alkali metal ion transporters (Gostinčar et al. 2014).

11.4 Concluding Remarks

A large part of our planet is cold, but despite this, little is known about cold-adapted yeasts that colonise these environments. Increasing scientific interest in these kinds of microorganisms is mainly aimed at their high biotechnological potential, while their ecology, diversity and environmental roles remain generally unknown.

One aspect that should be taken into consideration in future investigations is global warming. The Antarctic and Arctic polar regions are warming twice as fast as the rest of the globe. The thawing and melting that results can promote not only the release of stored genomic diversity, which can enrich terrestrial environments and oceans, but also the release of ancient animal carcasses, plant debris and various potentially pathogenic microorganisms, including yeasts. To date, cryospheric microbial-associated morbidity and mortality have been confined to anecdotal reports (Fiore et al. 1997; http://geographical.co.uk/nature/polar/item/1884-death-from-below-how-global-warming-is-reawakening-long-buried-diseases). However, it now appears that changes in the cryosphere might well precipitate a range of microbial threats (Revich et al. 2012; Edwards 2015).

Studies of yeasts in polar and subpolar environments have shown that these are dominated by psychrotolerant yeasts that can adapt to growth across a wide range of temperatures. Some of these can even grow at 37 °C, which is the most important prerequisite for pathogenicity in fungi (Gostinčar et al. 2010, 2011) and is exemplified by *A. melanogenum*, *Exophiala dermatitidis*, *C. parapsilosis* and *Na. liquefaciens*. As mentioned above, the previously polyphyletic genus *Cryptococcus* has been split in several genera, which include representatives of both opportunistic pathogens and environmental yeasts (de Garcia et al. 2012; Chen et al. 2014; Turchetti et al. 2015; Liu et al. 2015). Thus, new investigations into the opportunistic nature of individual species are needed.

Another potential threat for immunocompromised humans is represented by the poly-extremotolerant yeasts, which can adapt their physiology from cold to warm, to man-made indoor environments (Gostinčar et al. 2010, 2011). These mainly constitute the black yeasts (Gostinčar et al. 2011), which are characterised by their extraordinary phenotypic plasticity. This is exemplified by *A. pullulans*, which has been recovered from polar and non-polar extreme environments and which can also cause opportunistic infections (Salkin et al. 1986; Bolignano and Criseo 2003; Chan et al. 2012).

The cryosphere acts as storage for genomic diversity. Many new yeast species with psychrophilic or psychrotolerant aptitudes and particular physiological adaptations that ensure their survival under cold stress were recently described for polar and subpolar regions. Accordingly, a number of species that belong mainly to the Basidiomycota phylum, and to a lesser extent to the Ascomycota phylum that includes a few black yeast genera (e.g. *Aureobasidium* sp., *Exophiala* sp.), have been described to date (Tables 11.1 and 11.2). Their numbers will probably increase in the future following new research in these regions, particularly because in spite of the novel data reported more recently, the yeast diversity in polar and subpolar habitats remains understudied or undersampled. Specific and sometimes very adverse conditions can occur in these habitats, to apply selective pressure towards the survival of species with greater ability to colonise such extreme niches, which is also reflected in the isolation bias.

The new non-culturable metagenomic approaches used for studying microbial diversity, namely, next-generation sequencing (i.e. a 454 pyrosequencing platform initially and Illumina systems in more recent years), have allowed increasing numbers of sequences to be obtained and compared with the conventional isolation and identification methods. These can now be studied further with their ecological data. Indeed, most phylotypes were not cultured from the same samples, which implies that these samples harbour poorly documented, and as yet uncultured, yeast communities. In contrast to using DNA for the analysis of microbial diversity, co-sequencing of both DNA and RNA provides an assessment of the metabolically active community vs. potentially transient DNA from dead or inactive taxa in the environment (Baldrian et al. 2012).

There are also the technical bottlenecks to be overcome. The ‘big data’ obtained by these non-culturable metagenomic approaches have to be interpreted with caution, because the length of the sequences generated and the public sequence databases used for comparisons, which have been faced with the challenge of accurately placing sequences into a given reference tree, might fail to differentiate certain yeast taxa. As a consequence, the identification process is frequently restricted to the family or genus level (e.g. *Kluyveromyces* sp., *Saccharomyces* sp., *Pichia* sp., *Taphrina* sp., *Yarrowia* sp., *Apiotrichum* sp., *Bensingtonia* sp., *Lipomyces* sp., *Tremella* sp.; Tables 11.1 and 11.2). As suggested by Timling et al. (2014), the use of a narrower cut-off and more sensitive markers would detect further species groups in some taxa and might reveal more consistent distributions of the different species.

The changes in the global climate that mainly affect the polar areas of the world will change the ecology of many of the yeasts described in this report, potentially through changes in their behaviour and adaptation, and thence in the regions of the world that they populate.

References

Abyzov SS (1993) Microorganisms in the Antarctic ice. In: Friedman EI (ed) Antarctic microbiology. Wiley, New York, pp 265–295

Abyzov SS, Hoover RB, Imura S, Mitskevich IN, Naganuma T, Poglazova MN, Ivanov MV (2004) Use of different methods for discovery of ice-entrapped microorganisms in ancient layers of the Antarctic glacier. Adv Space Res 33:1222–1230

Alcaíno J, Cifuentes V, Baeza M (2015) Physiological adaptations of yeasts living in cold environments and their potential applications. World J Microbiol Biotechnol 31:1467–1473

Altunatmaz SS, Issa G, Aydin A (2012) Detection of airborne psychrotrophic bacteria and fungi in food storage refrigerators. Braz J Microbiol 43:1436–1443

Andrews JH, Spear RN, Nordheim EV (2002) Population biology of *Aureobasidium pullulans* on apple leaf surfaces. Can J Microbiol 48:500–513

Antony R, Sanyal A, Kapse N, Dhakephalkar PK, Thamban M, Nair S (2016) Microbial communities associated with Antarctic snow pack and their biogeochemical implications. Microbiol Res 192:192–202

Arenz BE, Blanchette RA (2011) Distribution and abundance of soil fungi in Antarctica at sites on the Peninsula, Ross Sea Region and McMurdo Dry Valleys. Soil Biol Biochem 43:308–315

Arenz BE, Held BW, Jurgens JA, Farrell RL, Blanchette RA (2006) Fungal diversity in soils and historic wood from the Ross Sea region of Antarctica. Soil Biol Biochem 38:3057–3064

Arenz BE, Held BW, Jurgens JA, Blanchette RA (2011) Fungal colonization of exotic substrates in Antarctica. Fungal Div 49:13–22

Atlas RM, di Menna ME, Cameron RE (1978) Ecological investigations of yeasts in Antarctic soils. Antarc Res Ser 30:27–34

Baharaeen S, Vishniac HS (1982) *Cryptococcus lupi* sp. nov., an Antarctic basidioblastomycete. Int J Syst Bacteriol 32:229–232

Baldrian P, Kolarik M, Stursova M, Kopecky J, Valaškova V, Vetrovsky T, Zifčakova L, Ridl J, Vlček C, Voriškova J (2012) Active and total microbial communities in forest soil are largely different and highly stratified during decomposition. ISME J 6:248–258

Barahona S, Yuivar Y, Socias G, Alcaíno J, Cifuentes V, Baeza M (2016) Identification and characterization of yeasts isolated from sedimentary rocks of Union Glacier at the Antarctica. Extremophiles 20:479–491

Baublis JA, Wharton RA Jr, Volz PA (1991) Diversity of micro-fungi in an Antarctic dry valley. J Basic Microbiol 31:3–12

Bellemain E, Davey ML, Kauserud H, Epp LS, Boessenkool S, Coissac E, Geml J, Edwards M, Willerslev E, Gussarova G, Taberlet P, Brochmann C (2013) Fungal palaeodiversity revealed using high-throughput metabarcoding of ancient DNA from arctic permafrost. Environ Microbiol 15:1176–1189

Benn DI, Evans DJA (1998) Glacier and glaciation. Arnold, London

Birgisson H, Delgado O, Garcia Arroyo L, Hatti-Kaul R, Mattiasson B (2003) Cold-adapted yeasts as producers of cold-active polygalacturonases. Extremophiles 7:185–193

Blaalid R, Davey ML, Kauserud H, Carlsen T, Halvorsen R, Høiland K, Eidesen PB (2014) Arctic root-associated fungal community composition reflects environmental filtering. Mol Ecol 23:649–659

Boetius A, Anesio AM, Deming JW, Mikucki JA, Rapp JZ (2015) Microbial ecology of the cryosphere: sea ice and glacial habitats. Nat Rev Microbiol 13:677–690

Bolignano G, Criseo G (2003) Disseminated nosocomial fungal infection by *Aureobasidium pullulans* var. *melanogenum*: a case report. J Clin Microbiol 41:4483–4485

Botnen S, Vik U, Carlsen T, Eidesen PB, Davey ML, Kauserud H (2014) Low host specificity of root-associated fungi at an Arctic site. Mol Ecol 23:975–985

Branda E, Turchetti B, Diolaiuti G, Pecci M, Smiraglia C, Buzzini P (2010) Yeast and yeast-like diversity in the southernmost glacier of Europe (Calderone Glacier, Apennines, Italy). FEMS Microbiol Ecol 72:354–369

Bridge PD, Newsham KK (2009) Soil fungal community composition at Mars Oasis, a southern maritime Antarctic site, assessed by PCR amplification and cloning. Fungal Ecol 2:66–74

Butinar L, Santos S, Spencer-Martins I, Oren A, Gunde-Cimerman N (2005) Yeast diversity in hypersaline habitats. FEMS Microbiol Lett 244:229–234

Butinar L, Spencer-Martins I, Gunde-Cimerman N (2007) Yeasts in high Arctic glaciers: the discovery of a new habitat for eukaryotic microorganisms. A van Leeuwenhoek 91:277–289

Butinar L, Strmole T, Gunde-Cimerman N (2011) Relative incidence of ascomycetous yeasts in arctic coastal environments. Microbial Ecol 61:832–843

Buzzini P, Margesin R (2014) Cold-adapted yeasts. Springer, Berlin

Buzzini P, Branda E, Goretti M, Turchetti B (2012) Psychrophilic yeasts from worldwide glacial habitats: diversity, adaptation strategies and biotechnological potential. FEMS Microbiol Ecol 82:217–241

Byrd RE (1938) Alone: the classic polar adventure. Island Press, Washington, DC

Cameron RE (1971) Antarctic soil microbial and ecological investigations. In: Quam LO, Porter HD (eds) Research in the Antarctic. AAAS, Washington, DC, pp 137–189

Carrasco M, Rozas JM, Barahona S, Alcaíno J, Cifuentes V, Baeza M (2012) Diversity and extracellular enzymatic activities of yeasts isolated from King George Island, the sub-Antarctic region. BMC Microbiol 12:251

Castoria R, de Curtis F, Lima G, Caputo L, Pacifico S, de Cicco V (2001) *Aureobasidium pullulans* (LS30) an antagonist of postharvest pathogens of fruits: study on its modes of action. Postharvest Biol Technol 22:7–17

Catellani P, Miotti Scapin R, Alberghini L, Radu IL, Giaccone V (2014) Levels of microbial contamination of domestic refrigerators in Italy. Food Control 42:257–262

Chan GF, Bamadhaj HM, Gan HM, Rashid NAA (2012) Genome sequence of *Aureobasidium pullulans* AY4, an emerging opportunistic fungal pathogen with diverse biotechnological potential. Eukaryot Cell 11:1419–1420

Chen SC, Meyer W, Sorrell TC (2014) *Cryptococcus gattii* infections. Clin Microbiol Rev 27:980–1024

Cheng KC, Demirci A, Catchmark JM (2011) Pullulan: biosynthesis, production, and applications. Appl Microbiol Biotechnol 92:29–44

Chi Z, Wang F, Yue L, Liu G, Zhang T (2009) Bioproducts from *Aureobasidium pullulans*, a biotechnologically important yeast. Appl Microbiol Biotechnol 82:793–804

Coleine C, Selbmann L, Ventura D'ALP, Onofri S, Zucconi L (2015) Fungal biodiversity in the alpine Tarfala Valley. Microorganisms 3:612–624

Connell LB, Redman R, Craig S, Scorzetti G, Iszard M, Rodriguez R (2008) Diversity of soil yeasts isolated from South Victoria Land, Antarctica. Microb Ecol 56:448–459

Connell LB, Redman R, Rodriguez R, Barrett A, Iszard M, Fonseca A (2010) *Dioszegia antarctica* sp. nov. and *Dioszegia cryoxerica* sp. nov., psychrophilic basidiomycetous yeasts from polar desert soils in Antarctica. Int J Syst Evol Microbiol 60:1466–1472

Connell LB, Rodriguez RR, Redman RS, Dalluge JJ (2014) Cold-adapted yeasts in Antarctic Deserts. In: Buzzini P, Margesin R (eds) Cold-adapted yeasts. Springer, Berlin, pp 75–98

Connell LB, Staudigel H (2013) Fungal diversity in a Dark Oligotrophic Volcanic Ecosystem (DOVE) on Mount Erebus, Antarctica. Biology 2(2):798–809

D'Elia T, Veerapaneni R, Theraisnathan V, Rogers SO (2009) Isolation of fungi from Lake Vostok accretion ice. Mycologia 101:751–763

Davey M, Blaalid R, Vik U, Carlsen T, Kauserud H, Eidesen PB (2015) Primary succession of *Bistorta vivipara* (L.) Delabre (Polygonaceae) root-associated fungi mirrors plant succession in two glacial chronosequences. Environ Microbiol 17:2777–2790

de García V, Brizzio S, Libkind D, Buzzini P, Van Broock M (2007) Biodiversity of cold-adapted yeasts from glacial meltwater rivers in Patagonia, Argentina. FEMS Microbiol Ecol 59:331–341

de Garcia V, Zalar P, Brizzio S, Gunde-Cimerman N, van Broock M (2012) *Cryptococcus* species (Tremellales) from glacial biomes in the southern (Patagonia) and northern (Svalbard) hemispheres. FEMS Microbiol Ecol 82:523–539

de Garcia V, Coelho MA, Maia TM, Ros LH, Vaz AM, Rosa CA, Sampaio JP, Goncalves P, van Broock M, Libkind D (2015) Sex in the cold: taxonomic reorganization of psychrotolerant yeasts in the order Leucosporidiales. FEMS Yeast Res 15:fov019

de Hoog GS, Zalar P, Urzì C, de Leo F, Yurlova NA, Sterflinger K (1999) Relationships of dothideaceous black yeasts and meristematic fungi based on 5.8S and ITS2 rDNA sequence comparison. Stud Mycol 43:31–37

di Menna ME (1966a) Three new species from Antarctic soil: *Candida nivalis*, *Candida gelida* and *Candida frigida* spp. nov. A van Leeuwenhoek 32:25–28

di Menna ME (1966b) Yeasts in Antarctic soil. A van Leeuwenhoek 32:29–38

Dreesens LL, Lee CK, Cary SC (2014) The distribution and identity of edaphic fungi in the McMurdo Dry Valleys. Biology 3:466–483

Duarte AWF, Dayo-Owoyemi I, Nobre FS, Pagnocca FC, Chaud LCS, Pessoa A, Felipe MGA, Sette LD (2013) Taxonomic assessment and enzymes production by yeasts isolated from marine and terrestrial Antarctic samples. Extremophiles 17:1023–1035

Duarte AWF, Passarini AWF, Delforno TP, Pellizzari FM, Cipro CVZ, Montone RC, Petry MV, Putzke J, Rosa LH, Sette LD (2016) Yeasts from macroalgae and lichens that inhabit the South Shetland Islands, Antarctica. Environ Microbiol Rep 8:874–885

Dujon B, Sherman D, Fischer G, Durrens P, Casaregola S, Lafontaine I, De Montigny J, Marck C, Neuveglise C, Talla E, Goffard N, Frangeul L, Aigle M, Anthouard V, Babour A, Barbe V, Barnay S, Blanchin S, Beckerich JM, Beyne E, Bleykasten C, Boisrame A, Boyer J, Cattolico L, Confanioleri F, De Daruvar A, Despons L, Fabre E, Fairhead C, Ferry-Dumazet H, Groppi A, Hantraye F, Hennequin C, Jauniaux N, Joyet P, Kachouri R, Kerrest A, Koszul R, Lemaire M, Lesur I, Ma L, Muller H, Nicaud JM, Nikolski M, Oztas S, Ozier-Kalogeropoulos-O, Pellenz S, Potier S, Richard GF, Straub ML, Suleau A, Swennen D, Tekaia F, Wesolowski-Louvel M, Westhof E, Wirth B, Zeniou-Meyer M, Zivanovic I, Bolotin-Fukuhara M, Thierry A, Bouchier C, Caudron B, Scarpelli C, Gaillardin C, Weissenbach J, Wincker P, Souciet JL (2004) Genome evolution in yeasts. Nature 430:35–44

Edwards A (2015) Coming in from the cold: potential microbial threats from the terrestrial cryosphere. Front Earth Sci 3:12

Edwards A, Douglas B, Anesio AM, Rassner SM, Irvine-Fynn TDL, Sattler B, Griffith GW (2013) A distinctive fungal community inhabiting cryoconite holes on glaciers in Svalbard. Fungal Ecol 6:168–176

Edwards A, Mur LA, Girdwood SE, Anesio AM, Stibal M, Rassner SM, Hell K, Pachebat JA, Post B, Bussell JS, Cameron SJ, Griffith GW, Hodson AJ, Sattler B (2014) Coupled cryoconite ecosystem structure-function relationships are revealed by comparing bacterial communities in alpine and Arctic glaciers. FEMS Microbiol Ecol 89:222–237

Farrell RL, Arenz BE, Duncan SM, Held BW, Jurgens JA, Blanchette RA (2011) Introduced and indigenous fungi of the Ross Island historic huts and pristine areas of Antarctica. Polar Biol 34:1669–1677

Fell JW, Hunter IL (1974) *Torulopsis austromarina* sp. nov. A yeast isolated from the Antarctic ocean. A van Leeuwenhoek 40:307–310

Ferrari BC, Zhang C, van Dorst J (2011) Recovering greater fungal diversity from pristine and diesel fuel contaminated sub-Antarctic soil through cultivation using both a high and a low nutrient media approach. Front Microbiol 2:217

Fiore DC, Mckee DD, Janiga MA (1997) Red snow: is it safe to eat? A pilot study. Wilderness Environ Med 8:94–95

Fonseca A, Boekhout T, Fell JW (2011) *Cryptococcus* Vuillemin (1901). In: Kurtzman CP, Fell JW, Boekhout T (eds) The yeasts: a taxonomic study, 5th edn. Elsevier, Amsterdam, pp 1661–1737

Fonseca A, Rodrigues MG (2011) *Taphrina* Fries (1832). In: Kurtzman CP, Fell JW, Boekhout T (eds) The yeasts: a taxonomic study, 5th edn. Elsevier, Amsterdam, pp 823–858

Frisvad JC (2008) Fungi in cold ecosystems. In: Margesin R, Schinner F, Marx J-C, Gerday C (eds) Psychrophiles: from biodiversity to biotechnology. Springer, Berlin, pp 137–156

Geml J, Morgado LN, Semenova TA, Welker JM, Walker MD, Smets E (2015) Long-term warming alters richness and composition of taxonomic and functional groups of arctic fungi. FEMS Microbiol Ecol 91:fiv095

Gesheva V, Negoita T (2012) Psychrotrophic microorganism communities in soils of Haswell Island, Antarctica, and their biosynthetic potential. Polar Biol 35:291–297

Gilichinsky D, Rivkina E, Bakermans C, Shcherbakova V, Petrovskaya L, Ozerskaya S, Ivanushkina N, Kochkina G, Laurinavichuis K, Pecheritsina S, Fattakhova R, Tiedje JM (2005) Biodiversity of cryopegs in permafrost. FEMS Microbiol Ecol 53:117–128

Gilichinsky DA, Wilson GS, Friedmann EI, McKay CP, Sletten RS, Rivkina EM, Vishnivetskaya TA, Erokhina LG, Ivanushkina NE, Kochkina GA, Shcherbakova VA, Soina VS, Spirina EV, Vorobyova EA, Fyodorov-Davydov DG, Hallet B, Ozerskaya SM, Sorokovikov VA, Laurinavichyus KS, Shatilovich AV, Chanton JP, Ostroumov VE, Tiedje JM (2007) Microbial populations in Antarctic permafrost: biodiversity, state, age, and implication for astrobiology. Astrobiology 7:275–311

Gittel A, Bárta J, Kohoutová I, Mikutta R, Owens S, Gilbert J, Schnecker J, Wild B, Hannisdal B, Maerz J, Lashchinskiy N, Capek P, Santrůčková H, Gentsch N, Shibistova O, Guggenberger G, Richter A, Torsvik VL, Schleper C, Urich T (2014) Distinct microbial communities associated with buried soils in the Siberian tundra. ISME J 8:841–853

Godinho VM, Furbino LE, Santiago IF, Pellizzari FM, Yokoya NS, Pupo D, Alves TMA, Junior PAS, Romanha AJ, Zani CL, Cantrell CL, Rosa CA, Rosa LH (2013) Diversity and bioprospecting of fungal communities associated with endemic and cold-adapted macroalgae in Antarctica. ISME J 7:1434–1451

Golubev WI (1998) New species of basidiomycetous yeasts, *Rhodotorula creatinovora* and *R. yakutica*, isolated from permafrost soils of Eastern-Siberian Arctic. Mykol Phytopathol 32:8–13

Goncalves VN, Carvalho CR, Johann S, Mendes G, Alves TMA, Zani CL, Junior PAS, Murta SMF, Romanha AJ, Cantrell CL, Rosa CA, Rosa LH (2015) Antibacterial, antifungal and antiprotozoal activities of fungal communities present in different substrates from Antarctica. Polar Biol 38:1143–1152

Goordial J, Davila A, Lacelle D, Pollard W, Marinova MM, Greer CW, DiRuggiero J, McKay CP, Whyte LG (2016) Nearing the cold-arid limits of microbial life in permafrost of an upper dry valley, Antarctica. ISME J 10:1613–1624

Gostinčar C, Turk M, Trbuha T, Vaupotič T, Plemenitaš A, Gunde-Cimerman N (2008) Expression of fatty-acid-modifying enzymes in the halotolerant black yeast *Aureobasidium pullulans* (de Bary) G. Arnaud under salt stress. Stud Mycol 61:51–59

Gostinčar C, Grube M, de Hoog S, Zalar P, Gunde-Cimerman N (2010) Extremotolerance in fungi: evolution on the edge. FEMS Microbiol Ecol 71:2–11

Gostinčar C, Grube M, Gunde-Cimerman N (2011) Evolution of fungal pathogens in domestic environments. Fungal Biol 115:1008–1018

Gostinčar C, Ohm RA, Kogej T, Sonjak S, Turk M, Zajc J, Zalar P, Grube M, Sun H, Han J, Sharma A, Chiniquy J, Ngan CY, Lipzen A, Barry K, Grigoriev IV, Gunde-Cimerman N

(2014) Genome sequencing of four *Aureobasidium pullulans* varieties: biotechnological potential, stress tolerance, and description of new species. BMC Genomics 15:549

Goto S, Sugiyama J, Iizuka H (1969) A taxonomic study of Antarctic yeasts. Mycologia 61:748–774

Gounot AM (1986) Psychrophilic and psychrotrophic microorganisms. Experientia 42:1192–1197

Guffogg SP, Thomas-Hall SR, Holloway P, Watson K (2004) A novel psychrotolerant member of the hymenomycetous yeasts from Antarctica: *Cryptococcus watticus* sp. nov. Int J Syst Evol Microbiol 54:275–277

Gunde-Cimerman N, Zalar P, de Hoog GS, Plemenitaš A (2000) Hypersaline waters in salterns-natural ecological niches for halophilic black yeasts. FEMS Microbiol Ecol 32:235–240

Gunde-Cimerman N, Sonjak S, Zalar P, Frisvad JC, Diderichsen B, Plemenitaš A (2003) Extremophilic fungi in Arctic ice: a relationship between adaptation to low temperature and water activity. Phys Chem Earth 28:1273–1278

Gunde-Cimerman N, Plemenitaš A, Buzzini P (2014) Change in lipids composition and fluidity of yeast plasma membrane as response to cold. In: Buzzini P, Margesin R (eds) Cold-adapted yeasts. Springer, Berlin, pp 225–242

Henriquez M, Vergara K, Norambuena J, Beiza A, Maza F, Ubilla P, Araya I, Chavez R, San-Martin A, Darias J, Darias MJ, Vaca I (2014) Diversity of cultivable fungi associated with Antarctic marine sponges and screening for their antimicrobial, antitumoral and antioxidant potential. World J Microbiol Biotechnol 30:65–76

Holdgate MW (1977) Terrestrial ecosystems in the Antarctic. Philos Trans R Soc Lond B Biol Sci 279:5–25

Irvine-Fynn TD, Edwards A (2014) A frozen asset: the potential of flow cytometry in constraining the glacial biome. Cytometry A 85:3–7

Jacobs MPH, Shafer JC (1964) Studies of fungi at Amundsen-Scott IGY South Polar Base. Arch Dermatol 89:177–183

Jacques N, Zenouche A, Gunde-Cimerman N, Casaregola S (2015) Increased diversity in the genus *Debaryomyces* from Arctic glacier samples. A van Leeuwenhoek 107:487–501

Jones GEB (1976) Recent advances in aquatic mycology. The Gresham Press, Old Woking, Surrey

Kachalkin AV (2010) New data on the distribution of certain psychrophilic yeasts in Moscow oblast. Microbiology 79:843–847

Kachalkin AV (2014) Yeasts of the White Sea intertidal zone and description of *Glaciozyma litorale* sp. nov. A van Leeuwenhoek 105:1073–1083

Kim HJ, Shim HE, Lee JH, Kang YC, Hur YB (2015) Ice-binding protein derived from *Glaciozyma* can improve the viability of cryopreserved mammalian cells. J Microbiol Biotechnol 25:1989–1996

Knowlton C, Veerapaneni R, Rogers SO (2013) Microbial analyses of ancient ice core sections from Greenland and Antarctica. Biology 2:206–232

Kochkina G, Ivanushkina N, Ozerskaya S, Chigineva N, Vasilenko O, Firsov S, Spirina E, Gilichinsky D (2012) Ancient fungi in Antarctic permafrost environments. FEMS Microbiol Ecol 82:501–509

Krishnan A, Alias SA, Wong CMVL, Pang K-L, Convey P (2011) Extracellular hydrolase enzyme production by soil fungi from King George Island, Antarctica. Polar Biol 34:1535–1542

Kurtzman CP (2011) Pichia E.C. Hansen (1904). In: Kurtzman CP, Fell JW, Boekhout T (eds) The yeasts: a taxonomic study, 5th edn. Elsevier, Amsterdam, pp 685–707

Kurtzman CP, Mateo RQ, Kolecka A, Theelen B, Robert V, Boekhout T (2015) Advances in yeast systematics and phylogeny and their use as predictors of biotechnologically important metabolic pathways. FEMS Yeast Res 15:fov050

Laich F, Vaca I, Chavez R (2013) *Rhodotorula portillonensis* sp. nov., a basidiomycetous yeast isolated from Antarctic shallow-water marine sediment. Int J Syst Evol Microbiol 63:3884–3891

Laich F, Chàvez R, Vaca I (2014) *Leucosporidium escuderoi* f.a., sp. nov., a basidiomycetous yeast associated with an Antarctic marine sponge. A van Leeuwenhoek 105:593–601

Libkind D, Moline M, Sampaio JP, van Broock M (2009) Yeasts from high-altitude lakes: influence of UV radiation. FEMS Microbiol Ecol 69:353–362

Liu XZ, Wang QM, Göker M, Groenewald M, Kachalkin AV, Lumbsch HT, Millanes AM, Wedin M, Yurkov AM, Boekhout T, Bai FY (2015) Towards an integrated phylogenetic classification of the Tremellomycetes. Stud Mycol 81:85–147

Loperena L, Soria V, Varela H, Lupo S, Bergalli A, Guigou M, Pellegrino A, Bernardo A, Calvino A, Rivas F, Batista S (2012) Extracellular enzymes produced by microorganisms isolated from maritime Antarctica. World J Microbiol Biotechnol 28:2249–2256

Ludley KE, Robinson CH (2008) "Decomposer" basidiomycota in Arctic and Antarctic ecosystems. Soil Biol Biochem 40:11–29

Lutz S, Anesio AM, Edwards A, Benning LG (2016) Linking microbial diversity and functionality of arctic glacial surface habitats. Environ Microbiol. doi:10.1111/1462-2920.13494

Ma LJ, Catranis CM, Starmer WT, Rogers SO (1999) Revival and characterization of fungi from ancient polar ice. Mycologist 13:70–73

Maccario L, Vogel TM, Larose C (2014) Potential drivers of microbial community structure and function in Arctic spring snow. Front Microbiol 5:413

Maraz A, Kovacs M (2014) Food spoilage by cold-adapted yeasts. In: Buzzini P, Margesin R (eds) Cold-adapted yeasts. Springer, Berlin, pp 497–532

Margesin R (2008) Effect of temperature on growth parameters of psychrophilic bacteria and yeasts. Extremophiles. 13:257–262

Margesin R, Fell JW (2008) *Mrakiella cryoconiti* gen. nov., sp. nov., a psychrophilic, anamorphic, basidiomycetous yeast from Alpine and Arctic habitats. Int J Syst Evol Microbiol 58:2977–2982

Margesin R, Fonteyne PA, Schinner F, Sampaio JP (2007) *Rhodotorula psychrophila* sp. nov., *Rhodotorula psychrophenolica* sp. nov. and *Rhodotorula glacialis* sp. nov., novel psychrophilic basidiomycetous yeast species isolated from alpine environments. Int J Syst Evol Microbiol 57:2179–2184

Martinez A, Cavello I, Garmendia G, Rufo C, Cavalitto S, Vero S (2016) Yeasts from sub-Antarctic region: biodiversity, enzymatic activities and their potential as oleaginous microorganisms. Extremophiles 20:759–769

Mittelbach M, Vannette RL (2017) Mutualism in yeasts. In: Buzzini P, Lachance MA, Yurkov AM (eds) Yeasts in natural ecosystems: ecology. Springer, Heidelberg, pp 155–178

Montes MJ, Belloch C, Galiana M, Garcia MD, Andres C, Ferrer S, Torres-Rodriguez JM, Guinea J (1999) Polyphasic taxonomy of a novel yeast isolated from Antarctic environment: description of *Cryptococcus victoriae* sp. nov. Syst Appl Microbiol 22:97–110

Nesje A, Dahl SV (2000) Glaciers and environmental change. Arnold, London

Newell SY, Fell JW (1970) The perfect form of a marine occurring yeast of the genus *Rhodotorula*. Mycologia 62:272–281

NSF International (2013) The Public Health and Safety Organization, Household Germ Study. https://www.nsf.org/newsroom_pdf/2013_germ_study_FOR-WEB-ONLY.pdf. Accessed 21 Nov 2016 (password protected)

Onofri S, Selbmann L, Zucconi L, Pagano S (2004) Antarctic microfungi as models for exobiology. Planet Space Sci 52:229–237

Onofri S, Zucconi L, Tosi S (2007) Continental Antarctic fungi. IHW-Verlag, Eching, Germany

Onofri S, Barreca D, Selbmann L, Isola D, Rabbow E, Horneck G, de Vera JPP, Hatton J, Zucconi L (2008) Resistance of Antarctic black fungi and cryptoendolithic communities to simulated space and Martian conditions. Stud Mycol 61:99–109

Osono T, Ueno T, Uchida M, Kanda H (2012) Abundance and diversity of fungi in relation to chemical changes in arctic moss profiles. Polar Sci 6:121–131

Ozerskaya SM, Kochkina GA, Ivanushkina NE, Knyazeva EV, Gilichinskii DA (2008) The structure of micromycete complexes in permafrost and cryopegs of the Arctic. Microbiology 77:482–489

Pathan AAK, Bhadra B, Begum Z, Shivaji S (2010) Diversity of yeasts from puddles in the vicinity of Midre Lovenbreen Glacier, Arctic and bioprospecting for enzymes and fatty acids. Curr Microbiol 60:307–314

Pavlova K, Angelova G, Savova I, Grigorova D, Kupenov L (2002) Studies of Antarctic yeast for b-glucosidase production. World J Microbiol Biotechnol 18:569–573

Pavlova K, Panchev I, Krachanova M, Gocheva M (2009) Production of an exopolysaccharide by Antarctic yeast. Folia Microbiol 54:343–348

Petrescu I, Lamotte-Brasseur J, Chessa JP, Ntarima P, Claeyssens M, Devreese B, Marino G, Gerday C (2000) Xylanase from the psychrophilic yeast *Cryptococcus adeliae*. Extremophiles 4:137–144

Pfeffer WT, Arendt AA, Bliss A, Bolch T, Cogley JG, Gardner AS, Hagen J-O, Hock R, Kaser G, Kienholz C, Miles ES, Moholdt G, Mölg N, Paul F, Radić V, Rastner P, Raup BH, Rich J, Sharp M (2014) The Randolph Glacier Inventory: a globally complete inventory of glaciers. J Glaciol 60:537–552

Pitt JI, Hocking AD (1999) Fungi and food spoilage, 2nd edn. Aspen Publishers, Gaithersburg, MD

Poli A, Anzelmo G, Tommonaro G, Pavlova K, Casaburi A, Nicolaus B (2010) Production and chemical characterization of an exopolysaccharide synthesized by psychrophilic yeast strain *Sporobolomyces salmonicolor* AL1 isolated from Livingston Island, Antarctica. Folia Microbiol 55:576–581

Price NP, Manitchotpisit P, Vermillion KE, Bowman MJ, Leathers TD (2013) Structural characterization of novel extracellular liamocins (mannitol oils) produced by *Aureobasidium pullulans* strain NRRL 50380. Carbohydr Res 370:24–32

Rämä T, Davey ML, Nordén J, Halvorsen R, Blaalid R, Mathiassen GH, Alsos IG, Kauserud H (2016) Fungi sailing the Arctic ocean: speciose communities in North Atlantic driftwood as revealed by high-throughput amplicon sequencing. Microb Ecol 72:295–304

Ray MK, Devi KU, Kumar GS, Shivaji S (1992) Extracellular protease from the antarctic yeast *Candida humicola*. Appl Environ Microbiol 58:1918–1923

Revich B, Tokarevich N, Parkinson AJ (2012) Climate change and zoonotic infections in the Russian Arctic. Int J Circumpolar Health 71:18792

Robinson CH (2001) Cold adaptation in Arctic and Antarctic fungi. New Phytol 151:341–353

Rovati JI, Pajot HF, Ruberto L, Mac Cormack W, Figueroa LIC (2013) Polyphenolic substrates and dyes degradation by yeasts from 25 de Mayo/King George Island (Antarctica). Yeast 30:459–470

Ruisi S, Barreca D, Selbmann L, Zucconi L, Onofri S (2007) Fungi in Antarctica. Rev Environ Sci Biotechnol 6:127–141

Salkin IF, Martinez JA, Kemna ME (1986) Opportunistic infection of the spleen caused by *Aureobasidium pullulans*. J Clin Microbiol 23:828–831

Sampaio JP (2011) *Rhodotorula* Harrison (1928). In: Kurtzman CP, Fell JW, Boekhout T (eds) The yeasts: a taxonomic study, 5th edn. Elsevier, Amsterdam, pp 1873–1927

Santiago IF, Soares MA, Rosa CA, Rosa LH (2015) Lichensphere: a protected natural microhabitat of the non-lichenised fungal communities living in extreme environments of Antarctica. Extremophiles 19:1087–1097

Scorzetti G, Petrescu I, Yarrow D, Fell JW (2000) *Cryptococcus adeliensis* sp. nov., a xylanase producing basidiomycetous yeast from Antarctica. A van Leeuwenhoek 77:153–157

Selbmann L, de Hoog GS, Mazzaglia A, Friedmann EI, Onofri S (2005) Fungi at the edge of life: cryptoendolithic black fungi from Antarctic desert. Stud Mycol 51:1–32

Selbmann L, de Hoog GS, Zucconi L, Isola D, Ruisi S, van den Ende AH, Ruibal C, De Leo F, Urzi C, Onofri S (2008) Drought meets acid: three new genera in a dothidealean clade of extremotolerant fungi. Stud Mycol 61:1–20

Selbmann L, Zucconi L, Onofri S, Cecchini C, Isola D, Turchetti B, Buzzini P (2014a) Taxonomic and phenotypic characterization of yeasts isolated from worldwide cold rock-associated habitats. Fungal Biol 118:61–71

Selbmann L, Turchetti B, Yurkov A, Cecchini C, Zucconi L, Isola D, Buzzini P, Onofri S (2014b) Description of *Taphrina antarctica* f.a. sp. nov., a new anamorphic ascomycetous yeast species associated with Antarctic endolithic microbial communities and transfer of four *Lalaria* species in the genus *Taphrina*. Extremophiles 18:707–721

Selbmann L, de Hoog GS, Zucconi L, Isola D, Onofri S (2014c) Black yeasts in cold habitats. In: Buzzini P, Margesin R (eds) Cold-adapted yeasts. Springer, Berlin, pp 173–189

Semenova TA, Morgado LN, Welker JM, Walker MD, Smets E, Geml J (2016) Compositional and functional shifts in arctic fungal communities in response to experimentally increased snow depth. Soil Biol Biochem 100:201–209

Shivaji S, Prasad GS (2009) Antarctic yeasts: biodiversity and potential applications. In: Satyanarayana T, Kunze G (eds) Yeast biotechnology: diversity and applications. Springer, Berlin, pp 3–16

Shtarkman YM, Kocer ZA, Edgar R, Veerapaneni RS, D'Elia T, Morris PF, Rogers SO (2013) Subglacial Lake Vostok (Antarctica) accretion ice contains a diverse set of sequences from aquatic, marine and sediment-inhabiting bacteria and eukarya. PLoS One 8:e67221

Siddiqui KS, Cavicchioli R (2005) Improved thermal stability and activity in the cold-adapted lipase B from *Candida antarctica* following chemical modification with oxidized polysaccharides. Extremophiles 9:471–476

Simon C, Wiezer A, Strittmatter AW, Daniel R (2009) Phylogenetic diversity and metabolic potential revealed in a glacier ice metagenome. Appl Environ Microbiol 75:7519–7526

Singh P, Singh SM, Tsuji M, Prasad GS, Hoshino T (2014) *Rhodotorula svalbardensis* sp. nov., a novel yeast species isolated from cryoconite holes of Ny-Alesund, Arctic. Cryobiology 68:122–128

Singh SM, Tsuji M, Gawas-Sakhalker P, Loonen MJJE, Hoshino T (2016) Bird feather fungi from Svalbard Arctic. Polar Biol 39:523–532

Soneda M (1961) On some yeasts from the Antarctic region. Special Publ Seto Marine Biol Lab Sirahama, Wakayama-Ken, Japan, pp 3–10

Takesako K, Ikai K, Haruna F, Endo M, Shimanaka K, Sono E, Nakamura T, Kato I, Yamaguchi H (1991) Aureobasidins, new antifungal antibiotics taxonomy, fermentation, isolation, and properties. J Antibiot (Tokyo) 44:919–924

Thomas-Hall SR, Watson K (2002) *Cryptococcus nyarrowii* sp. nov., a basidiomycetous yeast from Antarctica. Int J Syst Evol Microbiol 52:1033–1103

Thomas-Hall SR, Watson K, Scorzetti G (2002) *Cryptococcus statzelliae* sp. nov. and three novel strains of *Cryptococcus victoriae*, yeasts isolated from Antarctic soils. Int J Syst Evol Microbiol 52:2303–2308

Thomas-Hall SR, Turchetti B, Buzzini P, Branda E, Boekhout T, Theelen B, Watson K (2010) Cold adapted yeasts from Antarctica and the Italian Alps. Description of three novel species: *Mrakia robertii* sp. nov., *Mrakia blollopis* sp. nov. and *Mrakiella niccombsii* sp. nov. Extremophiles 14:47–59

Timling I, Walker DA, Nusbaum C, Lennon NJ, Taylor DL (2014) Rich and cold: diversity, distribution and drivers of fungal communities in patterned-ground ecosystems of the North American Arctic. Mol Ecol 23:3258–3272

Tsuji M (2016) Cold-stress responses in the Antarctic basidiomycetous yeast *Mrakia blollopis*. R Soc Open Sci 3:160106

Tsuji M, Fujiu S, Xiao N, Hanada Y, Kudoh S, Kondo H, Tsuda S, Hoshino T (2013) Cold adaptation of fungi obtained from soil and lake sediment in the Skarvsnes ice-free area, Antarctica. FEMS Microbiol Lett 346:121–130

Tsuji M, Yokota Y, Kudoh S, Hoshino T (2015) Comparative analysis of milk fat decomposition activity by *Mrakia* spp. isolated from Skarvsnes ice-free area, East Antarctica. Cryobiology 70:293–296

Tsuji M, Uetake J, Tanabe Y (2016) Changes in the fungal community of Austre Brøggerbreen deglaciation area, Ny-Ålesund, Svalbard, High Arctic. Mycoscience 57:448–451

Turchetti B, Buzzini P, Goretti M, Branda E, Diolaiuti G, D'Agata C, Smiraglia C, Vaughan-Martini A (2008) Psychrophilic yeasts in glacial environments of Alpine glaciers. FEMS Microbiol Ecol 63:73–83

Turchetti B, Thomas-Hall SR, Connell LB, Branda E, Buzzini P, Theelen B, Muller WH, Boekhout T (2011) Psychrophilic yeasts from Antarctica and European glaciers: description of *Glaciozyma* gen. nov., *Glaciozyma martinii* sp. nov. and *Glaciozyma watsonii* sp. nov. Extremophiles 15:573–586

Turchetti B, Selbmann L, Blanchette RA, Di Mauro S, Marchegiani E, Zucconi L, Arenz BE, Buzzini P (2015) *Cryptococcus vaughanmartiniae* sp. nov. and *Cryptococcus onofrii* sp. nov.: two new species isolated from worldwide cold environments. Extremophiles 19:149–159

Turk M, Plemenitaš A, Gunde-Cimerman N (2011) Extremophilic yeasts: plasma-membrane fluidity as determinant of stress tolerance. Fungal Biol 115:950–958

Turkiewicz M, Pazgier M, Kalinowska H, Bielecki S (2003) A cold-adapted extracellular serine proteinase of the yeast *Leucosporidium antarcticum*. Extremophiles 7:435–442

Turkiewicz M, Pazgier M, Donachie SP, Kalinowska H (2005) Invertase and a-glucosidase production by the endemic Antarctic marine yeast *Leucosporidium antarcticum*. Pol Polar Res 26:125–136

Uetake J, Yoshimura Y, Nagatsuka N, Kanda H (2012) Isolation of oligotrophic yeasts from supraglacial environments of different altitude on the Gulkana Glacier (Alaska). FEMS Microbiol Ecol 82:279–286

Vaca I, Faundez C, Maza F, Paillavil B, Hernandez V, Acosta F, Levican G, Martinez C, Chavez R (2013) Cultivable psychrotolerant yeasts associated with Antarctic marine sponges. World J Microbiol Biotechnol 29:183–189

Vaz ABM, Rosa LH, Vieira MLA, de Garcia V, Brandão LR, Teixeira LCRS, Moliné M, Libkind D, van Broock M, Rosa CA (2011) The diversity, extracellular enzymatic activities and photoprotective compounds of yeasts isolated in Antarctica. Braz J Microbiol 42:937–947

Vishniac HS (1985a) *Cryptococcus socialis* sp. nov. and *Cryptococcus consortionis* sp. nov., Antarctic basidioblastomycetes. Int J Syst Bacteriol 35:119–122

Vishniac HS (1985b) *Cryptococcus friedmannii*, a new species of yeast from the Antarctic. Mycologia 77:149–153

Vishniac HS (1996) Biodiversity of yeasts and filamentous microfungi in terrestrial Antarctic ecosystems. Biodivers Conserv 5:1365–1378

Vishniac HS (2002) *Cryptococcus tephrensis*, sp. nov., and *Cryptococcus heimaeyensis*, sp. nov.; new anamorphic basidiomycetous yeast species from Iceland. Can J Microbiol 48:463–467

Vishniac HS (2006a) A multivariate analysis of soil yeasts isolated from a latitudinal gradient. Microb Ecol 52:90–103

Vishniac HS (2006b) Yeast biodiversity in the Antarctic. In: Rosa C, Gabor P (eds) Biodiversity and ecophysiology of yeasts. Springer, Berlin, pp 419–440

Vishniac HS, Hempfling WP (1979) *Cryptococcus vishniacii* sp. nov., an Antarctic yeast. Int J Syst Bacteriol 29:153–158

Vishniac HS, Kurtzman CP (1992) *Cryptococcus antarcticus* sp. nov, and *Cryptococcus albidosimilis* sp. nov., basidioblastomycetes from antarctic soils. Int J Syst Bacteriol 42:547–553

Vishniac HS, Onofri S (2002) *Cryptococcus antarcticus* var. *circumpolaris* var. nov., a basidiomycetous yeast from Antarctica. A van Leeuwenhoek 83:231–233

Vishniac HS, Takashima M (2010) *Rhodotorula arctica* sp. nov., a basidiomycetous yeast from Arctic soil. Int J Syst Evol Microbiol 60:1215–1218

Wang QM, Yurkov AM, Göker M, Lumbsch HT, Leavitt SD, Groenewald M, Theelen B, Liu XZ, Boekhout T, Bai FY (2015) Phylogenetic classification of yeasts and related taxa within Pucciniomycotina. Stud Mycol 81:149–189

Xiao N, Inaba S, Tojo M, Degawa Y, Fujiu S, Hanada Y, Kudoh S, Hoshino T (2010) Antifreeze activities of various fungi and Stramenophilia isolated from Antarctica. N Am Fungi 5:215–220

Xin M, Zhou P (2007) *Mrakia psychrophila* sp. nov., a new species isolated from Antarctic soil. J Zhejiang Univ Sci B 8:260–265

Yurkov AM (2017) Temporal and geographic patterns in yeast distribution. In: Buzzini P, Lachance MA, Yurkov AM (eds) Yeasts in natural ecosystems: ecology. Springer, Heidelberg, pp 101–130

Yurkov A, Inacio J, Chernov IY, Fonseca A (2015a) Yeast biogeography and the effects of species recognition approaches: the case study of widespread basidiomycetous species from birch forests in Russia. Curr Microbiol 70:587–601

Yurkov A, Kachalkin AV, Daniel HM, Groenewald M, Libkind D, de Garcia V, Zalar P, Gouliamova DE, Boekhout T, Begerow D (2015b) Two yeast species *Cystobasidium psychroaquaticum* f.a. sp. nov. and *Cystobasidium rietchieii* f.a. sp. nov. isolated from natural environments, and the transfer of *Rhodotorula minuta* clade members to the genus *Cystobasidium*. A van Leeuwenhoek 107:173–185

Zalar P, Gunde-Cimerman N (2014) Cold-adapted yeasts in Arctic habitats. In: Buzzini P, Margesin R (eds) Cold-adapted yeasts. Springer, Berlin, pp 49–74

Zalar P, Gostinčar C, De Hoog GS, Uršič V, Sudhadham M, Gunde-Cimerman N (2008) Redefinition of *Aureobasidium pullulans* and its varieties. Stud Mycol 61:21–38

Zalar P, Novak M, De Hoog GS, Gunde-Cimerman N (2011) Dishwashers – A man-made ecological niche accommodating human opportunistic fungal pathogens. Fungal Biol 115:997–1007

Zhang T, Yao Y-F (2015) Endophytic fungal communities associated with vascular plants in the high Arctic zone are highly diverse and host-plant specific. PLoS One 10:e0130051

Zhang T, Zhang YQ, Liu HY, Su J, Zhao LX, Yu LY (2014) *Cryptococcus fildesensis* sp. nov., a psychrophilic basidiomycetous yeast isolated from Antarctic moss. Int J Syst Evol Microbiol 64:675–679

Zhang T, Wei XL, Zhang YQ, Liu HY, Yu LY (2015a) Diversity and distribution of lichen-associated fungi in the Ny-Ålesund Region (Svalbard, High Arctic) as revealed by 454 pyrosequencing. Sci Rep 5:14850

Zhang T, Wang NF, Zhang YQ, Liu HY, Yu LY (2015b) Diversity and distribution of fungal communities in the marine sediments of Kongsfjorden, Svalbard (High Arctic). Sci Rep 5:14524

Zucconi L, Selbmann L, Buzzini P, Turchetti B, Guglielmin M, Frisvad JC, Onofri S (2012) Searching for eukaryotic life preserved in Antarctic permafrost. Polar Biol 35:749–757

Chapter 12
Yeasts in Nonpolar Cold Habitats

Ciro Sannino, Giorgia Tasselli, Sara Filippucci, Benedetta Turchetti, and Pietro Buzzini

Abstract The main terrestrial cold areas (outside Antarctica and the Arctic and subarctic regions, viz., the Earth strip running approximately between 60°N and 60°S) are those covered by glaciers and permafrost soil, primarily confined in the Himalayas, Andes, and European high mountains and only little portions in other parts of the globe.

The study of microbial (prokaryotic and eukaryotic) biological diversity in nonpolar cold habitats represents a contribution to obtain a better defined picture of fungal diversity (including yeasts) inhabiting those ecosystems. The present chapter will provide an overview of both culturable and non-culturable yeast diversity found in nonpolar cold habitats. Yeasts found were identified as belonging to a number of ascomycetous and basidiomycetous species: among them, Basidiomycota dominated yeast diversity; in particular, species of the genera *Cystofilobasidium*, *Dioszegia*, *Filobasidium*, *Holtermanniella*, *Leucosporidium*, *Mrakia*, *Naganishia*, *Rhodotorula*, *Saitozyma*, *Solicoccozyma*, and *Vishniacozyma* were the most abundant. A number of new species were also found. Most of the yeast species are apparently ubiquitous in different geographical locations and exhibit some adaptation of their physiology and metabolism that increase cell protection against the damaging effects of low temperatures. Due to those physiological and metabolic adaptations, they could play an ecological role in nonpolar cold ecosystems, especially in relation to the in situ hydrolysis of complex organic macromolecules connected with the mineralization of organic matter.

Keywords Nonpolar habitats • Taxonomy • Diversity • Culturable and non-culturable yeasts

Contents

C. Sannino • G. Tasselli • S. Filippucci • B. Turchetti • P. Buzzini (✉)
Department of Agricultural, Food and Environmental Sciences, University of Perugia, Borgo XX Giugno 74, 06121 Perugia, Italy
e-mail: pietro.buzzini@unipg.it

P. Buzzini et al. (eds.), *Yeasts in Natural Ecosystems: Diversity*,
DOI 10.1007/978-3-319-62683-3_12

12.1 Introduction

More than 50% of Earth cold environments are localized in deep oceans, whereas cold terrestrial habitats, which are incredibly varied in their nature, cover approximately 25% of the world's land surface (Cavicchioli and Tortsen 2000; Buzzini and Margesin 2014). Although a clear definition of terrestrial cold habitats is difficult because their global distribution is not geographically and historically static, a consistent part of them overlaps with cryosphere, i.e., the portion of Earth surface where water occurs in solid forms, such as ice (lake, river, sea, etc.), snowpack, glaciers, and permafrost (Benn and Evans 1998; Nesje and Dahl 2000). Geological, chemical, and isotopic evidences indicate that in its history, Earth has experienced numerous periods of extensive glaciations. At present, approximately 11% of Earth's land surface is covered by ice (Hamilton et al. 2013). In many ecosystems, cold conditions are frequently associated with other limiting factors (e.g., low A_w and nutrient availability, high hydrostatic pressure, oxidative stress, and solar irradiation, etc.) which make such extreme habitats very inhospitable for life (Vincent 1988; Campbell and Claridge 2000).

One of the main factors controlling the presence of cold habitats in a given area is latitude. Antarctica (about 14 million km^2) and Arctic (about 4) are the most representative examples of terrestrial cold habitats, of which almost the totality of surface is covered by ice and snow: both areas include over 95% of Earth's cryosphere (Holdgate 1977). Below the Arctic Circle, there is the so-called subarctic region, which is an area that covers a ring of continental lands and islands surrounding the ice-covered Arctic Ocean, i.e., northern Asia (Russia, Siberia), Europe (Scandinavia), and North America (northern Canada, Alaska), and the large islands of Nova Zemlya (Russia), Spitsbergen (Norway), Iceland, and southern Greenland (Denmark). Generally, subarctic regions fall between 50°N and 70°N latitude, depending on the local climate (Benn and Evans 1998; Nesje and Dahl 2000).

Based on the inventory published in 2012 by the World Glacier Monitoring Service, the main terrestrial cold areas occurring approximately between 60°N and 60°S (outside Antarctica and the Arctic and subarctic regions) are those covered by glaciers, which are mainly confined in the Himalayas (approx. 114,800 km^2), Andes (25,500 km^2), European high mountains (3800 km^2), and New Zealand (1600 km^2). Other minor glacial areas are placed in Africa (6 km^2), North American Rocky Mountains (5 km^2), and New Guinea (3 km^2) (WGMS 2012).

Despite the well-known contribution of biotic and abiotic factors to global biogeochemical cycles, the composition and phylogenetic structure of active microbial communities sharing Earth glacial ecosystems are far to be fully described (Hamilton et al. 2013). Both the Arctic and Antarctic regions have been studied since the 1950s as important reservoirs of psychrophilic and psychrotolerant microorganisms. Early studies (Abyzov et al. 1982, 2001) found the presence of viable microorganisms in the deep ice cores drilled from Lake Vostok, Antarctica. The diversity of yeasts in polar and subpolar cold habitats is reported in Chap. 11 of this book.

Some decades later, the fungal diversity (including yeasts) of nonpolar cold worldwide regions attracted the interest of some researchers, and a number of studies were mainly focused on the cultivable portion of fungal microbiota (Margesin et al. 2002; Margesin 2009; Lazzaro et al. 2012; Buzzini et al. 2012; Buzzini and Margesin 2014; França et al. 2016). Nevertheless, the most recent studies have been performed using metagenomic approaches, in order to investigate the non-cultivable portion of microbial communities, thus giving a more complete picture of the diversity of the bacterial and fungal populations (including yeasts) sharing those ecosystems. Both culture-dependent and culture-independent investigations have revealed that nonpolar cold habitats can harbor diverse microbial communities (viz., Gram-positive and Gram-negative bacteria, archaea, yeasts, filamentous fungi, and microalgae). Some microbial taxa are ubiquitous in different geographical locations, suggesting their psychrophilic or psychrotolerant nature. A number of metagenomic studies investigated the presence of yeast operational taxonomic units (OTUs), describing them at the genus and species level, while in some others the fungal populations were studied only at the level of higher fungal taxa (i.e., classes, orders, or families), or even by simply discriminating bacterial from fungal populations, with no detailed mention on the presence of specific yeast genera and species. However, it is important to underline that yeast and filamentous fungi are both saprophytic organisms exhibiting similar physiological and ecological features. Therefore, the presence of living and metabolically active fungal taxa in a given nonpolar cold habitat could suggest the presence of chemical, physical, and ecological conditions favoring either survival or growth of both yeasts and filamentous fungi.

The present chapter will provide an overview of both culturable and non-culturable yeast diversity in nonpolar cold habitats.

12.2 Ecological Niches in Worldwide Nonpolar Cold Habitats

A consistent part of investigations on yeast diversity in nonpolar cold habitats concerned the study of microbial communities colonizing the array of habitats associated with the glacial systems, in particular sediments collected at the surface or beneath the ice masses (supraglacial and subglacial, respectively), ice cores, cryoconite holes, meltwater, snow, and soil (including permafrost). Although ice occurring in nonpolar glaciers cannot be considered as ancient as those occurring in polar and subarctic regions, ice cores from glaciers located in Asia, North and South America, and Europe should be considered extremely important, because they constitute a continuous record of chemical, physical, and biological information about climate evolution during the different ages. In particular, the reduced dimension and the general conditions of some nonpolar glaciers make them particularly subjected to global change (Shea et al. 2015).

Ice masses of glaciers may act as long-term repositories of viable microorganisms, including yeasts. Price (2000, 2007) demonstrated that biological activity may occur inside ice masses, although with extremely low metabolic rates. Indeed, due to its oligotrophic nature, most of the organic and inorganic elements available in situ are probably used mainly for repairing cellular and molecular damages than for reproducing. Although ice cores can be really regarded as extreme habitats because of their severe ecological conditions (a combination of subzero temperatures, high hydrostatic pressure, low A_w, and nutrient availability), a few peculiar features could apparently increase the chances of microbial life to overcome the adverse effects of such limiting condition: (a) the high hydrostatic pressure occurring at the basal ice layers can reach the pressure of the melting point, thus determining an increase of both temperature and A_w, and (b) the presence of unfrozen microlayers of water around solid impurities or gas bubbles (viz., "hydratation distance") enclosed into the ice cores, as well as around crevasses and water channels. These micro-amounts of water contain little concentrations of organic and inorganic elements, polymeric compounds, and particulate matter (Price 2000, 2007). This evidence allows to postulate that microbial populations could survive within ice cores, thus playing an important role in nutrient cycling inside glacier masses and ice sheets (Castello and Rogers 2005; D'Amico et al. 2006; Price 2007; Deming 2009).

Snow can be considered a typical oligotrophic niche occurring in nonpolar cold habitats. Snowfalls can form snow layers (labeled as "snowpacks") that accumulate in high altitudes where the climate allows cold weather conditions for extended periods during the year. Snow also contributes to increase the mass of glaciers in their accumulation zone. Due to its oligotrophic nature, the ecological role of snow can be substantially considered like a carrier because it provides inoculation of microbial cells and supplies little amounts of nutrients to cold surfaces where it falls (Stibal et al. 2007). Complex microbial consortia were found in snowpack covering the surface of a few glaciers located in the Alps and Andes (De Angelis and Gaudichet 1991; Amato et al. 2007; Turchetti et al. 2013).

Paterson (2002) classified waters occurring in glacial habitats on the basis of their origin: (a) streaming of meltwater on the glacier surface due to ice and snow ablation during summer, (b) percolation of superficial water into ice crevasses and channels, (c) ice melting induced by the high hydrostatic pressure in the basal ice layers, (d) water entrapped inside pockets and bubbles originated from ice formation, and (e) lakes and lagoons formed within the margins of the glacier or on the glacier mass. Meltwaters are considered an important connection way among different niches within the same glacier or even as a natural link between a glacial habitat and glacier forefield (Turchetti et al. 2014).

The term "cryoconite" (ice dust) refers to small (0.1–3 mm in diameter) dark-colored, wind-borne particles at the bottom of cryoconite holes, which include mineral components of soils, sediments, and organic components (i.e., litter, pollen, prokaryotic, and eukaryotic cells). Cryoconites are formed inside water-filled little holes on the glacial surface created as a consequence of the so-called black body effect, i.e., the warming due to solar radiation of the dark particles deposited on the glacier surface (because of their lower albedo) that provokes the accelerated melting of the ice below the cryoconite and thus the formation of the hole (Takeuchi et al. 2001; Cameron et al. 2012). Cryoconite holes are found worldwide in polar and nonpolar glaciers (MacDonell and Fitzsimons 2008; Miteva 2008; Anesio et al. 2009). Due to their eutrophic conditions, they could be considered "hot spots" of extremely complex prokaryotic and eukaryotic microbial consortia, including yeasts (Margesin et al. 2002; Hodson et al. 2008; Lee et al. 2011; Anesio and Laybourn-Parry 2012; Turchetti et al. 2014). In addition, they could serve as biological refuges during extended periods of subzero temperatures; members of the microbial communities that survive inside cryoconite holes could also ensure the "reinoculation" of the surrounding environment during warmer periods (Christner et al. 2003; Margesin and Miteva 2011; Singh and Singh 2012; Turchetti et al. 2014). Members of individual cryoconite communities were frequently found to be specific to individual locations, suggesting that they may be disseminated primarily via wind transportation and/or during glacial melt from adjacent surrounding environments (Christner et al. 2003; Cameron et al. 2012). On the other hand, long-distance wind transport may provide a mechanism for colonizing newly exposed areas after glacier retreats (Wharton et al. 1985).

Supraglacial and subglacial sediments occurring in glaciers are also considered eutrophic habitats, because they contain significant amounts of organic carbon, nitrogen, and phosphorous. Accordingly, they constitute an important reservoir of metabolically active heterotrophic prokaryotic and eukaryotic microbial communities that could play a significant role in some important biogeochemical activities (such as organic matter decomposition and nutrient cycling) at low temperature (Turchetti et al. 2008, 2013, 2014; Branda et al. 2010; Buzzini et al. 2012).

Although cold rock-associated ecosystems are characterized by an impressive mixture of stressing conditions (i.e., cold, extremely low a_w, ultra-oligotrophy, high solar irradiation, sometimes very acidic conditions, etc.) that make them very inhospitable for life, a few studies found the presence of complex microbial consortia colonizing such peculiar ecosystems (Onofri et al. 2004; Selbmann et al. 2014). The possible role of extremophilic fungal populations colonizing mineral surfaces in biogenic weathering and mineral diagenesis and dissolution has also been supposed (Gorbushina 2007; Finlay et al. 2009).

12.3 Yeasts in Cold Regions of Asia

The Himalayas are an Asian mountain range that separates the plains of the Indian subcontinent from the Tibetan Plateau. The Himalayan range has the Earth's highest peak (Mount Everest) and includes over a hundred mountains exceeding 7000 m a.s.l. in elevation. According to recent researches, most of the estimated 5500 glaciers in the Himalayan range will disappear or drastically reduce their volume approx. by 70–99% by 2100 as the consequence of the climate change, with catastrophic hydrological consequences (Shea et al. 2015). Despite the Himalayan range being certainly the largest nonpolar glacial habitat of the globe, the number of microbiological studies carried out is still very limited. This is probably due to the very difficult accessibility of the main glaciers present in that vast area, which severely limit the collection and the transportation of samples to the closest laboratories. Therefore, yeast diversity in the Himalayan range is still little studied.

Shivaji et al. (2008) isolated psychrophilic yeasts from the soil surrounding Roopkund Lake, a high-altitude (approx. 5000 m) glacial lake in the Uttarakhand Himalayan range (India). The species *Cryptococcus gastricus* (currently *Goffeauzyma gastrica*) and *Rhodotorula himalayensis* (*Phenoliferia himalayensis*) were found. More recently, a study identified a number of culturable yeast strains from rocky samples. A psychrotolerant *Cryptococcus* sp. strain (now reassigned to the genus *Sirobasidium* sp. according to the phylogenetic position of the strain as defined in the original paper) was found in the microbial endolithic communities colonizing igneous rocks collected at the south side of K2 Mountain (approx. 5000 m a.s.l.), Karakorum, Pakistan (Selbmann et al. 2014).

The Tibetan Plateau, also known in China as the Qinghai–Tibet Plateau or Himalayan Plateau, is the highest plateau in the world (approx. 2,500,000 km^2) with an average elevation of around 4500 m. Sometimes termed the Third Pole, the Tibetan Plateau contains thousands of glaciers that provide water to the surrounding regions. Both culture-dependent and culture-independent investigations have revealed that cold habitats of the Tibetan Plateau are colonized by complex microbial consortia (Hu et al. 2015a) containing some basidiomycetous yeast species, namely, *Cryptococcus adeliensis* (currently *Naganishia adeliensis*), *Cryptococcus albidus* (*Naganishia albida*), *Cryptococcus magnus* (*Filobasidium magnum*), *Cryptococcus nyarrowii* (*Holtermanniella nyarrowii*), *Cryptococcus tephrensis* (*Vishniacozyma tephrensis*), *Cryptococcus victoriae* (*Vishniacozyma victoriae*), *Dioszegia fristingensis*, *Leucosporidiella creatinivora* (*Leucosporidium creatinivorum*), *Rhodotorula glutinis*, *Rhodosporidium babjevae* (*Rhodotorula babjevae*), *Rhodotorula vanillica* (*Sampaiozyma vanillica*), *Rhodotorula ingeniosa* (*Sampaiozyma ingeniosa*), *Rhodotorula minuta* (*Cystobasidium minutum*), *Rhodotorula laryngis* (*Cystobasidium laryngis*), and *Rhodotorula mucilaginosa* (Li et al. 2012; Hu et al. 2014, 2015a, b).

Uetake et al. (2011) used both culture-dependent and culture-independent methods to study yeast abundance and diversity in snow and ice cores collected from a high-altitude glacier (approx. 4500 m a.s.l.) placed in Belukha Mountain, which is a massif that rises along the to the border between Russia and Kazakhstan. This is the highest mountain in Altai Mountains, which is a part of Altai-Sayan Mountains. 26S rRNA clonal direct amplification analysis (after melting) of ice cores revealed the presence of strains of the genus *Rhodotorula*. Based on the analysis of peaks in yeast concentration, the role of occasional melting as one of the factors driving yeast propagation from surface to deep snow layers has been hypothesized.

The Pamir Mountains are a mountain range in Central Asia at the junction of the Himalayas with the Tian Shan, Karakoram, Kunlun, Hindu Kush, and Hindu Raj ranges. They are among the world's highest mountains. The three highest mountains in the Pamir area are Ismoil Somoni Peak (7495 m a.s.l.), Ibn Sina Peak (7134 m a.s.l.), and Peak Korzhenevskaya (7105 m a.s.l.). In the Eastern Pamir, China's Kongur Tagh is the highest at 7649 m a.s.l. Babjeva and Reshetova (1971) explored the yeast diversity in cold soils of the Pamir Mountains. A number of species were found, namely, *Torulopsis candida* (*Candida saitoana*) and *Torulopsis famata* (*Debaryomyces hansenii*) among Ascomycota, *Cryptococcus laurentii* (currently *Papiliotrema laurentii*), *Cryptococcus luteolus* (*Hannaella luteola*), *Na. albida*, *Naganishia diffluens*, *Rh. glutinis*, *Rh. mucilaginosa*, *Rhodotorula aurantiaca* (*Buckleyzyma aurantiaca*), *Rhodotorula flava* (*Saitozyma flava*), *Sporobolomyces holsaticus* (*Sporobolomyces johnsonii*), and *Torulopsis aeria* (*Solicoccozyma aeria*) among Basidiomycota. The list of yeast species found in cold habitats of Asia is summarized in Table 12.1.

12.4 Yeasts in Cold Habitats of North America

Excluding Alaska and the northern part of Canada (all falling into the subarctic region), the Rocky Mountains can be considered the main North American cold habitats in which the diversity of yeasts and fungi was studied. They are the major mountain range in western North America (the highest peak is Mount Elbert, 4400 m a.s.l.). They are more than 4800 km long, from the north of British Columbia (West Canada) to New Mexico (southwest of USA), and contain a variety of cold habitats, including a number of glaciers.

The presence of active prokaryotic and eukaryotic OTUs, including some ascomycetous and basidiomycetous species, i.e., *Kloeckera lindneri*, *Taphrina vestergrenii*, *Sporobolomyces falcatus* (*Colacogloea falcata*), *Itersonilia perplexans*, *Kriegeria eriophori*, *Mastigobasidium intermedium* (*Leucosporidium intermedium*), *Malassezia pachydermatidis*, *Pseudozyma aphidis* (*Moesziomyces aphidis*), and *Sporobolomyces tsugae* (*Slooffia tsugae*), was found in samples collected at Robertson Glacier (Alberta, Canada). The phylogenetic structure of microbial consortia found in subglacial sediment indicated the presence of a greater diversity than that observed on the ice surface, thus confirming the hypothesis that subglacial habitats could be considered as endogenous eutrophic ecosystems

Table 12.1 Yeast species found in cold habitats of Asia

Species	Original taxonomic designation	Source					Locality					References
		S	Sn	R	Ic	Sp	TP	HI	KP	BG	PM	
Ascomycetous yeasts												
Candida saitoana	*Torulopsis candida*	■									■	Babjeva and Reshetova 1971
Debaryomyces hansenii	*Torulopsis famata*	■									■	Babjeva and Reshetova 1971
Basidiomycetous yeasts												
Buckleyzyma aurantiaca	*Rhodotorula aurantiaca*	■									■	Babjeva and Reshetova 1971
Cystobasidium laryngis *	*Rhodotorula laryngis*	■					■					Hu et al. 2014
Cystobasidium minutum *	*Rhodotorula minuta*	■					■					Hu et al. 2014
Cystofilobasidium infirmominiatum *		■					■					Hu et al. 2014
Dioszegia fristingensis *		■					■					Hu et al. 2014
*Filobasidium magnum**	*Cryptococcus magnus*	■					■					Hu et al. 2014
(*Filobasidium* sp.)	*Cryptococcus* sp.			■					■			Selbmann et al. 2014
Goffeauzyma gastrica	*Cryptoccoccus gastricus*	■						■				Shivaji et al. 2008
Hannaella luteola	*Cryptococcus luteolus*	■									■	Babjeva and Reshetova 1971
Holtermanniella nyarrowii *	*Cryptococcus nyarrowii*	■					■					Hu et al. 2014
Leucosporidium creatinivurum *	*Leucosporidiella creatinivora*	■					■					Hu et al. 2014
(*Naganishia* sp. *)	*Cryptococcus* sp.	■					■					Hu et al. 2014
Naganishia adeliensis	*Cryptococcus adeliensis*	■				■	■					Li et al. 2012; Hu et al. 2014
Naganishia albida #	*Cryptococcus albidus*	■					■				■	Babjeva and Reshetova 1971; Hu et al. 2014
Naganishia diffluens	*Cryptococcus diffluens*	■									■	Babjeva and Reshetova 1971
Papiliotrema laurentii	*Cryptococcus laurentii*	■									■	Babjeva and Reshetova 1971
(*Phenoliferia* sp.)	*Rhodotorula* sp.		■		■					■		Uetake et al. 2011

Rhodotorula sp.												Hu et al. 2014
Rhodotorula babjevae *	*Rhodosporidium babjevae*											Hu et al. 2014
Rhodotorula glutinis #												Babjeva and Reshetova 1971; Hu et al. 2014
Rhodotorula mucilaginosa												Babjeva and Reshetova 1971; Li et al. 2012; Hu et al. 2015
Saitozyma flava	*Cryptococcus flavus*											Babjeva and Reshetova 1971
Sampaiozyma ingeniosa *	*Rhodotorula ingeniosa*											Hu et al. 2014
Sampaiozyma vanillica *	*Rhodotorula vanillica*											Hu et al. 2014
Solicoccozyma aeria	*Torulopsis aeria*											Babjeva and Reshetova 1971
Sporobolomyces johnsonii	*Sporobolomyces holsaticus*											Babjeva and Reshetova 1971
Tilletiopsis albescens												Li et al. 2012
Vishniacozyma tephrensis *	*Cryptococcus tephrensis*											Hu et al. 2014
Vishniacozyma victoriae *	*Cryptococcus victoriae*											Hu et al. 2014

Genera in brackets = authors arbitrary new genus/species nomenclature designation based on recent literature for that strains/OTUs for which a clear phylogenetic position was defined in the original paper

S soil; *Sn* snow; *R* rock; *Ic* ice cores; *Sp* Spring; *TP* Tibetan Plateau, China; *HI* Himalaya, India; *KP* = Karakorum, Pakistan; *BG* Belukha glacier, Russia/Kazakhstan; *PM* Pamir Mountains, Central Asia

[a]Species found using culture-independent approach

[b]Species found using both culture-dependent and culture-independent approach

harboring microbial life and potentially impacting on biogeochemical cycles over extended periods of time (Hamilton et al. 2013).

A recent study on non-culturable fungal and bacterial successions along a glacier forefront of Lyman Glacier (approx. 1900 m a.s.l., North Cascade Mountain range, WA, USA) found the presence of *Cryptococcus skinneri* (currently *Phaeotremella skinneri*) in soil samples. Results indicated that many microbial taxa were non-randomly distributed across the glacier foreland (Brown and Jumpponen 2014). More recently, Turchetti et al. (2015) found the species *Cryptococcus onofrii* (*Naganishia onofrii*) among the eukaryotic culturable portion of the microbial community in a high mountain soil collected at Krumholtz, Colorado, USA.

The list of yeast species found in cold habitats of North America is reported in Table 12.2.

12.5 Yeasts in Cold Habitats of South America

South American cold habitats are substantially associated to Andean glacier ecosystems, which are subjected to various climatic regimes. The Andes are the longest continental mountain range and the world's highest outside of Asia. They are about 7000 km long, from 200 to 700 km wide and exhibit an average elevation of about 4000 m a.s.l. (Lliboutry 1998). The highest peak is Mount Aconcagua, which rises to an elevation of 6962 m a.s.l. A number of glaciers occur along the Andes: from the very small glaciers or ice masses ("glacierets") in North Chile to the approx. 13,000 km^2 glaciers in South Patagonia. Los Glaciares National Park and the Patagonian ice fields (Hielos Patagónicos) are the largest nonpolar ice masses in the Southern Hemisphere. Perito Moreno Glacier is located inside the Patagonian ice fields (Lliboutry 1998; Skvarca et al. 2003; Stuefer et al. 2007). The Andean glaciers reached their maximum extents in the Little Ice Age, while they have gone through a gradual retreat in more recent ages, with two periods of accelerated melt: the first in the late 1800s and the second, much larger and accelerated, in the late 1900s and early 2000s (Villarosa et al. 2008).

Like other glacial ecosystems, Patagonian glaciers harbor psychrophilic and psychrotolerant yeast life. A number of ecological studies have been carried out in the area of Nahuel Huapi National Park (NHNP), Patagonia, Argentina. NHNP exhibits a great variety of glacially formed oligotrophic to ultra-oligotrophic cold water bodies fed by glacier meltwater and sometimes surrounded by dense native forests (Quirós and Drago 1985; Díaz et al. 2000). A number of culturable ascomycetous and basidiomycetous yeasts were found in high-altitude lakes and lagoons of NHNP: *Vishn. victoriae*, *Leucosporidiella fragaria* (*Leucosporidium fragarium*), and *Rh. mucilaginosa* were the most frequently isolated species (Libkind et al. 2003, 2009; de García et al. 2007, 2010a, b; Brandão et al. 2011). Likewise, both psychrophilic and psychrotolerant Basidiomycota (in particular *Vishn. victoriae*) were observed in the ecosystems of Perito Moreno and Mount Tronador glaciers. A few newly described species were found in Patagonian cold habitats: *Cryptococcus*

Table 12.2 Yeast species found in cold habitats of North America

Species	Original taxonomic designation	Source				Locality		References
		S	Sd	Sn	Ch	RG	RM	
Ascomycetous yeasts								
Kloeckera lindneri *								Hamilton et al. 2013
Taphrina vestergrenii *								Hamilton et al. 2013
Basidiomycetous yeasts								
Colacogloea falcate *	*Sporobolomyces falcatus*							Hamilton et al. 2013
Itersonilia perplexans *								Hamilton et al. 2013
Kriegeria eriophori *								Hamilton et al. 2013
Leucosporidium intermedium *	*Mastigobasidium intermedium*							Hamilton et al. 2013
Leucosporidium sp. *								Hamilton et al. 2013
Malassezia pachydermatis *								Hamilton et al. 2013
Naganishia onofrii	*Cryptococcus onofrii*							Turchetti et al. 2015
Phaeotremella skinneri *	*Cryptococcus skinneri*							Brown and Jumpponen 2014
Moesziomyces aphidis *								Hamilton et al. 2013
Slooffia tsugae *	*Sporobolomyces tsugae*							Hamilton et al. 2013
Sporobolomyces sp. *								Hamilton et al. 2013

S soil; *Sd* sediments; *Sn* snow; *Ch* cryoconite holes; *RG* Robertson Glacier, Alberta, Canada; *RM* Rocky Mountains, USA

[a]Species found using culture-independent approach

fonsecae (now *Papiliotrema fonsecae*), *Cryptococcus psychrotolerans* (*Vishniacozyma psychrotolerans*), *Cryptococcus frias* (*Papiliotrema frias*), and *Cryptococcus tronadorensis* (*Fonsecazyma tronadorensis*). Cold-adapted yeasts were able to hydrolyze organic macromolecules at low temperatures, thus suggesting their ecological role as in situ organic matter decomposers and nutrient cyclers (de Garcia et al. 2012a, b).

An interesting metagenomic study on non-culturable microbiota performed on the cold desert of Atacama (Chile) has shown that mineral soils collected above 6000 m a.s.l. are dominated by basidiomycetous yeast taxa, which increased their relative abundance in soils subjected to frequent freeze–thaw cycles (Vimercati et al. 2016). On the other hand, Selbmann et al. (2014) found the presence of *Cryptococcus friedmannii* (*Naganishia friedmannii*) inside the eukaryotic culturable portion of a cryptoendolithic microbiota colonizing igneous rocks collected at 4550 m a.s.l. in the Salar de Uyuni desert, Andes, Bolivia.

Environmental factors influencing culturable yeast community in cold pristine Andean forests were studied by Mestre et al. (2014, 2016). Yeast community structure in the investigated forest sites varied in dependence to some environmental parameters, all of which were strongly shaped by the climate. The species *Cryptococcus podzolicus* (*Saitozyma podzolica*) was frequently isolated in moist and warmer soils, while *Trichosporon porosum* (*Apiotrichum porosum*) and *Cryptococcus phenolicus* (*Solicoccozyma phenolica*) were more prominent in cold mountain soils with higher pH and lower water content (Mestre et al. 2014, 2016).

Yeast species isolated from cold habitats of South America are resumed in Table 12.3.

12.6 Yeasts in Cold Habitats of Europe

European cold habitats (with the exclusion of Scandinavia, Iceland, and Greenland; see Chap. 11 of this book) are almost entirely confined to glaciers of the Alps, Pyrenees, and Apennines (Messerli 1980). The Alps are one of the great mountain range systems of Europe, whose length is approximately 1200 km. The whole Alpine area contains many peaks higher than 4000 m a.s.l. including the highest mountain in Europe (Mont Blanc, 4810 m a.s.l.). Alpine glaciers, which account only for a tiny part of the total Earth's ice masses, lost over one-third of their total surface in the last century and almost 50% by the early 2000s. It was predicted that in the Alps, most small glaciers (80% of total glacial coverage, which represents an important contribution to water resources of surrounding areas) could disappear in the next decades (Cannone et al. 2008). In this context, global warming may have a strong impact on Alpine mountains, and a dramatic scenario for the future, including complete deglaciation of entire European mountain ranges, can be depicted (Zemp et al. 2006; Pecci et al. 2008).

Habitats of both polluted and nonpolluted Alpine glaciers have recently been studied as sources of psychrophilic and psychrotolerant yeast life: overall,

Table 12.3 Yeast species found in cold habitats of South America

Species	Original taxonomic designation	Source					Locality					References
		S	R	Ic	Mw	Cw	PA	PLk	PLg	PR	CD	
Ascomycetous yeasts												
Ambrosiozyma sp.		■					■					Mestre et al. 2014
Candida sp.		■		■	■		■					de García et al. 2012b; Mestre et al. 2016
Candida ralunensis		■					■					Mestre et al. 2016
Candida maritima		■			■		■					de García et al. 2012b; Mestre et al. 2014, 2016
Candida mesenterica				■	■		■					de García et al. 2012b
Candida parapsilosis				■	■			■				Brandão et al. 2011
Candida railenensis		■					■					Mestre et al. 2014
Candida saitoana		■					■					Mestre et al. 2014, 2016
Candida sakè		■			■		■	■				Brandão et al. 2011; Mestre et al. 2014
Debaryomyces hansenii				■		■	■	■				de García et al. 2007; Brandão et al. 2011; de García et al. 2012b
Lachancea nothofagi		■					■					Mestre et al. 2014
Saccharomyces eubayanus		■					■					Mestre et al. 2014
Wickerhamomyces sp.		■					■					Mestre et al. 2014, 2016
Wickerhamomyces patagonicus				■			■					de García et al. 2010a, 2012b
Basidiomycetous yeasts												
Apiotrichum porosum	*Trichosporon porosum*	■					■					Mestre et al. 2014
Bannozyma yamatoana	*Bensingtonia yamatoana*			■			■					de García et al. 2012b
(*Curvibasidium* sp.)	*Rhodotorula* sp.	■					■					Mestre et al. 2014
Cutaneotrichosporon moniliiforme	*Trichosporon moniliiforme*	■					■					Mestre et al. 2014
(*Cystobasidium* sp.)	*Rhodotorula* sp.			■			■					de García et al. 2012b

(continued)

Table 12.3 (continued)

Cystobasidium laryngis	*Rhodotorula laryngis*											de García et al. 2007; Brandão et al. 2011
Cystobasidium minutum	*Rhodotorula minuta*											Libkind et al. 2003
Cystobasidium pinicola	*Rhodotorula pinicola*											Brandão et al. 2011
Cystofilobasidium capitatum												Libkind et al. 2003; Brandão et al. 2011; Mestre et al. 2014
Cystofilobasidium infirmominiatum												Libkind et al. 2003; Brandão et al. 2011; Mestre et al. 2014
Cystofilobasidium lacus-mascardii												Libkind et al. 2009
Cystofilobasidium macerans												de García et al. 2007
Dioszegia butyracea												de García et al. 2012b
Dioszegia crocea												de García et al. 2007, 2012b
Dioszegia fristingensis												de García et al. 2007, 2012b
Dioszegia hungarica												de García et al. 2007; Brandão et al. 2011
Filobasidium wieringae	*Cryptococcus wieringae*											Brandão et al. 2011; de García et al. 2012b
Filobasidium stepposum	*Cryptococcus stepposus*											Brandão et al. 2011
Filobasidium magnum	*Cryptococcus magnus*											Brandão et al. 2011
Fonsecazyma tronadorensis	*Cryptococcus tronadorensis*											de García et al. 2012a
Gelidatrema spencermartinsiae	*Cryptococcus spencermartinsiae*											de García et al. 2010b, 2012b
Goffeauzyma gastrica	*Cryptococcus gastricus*											Mestre et al. 2014
(*Heterocephalacria* sp.)	*Cryptococcus* sp.											de García et al. 2012b
Holtermanniella sp.												de García et al. 2012b; Mestre et al. 2016
Holtermanniella festucosa	*Cryptococcus festucosus*											de García et al. 2007, 2012b; Brandão et al. 2011
Holtermanniella takashimae												Mestre et al. 2016
Holtermanniella wattica	*Cryptococcus watticus*											Mestre et al. 2014
Itersonilia pannonica	*Udeniomyces pannonicus*											de García et al. 2007, 2012b
(*Kwoniella* sp.)	*Cryptococcus* sp.											de García et al. 2012b
Kwoniella dendrophila	*Bullera dendrophila*											Brandão et al. 2011
Kwoniella heveanensis	*Cryptococcus heveanensis*											Brandão et al. 2011
Leucosporidium creatinivorum	*Leucosporidiella creatinivora*											de García et al. 2007
Leucosporidium fragarium	*Leucosporidiella fragaria*											de García et al. 2007

Leucosporidium intermedium	*Mastigobasidium intermedium*			■			■					de García et al. 2012b
Mrakia sp.				■			■					de García et al. 2012b
Mrakia aquatica	*Mrakiella aquatica*			■			■					de García et al. 2012b
Mrakia frigida	*Leucosporidium frigidum*				■		■					de García et al. 2007
Mrakia robertii				■			■					de García et al. 2012b
(*Naganishia* sp.*)	*Cryptococcus* sp.	■					■					Mestre et al. 2014; Vimercati et al. 2016
Naganishia adeliensis	*Cryptococcus adeliensis*					■		■				de García et al. 2007; Brandão et al. 2011
Naganishia bhutanensis	*Cryptococcus bhutanensis*	■					■					Mestre et al. 2014
Naganishia diffluens	*Cryptococcus diffluens*					■		■				Brandão et al. 2011
Naganishia friedmannii	*Cryptococcus friedmannii*		■								■	Selbmann et al. 2014
Naganishia globosa	*Cryptococcus saitoi*				■			■				Brandão et al. 2011
Naganishia globosa/ friedmannii	*Cryptococcus saitoi/friedmannii*	■					■					Mestre et al. 2014
(*Papiliotrema* sp.)	*Cryptococcus* sp.				■		■					de García et al. 2012b
Papiliotrema frias	*Cryptococcus frias*				■		■					de García et al. 2012a
Papiliotrema laurentii	*Cryptococcus laurentii*				■	■	■		■	■		de García et al. 2007
(*Phenoliferia* sp.)	*Rhodotorula* sp.				■		■					de García et al. 2012b
Piskurozyma cylindrica	*Cryptococcus cylindricus*	■			■		■			■		de García et al. 2007; Mestre et al. 2016
Piskurozyma filicatus	*Cryptococcus filicatus*	■					■					Mestre et al. 2014
Rhodosporidiobolus colostri	*Rhodotorula colostri*	■			■	■	■	■	■	■		de García et al. 2007; Brandão et al. 2011; Mestre et al. 2014
Rhodotorula babjevae	*Rhodosporidium babjevae*					■		■	■			Libkind et al. 2003
Rhodotorula diobovata	*Rhodosporidium diobovatum*				■			■	■			Brandão et al. 2011
Rhodotorula kratochvilovae	*Rhodosporidium kratochvilovae*					■		■	■			Libkind et al. 2003; de García et al. 2007
Rhodotorula mucilaginosa				■		■	■	■	■			Libkind et al. 2003; de García et al. 2007; Brandão et al. 2011; de García et al. 2012b
Saitozyma podzolica	*Cryptococcus podzolicus*	■					■					Mestre et al. 2014
Solicoccozyma aeria	*Cryptococcus* cf. *aerius*	■					■					Mestre et al. 2014
Solicoccozyma phenolica	*Cryptococcus phenolicus*	■					■					Mestre et al. 2014
(*Solicoccozyma* sp.)	*Cryptococcus* sp.	■					■					Mestre et al. 2014
Solicoccozyma terrea	*Cryptococcus terreus*	■					■					Mestre et al. 2014

(continued)

Table 12.3 (continued)

Solicoccozyma terricola	*Cryptococcus terricola*	■		■			■					de García et al. 2012b; Mestre et al. 2014
Sporobolomyces metaroseus	*Sporidiobolus metaroseus*					■		■				Libkind et al. 2003
Sporobolomyces ruberrimus				■		■	■	■	■			Libkind et al. 2003; de Garcia et al. 2007, 2012b
Sporobolomyces salmonicolor	*Sporidiobolus salmonicolor*					■		■				Libkind et al. 2003
Tausonia pullulans	*Guehomyces pullulans*	■		■	■		■	■				Brandão et al. 2011; de García et al. 2012b; Mestre et al. 2014, 2016
Trichosporon sp.		■					■					Mestre et al. 2014, 2016
Udeniomyces megalosporus				■			■					de García et al. 2012b
Udeniomyces pyricola				■			■					de García et al. 2012b
Vanrija albida	*Asterotremella albida*, *Sporobolomyces albidus*	■					■					Mestre et al. 2016
Vishniacozyma foliicola	*Cryptococcus folliicola*				■		■					de García et al. 2012a
(*Vishniacozyma* sp.)	*Cryptococcus* sp.				■		■					de García et al. 2012b
Vishniacozyma tephrensis	*Cryptococcus tephrensis*				■			■				Brandão et al. 2011
Vishniacozyma victoriae	*Cryptococcus victoriae*			■	■	■	■	■		■		Brandão et al. 2011; de García et al. 2012a, 2012b
Yeast-like organisms												
Aureobasidium pullulans				■	■		■	■				Brandão et al. 2011; de García et al. 2012b
Phaeococcomyces sp.				■			■					de García et al. 2012b

Genera in brackets = authors arbitrary new genus/species nomenclature designation based on recent literature for that strains/OTUs for which a clear phylogenetic position was defined in the original paper

S soil; *R* rock; *Ic* ice cores; *Mw* meltwater; *Cw* cold water; *Cr* cold river; *PA* Patagonian Andes, Argentina; *PLk* Patagonian cold lakes, Argentina; *PLg* Patagonian cold lagoons, Argentina; *PR* Patagonian cold rivers, Argentina; *CD* Cold desert, Bolivia

[a]Species found using culture-independent approach

Basidiomycota dominated yeast diversity (Bergauer et al. 2005; Buzzini et al. 2005; Krallish et al. 2006; Turchetti et al. 2008; Brunner et al. 2011). The species *Cryptococcus gilvescens* (currently *Goffeauzyma gilvescens*) accounted for over 50% of the total yeasts isolated from Italian Alpine glaciers (Turchetti et al. 2008), while a few novel species were described: *Mrakia robertii* and *Glaciozyma watsonii* (Thomas-Hall et al. 2010; Turchetti et al. 2011). The relationships between some abiotic parameters (pH, dry weight, organic carbon, nitrogen, and phosphorous) on culturable yeast diversity was investigated in two neighboring Alpine glaciers placed on the slopes of Mont Blanc (Italy and France). Overall, the most represented species were *Na. adeliensis*, *Cryptococcus antarcticus* (*Naganishia antarctica*), *Goff. gastrica*, *Goff. gilvescens*, *Cryptococcus terricola* (*Solicoccozyma terricola*), *Vishn. victoriae*, *Mr. robertii*, *Rhodotorula glacialis* (*Phenoliferia glacialis*), and *Rhodotorula psychrophenolica* (*Phenoliferia psychrophenolica*) (Turchetti et al. 2013). Some species have been regularly observed in worldwide cold environments, while some others are regarded as ubiquitous ones, e.g., the dimorphic yeast-like fungus *Aureobasidium pullulans*, *Cryptococcus chernovii* (*Filobasidium chernovii*), *Cryptococcus diffluens* (*Nag. diffluens*), and *Cyst. laryngis* (Buzzini et al. 2012). Principal component analysis (PCA) revealed that organic carbon, nitrogen, and phosphorous are selectively related to culturable yeast abundance and diversity (Turchetti et al. 2013). A few novel species were also described: *Cryptococcus vaughanmartiniae* (*Naganishia vaughanmartiniae*) and *Na. onofrii* (Turchetti et al. 2015).

Although some studies on microbiota occurring in cryoconite holes found the presence of a composite microbiota (Christner et al. 2003; Anesio et al. 2009; Stibal et al. 2012), there is still little information on the role of yeasts in such habitats (Margesin et al. 2002; Singh and Singh 2012; Turchetti et al. 2014). Alpine cryoconites of an Austrian glacier were found to contain both culturable aerobic heterotrophic bacteria and yeasts able to hydrolyze organic macromolecules: *Goff. gilvescens* was the dominant species (Margesin et al. 2002, 2005; Bergauer et al. 2005; Turchetti et al. 2014). The description of novel yeast species, namely, *Phen. glacialis*, *Phen. psychrophenolica*, and *Mrakiella cryoconiti* (*Mrakia cryoconiti*), from the Alpine glacier cryoconite indicates the presence of a typical microbial community (Margesin et al. 2007; Margesin and Fell 2008; Margesin and Miteva 2011).

The effect of altitude (from approx. 500 to 2000 m) and season on abundance and diversity of the culturable bacterial and yeast communities were studied in South Tyrolean soils. Yeast diversity was essentially season dependent, while site- and altitude-specific effects on yeast diversity were only detected in spring. Species belonging to the classes Dothideomycetes, Saccharomycetes Tremellomycetes, and Mycrobotryomycetes were observed. The yeast species were predominantly Basidiomycota: *Ap. porosum* and *Sol. terricola* accounted for 15–55 and 25–75%, respectively, of the yeast diversity cultivated at 4 °C and a consistent part (5–40% and 5–45%, respectively) of that observed at 20 °C. On the contrary, *Sa. podzolica* was the most abundant species isolated at 20 °C. Some species were exclusively found at 4 °C, thus exhibiting a psychrophilic habit: *Bullera alba*,

Cystofilobasidium capitatum, *Dioszegia hungarica*, *Itersonilia pannonica*, *Mrakia gelida*, *Rhodosporidiobolus colostri*, and *Vishn. victoriae*. On the contrary, the species *Babjeviella inositovora*, *Deb. hansenii*, *Vanrija musci*, and *Yamadamyces rosulatus* were found exclusively at 20 °C (França et al. 2016).

The considerable retreat of the glaciers due to global warming caused by ongoing climate change annually releases a growing portion of newly exposed soil. Microbial prokaryotic and eukaryotic populations colonizing such new soils have been studied in recent years through metagenomic approaches. Kuhnert et al. (2012) investigated the yeast population of silty soils collected at the forefront of the Rotmoosferner glacier (approx. 2300 m a.s.l.), which constitutes a primary successional site in the Ötz valley in the Stubaier Alpen (Austrian Central Alps). The most abundantly detected fungus (22% of all OTUs) was the basidiomycetous yeast *Na. albida*, followed by the aquatic hyphomycete *Tetracladium maxilliforme* (9%). Saprophytic fungi (50%) and basidiomycetous yeasts (31%) dominated winter fungal communities (Kuhnert et al. 2012).

Likewise, Rime et al. (2015, 2016a, b) found basidiomycetous OTUs associated to the genera *Naganishia*, *Solicoccozyma* (both former *Cryptococcus*), *Leucosporidium*, *Malassezia*, and *Mrakia* in ice, sediments, and meltwater stream leaving the glacier forefront of the Damma Glacier (Switzerland). *Leucosporidium* and *Dioszegia* were also found in snow and sediments, while OTUs associated to the former phenotypic genera *Cryptococcus* and *Rhodotorula* were also found. The yeast population structure observed in snow and sediments were similar, despite the fact that these habitats were not always spatially connected. Interestingly, Lazzaro et al. (2015) correlated the degradation/oxidation profiles of polymeric substrates (BIOLOG) of snow samples collected from 1930 to 2519 m a.s.l. at the Tiefen Glacier (Canton Uri, Switzerland) to the presence of cold-adapted lipolytic microorganisms, including yeasts.

The presence of permafrost was found in some Alpine areas (Boeckli et al. 2012). Prokaryotic and eukaryotic microbiota sharing long-term alpine permafrost collected at the Muot da Barba Peider (East Switzerland) was studied with non-culturable techniques. The microbiota showed the presence of the genera *Mrakia* and *Leucosporidium.* Some OTUs were also associated to the former phenotypic genera *Cryptococcus* and *Rhodotorula*, together with some lichenized fungi (Frey et al. 2016). Finally, Selbmann et al. (2014) found psychrophilic and psychrotolerant culturable yeasts of the species *Na. antarctica*, *Cryptococcus albidosimilis* (*Naganishia albidosimilis*), and *Rhodotorula pinicola* (*Cystobasidium pinicola*) in Alpine endolithic microbial communities of rocks collected at Mount Rosa (Italian Alps).

Little is known so far about the eukaryotic microbiota colonizing cold ecosystems of the Pyrenees. García-Descalzo et al. (2013) used molecular 18S rRNA-based approaches to characterize some of the microbial eukaryotic populations associated with a few French Pyrenean glaciers. The presence of OTUs associated to the species *Bensingtonia yamatoana* (currently *Bannozyma yamatoana*) was found in ice cores collected at Aneto, Maladeta, Monte Perdido, and Literola glaciers.

The Apennines are a mountain range consisting of parallel smaller chains extending around 1200 km along the peninsular Italy. The highest peak (Corno Grande, Mount Gran Sasso, 2912 m a.s.l.) is partially covered by the Calderone glacier. With the substantial loss of the Corral de la Veleta glacier (Sierra Nevada, Pyrenees, Spain) in the early 1900s, it became the southernmost European glacier (Pecci et al. 2008). A recent study investigated the culturable psychrophilic and psychrotolerant yeast and yeast-like diversity in superficial and deep-piping sediments, ice cores, and meltwaters of the Calderone glacier. Yeast isolates were identified as belonging to a number of ascomycetous and basidiomycetous species. Overall, the most represented species were *Goff. gastrica*, *Phen. psychrophenolica*, and *Vishn. victoriae* (Branda et al. 2010). The new species *Glaciozyma martini* was described (Turchetti et al. 2011). Likewise, Selbmann et al. (2014) identified some yeasts isolated from endolithic microbial communities colonizing limestone rocks collected at Corno Piccolo (approx. 2500 m a.s.l., Mount Gran Sasso, Apennines, Italy) as belonging to the genera *Debaryomyces* and *Cystobasidium*.

The list of yeast species isolated from cold habitats of Europe is summarized in Table 12.4.

12.7 A Few Hypotheses on the Ecological Significance of Yeasts in Nonpolar Glaciers

Psychrophilic and psychrotolerant microorganisms (including yeasts) that colonize cold ecosystems are subjected to some adaptation of their physiology and metabolism, namely, the production of cold-active enzymes, antifreezing compounds, and extracellular polymers that increase cell protection against the damaging effects of near-to-zero temperatures. Besides, they preserve the fluidity and plasticity of their cytoplasm membranes until subzero temperatures by an increased percentage of unsaturated fatty acids and by a decrease in the average fatty acid chain length (Siddiqui and Cavicchioli 2006; Buzzini et al. 2012; Gunde-Cimerman et al. 2014).

It has been estimated that the nutrient input for microbial life in nonpolar glacier ecosystems (i.e., carbon, nitrogen, phosphorous, and oligoelements) derives primarily from wind, rainfall and snowfall, and alluvial deposition of allochthonous particulate materials, which can supply both organic and inorganic compounds. At the same time, microbial viable cells including yeasts can be transported on cold surfaces by air currents, snowfalls, and rainfalls. The combination of these two abiotic and biotic components, together with a supposed differential ability of different taxa either to survive in latent state or to grow in such habitats as metabolically active organisms responsible for certain natural processes, could justify the existence of dissimilarities on abundance and taxonomic composition of microbial consortia found in different nonpolar glacier habitats (Xiang et al. 2009; Anesio and Laybourn-Parry 2012; Buzzini et al. 2012; Turchetti et al. 2014).

Table 12.4 Yeast species found in cold habitats of Europe

Species	Original taxonomic designation	Source								Locality							References
		S	Sd	Sn	R	Ic	Mw	Air	Ch	ST	IA	FA	SA	AA	AP	FP	
Ascomycetous yeasts																	
Babjeviella inositovora		■								■							França et al. 2016
Candida oregonensis		■								■							França et al. 2016
Candida santamariae		■	■							■	■						Branda et al. 2010; Turchetti et al. 2013; França et al. 2016
Candida sp.			■	■							■	■					Turchetti et al. 2013
Debaryomyces hansenii		■			■					■					■		Selbmann et al. 2014; França et al. 2016
Schizoblastosporion starkeyi-henricii		■								■							França et al. 2016
Basidiomycetous yeasts																	
Apiotrichum dulcitum	*Trichosporon dulcitum*	■								■							Bergauer et al. 2005
Apiotrichum porosum	*Trichosporon porosum*	■								■							França et al. 2016
Bannozyma yamatoana *	*Bensingtonia yamatoana*					■										■	Garcia-Descalzo et al. 2013
Buckleyzyma aurantiaca	*Rhodotorula aurantiaca*							■				■					Turchetti et al. 2013
Bullera alba	*Bulleromyces albus*	■	■	■						■	■	■					Turchetti et al. 2013; França et al. 2016
Cryptococcus sp. **		■			■						■		■				Selbmann et al. 2014; Rime et al. 2015
Curvibasidium pallidicorallinum		■								■							França et al. 2016
Cutaneotrichosporon moniliiforme	*Trichosporon moniliiforme*	■								■							França et al. 2016
Cystobasidium laryngis	*Rhodotorula laryngis*		■	■		■					■	■				■	Turchetti et al. 2008, 2013; Branda et al. 2010
Cystobasidium pinicola	*Rhodotorula pinicola*				■						■						Selbmann et al. 2014

(*Cystobasidium* sp.)	*Rhodotorula* sp.		■		■						■				■		Turchetti et al. 2013; Selbmann et al. 2014
Cystofilobasidium capitatum	*Rhodosporidium capitatum*	■	■				■			■	■				■		Branda et al. 2010; Turchetti et al. 2013; França et al. 2016
Cystofilobasidium infirmominiatum				■								■					Turchetti et al. 2013
Cystofilobasidium macerans	*Cryptococcus macerans*	■	■	■			■			■	■	■			■		Branda et al. 2010; Turchetti et al. 2013; França et al. 2016
Dioszegia crocea			■	■								■			■		Branda et al. 2010; Turchetti et al. 2013
Dioszegia fristingensis				■								■					Turchetti et al. 2013
Dioszegia hungarica		■		■			■			■	■	■					Turchetti et al. 2013; França et al. 2016
Dioszegia sp. **			■	■		■	■				■	■	■				Turchetti et al. 2013; Rime et al. 2016a, 2016b
Erythrobasidium hasegawianum			■												■		Branda et al. 2010
Filobasidium chernovii	*Cryptococcus chernovii*			■								■					Turchetti et al. 2013
Filobasidium magnum	*Cryptococcus magnus*			■								■					Turchetti et al. 2013
Filobasidium oeirense	*Cryptococcus oeirensis*		■	■								■			■		Branda et al. 2010; Turchetti et al. 2013
(*Filobasidium* sp.)	*Cryptococcus* sp.	■	■	■						■	■	■			■		Branda et al. 2010; Turchetti et al. 2013; França et al. 2016
Filobasidium stepposum	*Cryptococcus stepposus*		■	■								■			■		Branda et al. 2010; Turchetti et al. 2013
Filobasidium wieringae	*Cryptococcus wieringae*		■	■								■			■		Branda et al. 2010; Turchetti et al. 2013
(*Gelidatrema* sp.)	*Cryptococcus* sp.		■	■						■	■	■					Turchetti et al. 2013
Glaciozyma martinii			■												■		Turchetti et al. 2011
Glaciozyma watsonii			■								■						Turchetti et al. 2011
Goffeauzyma gastrica	*Cryptococcus gastricus*	■	■	■			■			■	■	■			■		Branda et al. 2010; Turchetti et al. 2013; França et al. 2016
Goffeauzyma gilvescens	*Cryptococcus gilvescens*		■	■		■	■				■	■					Turchetti et al. 2008, 2013
(*Heterocephalacria* sp.)	*Cryptococcus* sp.		■							■	■	■					Turchetti et al. 2013
Holtermanniella festucosa	*Cryptococcus festucosus*	■	■							■	■	■			■		Branda et al. 2010; Turchetti et al. 2013; França et al. 2016
Holtermanniella takashimae		■	■	■			■			■	■	■			■		Branda et al. 2010; Turchetti et al. 2013; França et al. 2016

(continued)

Table 12.4 (continued)

Holtermanniella wattica	*Cryptococcus watticus*																Branda et al. 2010; Turchetti et al. 2013; França et al. 2016
Itersonilia pannonica	*Udeniomyces pannonicus*																Turchetti et al. 2013; França et al. 2016
(*Krasilnikovozyma* sp.)	*Mrakia* sp.																França et al. 2016
Leucosporidium sp. *																	Rime et al. 2015
Leucosporidium creatinivorum	*Leucosporidiella creatinivora, Rhodotorula creatinivora*																Bergauer et al. 2005; Turchetti et al. 2008, 2013
Leucosporidium drummii																	França et al. 2016
Leucosporidium intermedium	*Mastigobasidium intermedium*																Bergauer et al. 2005; Branda et al. 2010
Leucosporidium scottii																	de García et al. 2015
Leucosporidium sp.**																	Turchetti et al. 2013; Rime et al. 2016a, 2016b
Leucosporidium yakuticum	*Leucosporidiella yakutica*																de García et al. 2015
Malassezia sp. *																	Rime et al. 2016a, 2016b
Mrakia sp.*																	Rime et al. 2015
Mrakia aquatica	*Mrakiella aquatica; Cryptococcus aquaticus*																Branda et al. 2010; Turchetti et al. 2013
Mrakia blollopis																	Branda et al. 2010
Mrakia cryoconitii	*Mrakiella cryoconiti*																Margesin and Fell 2008; Branda et al. 2010; Turchetti et al. 2013
Mrakia frigida																	Margesin et al., 2002
Mrakia gelida																	Turchetti et al. 2008; Branda et al. 2010; Turchetti et al. 2013; França et al. 2016
Mrakia niccombsii	*Mrakiella niccombsii*																Turchetti et al. 2013
Mrakia psychrophila																	Turchetti et al. 2013
Mrakia robertii																	Thomas-Hall et al. 2010; Turchetti et al. 2013
Naganishia adeliensis	*Cryptococcus adeliensis*																Branda et al. 2010; Turchetti et al. 2013
Naganishia albida *	*Cryptococcus albidus*																Turchetti et al. 2013; França et al. 2016; Kuhnert et al. 2012
Naganishia albidosimilis	*Cryptococcus*																Branda et al. 2010; Turchetti et al. 2013;

	albidosimilis		■	■	■		■				■	■			■		Selbmann et al. 2014
Naganishia antarctica	*Cryptococcus antarcticus*		■				■				■						Turchetti et al. 2013
Naganishia antarctica	*Cryptococcus antarcticus*				■						■						Selbmann et al. 2014
Naganishia diffluens	*Cryptococcus diffluens*	■		■						■	■						Turchetti et al. 2013; França et al. 2016
Naganishia friedmannii	*Cryptococcus friedmannii*						■				■						Turchetti et al. 2013
Naganishia globosa	*Cryptococcus saitoi*		■			■					■				■		Turchetti et al. 2008; Branda et al. 2010
Naganishia onofrii	*Cryptococcus onofrii*			■			■				■	■					Turchetti et al. 2013
Naganishia vaughanmartiniae	*Cryptococcus vaughanmartiniae*		■			■					■	■			■		Branda et al. 2010; Turchetti et al. 2008, 2013
Papiliotrema frias	*Cryptococcus frias*		■												■		Branda et al. 2010
Papiliotrema laurentii	*Cryptococcus laurentii*			■								■					Turchetti et al. 2013
Phenoliferia glacialis	*Rhodotorula glacialis*	■	■			■	■		■		■			■			Margesin et al. 2007; Turchetti et al. 2008, 2013
Phenoliferia psychrophenolica	*Rhodotorula psychrophenolica*	■	■	■					■		■	■		■	■		Margesin et al. 2007; Turchetti et al. 2008, 2013; Branda et al. 2010
Phenoliferia psychrophila	*Rhodotorula psychrophila*	■												■			Margesin et al. 2007
(*Phenoliferia* sp.)	*Rhodotorula* sp.		■								■						Turchetti et al. 2013
(*Piskurozyma* sp.)	*Cryptococcus* sp.	■								■							França et al. 2016
Rhodosporidiobolus colostri	*Rhodotorula colostri*	■	■				■			■	■				■		Branda et al. 2010; Turchetti et al. 2013; França et al. 2016
Rhodosporidiobolus lusitaniae	*Rhodosporidium lusitaniae*	■													■		Bergauer et al. 2005
Rhodotorula sp. **			■	■									■				Rime et al. 2016a, 2016b
Rhodotorula bacarum			■								■						Turchetti et al. 2008
Rhodotorula sp.		■								■							França et al. 2016
Saitozyma podzolica	*Cryptococcus podzolicus*	■								■							França et al. 2016
Sampaiozyma ingeniosa	*Rhodotorula ingeniosa*	■												■			Bergauer et al. 2005
Solicoccozyma terricola	*Cryptococcus terricola/Cryptococcus terriculus*	■	■							■	■			■			Bergauer et al. 2005; Turchetti et al. 2008, 2013; França et al. 2016

(continued)

Table 12.4 (continued)

Solicoccozyma fuscescens	*Cryptococcus fuscescens*	■	■							■	■						Turchetti et al. 2013; França et al. 2016
Solicoccozyma phenolica	*Cryptococcus phenolicus*	■								■							França et al. 2016
Solicoccozyma terrea	*Cryptococcus terreus*	■	■							■	■			■			Bergauer et al. 2005; Turchetti et al. 2013; França et al. 2016
Solicoccozyyma aeria	*Cryptococcus aerius*		■	■							■	■			■		Branda et al. 2010; Turchetti et al. 2013
Sporobolomyces roseus		■	■	■						■	■	■		■	■		Bergauer et al. 2005; Branda et al. 2010; Turchetti et al. 2013; França et al. 2016
Tausonia pullulans	*Guehomyces pullulans*	■	■				■			■	■				■		Branda et al. 2010; Turchetti et al. 2013; França et al. 2016
Ustilentyloma graminis	*Rhodotorula hordea*		■								■						Turchetti et al. 2013
Vanrija musci	*Cryptococcus musci*	■								■							França et al. 2016
Vishniacozyma dimennae	*Cryptococcus dimennae*		■												■		Branda et al. 2010
Vishniacozyma tephrensis	*Cryptococcus tephrensis*		■	■							■	■			■		Branda et al. 2010; Turchetti et al. 2013
Vishniacozyma victoriae	*Cryptococcus victoriae*	■	■	■		■				■	■	■			■		Branda et al. 2010; Turchetti et al. 2013; França et al. 2016
Vishniacozyma carnescens	*Cryptococcus carnescens*		■								■						Turchetti et al. 2013
Yamadamyces rosulatus	*Rhodotorula rosulata*	■								■							França et al. 2016
Yeast-like organisms																	
Aureobasidium sp.			■								■						Turchetti et al. 2013
Aureobasidium pullulans		■	■	■			■	■		■	■	■	■		■		Turchetti et al. 2008; Branda et al. 2010; Brunner et al. 2011; Turchetti et al. 2013; França et al. 2016
Exophiala dermatitidis			■				■								■		Branda et al. 2010

Genera in brackets = authors arbitrary new genus/species nomenclature designation based on recent literature for that strains/OTUs for which a clear phylogenetic position was defined in the original paper

S soil; *Sd* sediment; *Sn* snow; *R* rock; *Ic* ice cores; *Mw* meltwater; *A* air; *Ch* cryoconites holes; *ST* South Tyrolean Alps; *IA* Italian Alps; *FA* French Alps; *SA* = Swiss Alps; *AA* Austrian Alps; *AP* Apennines, Italy; *FP* French Pyrenees

[a]Species found using culture-independent approach

[b]Species found using both culture-dependent and culture-independent approaches

The chemical and physical conditions exhibited by sediments and cryoconite holes could probably support an in situ microbial growth. In this context, the heterotrophic life of psychrophilic and psychrotolerant yeasts could suggest their ability to degrade efficiently some recalcitrant molecules and could be considered as responsible of some biogeochemical activities occurring in nonpolar glacial habitats. On the contrary, the oligotrophic (or even ultra-oligotrophic) nature of ice, snow, and meltwater could only allow a mere survival of yeasts in such habitats (Anesio and Laybourn-Parry 2012; Margesin et al. 2002; Turchetti et al. 2008, 2014; Branda et al. 2010; Buzzini et al. 2012).

Under the light of the above considerations, which are apparently supported by results reported by Turchetti et al. (2008, 2013), a few hypotheses on the ecological role of psychrophilic and psychrotolerant yeast life in nonpolar glaciers have been recently postulated (Turchetti et al. 2014): (a) not all species of yeast could be equally able to grow in sediments and cryoconite holes because they could exhibit a differential fitness to carbon, nitrogen, phosphorous, oligoelements, etc., as supposed by Turchetti et al. (2013); (b) this could lead to an in situ selective enrichment of some species in sediments and cryoconite holes at the expense of the others; (c) on the contrary, due to their significant oligotrophy, ice, snow, and meltwater could support only a survival of psychrophilic and psychrotolerant yeasts, thus substantially acting as "hibernation zones"; (d) meltwater could act as a driving vector by allowing the slow percolation of viable yeast cells from supraglacial sediments to unfrozen subglacial sediments passing throughout englacial ice through ice cracks, crevasses, and water channels (Turchetti et al. 2014).

12.8 Concluding Remarks

Investigations on the ecology of nonpolar cold habitats have revealed that microbiota is dynamically involved in biogeochemical transformations (Anesio and Laybourn-Parry 2012; Margesin et al. 2002; Turchetti et al. 2008, 2014; Branda et al. 2010; Buzzini et al. 2012). Psychrophilic and psychrotolerant yeasts can be considered as an active part of this community, because they have developed a complex array of physiological and metabolic adaptations that enhance their possibility to survive and, in some cases, to grow in glacial habitats. Therefore, although global warming due to climate change apparently predicts a strong influence on psychrophilic and psychrotolerant microbial communities of cold habitats, it could play a fundamental role in nonpolar cold ecosystems, especially in relation to the in situ hydrolysis of complex organic macromolecules connected with the mineralization of organic matter (Brunner et al. 2011).

A few authors (Price and Sowers 2004) underlined that the physiological characters of microorganisms cultured under controlled laboratory-simulated conditions do not necessarily reflect their in situ performances. Accordingly, the verification of the real impact of metabolic activity of psychrophilic and psychrotolerant bacteria, yeasts, and fungi in biogeochemistry of cold ecosystems is one of the most intriguing environmental questions for future studies. The functional diversity (including ecological interactions) of yeasts in nonpolar cold habitats may also be characterized on the basis of their physiological traits. However, although most of them are belonging to species acknowledged for their versatile metabolic traits and high resistance to stress associated with the cold environment (e.g., long-term freezing, freeze–thaw cycles, low Aw and nutrient availability, high solar irradiation), the ecology of yeasts in nonpolar cold habitats is still far to be fully explored.

Considering the fast progress in metagenomic studies and the continuous decreasing price of such techniques, it is possible to expect future studies related not only to the taxonomic position of the microorganisms colonizing the different habitats but also to the ecological role of them. Future metagenomic studies would make possible to determine the type and the intensity of in situ yeast metabolism, thus allowing speculations on their real impact on the biogeochemical cycles.

References

Abyzov SS, Bobin NE, Koudryashov BB (1982) Quantitative assessment of microorganisms in microbiological studies of Antarctic glaciers. Biol Bull Acad Sci USSR 9:558–564

Abyzov SS, Mitskevich IN, Poglazova MN, Barkov NI, Lipenkov VY, Bobin NE, Koudryashov BB, Pashkevich VM, Ivanov MV (2001) Microflora of the basal strata at Antarctic ice core above the Vostok Lake. Adv Space Sci 28:701–706

Amato P, Hennebelle R, Magand O, Sancelme M, Delort AM, Barbante C, Boutron C, Ferrari C (2007) Bacterial characterisation of the snow cover at Spitsbergen, Svalbard. FEMS Microbiol Ecol 59:255–264

Anesio AM, Laybourn-Parry J (2012) Glaciers and ice sheets as a biome. Trends Ecol Evol 27: 219–225

Anesio AM, Hodson AJ, Fritz A, Psenner R, Sattler B (2009) High microbial activity on glaciers: importance to the global carbon cycle. Global Change Biol 15:955–960

Babjeva IP, Reshetova IS (1971) Yeasts in Pamir soils (in Russian). Moscow Univ Soil Sci Bull 2: 80–86

Benn DI, Evans DJA (1998) Glaciers and glaciation. Arnold, London

Bergauer P, Fonteyne PA, Nolard N, Schinner F, Margesin R (2005) Biodegradation of phenol and phenol-related compounds by psychrophilic and cold-tolerant alpine yeasts. Chemosphere 59: 909–918

Boeckli L, Brenning A, Gruber S, Noetzli J (2012) Permafrost distribution in the European Alps: calculation and evaluation of an index map and summary statistics. Cryosphere 6:807–820

Branda E, Turchetti B, Diolaiuti G, Pecci M, Smiraglia C, Buzzini P (2010) Yeast and yeast-like diversity in the southernmost glacier of Europe (Calderone Glacier, Apennines, Italy). FEMS Microbiol Ecol 72:354–369

Brandão RL, Libkind D, Vaz BMA, Espirito Santo LC, Moliné M, de García V, van Broock M, Rosa C (2011) Yeasts from an oligotrophic lake in Patagonia (Argentina): diversity, distribution and synthesis of photoprotective compounds and extracellular enzymes. FEMS Microbiol Ecol 76:1–13

Brown SP, Jumpponen A (2014) Contrasting primary successional trajectories of fungi and bacteria in retreating glacier soils. Mol Ecol 23:481–497

Brunner I, Plotze M, Rieder S, Zumsteg A, Furrer G, Frey B (2011) Pioneering fungi from the Damma glacier forefield in the Swiss Alps can promote granite weathering. Geobiology 9: 266–279

Buzzini P, Margesin R (2014) Cold-adapted yeasts. Springer, Berlin

Buzzini P, Turchetti B, Diolaiuti G, D'Agata C, Martini A, Smiraglia C (2005) Culturable yeasts in meltwaters draining from two glaciers in the Italian Alps. Ann Glaciol 40:119–122

Buzzini P, Branda E, Goretti M, Turchetti B (2012) Psychrophilic yeasts from worldwide glacial habitats: diversity, adaptation strategies and biotechnological potential. FEMS Microbiol Ecol 82:217–241

Cameron KA, Hodson AJ, Osborn AM (2012) Structure and diversity of bacterial, eukaryotic and archaeal communities in glacial cryoconite holes from the Arctic and the Antarctic. FEMS Microbiol Lett 82:254–267

Campbell IB, Claridge GGC (2000) Soil temperature, moisture and salinity patterns in Transantarctic Mountain cold desert ecosystems. In: Davidson W, Howard-Williams C, Broady P (eds) Antarctic ecosystems: models for wider understanding. Caxton Press, Christchurch, pp 2332–2340

Cannone N, Diolaiuti G, Guglielmin M, Smiraglia C (2008) Accelerating climate change impacts on alpine glacier forefield ecosystems in the European Alps. Ecol Appl 18:637–648

Castello JD, Rogers SO (2005) Life in ancient ice. Princeton University Press, Princeton

Cavicchioli R, Tortsen T (2000) Extremophilic. In: Lederberg J (ed) Encyclopedia of microbiology, vol 2, 2nd edn. Academic Press, London, pp 317–337

Christner BC, Krivko BH II, Reeve JN (2003) Molecular identification of bacteria and eukarya inhabiting an Antarctic cryoconite hole. Extremophiles 7:177–183

D'Amico S, Collins T, Marx JC, Feller G, Gerday C (2006) Psychrophilic microorganisms: challenges for life. EMBO Rep 7:385–389

De Angelis M, Gaudichet A (1991) Saharan dust deposition over Mont Blanc (French Alps) during the last 30 years. Tellus 43B:61–75

de García V, Brizzio S, Libkind D, Buzzini P, van Broock M (2007) Biodiversity of cold-adapted yeasts from glacial meltwater rivers in Patagonia, Argentina. FEMS Microbiol Ecol 59: 331–341

de García V, Brizzio S, Libkind D, Rosa CA, van Broock M (2010a) *Wickerhamomyces patagonicus* sp. nov., an ascomycetous yeast species from Patagonia, Argentina. Int J Syst Evol Microbiol 60:1693–1696

de García V, Brizzio S, Russo G, Rosa CA, Boekhout T, Theelen B, Libkind D, van Broock M (2010b) *Cryptococcus spencermartinsiae* sp. nov., a basidiomycetous yeast isolated from glacial waters and apple fruits. Int J Syst Evol Microbiol 60:707–711

de García V, Zalar P, Brizzio S, Gunde-Cimerman N, van Broock M (2012a) *Cryptococcus* species (Tremellales) from glacial biomes in the southern (Patagonia) and northern (Svalbard) hemispheres. FEMS Microbiol Ecol 82:523–539

de García V, Brizzio S, van Broock M (2012b) Yeasts from glacial ice of Patagonian Andes, Argentina. FEMS Microbiol Ecol 82:540–550

de García V, Coelho MA, Maia TM, Rosa LH, Vaz AM, Rosa CA, Sampaio JP, Gonçalves P, van Broock M, Libkind D (2015) Sex in the cold: taxonomic reorganization of psychrotolerant yeasts in the order Leucosporidiales. FEMS Yeast Res 15:fov019

Deming JW (2009) Extremophiles: cold environments. In: Schaechter M (ed) Encyclopedia of microbiology, 3rd edn. Academic Press, Oxford, pp 147–158

Díaz M, Pedrozo F, Baccala N (2000) Summer classification of Southern Hemisphere temperate lakes (Patagonia, Argentina). Lakes Reserv 5:213–229

Finlay R, Wallander H, Smits M, Holmstrom S, Van Hees P, Lian B, Rosling A (2009) The role of fungi in biogenic weathering in boreal forest soils. Fungal Biol Rev 23:101–106

França L, Sannino C, Turchetti B, Buzzini P, Margesin R (2016) Seasonal and altitudinal changes of culturable bacterial and yeast diversity in Alpine forest soils. Extremophiles 20:855–873

Frey B, Rime T, Phillips M, Stierli B, Hajdas I, Widmer F, Hartmann M (2016) Microbial diversity in European alpine permafrost and active layers. FEMS Microbiol Ecol 92:fiw018

García-Descalzo L, García-López E, Postigo M, Baquero F, Alcazar A, Cid C (2013) Eukaryotic microorganisms in cold environments: examples from Pyrenean glaciers. Front Microbiol 4: 1–14

Gorbushina AA (2007) Life on the rocks. Environ Microbiol 9:1613–1631

Gunde-Cimerman N, Plemenitaš A, Buzzini P (2014) Changes in lipids composition and fluidity of yeast plasma membrane as response to cold. In: Buzzini P, Margesin R (eds) Cold-adapted yeasts. Springer, Berlin, pp 225–242

Hamilton TL, Peters JW, Skidmore ML, Boyd ES (2013) Molecular evidence for an active endogenous microbiome beneath glacial ice. Int Soc Microb Ecol 7:1402–1412

Hodson A, Anesio AM, Trantner M, Fountain A, Osborn M, Priscu J, Laybourn-Parry J, Sattler B (2008) Glacial ecosystems. Ecol Monograph 78:41–67

Holdgate MV (1977) Terrestrial ecosystems in the Antarctic. Philos Trans Royal Soc B 279:5–25

Hu W, Zhang Q, Li D, Cheng G, Mu J, Wu Q, Niu F, An L, Feng H (2014) Diversity and community structure of fungi through a permafrost core profile from the Qinghai-Tibet Plateau of China. J Basic Microbiol 54:1331–1341

Hu W, Zhang Q, Tian T, Cheng G, An L, Feng H (2015a) The microbial diversity, distribution, and ecology of permafrost in China: a review. Extremophiles 19:693–705

Hu H, Yan F, Wilson C, Shen Q, Zheng X (2015b) The ability of a cold-adapted *Rhodotorula mucilaginosa* strain from Tibet to control blue mold in pear fruit. Antonie van Leeuwenhoek 108: 1391–1404

Krallish I, Gonta S, Savenkova L, Bergauer P, Margesin R (2006) Phenol degradation by immobilized cold-adapted yeast strains of *Cryptococcus terreus* and *Rhodotorula creatinivora*. Extremophiles 10:441–449

Kuhnert R, Oberkofler I, Peintner U (2012) Fungal growth and biomass development is boosted by plants in snow-covered soil. Microb Ecol 64:79–90

Lazzaro A, Brankatschk R, Zeyer J (2012) Seasonal dynamics of nutrients and bacterial communities in unvegetated alpine glacier forefields. Appl Soil Ecol 53:10–22

Lazzaro A, Wismer A, Schneebeli M, Erny I, Zeyer J (2015) Microbial abundance and community structure in a melting alpine snowpack. Extremophiles 19:631–642

Lee YM, Kim S-Y, Jung J, Kim EH, Cho KH, Schinner F, Margesin R, Hong SG, Lee HK (2011) Cultured bacterial diversity and human impact on alpine glacier cryoconite. J Microbiol 49: 355–362

Li SL, Lin Q, Li XR, Xu H, Yang YX, Qiao DR, Cao Y (2012) Biodiversity of the oleaginous microorganisms in Tibetan Plateau. Brazilian J Microbiol:627–634

Libkind D, Brizzio S, Ruffini A, Gadanho M, van Broock MR, Sampaio JP (2003) Molecular characterization of carotenogenic yeasts from aquatic environments in Patagonia, Argentina. A van Leeuwenhoek 84:313–322

Libkind D, Moline M, Sampaio JP, van Broock M (2009) Yeasts from high-altitude lakes: influence of UV radiation. FEMS Microbiol Ecol 69:353–362

Lliboutry L (1998) Glaciers of South America: glaciers of Chile and Argentina. In: Williams RS, Ferrigno JG (eds) Satellite image atlas of glaciers of the world. US Geological Survey, Denver, pp 1109–1119

MacDonell S, Fitzsimons S (2008) The formation and hydrological significance of cryoconite holes. Prog Phys Geogr 32:595–610

Margesin R (2009) Effect of temperature on growth parameters of psychrophilic bacteria and yeasts. Extremophiles 13:257–262

Margesin R, Fell JW (2008) *Mrakiella cryoconiti* gen. nov., sp. nov., a psychrophilic, anamorphic, basidiomycetous yeast from alpine and arctic habitats. Int J Syst Evol Microbiol 58:2977–2982

Margesin R, Miteva V (2011) Diversity and ecology of psychrophilic microorganisms. Res Microbiol 162:346–361

Margesin R, Zacke G, Schinner F (2002) Characterization of heterotrophic microorganisms in Alpine glacier cryoconite. Arctic Antarct Alp Res 34:88–93

Margesin R, Fauster V, Fonteyne PA (2005) Characterization of cold-active pectate lyases from psychrophilic *Mrakia frigida*. Lett Appl Microbiol 40:453–459

Margesin R, Fonteyne PA, Schinner F, Sampaio JP (2007) Novel psychrophilic basidiomycetous yeasts from Alpine environments: *Rhodotorula psychrophila* sp. nov., *Rhodotorula psychrophenolica* sp. nov. and *Rhodotorula glacialis* sp. nov. Int J Syst Evol Microbiol 57:2179–2184

Messerli B (1980) Mountain glaciers in the Mediterranean Area and in Africa, World Glacier Inventory, Proc Riederal Workshop, IAHS, pp 197–211

Mestre MC, Fontenla S, Rosa CA (2014) Ecology of cultivable yeasts in pristine forests in northern Patagonia (Argentina) influenced by different environmental factors. Can J Microbiol 60:1–12

Mestre MC, Fontenla S, Bruzone MC, Fernández NV, Dames J (2016) Detection of plant growth enhancing features in psychrotolerant yeasts from Patagonia (Argentina). J Basic Microbiol 56:1–9

Miteva V (2008) Bacteria in snow and glacier ice. In: Margesin R, Schinner F, Marx JC, Gerday C (eds) Psychrophiles: from biodiversity to biotechnology. Springer, Berlin, pp 31–50

Nesje A, Dahl SO (2000) Glaciers and environmental changes. Arnold, London

Onofri S, Selbmann L, Zucconi L, Pagano S (2004) Antarctic microfungi as models for exobiology. Planet Space Sci 52:229–237

Paterson WSB (2002) The physics of glaciers. Butterworth-Heinemann/Elsevier Science, Burlington

Pecci M, D'Agata C, Smiraglia C (2008) Ghiacciaio del Calderone (Apennines, Italy): the mass balance of a shrinking Mediterranean glacier. Geogr Fis Din Quat 31:55–62

Price PB (2000) A habitat for psychrophiles in deep, Antarctic ice. Proc Natl Acad Sci USA 97: 1247–1251

Price PB (2007) Microbial life in glacial ice and implications for a cold origin of life. FEMS Microbiol Ecol 59:217–231

Price PB, Sowers T (2004) Temperature dependence of metabolic rates for microbial growth, maintenance, and survival. Proc Natl Acad Sci USA 101:4631–4636

Quirós R, Drago E (1985) Relaciones entre variables físicas, morfométricas y climáticas en lagos patagónicos. Rev Asoc Cs Nat Litoral 16:181–199

Rime T, Hartmann M, Brunner I, Widmer F, Zeyer J, Frey B (2015) Vertical distribution of the soil microbiota along a successional gradient in a glacier forefield. Mol Ecol 24:1091–1108

Rime T, Hartmann H, Frey B (2016a) Potential sources of microbial colonizers in an initial soil ecosystem after retreat of an alpine glacier. Int Soc Microb Ecol:1–17

Rime T, Hartmann M, Stierli B, Anesio AM, Frey B (2016b) Assimilation of microbial and plant carbon by active prokaryotic and fungal populations in glacial forefields. Soil Biol Biochem 98:30–41

Selbmann L, Zucconi L, Onofri S, Cecchini C, Isola D, Turchetti B, Buzzini P (2014) Taxonomic and phenotypic characterization of yeasts isolated from worldwide cold rock-associated habitats. Fungal Biol 118:61–71

Shea JM, Immerzeel WW, Wagnon P, Vincent C, Bajracharya S (2015) Modelling glacier change in the Everest region, Nepal Himalaya. Cryosphere 9:1105–1128

Shivaji S, Bhadra B, Rao RS, Pradhan S (2008) *Rhodotorula himalayensis* sp. nov., a novel psychrophilic yeast isolated from Roopkund Lake of the Himalayan mountain ranges, India. Extremophiles 12:375–381

Siddiqui KS, Cavicchioli R (2006) Cold-adapted enzymes. Annu Rev Biochem 75:403–433

Singh P, Singh SM (2012) Characterization of yeast and filamentous fungi isolated from cryoconite holes of Svalbard, Arctic. Polar Biol 35:575–583

Skvarca L, Raup B, De Angelis H (2003) Recent behaviour of Glaciar Upsala, a fast-flowing calving glacier in Lago Argentino, southern Patagonia. Ann Glaciol 36:184–188

Stibal M, Elster J, Sabacka M, Kastovska K (2007) Seasonal and diel changes in photosynthetic activity of the snow alga *Chlamydomonas nivalis* (Chlorophyceae) from Svalbard determined by pulse amplitude modulation fluorometry. FEMS Microbiol Ecol 59:265–273

Stibal M, Sabacka M, Zarsky J (2012) Biological processes on glacier and ice sheet surfaces. Nat Geosci 5:771–774

Stuefer M, Rott H, Skvarca P (2007) Glaciar Perito Moreno, Patagonia: climate sensitivities and characteristics preceding 2003/04 and 2005/06 damming events. J Glaciol 53:3–16

Takeuchi N, Kohshima S, Seko K (2001) Structure, formation and darkening process of albedo-reducing material (cryoconite) on a Himalayan glacier: a granular algal mat growing on the glacier. Arct Antarct Alp Res 33:115–122

Thomas-Hall SR, Turchetti B, Buzzini P, Branda E, Boekhout T, Theelen B, Watson K (2010) Cold adapted yeasts from Antarctica and the Italian Alps. Description of three novel species: *Mrakia robertii* sp. nov., *Mrakia blollopis* sp. nov. and *Mrakiella niccombsii* sp. nov. Extremophiles 14:47–59

Turchetti B, Buzzini P, Goretti M, Branda E, Diolaiuti G, D'Agata C, Smiraglia C, Vaughan-Martini A (2008) Psychrophilic yeasts in glacial environments of Alpine glaciers. FEMS Microbiol Ecol 63:73–83

Turchetti B, Thomas-Hall SR, Connell LB, Branda E, Buzzini P, Theelen B, Muller WH, Boekhout T (2011) Psychrophilic yeasts from Antarctica and European glaciers: description of *Glaciozyma* gen. nov., *Glaciozyma martinii* sp. nov. and *Glaciozyma watsonii* sp. nov. Extremophiles 15:573–586

Turchetti B, Goretti M, Branda E, Diolaiuti G, D'Agata C, Smiraglia C, Onofri A, Buzzini P (2013) Influence of abiotic variables on culturable yeast diversity in two distinct Alpine glaciers. FEMS Microbiol Ecol 86:327–340

Turchetti B, Goretti M, Buzzini P, Margesin R (2014) Cold-adapted yeasts in Alpine and Apennine Glaciers. In: Buzzini P, Margesin R (eds) Cold-adapted yeasts. Springer, Berlin, pp 99–122

Turchetti B, Selbmann L, Blanchette RA, Di Mauro S, Marchegiani E, Zucconi L, Arenz BE, Buzzini P (2015) *Cryptococcus vaughanmartiniae* sp. nov. and *Cryptococcus onofrii* sp. nov.: two new species isolated from worldwide cold environments. Extremophiles 19:149–159

Uetake J, Kohshima S, Nakazawa F, Takeuchi N, Fujita K, Miyake T, Narita H, Aizen V, Nakawo M (2011) Evidence for propagation of cold-adapted yeast in an ice core from a Siberian Altai Glacier. J Geophys Res 116:G01019

Villarosa G, Outes V, Masiokas M, Villalba R, Rivas S (2008) El Monte Tronador: historias de hielo y fuego. In: Rastelli D (ed) Sitios de Interés Geológico de la República Argentina. SEGEMAR (Servicio de Geología y Minería de la República Argentina), Buenos Aires, pp 627–641

Vimercati L, Hamsher S, Schubert Z, Schmidt SK (2016) Growth of high-elevation *Cryptococcus* sp. during extreme freeze–thaw cycles. Extremophiles 20:579

Vincent CF (1988) Microbial ecosystems of Antarctica. Cambridge University Press, Cambridge

Wharton JRA, McKay CP, Simmons GM, Parker BC (1985) Cryoconite holes on glaciers. Bioscience 8:499–503

World Glacier Monitoring Service (WGMS) (2012) Fluctuations of glaciers 2005–2010. ICSU (WDS) – IUGG (IACS) – UNEP – UNESCO – WMO, Staffel Druck AG, Zurich

Xiang SR, Shang TC, Chen Y, Yao TD (2009) Deposition and post deposition mechanisms as possible drivers of microbial population variability in glacier ice. FEMS Microbiol Ecol 70: 165–176

Zemp M, Haeberli W, Hoelzle M, Paul F (2006) Alpine glaciers to disappear within decades? Geophys Res Lett 33:L13504

Chapter 13
Yeasts in Insects and Other Invertebrates

Meredith Blackwell

Abstract Minute organisms, including yeasts with diverse physiological capabilities, make possible the existence of arthropods, especially insects, the most speciose group of organisms on Earth. The yeast growth form occurs throughout most of the fungal kingdom and is often associated with insects. Fungi and insects evolved together in the same habitats where casual associations certainly occurred early in their shared geological history and yeast attractants for insects developed over their lives together. Examples of their interactions range from accidental dispersal and the use of yeasts as food to obligate mutualisms described in this chapter. These include discussions of yeast-like symbionts, the use of the wasp gut for yeast outcrossing, extension of endophyte life cycles to include dispersal by insects, the advantage of yeasts in the diet of many insects such as blood-sucking dipterans, and the yeast-like germination of phoretic fungi in ephemeral habitats. Future studies of yeast–insect associations will continue to include species discovery but also approach theoretical questions of sexual and asexual reproduction, host specificity, host switching, advantages of horizontal and vertical dispersal, and studies that include entire interactive communities.

Keywords Erotylidae • Transitions to mutualism • Hemiptera • Ophiocordycipitaceae • Tenebrionidae

Contents

M. Blackwell (✉)
Department of Biological Sciences, Louisiana State University, Baton Rouge, LA 70803, USA

Department of Biological Sciences, University of South Carolina, Columbia, SC 29208, USA
e-mail: mblackwell@lsu.edu

P. Buzzini et al. (eds.), *Yeasts in Natural Ecosystems: Diversity*,
DOI 10.1007/978-3-319-62683-3_13

13.1 Introduction

Haeckel (1899) made an early attempt to popularize small organisms for the general public, but recognition of microbes as important to Earth's environments is only now being widely recognized (Hug et al. 2016). Price (1988), who called for more interest in microbes, chided "Noah's Ark Ecology." He emphasized evidence that the greatest energy transfer in food webs was not by large, familiar animals and plants but by organisms smaller than "bird food." Price's consideration of food webs in a phylogenetic context led him to express an essential principle, "ecology recapitulates phylogeny." Pimentel et al. (1992) estimated that 90% of the animal and plant species present in certain regions are arthropods and microbes, a vast majority of the species on Earth. More recently estimates based on a universal dominance scaling law proposed the existence of as many as 1 trillion (10^{12}) species of microbes, including bacteria, archaea, and microscopic fungi (Locey and Lennon 2016).

Fungal biologists, however, have been interested in fungal associations with other organisms for many years, but recently general attention to microbes has increased due to the human microbiome project (Gill et al. 2006; Turnbaugh et al. 2007; Rodrigues Hoffmann et al. 2016). It is common to see eye-catching headlines in popular literature touting large numbers of microbes on different parts of the human body and T-shirts declaring we are "10% human, 90% bacteria," and although new estimates downplay the number of microbial cells in a person, the numbers are still impressive (Sender et al. 2016). Anthropocentric attention has brought interest to the study of all microbiomes, and the development of high-throughput sequencing, metagenomic analysis, and other methods has benefitted the study of fungi (Hoffmann et al. 2013). Questions, including those focused on all yeast–insect associations and interactions discussed here, should be invigorated by the improved tools of high-throughput sequencing and other methods to provide answers to who, what, when, where, why, and how yeasts and their host organisms became associated; eventually the view will be broadened to consider entire microbial communities (Ganter 2006; Douglas and Werren 2016).

The period from the late nineteenth century until the present has been a time of discovery of associations among microbes and insects (Buchner 1953). The translation of Buchner's (1965) classic work has made the information widely available to English speakers. Kurtzman and Robnett (1998) developed methods not only to identify yeasts rapidly but also to begin to arrange them phylogenetically. Ganter (2006) remedied his perceptive concern for the lack of current discussion of fungi, especially yeasts associated with insects, among biologists by his thorough presentation of the subject. The following list of events marks progress in the study of fungal–invertebrate animal associations, the focus of this chapter, and it provides perspective for the discussion. Items marked by asterisk indicate reviews on the topic:

- 1835—Agostino Bassi di Lodi studied first animal (insect) disease, caused by a fungal agent later named for him (*Beauveria bassiana*).
- 1857—Pasteur determined microbial agents of fermentation.
- 1870—Pasteur saved the French silk industry when he discovered a microsporidian fungus (*Nosema bombycis*), the agent of pébrine disease of silkworms.
- 1952—*The Yeasts: A Taxonomic Study*, first edition (Lodder and Kreger-van Rij 1952).
- 1953*—Publication of *Endosymbiose der Tiere mit Pflanzlichen Mikroorganismen*, a detailed overview of the nineteenth- and twentieth-century literature (Buchner 1953).
- 1965*—Publication of the English translation of Buchner's work on endosymbiotic microbes of animals (Buchner 1965).
- 1970—*The Yeasts: A Taxonomic Study*, second edition (Lodder 1970).
- 1972—Phaff and Heed begin to publish on cactophilic yeast communities.
- 1976—Heed, Fogleman, Starmer, Lachance, and Rosa et al. begin to publish on yeasts associated with insects (see Chap. 8 of this book).
- 1984—*The Yeasts: A Taxonomic Study*, third edition (Kreger van Rij 1984).
- 1988—Lachance and Starmer et al. begin to publish on yeasts associated with ephemeral flowers, bees, and *Drosophila.*
- 1996—Publication of the first eukaryote complete genomic sequence, *Saccharomyces cerevisiae* (Goffeau et al. 1996).
- 1998—*The Yeasts: A Taxonomic Study*, fourth edition (Kurtzman and Fell 1998).
- 1998—Kurtzman and Robnett publish on large subunit ribosomal LSU rDNA (large subunit ribosomal DNA) as a marker and develop a database of yeast sequences (Kurtzman and Robnett 1998).
- 2004—Suh, Nguyen, Urbina, and Blackwell et al. begin to publish on insect gut yeasts.
- 2005*—Vega and Dowd publish often cited chapter on the role of yeasts as insect endosymbionts (Vega and Dowd 2005).
- 2005—Boekhout reports that only 6% of the CBS-KNAW Fungal Biodiversity Centre collection yeasts are from insect sources (Boekhout 2005).
- 2006—Rosa, Cadete, and Lachance begin to publish on yeasts in decaying wood (see Chap. 9 of this book).

- 2006*—Publication of *Biodiversity and Ecophysiology of Yeasts*, Springer, Berlin [includes Ganter (2006) review of yeasts and invertebrate associations] (Rosa and Péter 2006).
- 2011*—*The Yeasts. A Taxonomic Study*, fifth edition (Kurtzman et al. 2011); includes Chap. 6, Yeast Ecology (Starmer and Lachance 2011).
- 2016—First yeast genome-wide study includes some insect-associated yeasts (Riley et al. 2016).
- 2016—Backbone phylogeny of Saccharomycotina published using genome-scale data provided more stable family-level classification (Shen et al. 2016).
- 2016*—Eleven years of increase in insect-associated yeasts accessioned in the CBS—KNAW Fungal Biodiversity Centre collection rises to 7.25% of total with more to be cataloged (see Chap. 15 of this book).

Ganter (2006) distinguished between associations, recognized by established co-occurrence, and interactions, in which a functional relationship has been "established or strongly supported by observational data"; in this chapter association will be used in a general sense, and interaction when a known benefit to a participant is referred to in an interaction. Most of the interactions between yeasts and insects discussed in this chapter are mutualistic, and they probably benefit, not harm, the participants in the association. In many cases, however, we do not understand the basis for the close associations. Diffuse mutualisms are not species specific (Johnstone and Bshary 2002; Starmer and Lachance 2011). Douglas (2009) considered these associations to be some of the more interesting ecologically, because they introduce variables in populations. Some of the associations are casual, while in others organisms interact in a fixed and sometimes obligate manner that affects the fitness of one or both participants; the partners may not benefit to an equal extent in asymmetric interactions. The presence of several classes of microbes (e.g., yeasts and bacteria) in the gut of individual insects is common, perhaps with collaborative roles that are not always considered in the study of the associations (Chandler et al. 2012; Ceja-Navaro et al. 2014, 2015; Douglas and Werren 2016). Certain organisms involved in a mutualism (e.g., multipartite mutualisms), such as a third participant, may be unknown. In a few cases, insects (e.g., planthoppers, roaches, termites) may rely on bacterial mutualists in closely related species but then switch to a fungus under new pressures (Engel and Moran 2013). For example, major switches in symbionts have occurred over evolutionary time coinciding with changes in diet as in roach and termite lineages (Brune and Dietrich 2015).

The work reviewed in this chapter is heavily biased toward terrestrial insects, especially Coleoptera (beetles), Diptera (flies), and a few Hemiptera (planthoppers, leafhoppers, aphids), and their ascomycete yeast associates, because these organisms have been the subject of the most research to date. Perhaps this focus is justified because estimates suggest that one of four animals on Earth is a beetle (Grimaldi and Engel 2005). Ganter (2006) covered the topic in great detail, and that material is not repeated in this chapter except when background information is needed. Another source of a list of specific associations is available on line

(Urubschurov and Janczyk 2011). Other chapters in this book on specific habitats in which yeasts and insects are closely associated include Chap. 8 ("Yeasts in Cactus and Tropical Fruit Habitats"), Chap. 9 ("Yeasts Associated with Decomposing Plant Material and Rotting Wood"), and Chap. 15 ("Census of Yeasts Isolated from Natural Ecosystems and Conserved in Worldwide Collections"). Besides, the book *Yeasts in Natural Ecosystems: Ecology* (Buzzini et al. 2017) includes a few chapters focused on mutualism in yeasts (Mittelbach and Vannette 2017), parasitism in yeasts (Begerow et al. 2017), and antagonistic interactions and killer yeasts (Klassen et al. 2017).

13.1.1 A Paradox of Success

Insects and other arthropods are unable to synthesize many of the essential nutrients and other resources that they require to maintain life (e.g., amino acids, sterols). A paradox exists: the world's most speciose group of organisms on Earth diverged to fill many habitats, often by relying on resources derived from bacteria and fungi. Douglas (2009) pointed out that distinguishing between microbes as food (e.g., ambrosia of beetles, fungal gardens of attine ants) versus symbionts performing other functions is not always clear, but what is certain is that "foodstuffs such as plant sap, vertebrate blood and sound wood, would be unavailable to insects without nutritional input from microorganisms." For example, vitamins for physiological functions and enzymes for degradation of plant cell walls and detoxification of plant secondary metabolites often are acquired from microbes, including yeasts and yeast-like fungi (Vega and Dowd 2005; Ganter 2006; Douglas 2009; Starmer and Lachance 2011). Bacteria may be more effective than fungi as mutualists in providing enzymes for degradation of plant cell walls and fixing nitrogen. Fungi, however, can also supply vitamins, recycle nitrogen in certain cases, and clearly have an advantage in providing sterols that are not synthesized by bacteria.

13.1.2 How Long Has This Been Going On? Dating the Associations

How long have invertebrate animals and microbes been associated? The question is important to consider because we want to understand the evidence for diversification of speciose groups. Based on available evidence, well-established groups of fungi, insects, and plants (including woody plants) could have been living closely in the same habitats by the end of the Carboniferous [~300 million years ago (Ma)] or even earlier (Taylor et al. 2005). Estimates of the ages of fungal taxa and insect orders have been based on DNA, morphology, the fossil record, and environmental data (Lücking et al. 2009; Misof et al. 2014; McKenna et al. 2015; Chang et al.

2015; Toussaint et al. 2017). Misof et al. (2014) dated the origin of insects at Early Ordovician [~479 Ma) and insect flight, Early Devonian (~406 Ma). Based on the illustration (Fig. 1 reported by Misof et al. 2014), estimates suggest that by 350–200 Ma insect orders associated with yeasts today were already established: Isoptera (termites) and Blattodea (Legendre et al. 2015), Hemiptera (bugs, but not necessarily including all members in Fulgoromorpha), Hymenoptera (ants and others), Neuroptera (lacewings), Lepidoptera (moths and butterflies), and the earliest of several divergences of Diptera (true flies) (Wiegmann et al. 2011). Toussaint et al. (2017) provided dates for many groups of beetles based on new fossil calibrations that are generally earlier. Their estimates for crown groups included *holometabolous* insects (those with egg, larva, pupa, and adult stages) before the Carboniferous about ~385 Ma, Coleoptera in Mid-Carboniferous (~325 Ma), and the four extant beetle suborders, including Polyphaga in which most of the fungus-feeding beetles are classified, by Late Carboniferous to Early Permian (~300 Ma). They estimated that many clades (series and superfamily level) survived the mass extinction at the end of the Permian (~250 Ma). The crown groups of beetles important to this discussion (plant-associated families Curculionidae, Cerambycidae, and Chrysomelidae, certain cucujoid families, e.g., Nitidulidae, Erotylidae with fungus-feeding lineages, and Tenebrionidae) were present by the end of the Jurassic (200–150 Ma). These dates are earlier than those given previously for beetles (Hunt et al. 2007; McKenna et al. 2015), but they predate the earliest known fossils available for the different groups and are in better agreement with what is known of radiations of other major groups of organisms.

Dates suggested for fungal evolution vary widely, in part because of the lack of reliable fossils and calibration points. Fossil yeasts have not been reported, even when fossils of insect groups known to harbor them are found, but ascomycetous yeasts certainly are older than the 35- to 45-million-year-old "fossil" yeasts advertised on the Internet for recent brewing ventures. Lücking et al. (2009) relied on previous publications and data from Schoch et al. (2009) to estimate the age of fungal divergences. Based on their most conservative, recalibrated phylogenetic tree ages, they suggested that the divergence of Ascomycota and Basidiomycota coincided with the appearance of nonvascular plants by 500 Ma; divergence of the Saccharomycotina and other Ascomycota, about 400 Ma; the base of Pezizomycotina, about 400–520 Ma; and the Pezizomycotina diversification, 320–400 Ma, corresponding with vascular plant diversification (see Table III and Fig. 2 reported by Lücking et al. 2009). These estimates indicate that yeast–insect associations, including those involving the use of yeasts as food, could be much older than 300 million years. Although yeast feeding has not been recognized in fossil insects, a staphylinid from Burmese amber (~99 Ma) has a spore brush on part of the maxilla, a known spore-feeding mouthpart specialization, as well as possible spores in the gut; this fossil is the earliest known beetle with evidence of a spore-feeding habit (Cai et al. 2016). Analysis of insect mouthparts and gut contents provides evidence of ecological associations with extant insects as well as fossils (Robertson personal communication 2014). Information is available on symbiotic-farming associations of beetles, ants, and termites. The symbioses apparently appear rather late compared to what is surmised about the lineages. Jordal and Cognato (2012) used calibrated molecular divergence rates based on updated fossil records to suggest that the fungus-farming habit in bark and ambrosia beetles arose independently at least ten times with origins about 21 Ma–50 Ma, at least 50 million years after the proposed origin of the subfamily Scolytinae (100–120 Ma). Furthermore, they proposed the origin of farming occurred during two periods of global warming in the Cenozoic. The 50 Ma date was similar to the proposed origins of attine ant associations with Leucocoprineae basidiomycetes (Schultz and Brady 2008) and a somewhat younger origin for African termites and their associated *Termitomyces* spp. of 31 Ma (19–49 Ma). There has been some host switching among termites within *Termitomyces* spp. (Nobre et al. 2011), but one attine lineage developed an association with a completely different group of basidiomycetes, coral fungi in Pterulaceae, about 10–20 Ma (Schultz and Brady 2008).

Yeasts may have been included in the diet of insects much earlier but with little specificity and without vertical transmission of clones as happens in the farming associations (e.g., ants). Elaborate behavioral traits would not have been required for transmission, although morphological adaptations such as gastric caeca and mycangia have developed in some insects. Certain yeast clades (e.g., *Suhomyces*, *Teunomyces*) may have diversified with beetles that favor mushroom diets, but there is only specificity at the level of genus or above. Exceptions are noted in discussions below.

13.1.3 Diversity of Yeasts and Yeast-Like Fungi in Animal Associations

As mentioned before (Blackwell 1984), the observations of Otto von Münchhausen, the eighteenth-century correspondent of Linnaeus, revealed that fungi were the dwellings of animals (Ramsbottom 1953; Findlay 1982). Since the time of Münchhausen, we have learned that some fungi are indeed the houses of animals, the insects within which yeast and yeast-like fungi dwell. It took several hundred years and the development of new molecular methods to determine the general nature of fungi (Alexopoulos et al. 1996; Hibbett et al. 2007), and although more is to be learned about the earliest-diverging lineages, we now have a good understanding of fungal limits (Grossart et al. 2016). Although yeast species are difficult to recognize solely on the basis of morphology, their recognition as fungi was followed rather quickly by classification in different taxonomic groups based largely on ecology, physiology, cell wall composition, guanine–cytosine (G-C) content of nuclear DNA, and in the case of some basidiomycetes by morphology (Suh et al. 2006b; Kurtzman and Robnett 1998; Barnett 2004). Many yeasts are associated with other organisms, and the essential work of determining the species continues (Buchner 1965; Ganter 2006; Starmer and Lachance 2011; Blackwell

2017). There is a high probability that many additional associations will be discovered with the use of high-throughput sequencing.

Yeasts are taxonomically diverse, and the terms "yeast" and "yeast-like" refer to a fungal growth form that includes budding in asexual reproduction, although other types of asexual reproduction also may occur. A number of fungi are dimorphic, and in the same life cycle, they may grow as yeasts in association with insects (e.g., in the mycangia of many ambrosia beetles) and as hyphal filaments when free of the insects (e.g., in a wood substrate). Fungal groups with yeast-like growth forms include Ascomycota, Basidiomycota, and some zygospore-producing Mucoromycotina, although clear evidence of invertebrate animal association is not known for species of Mucoromycotina. Genome studies provided evidence that the irregular occurrence of the yeast growth form throughout the fungal kingdom is due to evolutionary convergence regulated by Zn-cluster transcription factors that regulate yeast–filamentous growth switches over time (Nagy et al. 2014). The evolutionary event that allowed for switches in growth form occurred early in the divergence of the fungal lineage, accounting for the sporadic appearance of the yeast growth form throughout the fungal tree, often in association with insects and other animals.

The yeasts (e.g., *Saccharomyces* spp.) and yeast-like fungi (e.g., *Symbiotaphrina* spp.) discussed below are mostly ascomycetes, but some basidiomycetes (e.g., *Trichosporon* spp. and new genera of Tremellomycetes previously placed in *Cryptococcus*—see below; Liu et al. 2016) also are associated with insects and other invertebrates. Ascomycetous and basidiomycetous yeasts are ecologically very different from each other, and the differences help to explain the more common occurrence of ascomycetous yeasts with animals. Ascomycetes usually have specialized niches rich in organic carbon (copiotrophic) where they are fermentative, often fragrant, and associated with other organisms, including invertebrate animals that disperse them. Basidiomycetous yeasts differ by their use of a wider variety of carbon compounds at low concentrations (oligotrophic), and they are less often fermentative, fragrant, or animal associated. Carotenoid pigments, forcibly discharged basidiospores, and clamp connections, when present in some taxa, allow them to be distinguished easily as basidiomycetes (Lachance and Starmer 1998; Suh et al. 2006b).

Ascomycetes are divided into three subphyla based on morphology and DNA analyses: Taphrinomycotina, Saccharomycotina, and Pezizomycotina. None of the Taphrinomycotina is known to be intimately associated with invertebrate animals, although occasional dispersal of yeast cells may occur. Members of Saccharomycotina (e.g., *Suhomyces*, *Teunomyces*) have extensive associations with animals. Several yeast forms evolved within the filamentous ascomycetes, Pezizomycotina (e.g., *Symbiotaphrina*), and while these taxa are not diverse in numbers, they are essential associates of certain insects, and studies of their genomes have given us data to provide new clues to the way in which strict symbioses evolved. Some arthropod-associated fungi with yeast states also occur among the three major subphyla of basidiomycetes. Pucciniomycotina is a group that most mycologists think of as plant parasites, but it is much more diverse and contains some biotrophic parasites of insects (e.g., *Septobasidium* spp.). Other yeasts are classified in one clade of Agaricomycotina (Tremellomycetes: new genera previously placed in *Cryptococcus*—see below; Liu et al. 2016) and in Ustilaginomycotina (e.g., *Sporisorium*) (Alexopoulos et al. 1996; Hibbett et al. 2007; Kurtzman et al. 2011; Begerow et al. 2014). Ganter (2006) discussed yeast–insect reports (1966–2004) characterized by habitat as well as taxonomic group. His records included a variety of productive plant-associated habitats, including decaying wood, basidiocarps, and the insects associated with these habitats (e.g., Platypodinae, Scolytinae, Biphyllidae, Bostrichidae, Buprestidae, Cerambycidae, Languriidae, Scarabaeidae), and his work should be consulted for detail.

13.1.4 Finding a Partner: Yeast and Animal Associations

Animals may be attracted by taste, feel, or sight. For example, certain insects are attracted to bright orange and yellow colors, and flies have been reported to be attracted to the fungus *Neurospora* growing in a bright mass, or they may react to the sweetness of ergot (Hawker 1957; Witzgall et al. 2012). Insects may use color only (Ernst et al. 2016) or odor at far distances and color in close quarters (Streinzer et al. 2009); many insects rely heavily on odor alone. Yeasts and other fungi produce volatile organic compounds (VOCs), which act as semiochemicals to signal other organisms, especially insects (Davis et al. 2013). The volatiles may reveal the presence of "nutrient resources, competitors, predators, potential mates, and habitat suitability" (Price et al. 2011), and the strength of the signal varies in some cases targeting only one to few organisms or in other cases attracting more broadly to include parasites. The effect of yeast volatile compounds was clearly demonstrated to me the day I inhaled the odor of bananas near the lab transfer chamber, unusual for our odorless (to me) mushroom-feeding beetle yeasts, including a number from Nitidulidae. A check of accession numbers revealed that these yeasts were in fact not from mushroom-feeding beetles but from nitidulid beetles that had been baited with fruit.

Associations between yeasts and invertebrates vary from very casual with the fungi having the advantage of being dispersed to fresh substrates by an animal accidently contacting the cells. Scanning electron micrographs show the variety of fungal diaspores attached to setae and resting in the depressions of arthropod exoskeletons (Blackwell 1984), and casual associations are not regarded as symbioses. Some yeasts are ingested, and after ingestion they are carried internally in the gut of insects; a few cells often survive gut passage and dispersal to new habitats. Douglas (2009) pointed out the benefits of microbial resources to nutrition, including the provision of sterols, amino acids, vitamins, and enzymes for digestion and detoxification of plant materials. She cautioned that some benefits ascribed to microbes have not been tested. For example, some animals, including insects, produce intrinsic cellulases, which have been hard to detect. Also, it is not clear that B-complex vitamins are always supplied to animals unless experimental work and genome annotation support the conclusion.

Studies focusing on yeasts associated with insect pollinators for dispersal have shown that a variety of ascomycetous yeasts are present in nectar, an osmophilic habitat relatively low in competition. Nectar is a well-known habitat for these yeasts, many of which are dispersed by the pollinators (Lachance and Starmer 1998; Andreadis et al. 2015). In a study that compared nectar composition, volume, and pollinators, Mittelbach et al. (2015) found a surprising number of basidiomycetous yeasts present in flowers with nectars high in monosaccharides rather than sucrose and visited by birds, indicating that nectar composition and type of vector are factors that lead to patterns of diversity (Mittelbach et al. 2015). Andréadis et al. (2015) were successful in using yeast cultures (four species of *Metschnikowia* and *Vishniacozyma tephrensis* (former *Cryptococcus tephrensis*)) to bait insects in an organic apple orchard. A diverse group of 93 arthropods from 15 orders, mostly insects, especially flies, were attracted to specific yeasts. Differences in taxa and their abundance were clear in almost all cases. The results suggested that attraction by yeasts and their volatile compounds could be the basis of effective detection and control of pest species.

Specificity has been assumed from observations of repeated associations and experimental studies (Andréadis et al. 2015). In some insects, yeasts are retained for long periods in structures such as mycangia and gut caeca. Often the associations do not show a high degree of species specificity except at the level of a clade as mentioned above or even unrelated groups of species with common ecological traits. The assumption is that evolution proceeds from casual associations with close contact in nature and progresses to a point of no return when a fungus and animal interact to a point that they can no longer live separately in nature, and sometimes evidence exists for such a progression in the examples that follow.

13.2 Examples of Associations

13.2.1 Virulent Pathogen to Obligate Mutualist

Obligate interactions of insects and microbial symbionts are common, but most of the microbes are bacteria and less often fungi. Some lineages, however, are exceptions. The usual symbionts that occur throughout a number of clades of Hemiptera are bacteria (Moran et al. 2008), but in a few species belonging to several clades (aphids, planthoppers, leafhoppers), fungi have replaced bacteria. Planthoppers have sucking mouthparts designed for acquiring phloem sap from plants, a diet that may be poor in nutrients without microbial supplements. The yeast-like symbionts (YLSs) of hemipterans have never been found free living in nature, and the host and microbial organisms cannot be grown separately, characteristics of obligate interactions. The most studied YLS is that of the brown planthopper, *Nilaparvata lugens*. The YLS has gone nameless until recently when the species was called *Entomomyces delphacidicola* (Fan et al. 2015), but apparently the taxon has not been described formally according to the rules of nomenclature. The naming may better await broader taxon sampling, however, because of possible polyphyly as noted below.

The YLSs are intracellular (endosymbiotic), located within specialized cells (mycetocytes) in the fat body, a differentiated cluster of cells connected to the gut, of planthoppers, aphids, and leafhoppers. The YLSs are present in all stages of the insect life history from egg to adult, and vertical transmission from one generation to another is ensured by transmission from the female parent to the egg by transovarial infection. As mentioned earlier, the brown planthopper, *N. lugens*, has been most studied, so generalizations may not always be accurate. During the process yeast cells are released from the mycetocytes to the hemocoel, then to cells surrounding the oocytes, and from that point to the egg surface where the hatching larvae acquire the yeasts as they chew out of the eggs (Buchner 1965; Cheng and Hou 2001). The relationship between *N. lugens* and its symbionts has been an object of investigation for more than 150 years because the insect is a destructive pest of cultivated rice. Although the yeasts cannot be grown in vitro, a method of harvesting cells by density gradient centrifugation has provided material for study (Noda and Omura 1992).

Various researchers helped to develop the planthopper–YLS interaction as a model system in which the YLS provides essential functions for the insect, allowing for survival on a low nutrient diet (Sasaki et al. 1996; Hongoh and Ishikawa 1997, 2000; Noda and Koizumi 2003). Studies of complete genomes of the planthopper, the YLS, a newly discovered bacterial associate, and rice provide insight, not only into the interactions of the organisms, but also into the evolutionary steps leading to obligate symbiosis. The brown planthopper genome size and G-C content are similar to that of the symbiont-associated palm aphid (*Cerataphis brasiliensis*) YLS, but the genome is somewhat reduced and has a higher G-C content when compared to many other ascomycetes (Xue et al. 2014; Fan et al. 2015). The YLS contributes enzymes for synthesis of essential amino acids, nitrogen storage and recycling, and sterols to supplement the poor nutritional quality of the highly specialized rice phloem sap diet of the YLS. Relatively large gene clusters for carbohydrate and amino acid metabolism probably reflect carbohydrate utilization and amino acid biosynthesis functions performed by the YLS (Xue et al. 2014; Fan et al. 2015; Wan et al. 2016). The bacterial associate synthesizes B-complex vitamins for the multipartite association (Xue et al. 2014). The loss of the YLS MAT loci involved in mating results in loss of sexual reproduction (Fan et al. 2015, also see below for apparent asexual reproduction in another yeast group associated with insects).

Not only did the YLS not have a formal name, but until the advent of DNA sequencing, it could not be placed among its close relatives. Early studies using phylogenetic analysis of rDNA placed both planthopper and aphid YLSs among filamentous ascomycetes in Pezizomycotina (Noda et al. 1995; Fukatsu and Ishikawa 1996); later the YLSs were placed near species of necrotrophic insect pathogens (Suh et al. 2001). A major revision of most taxa previously classified in Clavicipitales now places the YLSs in Ophiocordycipitaceae (Hypocreales), a group that includes *Ophiocordyceps sinensis*, renowned in herbal medicine, and *Elaphocordyceps* spp., parasites of hypogeous mycorrhizal ascomycetes, *Elaphomyces* spp. (Sung et al. 2007; Spatafora et al. 2007).

Dramatic symbiont changes have been observed in several hosts, including the roach–termite lineage. *Blattabacterium* sp. resides in mycetocytes of the common ancestor of roaches and termites, but soon after the divergence of termites, the bacterium was lost in all but the earliest-diverging termites, likely due to a change in diet to wood (Sabree and Moran 2014). A similar switch from prokaryotic to eukaryotic symbionts appears to have occurred independently in Hemiptera on several occasions (Nishino et al. 2016), perhaps driven by different dietary needs. The YLSs evolved from within *Ophiocordyceps* on at least two independent occasions, one from within *Ophiocordyceps* to a subclade that contained parasites of beetle grubs, ants, and YLS of planthoppers (*N. lugens* and *Sogatella furcifera*) plus an aphid [*Tuberaphis* (= *Hamiltonaphis*) styraci]; the second subclade contained *Ophiocordyceps* parasites of caterpillars (including *Oph. sinensis*) and YLS of eared leafhoppers (Cicadellidae, Ledrinae) (Nishino et al. 2016). Based on analyses from extant organisms, the phylogenetic evidence suggests stepwise evolutionary transitions with an ancestor containing two bacterial symbionts, then one of the bacteria retained and the other replaced by a fungus, and eventual loss of the remaining bacterium to a fungus symbiont (the YLS) only in the evolutionary history of leafhoppers (Nishino et al. 2016). Although the aphid and planthopper YLS clade and the leafhopper YLS clade are both derived from within *Ophiocordyceps* fungal parasites, they appear to have diverged independently. The switch to the fungus offers an advantage over bacterial symbionts, because, although both types of microbes have uricase genes used in nitrogen recycling, bacteria do not synthesize sterols, essential requirements for insect metabolism (Gibson and Hunter 2010; Fan et al. 2015).

The concept of host switching (as "host jumping") has been used in fungal biology to explain the interkingdom host shifts observed in the *Ophiocordyceps* lineage in evolution from an insect to an *Elaphomyces* host in a common hypogeous environment with connections to plant roots (Nikoh and Fukatsu 2000), and other examples of dramatic host switches also occur in the Hypocreales (Kepler et al. 2012). In a discussion of host–parasite evolution, Araujo et al. (2015) gave evidence that host switching is far more common than co-speciation. In fact, there are few if any examples of co-speciation among fungi. The *Ophiocordyceps* lineage example satisfied the first requirement stipulated by Araujo et al. (2015), opportunity, by temporal and spatial coexistence amid plant roots. Exactly what has been the evolutionary pathway to go another step to overcome changes from virulent filamentous pathogen to obligate YLSs of insects such as planthoppers? Does loss of virulence involve a more efficient immune system of the new host? Does the host switch occur rapidly enough for the newly infected host to survive pathogenicity? Accelerated evolution of a parasite occurs when it changes to a mutualistic lifestyle (Lutzoni and Pagel 1997), and changes in bacterial effects have sometimes been observed in relatively time short periods, perhaps generational time. Weeks et al. (2007) found evidence that cytoplasmic incompatibility in *Drosophila* occurred in matings between males carrying a certain *Wolbachia* infection and females that were either uninfected or infected with a different *Wolbachia*. The fecundity rate (actual reproductive rate), believed to be the only effect on the newly infected population, was measured after 20 years. Evolutionary changes in the parasite *Wolbachia* caused a reduction of an original fecundity disadvantage of 20%, and in less than 20 years, the disadvantage had become an advantage of 10%. The evidence also supported a more encompassing hypothesis that evolution of a mutualism occurs through maternal vertical transmission rather than by horizontal transmission (Weeks et al. 2007).

But could changes in even a relatively short generational time frame allow for coexistence long enough for the virulent parasite to develop a mutualistic life with the planthopper? Perhaps the first steps in the rapid shift from virulent pathogen to obligate mutualist could occur almost instantaneously as has happened with some virus-infected filamentous fungi, for example, hypovirulence of the agent of chestnut blight, *Cryphonectria parasitica*. Immediate change from tree-killing parasite to hypovirulent parasite comes by infection with a virus (Alexopoulos et al. 1996; Anagnostakis 2001). The process occurs naturally when fungal strains are vegetatively compatible, and hypovirulent (virus +) strains fuse with the virulent (virus –) strains, and the virus migrates throughout the cytoplasm converting the virulent strain to hypovirulent, resulting in an infection that does not kill the tree as a result of the now hypovirulent infection. The phylogenetic evidence of Nishino et al. (2016) suggesting amelioration by a bacterium could have aided in the transition of the last step in the development of the mutualism, suggests a way in which the virulent pathogen to YLS could have evolved quickly. Such a progression has also been described as a "stepping-stone" process (Araujo et al. 2015).

13.2.2 Fungi as a Unique Habitat for Insects and Yeasts

Fungus feeding is important to many insects and may be a primitive feeding habit. Lawrence (1989) estimated that half of all beetle families are primarily mycophagous or dependent upon plant material that has been altered by fungal action, a statement reiterated by others (Hammond and Lawrence 1989; Leschen and Buckley 2007; Cai et al. 2016). However, more strictly speaking, about thirty families of beetles are associated with basidiocarps and other fungal fruiting bodies, upon which they feed and breed exclusively. More than 250 beetle species were collected from 2660 basidiocarps of a polypore species in an extensive study in eastern Canada (Gilbertson, 1984). In another study, 136 taxa from 30 different beetle families were reported from *Pleurotus* spp. in North America (Cline and Leschen 2005). Some beetles specialize only on specific parts of a basidiocarp, such as spores or context, and differentiate certain types of context tissue characterized by "hardness" and texture. For example, spore-feeding beetles in three different families of superfamily Staphylinoidea had convergent morphological features linked to the spore-feeding trait (Betz et al. 2003). A group of polypores including species of *Phellinus*, *Inonotus*, and *Phaeolus* judged to have the same "hardness" were inhabited by the same context-eating ciid (Ciidae) beetle species (Lawrence 1973, 1989). Mouthparts have been selected for feeding, and the changes in morphology can be used to track evolutionary changes across insect lineages corresponding to changes in diet in the fossil record (Leschen and Buckley 2007; Cai et al. 2016).

As a distinct major lineage of organisms, fungi differ in a number of ways from other organisms including plants. Many beetles associated with plant nutritional sources have mouthparts that allow them to penetrate living cells, degrade recalcitrant plant cell walls, or more often use microbes to degrade the cell walls. As mentioned above, the ability to determine feeding substrates based on mouthpart morphology underscores the possibility of being able to interpret spore-feeding behavior of insects from fossils in order to search for associated yeasts. Reliance on a single food source such as mushrooms and polypores throughout their lives draws attention to the possibility of an inadequate diet over the long term. It is not clear that insects could subsist on basidiocarp tissues alone, but microbes that inhabit the gut, sometimes in midgut caeca (McHugh et al. 1997), could provide additional nutrients. Suggestions that yeasts supply vitamin supplements and a variety of enzymes have been supported by findings of a variety of B-complex vitamins and assimilation of a wide variety of carbon substrates, including trehalose. Evolution of caeca in the gut of these beetles points to a benefit for the insect, but the exact basis of the association remains uncertain. The dispersal benefit for the microbes is extremely important for their survival, because fungi (e.g., wood decayers) and insects (e.g., fungus feeders) modify or completely destroy their substrates in a relatively short time and need a new, similar habitat.

A large percentage of the yeasts isolated from the gut of fungus-feeding beetles were previously unknown taxa, some more than 40 base pairs of LSU rDNA different from the closest known species in GenBank. The principal clades in the fungus-feeding beetle guts included *Suhomyces* species (previously known as the *Candida tanzawaensis* clade), *Teunomyces* species (previously, *Candida kruisii* clade), *Meyerozyma guilliermondii* clade (previously *Pichia guilliermondii*), and the *Candida mesenterica* and Trichomonascaceae clades. The insects showed no preference in a variety of wood-decaying basidiomycetous fruiting bodies indicating that as in the case of yeasts from ephemeral flowers and desert cacti, the gut yeasts are more closely tied to the host insect than to the specific food substrate used by the insect (Blackwell 2017). This fact indicates that the association likely benefits both insects and fungi, and that, as mentioned just above, dispersal of yeasts to a new habitat is of great importance. Often yeasts can be isolated from frass, and germination of cells on agar indicates survival after gut passage. This fact points to dispersal success inside an insect, but identification of the insect associate may not be possible (Imanishi et al. 2008).

Yeasts are more commonly associated with beetles having certain ecological and morphological similarities. Fungus-feeding beetles may possess gastric caeca located at the anterior part of the midgut, and these have been well characterized in beetles such as the fungus-feeding *Megalodacne heros* (Erotylidae; McHugh et al. 1997). Caeca and mycangia also are found in Anobiidae, Buprestidae, Cerambycidae, Cucujoidea, and Tenebrionidae, many of which are fungus feeders. The deep pockets of gastric caeca may help to retain gut yeasts even during molts in larval beetles (Engel and Moran 2013). Moreover, some beetles exhibit parental care, providing an opportunity for possible vertical transmission to offspring, including fungus-feeding Erotylidae (e.g., *Pselaphacus* spp.) and Tenebrionidae. Other families with parental care known to have fungus and yeast associations include Curculionidae (Scolytinae, Platypodinae), Passalidae, and Scarabaeidae. Evidence of consistent yeast-beetle associations has come from repeated isolation of the same yeast species from the same beetle species over a 5-year period at five different localities in southeastern Louisiana, the association of the same yeast and beetle species along the eastern and gulf coasts of the United States from Vermont to Georgia and Louisiana, and isolation of a yeast from both larvae and adults of a beetle species (Blackwell and Suh, unpublished data). Although species-specific associations do occur on occasion and vertical transmission may assure such an association over the short term, specificity is usually more common at the level of yeast clade.

Fungus substrates were chosen for study in my own work, because yeasts previously had been reported from mushrooms (Prillinger 1987), and the substrate offered an easy target to resample certain insects to determine repeated associations (Suh and Blackwell 2005; Suh et al. 2005). Several clades of yeasts were dominant among the taxa isolated from fungus-feeding beetles. The yeast strain, CBS 7422, had a long and interesting history since it was first isolated from the moss *Polytrichum commune* in Kanagawa Prefecture, Mount Tanzawa, Japan, by Nakase in 1966. Nakase refrained from describing the single strain for 22 years, while he waited unsuccessfully to isolate the yeast a second time before describing it as *C. tanzawaensis* (Nakase et al. 1988). Thirteen more years passed before it or any close relative was found. Kurtzman (2001) described six new related species from a variety of substrates including insect frass, rotted wood, and mushrooms, mostly from the central United States but also one from South Africa. *Suhomyces kilbournensis* was described most recently from maize kernels collected in Illinois, USA, and soil in Mexico (Kurtzman et al. 2016). In a study of yeasts from insect guts, sixteen more species in what is now known as the *Suhomyces tanzawaensis* clade were described from Panama and the southeastern United States in association with fungus-feeding beetles (Suh et al. 2004a).

Suhomyces spp. are common in the gut of a variety of fungus-feeding beetles. Out of about 650 strains from the gut of fungus-feeding beetles, 30% were members of this clade, and 85% of these were isolated from only two beetle families, Tenebrionidae and Erotylidae. While the clade is widespread in association with fungus-feeding beetles, the species may have more restricted distributions. Among the 16 novel species described, *Suhomyces panamericana*, *Suhomyces bribrorum*, and *Suhomyces anneliseae* (Table 13.1) have been found to have broad ranges

Table 13.1 Localities where species of *Suhomyces* and *Teunomyces* were isolated from the gut of fungus-feeding beetles (Blackwell, Luangsa-ard, Nguyen, Suh and Urbina, unpublished data)

Panama	SE United States	Thailand
Suhomyces anneliseae	*Suhomyces anneliseae*	*Suhomyces anneliseae* (0[a])
Suhomyces panamericana	*Suhomyces panamericana*	*Suhomyces* cfr. *panamericana* (13–15)
Suhomyces bribrorum	*Suhomyces bribrorum*	–
Suhomyces ambrosiae	*Suhomyces ambrosiae*	–
Suhomyces guaymorum	–	–
Suhomyces bokatorum	–	–
Suhomyces kunorum	–	–
Suhomyces terraborum	–	*Suhomyces* cfr. *terraborum* (7)
Suhomyces emberorum	–	–
Suhomyces wounanorum	–	–
Suhomyces maxi	–	–
Suhomyces taliae	–	–
–	*Suhomyces yuchorum*	–
–	*Suhomyces chickasaworum*	*Suhomyces* cfr. *chickasaworum* (0–11)
–	*Suhomyces choctaworum*	–
–	*Suhomyces bolitotheri*	–
–	*Suhomyces atakaporum*	*Suhomyces* cfr. *atakaporum* (16–30)
–	–	*Suhomyces tanzawaensis* (0–2)
–	–	*Suhomyces* sp. ST-431 (7)
–	–	*Suhomyces* sp. ST-370 (17–19)
Teunomyces panamensis	–	–
Teunomyces barrocoloradensis	–	–
Teunomyces stri	–	–
Teunomyces aglyptinia	–	–
Teunomyces gatunensis		–
	Teunomyces pallodes	–
	Teunomyces tritomae	–
	Teunomyces lycoperdinae	–
	Teunomyces atbi	–

[a](n) indicates number of base pairs difference in LSU rDNA from type strain

evidenced by their isolation from a variety of beetles in Panama, the United States (Suh et al. 2004a), and Thailand (Blackwell et al. unpublished data).

Su. anneliseae has been isolated from beetles in seven families of fungus-feeding beetles. In addition *Suhomyces ambrosiae* reported by Kurtzman (2001) from a number of localities in the central and northeastern United States also was isolated in Panama and the southeastern United States (Table 13.1). The basis of the specific relationship between fungus-feeding beetles and *Suhomyces* clade yeasts is not clear, but as mentioned above, the results of metabolic tests indicate that yeasts produce several B vitamins and a wide variety of carbohydrate-degrading enzymes that may be of use to the beetle hosts. It is of interest that the five species that were not isolated from the gut of beetles are in a subclade distinct from the 19 beetle-associated species.

Crosses of related strains of *Suhomyces* have not produced ascospores in culture (Suh et al. 2004a; Kurtzman et al. 2016). Little information from genomes has bearing on the topic, except that the genome of the type of *Su. tanzawaensis* does have a heterothallic *MAT* locus organization with only one *MAT* idiomorph present in the single strain examined (Riley et al. 2016). It is possible that the other mating type will be found at some point. An interesting finding was that many gut isolations recovered only a single species of yeast. The isolation of single species could be an artifact of the methods of isolation, including choice of medium or purification, so that rare yeasts were overgrown or the result of inhospitable insect gut conditions, although an effort was made to avoid these possibilities. In an effort to explain this finding, we looked for inhibition of growth of *Suhomyces* strains, and it is possible that some of the yeasts may inhibit the growth of related strains (Blackwell and Lu, unpublished data).

Species from a second previously rarely isolated yeast clade have been placed in the genus *Teunomyces*, although some strains have been found in habitats other than fungus-feeding beetles. The clade members are known widely from the northern and southern United States, Greece and Crete, Panama, and China (Table 13.1) (Kurtzman et al. 2016). *Teunomyces kruisii* and CBS 9453 (now *Teunomyces cretensis*) were the only members of the clade known until 2006. Of 650 fungus-feeding beetles dissected, about 100 isolates, the majority of which were from beetles in the Nitidulidae, especially *Pallodes*, were isolated, and nine new species were described (Suh et al. 2006a). There are no records of clade members from Thailand, but few nitidulids were collected, perhaps due to drought conditions (Blackwell et al. unpublished data). As in the case of *Suhomyces*, strains of *Teunomyces* have never produced ascospores in culture, despite numerous attempts at crossing closely related strains under different conditions (Suh et al. 2006a; Kurtzman et al. 2016).

A life away from fungus-feeding beetles has been a possibility for species belonging to the genera *Suhomyces* and *Teunomyces*, because strains have been isolated from frass and, more recently, free living associated with trees (Sylvester et al. 2015); isolation from the beetle gut, however, always has had a higher probability of success. A study using 454 sequencing of the LSU rDNA marker detected species of *Suhomyces* and *Teuomyces* in the gut of *Drosophila* spp. visiting *Russula* spp. at Stony Brook, NY, USA (Chandler et al. 2012). These species were distinct from all other yeasts analyzed from 13 other non-fungal substrates (mostly from California). Furthermore, yeasts common to plant-associated habitats were absent in the fungus-associated flies. Bacterial populations from the fungi also were distinctive, and bacteria were more diverse than yeasts in all substrates sampled including those from *Russula*.

The fact that communities centered around a fungal resource are distinct from plant-based communities is of interest but not surprising, because the organisms that specialize on fungal substrates would require distinctive physiology for efficient use of a fungal resource. Most fungi, including mushrooms, differ dramatically from plants in cell wall composition (e.g., chitin and glucan, not cellulose, hemicellulose, and lignin), energy resource (storage as glycogen and lipids, not starch), and sterols as precursors of cell membranes (ergosterol, not 24-alkyl-Δ(5)–sterols) (Alexopoulos et al. 1996; Weete et al. 2010). Starmer (1981) noted another difference: yeasts from relatively low-quality mushrooms had a higher catabolic activity than those from other yeast communities. Contributions by different yeasts to the fitness of *Drosophila* spp. have been determined (e.g., Anagnostou et al. 2010), and the differences in yeast communities from different habitats probably have been significant in *Drosophila* radiations (Broderick and Lemaitre 2012). The discovery of both *Suhomyces* spp. and *Teunomyces* spp. associated with drosophilids adds a new dimension to the life cycle of yeasts that should be integrated with the life histories of the flies. Are the yeasts simply acquired from the surface of the fungus where the drosophilids feed along with beetles, or does some other interaction with beetles occur?

13.2.3 Yeasts and Wood-Ingesting Insects

Several families of beetles (Passalidae, Cerambycidae, Buprestidae, and a Tenebrionidae species *(Phrenapates bennetti)* ingest wood as they tunnel through several year-old white-rotted logs. The most common yeasts associated with wood-ingesting beetles are members of the *Scheffersomyces* and *Spathaspora* clades. Many of the yeasts are marked by a rare attribute, xylose fermentation, among other properties; xylose-fermenting yeasts are consistently present in termites, wood roaches, and other animals that feed in dead wood (Prillinger et al. 1996; Suh et al. 2003; Urbina et al. 2013b). In addition several other clades of xylose-assimilating yeasts have also been present, including species of *Lodderomyces* and basidiomycetous species of *Trichosporon* and Tremellomycetes genera previously included within *Cryptococcus* (Urbina et al. 2013b; see Liu et al. 2016). Comparisons of gut yeasts from fungus-feeding beetles with those of wood-ingesting beetles show that very different communities are associated with each group of yeasts, in part due to physiological adaptations for use of nutrients and survival of the gut physical conditions.

The passalid gut is distinctly compartmentalized, similar to that of the termite gut but larger (Nardi et al. 2006). In addition to the foregut and midgut, the hindgut is divided into several different compartments that are morphologically and physiologically distinct. Parabasalid flagellates, related to termite hindgut organisms, occupy part of the anterior hindgut. The enlarged anterior hindgut has the lowest O_2 availability in the gut, and bacteria are attached to the convoluted gut walls. Although the bacteria in the anterior hindgut are not diverse in species numbers, the taxa present and gut physical conditions indicate that it is the site of decomposition of lignocellulosic materials, N_2 fixation, pH_2 regulation, and CH_4 production (Ceja-Navarro et al. 2014). The elongated posterior hindgut contains other bacteria, fungi, and gut organisms previously placed in a group known as trichomycetes (Lichtwardt et al. 1999). Yeasts are attached to the posterior end of the hindgut, and xylose-fermenting yeasts are always present in both larvae and adults. Adult beetles provide some level of parental care by provisioning the gallery walls with a mixture of chewed wood and frass. The frass contains yeasts, so that the hindgut microbial community can be renewed constantly after molting. Any advantage of these yeasts to the beetle is not clear. Although fermentation can increase the nutrient value of a poor diet and enhance the anaerobic conditions in the gut favoring the bacterial communities, the far posterior position of the yeasts may not promote an advantage because it is not clear how much absorption occurs in that part of the hindgut in insects.

Odontotaenius disjunctus (Passalidae) harbors the xylose-fermenting species, *Scheffersomyces stipitis*, in the hindgut. The yeast has a holdfast by which it attaches to the hindgut wall (Suh et al. 2004b; Nardi et al. 2006). A holdfast, rarely observed as a yeast adaptation, may indicate that selection has been in favor of the yeast that benefits from an advantageous position to acquire nutrients in a dynamic gut region (Nardi et al. 2006). All attempts to cure the beetles of yeasts failed so that experimental studies could not be pursued (Gross 2010). The species *Scheffersomyces cryptocercus* has been isolated several times in association with a wood roach, *Cryptocercus* sp. (Urbina et al. 2013a), and apparently free living from the bark of forest trees (Sylvester et al. 2015). A different xylose-fermenting yeast, *Spathaspora passalidarum*, also has been isolated from the gut of *O. disjunctus* (Nguyen et al. 2006b). *Scheff. stipitis* and *Spath. passalidarum* both form ascospores in culture, although *Spath. passalidarum* has more specific requirements, notably lower temperature (Nguyen et al. 2006b). Sequencing of a single isolate each of *Scheff. stipitis* and *Spath. passalidarum* showed that *Scheff. stipitis* has a homothallic *MAT* locus organization with both *MAT*a and *MAT*α present, but *Spath. passalidarum* has a heterothallic organization with a single idiomorph present (Riley et al. 2016). Related asexual species of both yeasts have been found in rotting wood, and some of them associated with wood-ingesting beetles also have been isolated directly from wood (Cadete et al. 2013; Morais et al. 2013). In tropical localities (Brazil, Guatemala, Panama, and Thailand), *Sc. stipitis* and related yeasts appear to be more species diverse. The list of xylose-fermenting yeasts from rotted wood, beetles, and termites grows longer and now includes species of *Sugiyamaella* and other plant cell wall-degrading yeasts of biotechnological potential (Jeffries et al. 2007; Urbina and Blackwell 2012; Morais et al. 2013; Suh et al. 2013; Urbina et al. 2013b; Cadete et al. 2016; Handel et al. 2016; Riley et al. 2016).

13.2.4 Basidiomycetes Also Inhabit the Gut of Insects

As mentioned earlier the work of Prillinger (1987) and Prillinger et al. (1993, 1996) was the stimulus to look at insects associated with mushrooms and other fungi in order to determine if insects might be active dispersers of yeasts. The possibility was that because the yeasts he isolated were from a mushroom parasitic on other mushrooms, an entire community of organisms might be dispersed between mushroom islands by insects. The basidiomycetous yeasts appeared infrequently (Prillinger et al. 1993), and we also isolated basidiomycetes much less often than ascomycete yeasts by our dissection and streaking methods. Among the Tremellomycetes (Agaricomycotina), species in the genera *Trichosporon*, *Bullera*, and others previously placed in *Cryptococcus* (Liu et al. 2016), including *Vanrija*, *Rhynchogastrema*, *Cystofilobasidium*, *Naganishia*, *Papiliorama*, *Piskurozyma*, *Filobasidium*, *Vanrija*, and *Vishniacozyma*, were isolated in Panama in association with fungus-feeding beetles in a variety of families (Curculionidae, Endomychidae, Nitidulidae, Passalidae, Scaphididae, and Scaraebidae). Ballistosporic yeasts (e.g., *Vanrija humicola*, former *Cryptococcus humicolus*) and relatives of those first reported by Prillinger et al. (1993) from species of *Asterophora* were occasionally found in the same beetles. Other basidiomycetous yeasts only rarely isolated from the gut of beetles included ballistosporic yeasts classified in Pucciniomycotina (e.g., previously classified in the genera *Sporobolomyces*, *Rhodosporidium*, *Rhodotorula*) and members of Ustilaginomycotina (*Pseudozyma*, *Sporisorium*).

Several basidiomycetous yeasts isolated from insect guts are known to interact with other fungi. One new species, *Trichosporon insectorum*, was isolated from the guts of a scarab and a passalid beetle from Gamboa and Barro Colorado Island, Panama, respectively, but also from artisanal cheese where it may have been insect or mite dispersed. Interestingly, all three strains possess an identical killer factor that is active against certain strains of *Cryptococcus neoformans* and *Cryptococcus gattii*. Another basidiomycete yeast from a termite hindgut, *Apiotrichum* (*Trichosporon*) *mycotoxinivorans*, has been reported to detoxify the mycotoxins, ochratoxin A, and zearalenone (Molnar et al. 2004).

13.2.5 Sex in the Gut of Wasps

To date, a great deal of time has been invested in looking for potential symbiotic associations in nature. There is, however, relatively little known about the activities of fungal associates within a host in nature. *S. cerevisiae* is one of the best-known fungi in the world because of its value to scientific research and its industrial importance, but despite the sophisticated knowledge of genomics and molecular biology of the yeast, relatively little was known of the complete life cycle in nature. Cavalieri and colleagues (Stefanini et al. 2012, 2016) determined that the gut of adult female social wasps (*Vespa crabro* and *Polistes* spp.) was a specialized niche for overwintering by *S. cerevisiae*, one in which conditions for outbreeding existed. Their experimental work indicated that the wasp gut also provided an environment conducive to outcrossing not just for *S. cerevisiae* but also for interspecific hybridization with its close relative, *Saccharomyces paradoxus*, both contributing to high genetic diversity. Their experiments showed that pure *S. paradoxus* ascospores were unable to survive gut passage, but hybrid spores remained viable.

Stefanini et al. (2016) have filled a gap in our knowledge of the life history of *S. cerevisiae* not only by determining that the yeasts overwinter in the wasp gut but also by determining that the spores are dispersed to the wasp progeny and then to the fruits in the wild or in vineyards. The wasp-yeast interaction provides a new role for an insect in providing a means of overwintering, outcrossing to increase diversity, and eventually dispersing the yeast. If gut passage of *S. cerevisiae* is compared to the yeasts of fungus-feeding beetles in *Suhomyces* and *Teunomyces*, the life histories are very different, primarily in the dramatic increase in genetic diversity of *S. cerevisiae* in the wasp gut and apparent lack of or only rare sex in the fungus-feeding beetles (see above), two very different life-history strategies.

13.2.6 YLSs of Anobiid Beetles

Symbiotaphrina buchneri and *Symbiotaphrina kochii* have long been known (Escherich 1900; van der Walt 1961) as YLSs associated with *Stegobium paniceum*, the drugstore beetle, and *Lasioderma serricorne*, the cigarette beetle (Anobiidae), respectively. However, unlike the YLS of planthoppers, *Symbiotaphrina* spp. grow readily in vitro, and they have been discovered growing independently in nature (Martin et al. 2015). The yeasts occupy caeca at the anterior end of the insect midgut and do not undergo sexual reproduction (Gams and von Arx 1980). The beetles eat processed plant material that often contains toxic secondary metabolites, and they have been found living in sacks of flour, cigarette tobacco shreds in packs, and other harsh, arid environments. The females pass yeast cells to their offspring by smearing them on the surface of the eggs. As the larvae chew out of the egg, they ingest some of the cells from the shell that will eventually populate the midgut caeca (Jurzitza 1979). Some of the yeast cells reproduce in the larval gut, but most of them never move from the gut into the caeca, so that part of the population is not passed on to the offspring vertically, an important observation that is mentioned again below (Frank 1996). Uninfected beetles are reported to die prematurely, while symbiont-infected beetle hosts survive. The beetles obtain amino acids and B-complex vitamins from the YLS (Nardon and Grenier 1989). The beetles also obtain ergosterol and 5-dihydroergosterol from the YLSs that they convert to 7-dehydrocholesterol (Pant and Fraenkel 1950; Nasir and Noda 2003). Experimental evidence from cultural studies indicates that *Symb. kochii* synthesizes broad-spectrum detoxifying enzymes (e.g., ester hydrolase, glucosidase, phosphatase, and glutathione transferase) to detoxify a variety of harmful natural and human-made products, including mycotoxins, insecticides, and herbicides, including parathion. Some of the compounds tested were used as a sole carbon source by the yeasts (Shen and Dowd 1991; Dowd 1992).

The phylogenetic placement of *Symbiotaphrina* spp. was impossible to determine until molecular methods became available. As the name suggests, the genus first was assumed to be a relative of early diverging ascomycetes (Taphrinomycotina) and then Saccharomycotina and, finally, based on small subunit rDNA, among Pezizomycotina in an uncertain position because it was not close to any sequences in GenBank (Noda et al. 1995; Jones and Blackwell 1996; Noda and Kodama 1996). More recently, Gazis et al. (2012) discovered a new lineage of the Pezizomycota radiation, the Leotiomyceta that contains *Xylona heveae* (Xylonomycetales), an endophyte isolated from *Hevea* spp. (rubber trees) in Amazonian Brazil. In a second study, Gazis et al. (2016) used a concatenated alignment of ITS (internal transcribed spacer) and LSU rDNA with well-chosen taxa for comparison to place the Xylonomycetales as a sister to Symbiotaphrinales, finally placing the YLS among close relatives. In addition, when the genome of *Xyl. heveae* was compared with those of other endophytes and plant-associated ascomycetes, there were fewer carbohydrate-active enzymes (CAZymes) for degradation of plant cell walls encoded, and the genome had more in common with animal-associated fungi that those associated with plants (compare with planthopper enzymes above). Thus, the ability of the endophytic *Xyl. heveae* to enter plants by direct penetration is limited, and Gazis et al. (2016) suggested that the species could be associated with an insect vector. Another unexpected finding from the study was the discovery of two strains of *Symb. kochii* in GenBank originally isolated as endophytes. The possibility that usually vertically transmitted symbionts have some cells that are not transmitted directly to the host offspring may benefit the symbionts. Frank (1996) included the life history of the drug-store beetle in his discussion of alternative horizontal dispersal as an evolutionary strategy favoring symbionts in order to avoid competition with similar genotypes.

13.2.7 Extending Endophyte Life Histories to Include Insects

Many fungi regularly produce budding cells in their life cycles. As Nagy et al. (2014) indicated, a genetic tool kit for directing changes from yeast to filamentous growth may have originated soon after the divergence of the fungal lineage. "Latent homology" was proposed to explain the switches in growth form in which yeasts appear throughout the fungi (Nagy et al. 2014). Some of the yeast-like fungi may have life cycles as endophytes (Vega 2008, Sasan and Bidochka 2012), including "multifunctional lifestyles" as insect pathogens (Barelli et al. 2016). Certain arthropod-associated yeasts have been reported as endophytes, including a mite-associated species of *Meira* (Sánchez Márquez et al. 2008). In addition to the YLS of anobiid beetles discussed above, other fungi with well-known arthropod associations also may live as endophytes (Martin et al. 2015).

The life history of *Septobasidium* species is interesting for their mutualistic associations with scale insects (Coccoidea). The genus is a member of the large, diverse clade of rust fungi (Pucciniomycotina), and it has a worldwide tropical to temperate distribution. The life histories of many species in culture include basidiospores that are forcibly discharged and germinate by undergoing several rounds of budding. Eventually hyphae grow out, and in nature a hyphal mat with a velvety or felt-like surface develops (Henk 2005; Henk and Vilgalys 2007; Henk personal communication). The interior of the colony has a series of chambers and tunnels, and a few to entire colonies of the insects are parasitized by fungal haustoria, typical of biotrophic parasites. The parasitized scale insects support the growth of the fungus, and in return the other scale insects are protected from parasitoid wasps and desiccation. The life cycle associated with scale insects was known for species of *Septobasidium* for more than a century (Couch 1938). Now, however, two strains of *Septobasidium* sp. have been discovered in wild trees of *Hevea* spp. and more than 30 environmental ITS sequences in GenBank may be endophytes (Martin et al. 2015). It is not clear how the endophytes fit into the scale insect-associated life history, but this finding deserves investigation.

13.2.8 Supplements for Blood-Feeding Insects

Some insects (e.g., tsetse flies, bat flies, kissing bugs, lice) are blood feeders for their entire lives, and they often harbor microbial symbionts, mostly bacteria including W*igglesworthia* spp. The bacteria are important because blood diets are deficient in essential B vitamins and some amino acids, and the bacteria help to overcome the dietary deficiencies (Douglas 2009; Gusmão et al. 2010). More recently, both laboratory-reared and wild mosquito populations have been reported to harbor a variety of yeast species in different stages of the insect life cycle (Ignatova et al. 1996; Gusmão et al. 2010). Attention to gut organisms in mosquitoes has increased with the spread of arthropod-borne diseases including malaria and a number of viruses (e.g., dengue, West Nile, Chikungunya, and Zika). The finding that has attracted new attention is that *Wickerhamomyces anomalus*, a mosquito gut yeast, produces an inhibitor of the β-1,3-glucans in the spore walls of several protistan parasites, including an agent of malaria, *Plasmodium berghei*. Inhibition of the protistan spores occurs in vitro but also in the mosquito gut where inhibition is up to 90% greater than in the controls (Ricci et al. 2011; Cappelli et al. 2014; Valzano et al. 2016; van Tol and Dimopoulos 2016).

13.2.9 Beetles that Farm

Beetles are one of the three groups of farming insects (the other two being termites and ants; Mueller and Gerardo 2002), but the beetles that acquired this habit are not a monophyletic group. Instead farming evolved independently in various weevils (Curculionidae) in the subfamilies Platypodinae and Scolytinae, with most ambrosia beetles found throughout Platypodinae and in two derived tribes of Scolytinae (Corthylini and Xyleborini). Host beetles may have simple pits in their exoskeletons or the specialized glandular structures (e.g., mycangia of many ambrosia beetles) in which dimorphic fungi grow as yeasts and are nourished as they are transported to new substrates (Harrington 2005; Six 2012; Mayers et al. 2015; Lu et al. 2016). Harrington (2005) hypothesized that certain beetles might have escaped to the xylem (wood) to avoid competition in bark habitats. Perhaps this was the perfect strategy and the reason for estimates of more than 3400 ambrosia beetles (Batra 1967; Harrington 2005); the fungal associates of very few of the known species of ambrosia fungi (1–5%) have been characterized (Harrington 2005; Hulcr and Stelinski 2017).

The beetles cultivate fungi that have been a cause of taxonomic uncertainty because convergent evolution of the taxonomic characters likely was convergent for insect dispersal (Spatafora and Blackwell 1993). In the past, sexually reproducing species were placed in a single genus, *Ceratocystis*, and sometimes even combined with taxa in Saccharomycotina based primarily on spore shape (Kurtzman 2011; Starmer and Lachance 2011). Currently, these fungi are classified with their asexual relatives (of which there are many) based on DNA analysis in Pezizomycotina as members of the Ophiostomatales, *Ophiostoma* clade (e.g., *Raffaelea* spp., de Beer et al. 2014), and Microascales, *Ceratocystis* clade (*Ambrosiella* and *Meredithiella*, Harrington et al. 2014; Mayers et al. 2015). Although these fungi are not usually counted among the "yeasts," they provide an appropriate example of the way in which many fungi associated with animals have yeast stages that appear over and over, even for short periods during the life cycle (see above, Nagy et al. 2014). Yeast stages are found in most ambrosia fungi, and the fungi have yeast-like budding of ascospores; species in Ceratocystidaceae, however, produce arthrospore-like cells rather than budding cells (Mayers et al. 2015).

Less is known about associations between these beetles and yeasts (Saccharomycotina). A number of yeast species have been reported in associations with the filamentous ambrosial fungi and ambrosia beetles, although some are from frass and some of the beetles were not identified. These genera include species of *Ambrosiozyma*, *Cyberlindnera*, *Dipodascus*, *Lindnera*, *Lipomyces*, *Ogataea*, *Pichia*, *Saccharomycopsis*, and other yeasts (Kurtzman et al. 2011; Davis 2015). Some of these yeasts have hat-shaped ascospores, apparently a convergent character common among some filamentous ascomycetes in the same habitats (Kurtzman et al. 2011). Although many yeasts probably do not form specific associations with certain beetles, a few such as *Kuraishia capsulata* and *Ogataea pini* occur repeatedly with many beetles, and the yeasts are probably important in their communities for production of pheromones (Davis 2015).

13.2.10 Phoresy, Yeasts Ready to Roll

Coprophilous fungi (i.e., fungi which grow on animal feces) are generally thought to exhibit "gut passage syndrome" in their life cycles, with spores surviving or even requiring heat and enzyme treatments as they travel through the gut of a mammal. After the spores are deposited in dung, they germinate and grow and their reproductive structures are oriented toward light for forcible discharge. To overgeneralize about the intriguing gut passage syndrome, however, is to neglect another type of life cycle in dung and other quickly deteriorating habitats, such as decaying plant material and beached wrack (Blackwell and Malloch 1989). Many of these dung fungi are mycoparasites and dependent on a fungal host and beetle and mite dispersers, and the timing of spore maturity and disperser migration is precisely timed for a move to a new dung pile. At maturity species of *Pyxidiophora* (Laboulbeniomycetes) produce ascospores that are passively discharged, oozing to the tip of a long-necked perithecium where they adhere to phoretic mites that then attach to beetles for transport to a fresh habitat, an example of hyperphoresy (Blackwell and Malloch 1989). Yeast cells are produced from the ascospores, which multiply the products of meiosis, and can be spread rapidly over the surface of the new substrate (Blackwell et al. 1986). The increase in number of cells by the time of arrival at the new substrate probably also increases the possibility of contact with the host fungus and speeds up the life cycle that must be completed before desiccation of the substrate (Blackwell and Malloch 1989). Species of *Kathistes* (Ophiostomatales) are poorly known, but in Ontario, Canada, they are dispersed by phoretic mites of parasitic flies of moose, another case of hyperphoresy. These species also have yeast cells that develop from the ascospores, again providing for multiplication of the meiotic products (Malloch and Blackwell 1990).

The holdfast of *Scheff. stipitis* that attaches to the hindgut of beetles, a part of the gut from which the cells could otherwise easily be eliminated, was discussed above. Only a few other yeasts are known to produce holdfasts by which they attach for dispersal (Kerrigan and Rogers 2003, 2013). The pseudohyphal filaments of several species of *Botryozyma* attach to nematodes by bifurcate basal holdfast cells distinct from those of *Scheff. stipitis*, and the filaments often are present in large numbers on nematodes phoretic on long-horned beetles, *Saperda calcarata* (Cerambycidae) in trembling aspen, another example of hyperphoresy. Because of the way we isolate yeasts in culture, they rarely are observed in situ, and it may be possible that other species could have holdfasts if we observed them in nature.

13.3 Concluding Remarks

More yeast–insect associations remain to be discovered. There already are many additional reports scattered throughout the literature of yeasts associated with a variety of insects. Examples include reports of lacewings and caddis flies (Nguyen et al. 2006a), dung beetles (Górz and Boroń 2016), codling moths (Witzgall et al. 2012), as well as many others that need more focused study. Social insects (termites, ants, bees, and wasps) are known for their reliance on microbes that play vital roles in their life cycles. Subsocial insects include a number of beetles that exhibit parental care [Erotylidae (e.g., *Pselaphacus* spp.), Silphidae, Chrysomelidae (Cassidinae), Carabidae, Lampyridae, Passalidae, Scarabaeidae, Staphylinidae, Curculionidae (Scolytinae, Platypodinae), and Tenebrionidae (*Phrenapates* spp.)] that also are potential hosts. Although these insects lack the specialized castes of social insects, they often live in large groups with overlapping generations and care for the young. Younger adult colony members may help perform duties connected with rearing young. Mycangia and gastric caeca (e.g., anobiids, cerambycids, buprestids, cucujoids, tenebrionoids) signal the presence of microbial symbionts.

High-throughput methods will help to discover new associations, but cultures in hand will still be important to determine biological traits. Genomes and other data will help to understand the kind of interactions that occur. Some insects change diets as they mature, and these organisms offer an opportunity to track microbial changes with changes in diet. Interest in diet changes over lineages and examination of the extant hosts, and microbes throughout these lineages may help to understand evolution. Fossils can be targeted for more records (Leschen and Buckley 2007; Cai et al. 2016). The basis of many of the interactions between yeasts and insects has been hypothesized, but seldom tested (but see Nasir and Noda 2003).

A number of theoretical considerations remain to be studied, and emphasis on yeast and animal associations could help to provide illumination on the following points:

- Interactions are rarely restricted to two participants, and unknown participants or multipartite interactions may have important consequences for the associations. A broader view to consider entire microbial communities should be pursued (Douglas and Werren 2016).
- The discovery of new microbial symbionts may be important to the way in which microbes function to benefit the host (Xue et al. 2014). Host switching and host-habitat hypotheses may be relevant in some cases (Nikoh and Fukatsu 2000; Araujo et al. 2015).
- Life-history studies may expand the involvement of insects in questions of horizontal and vertical transmission of microbes; is vertical transmission rare among most microbial associations as has been suggested from restricted examples (Boomsma and Aanen 2009)? Do both commonly occur in the same association? Can specialists be generalists (Agosta et al. 2010)?
- Is lack of sexual reproduction better for some symbionts as theory might suggest (Xue et al. 2014; Fan et al. 2015)? How does asexual reproduction affect associations over the long term (Moran 1996)?
- Is there confirmation that symbiosis is a "key driver of insect physiological processes, ecological interactions, evolutionary diversification, and impacts on humans," as Klepzig et al. (2009) suggest?

- Yeast–insect associations are excellent systems for studying applications for agriculture (Witzgall et al. 2012), medicine (Valzano et al. 2016), and evolutionary research.
- Studies resulting from the large amounts of funding available for human genomes can provide newer methods and background information for the study of microbe and insect associations (Putignani et al. 2014).

In the summer of 1970, I worked at Houston NASA during the Apollo 13 mission. The aborted moon launch left us without the anticipated moon rock samples to use in assays to ensure that pathogens had not come to Earth from the moon. To occupy ourselves, my colleagues and I attempted to cultivate symbiotic bacteria from roaches using several kinds of media including at the time sophisticated insect culture media. We never made up for the insect component of the interaction, and at last I understand why our attempts were unsuccessful (Engel and Moran 2013).

Acknowledgments This chapter is dedicated to Hansjörg Prillinger, who inspired this study by his work on ecology and physiology involving mushroom-associated yeasts.

I appreciate fruitful discussions with Fernando E. Vega, who helped to improve the manuscript with his insightful editorial comments. Friends who generously answered my questions and provided references include William Eberhard, Daniel A. Henk, Richard A. B. Leschen, Jennifer Luangsa-ard, Joseph V. McHugh, Nhu H. Nguyen, James A. Robertson, Joseph W. Spatafora, Sung-Oui Suh, Margaret K. Thayer, and Hector Urbina.

References

Agosta SJ, Janz N, Brooks DR (2010) How specialists can be generalists: resolving the "parasite paradox" and implications for emerging infectious disease. Fortschr Zool 27:151–162

Alexopoulos CJ, Mims CW, Blackwell M (1996) Introductory mycology, 4th edn. Wiley, New York

Anagnostakis SL (2001) American chestnut sprout survival with biological control of the chestnut-blight fungus population. For Ecol Manag 152:225–233

Anagnostou C, LeGrand EA, Rohlfs M (2010) Friendly food for fitter flies? – Influence of dietary microbial species on food choice and parasitoid resistance in *Drosophila*. Oikos 119:533–541

Andreadis SS, Witzgall P, Becher PG (2015) Survey of arthropod assemblages responding to live yeasts in an organic apple orchard. Front Ecol Evol 3:121

Araujo SBL, Braga MP, Brooks DR, Agosta SJ, Hoberg EP, von Hartenthal FW, Boege WA (2015) Understanding host-switching by ecological fitting. PLoS One 10:e0139225

Barelli L, Moonjely S, Behie SW, Bidochka MJ (2016) Fungi with multifunctional lifestyles: endophytic insect pathogenic fungi. Plant Mol Biol 90:657–664

Barnett JA (2004) History of research on yeasts: 8 Taxonomy. Yeast 2004(21):1141–1193

Batra LR (1967) Ambrosia fungi: extent of specificity to ambrosia beetles. Science 153:193–195

Begerow D, Schafer AM, Kellner R, Oberwinker D, Bauer R (2014) The Mycota, Vol. VII, Part A: systematics and evolution. Ustilaginomycotina, 2nd edn. Springer, Berlin, pp 295–329

Begerow D, Kemler M, Feige A, Yurkov AM (2017) Parasitism in yeasts. In: Buzzini P, Lachance MA, Yurkov AM (eds) Yeasts in natural ecosystems: ecology. Springer, Heidelberg, pp 179–210

Betz O, Thayer MK, Newton AF (2003) Comparative morphology and evolutionary pathways of the mouthparts in spore-feeding Staphylinoidea (Coleoptera). Acta Zool 84:179–238

Blackwell M (1984) Myxomycete life cycles and arthropod spore dispersers. In: Wheeler Q, Blackwell M (eds) Fungus-insect relationships: perspectives in ecology and evolution. Columbia University Press, New York, pp 67–90

Blackwell M. (2017) Made for each other: ascomycete yeasts and insects. The Fungal Kingdom. Microbiol Spectr 5(3). FUNK-0081-2016. ASM Press, Washington, DC

Blackwell M, Malloch D (1989) *Pyxidiophora:* life histories and arthropod associations of two species. Can J Bot 67:2552–2562

Blackwell M, Bridges JR, Moser JC, Perry TJ (1986) Hyperphoretic dispersal of a *Pyxidiophora* anamorph. Science 232:993–995

Boekhout T (2005) Biodiversity: gut feeling for yeasts. Nature 434:449–451

Boomsma JJ, Aanen DK (2009) Rethinking crop-disease management in fungus-growing ants. Proc Natl Acad Sci U S A 106:17611–17612

Broderick NA, Lemaitre B (2012) Gut-associated microbes of *Drosophila melanogaster*. Gut Microbes 3:307–321

Brune A, Dietrich C (2015) The gut microbiota of termites: digesting the diversity in the light of ecology and evolution. Annu Rev Microbiol 69:145–166

Buchner P (1953) Endosymbiose der Tiere mit Pflanzlichen Mikroorganismen Vol. 12, Birkhäuser, Basel/Stuttgart

Buchner P (1965) Endosymbiosis of animals with plant microorganisms. Wiley, New York

Buzzini P, Lachance MA, Yurkov AM (2017) Yeasts in natural ecosystems: ecology. Springer, Heidelberg,

Cadete RM, Melo MA, Zilli JE, Vital MJS, Mouro A, Prompt AH, Gomes FCO, Stambuk BU, Lachance M-A, Rosa CA (2013) *Spathaspora brasiliensis* sp. nov., *Spathaspora suhii* sp. nov., *Spathaspora roraimanensis* sp. nov. and *Spathaspora xylofermentans* sp. nov., four novel D-xylose-fermenting yeast species from Brazilian Amazonian forest. A van Leeuwenhoek 103:421–431

Cadete RM, de las Heras AM, Sandström AG, Ferreira C, Gírio F, Gorwa-Grauslund M-F, Rosa CA, Fonseca C (2016) Exploring xylose metabolism in *Spathaspora* species: XYL1.2 from *Spathaspora passalidarum* as the key for efficient anaerobic xylose fermentation in metabolic engineered *Saccharomyces cerevisiae*. Biotechnol Biofuels 9:167

Cai C, Newton AF, Thayer MK, Leschen RAB, Huang D (2016) Specialized proteinine rove beetles shed light on insect–fungal associations in the Cretaceous. Proc R Soc B 283:20161439

Cappelli A, Ulissi U, Valzano M, Damiani C, Epis S, Gabrielli MG, Conti S, Polonelli L, Bandi C, Favia G, Ricci I (2014) A *Wickerhamomyces anomalus* killer strain in the malaria vector *Anopheles stephensi*. PLoS One 9:e95988

Ceja-Navarro JA, Nguyen NH, Karaoz U, Gross SR, Herman DJ, Andersen GL, Bruns TD, Pett-Ridge J, Blackwell M, Brodie EL (2014) Compartmentalized microbial composition, oxygen gradients and nitrogen fixation in the gut of *Odontotaenius disjunctus*. ISME J 8:6–18

Ceja-Navarro JA, Vega FE, Karaoz U, Hao Z, Jenkins S, Lim HC, Kosina P, Infante F, Northen TR, Brodie EL (2015) Gut microbiota mediate caffeine detoxification in the primary insect pest of coffee. Nat Commun 6:7618

Chandler JA, Eisen JA, Kop A (2012) Yeast communities of diverse *Drosophila* species: comparison of two symbiont groups in the same hosts. Appl Environ Microbiol 78:7327–7336

Chang Y, Wang S, Sekimoto S, Aerts AL, Choi C, Clum A, LaButti KM, Lindquist EA, Ngan CY, Ohm RA, Salamov AA, Grigoriev IV, Spatafora JW, Berbee ML (2015) Phylogenomic analyses indicate that early fungi evolved digesting cell walls of algal ancestors of land plants. Genome Biol Evol 7:1590–1601

Cheng DJ, Hou RF (2001) Histological observations on transovarial transmission of a yeast-like symbiote in *Nilaparvata lugens* Stål (Homoptera, Delphacidae). Tissue Cell 33:273–279

Cline AR, Leschen RAB (2005) Coleoptera associated with the oyster mushroom, *Pleurotus ostreatus* Fries, in North America. Southeast Nat 4:409–420

Couch JN (1938) The genus *Septobasidium*. Univ of North Carolina Press, Chapel Hill

Davis TS (2015) The ecology of yeasts in the bark beetle holobiont: a century of research revisited. Microb Ecol 69:723–732

Davis TS, Crippen TL, Hofstetter RW, Tomberlin JK (2013) Microbial volatile emissions as insect semiochemicals. J Chem Ecol 39:840–859

de Beer ZW, Duong TA, Barnes I, Wingfield BD, Wingfield MJ (2014) Redefining *Ceratocystis* and allied genera. Stud Mycol 79:187–219

Douglas AE (2009) The microbial dimension in insect nutritional ecology. Funct Ecol 23:38–47

Douglas AE, Werren JH (2016) Holes in the hologenome: why host-microbe symbioses are not holobionts. MBio 7:e02099-15

Dowd PF (1992) Insect fungal symbionts: a promising source of detoxifying enzymes. J Ind Microbiol 9:149–161

Engel P, Moran NA (2013) The gut microbiota of insects – diversity in structure and function. FEMS Microbiol Rev 37:699–735

Ernst CM, Loboda S, Buddle CM (2016) Capturing northern biodiversity: diversity of arctic, subarctic and north boreal beetles and spiders are affected by trap type and habitat. Insect Conserv Divers 9:63–73

Escherich K (1900) Über das regelmäßige Vorkommen von Sproßpilzen in dem Darmepithel eines Käfers. Biol ZBL 20:350–358

Fan H-W, Noda H, Xie H-Q, Suetsugu Y, Zhu Q-H, Zhang C-X (2015) Genomic analysis of an ascomycete fungus from the rice planthopper reveals how it adapts to an endosymbiotic lifestyle. Genome Biol Evol 7:2623–2634

Findlay WPK (1982) Fungi: folklore, fiction, and fact. Mad River Press, Eureka, CA

Frank SA (1996) Host symbiont conflict over the mixing of symbiotic lineages. Proc R Soc Lond B 263:339–344

Fukatsu T, Ishikawa H (1996) Phylogenetic position of yeast-like symbiont of *Hamiltonaphis styraci* (Homoptera, Aphididae) based on 18S rDNA sequence. Insect Biochem Mol Biol 26:383–388

Gams W, von Arx JA (1980) Validation of *Symbiotaphrina* (Imperfect Yeasts). Persoonia 10:542–543

Ganter PF (2006) Yeast and invertebrate associations. In: Rosa CA, Péter G (eds) Biodiversity and ecophysiology of yeasts. The Yeast Handbook. Springer, Berlin, Heidelberg, pp 303–370

Gazis R, Miadlikowska J, Lutzoni F, Arnold AE, Chaverri P (2012) Culture-based study of endophytes associated with rubber trees in Peru reveals a new class of Pezizomycotina: Xylonomycetes. Mol Phylogenet Evol 65:294–304

Gazis R, Kuo A, Riley R, LaButti K, Lipzen A, Lin J, Amirebrahimi M, Hesse CN, Spatafora JW, Henrissat B, Hainaut M, Grigoriev IV, Hibbett DS (2016) The genome of *Xylona heveae* provides a window into fungal endophytism. Fungal Biol 120:26–42

Gibson CM, Hunter MS (2010) Extraordinarily widespread and fantastically complex: comparative biology of endosymbiotic bacterial and fungal mutualists of insects. Ecol Lett 13:223–234

Gilbertson RL (1984) Relationships between insects and wood-rotting Basidiomycetes. In: Wheeler Q, Blackwell M (eds) Fungus–insect relationships: perspectives in ecology and evolution. Columbia University Press, New York, pp 130–165

Gill SR, Pop M, Deboy RT, Eckburg PB, Turnbaugh PJ, Samuel BS, Gordon JI, Relman DA, Fraser-Liggett CM, Nelson KE (2006) Metagenomic analysis of the human distal gut microbiome. Science 312:1355–1359

Goffeau A, Barrell BG, Bussey H, Davis RW, Dujon B, Feldmann H, Galibert F, Hoheisel JD, Jacq C, Johnston M, Louis EJ, Mewes HW, Murakami Y, Philippsen P, Tettelin H, Oliver SG (1996) Life with 6000 genes. Science 274(546):563–567

Górz A, Boroń P (2016) The yeast fungus *Trichosporon lactis* found as an epizoic colonizer of dung beetle exoskeletons. Microb Ecol 71:422–427

Grimaldi D, Engel MS (2005) Evolution of the insects. Cambridge Univ Press, Cambridge

Gross SR (2010) A study of passalid beetle prokaryote prokaryote and yeast gut symbionts. MS Thesis, Louisiana State University. http://etd.lsu.edu/docs/available/etd-07092010-102353/unrestricted/GrossThesis.pdf

Grossart H-P, Wurzbacher C, James TY, Kagami M (2016) Discovery of dark matter fungi in aquatic ecosystems demands a reappraisal of the phylogeny and ecology of zoosporic fungi. Fungal Ecol 19:28–38

Gusmão DS, Santos AV, Marini DC, Bacci M Jr, Berbert-Molina MA, Lemos FJA (2010) Culture-dependent and culture-independent characterization of microorganisms associated with *Aedes aegypti* (Diptera: Culicidae) (L.) and dynamics of bacterial colonization in the midgut. Acta Trop 115:275–281

Haeckel E (1899) Kunstformen der Natur. Bibliographisches Institut, Leipzig and Wien

Hammond PM, Lawrence JF (1989) Mycophagy in insects: a summary. In: Wilding N, Collins NM, Hammond PM, Webber JF (eds) Insect–fungus interactions. 14th Symposium of the Royal Entomological Society of London. Academic Press, London, pp 275–324

Handel S, Wang T, Yurkov A, König H (2016) *Sugiyamaella mastotermitis* sp. nov. and *Papiliotrema odontotermitis* f.a., sp. nov. from the gut of the termites *Mastotermes darwiniensis* and *Odontotermes obesus*. Int J Syst Evol Microbiol 66:4600–4608

Harrington TC (2005) Ecology and evolution of mycophagous bark beetles and their fungal partners. In: Vega FE, Blackwell M (eds) Insect-fungal associations: ecology and evolution. Oxford University Press, New York, pp 257–291

Harrington TC, McNew D, Mayers C, Fraedrich SW, Reed SE (2014) *Ambrosiella roeperi* sp. nov. is the mycangial symbiont of the granulate ambrosia beetle, *Xylosandrus crassiusculus*. Mycologia 106:835–845

Hawker LE (1957) Ecological factors and the survival of fungi. Symp Soc Gen Microbiol 7:238–258

Henk DA (2005) New species of *Septobasidium* from southern Costa Rica and the southeastern United States. Mycologia 97:908–913

Henk DA, Vilgalys R (2007) Molecular phylogeny suggests a single origin of insect symbiosis in the Pucciniomycetes with support for some relationships within the genus *Septobasidium*. Am J Bot 94:1515–1526

Hibbett DS, Binder M, Bischoff JF, Blackwell M, Cannon PF, Eriksson OE, Huhndorf S, James T, Kirk PM, Lücking R, Thorsten Lumbsch H, Lutzoni F, Matheny PB, McLaughlin DJ, Powell MJ, Redhead S, Schoch CL, Spatafora JW, Stalpers JA, Vilgalys R, Aime MC, Aptroot A, Bauer R, Begerow D, Benny GL, Castlebury LA, Crous PW, Dai YC, Gams W, Geiser DM, Griffith GW, Gueidan C, Hawksworth DL, Hestmark G, Hosaka K, Humber RA, Hyde KD, Ironside JE, Kõljalg U, Kurtzman CP, Larsson KH, Lichtwardt R, Longcore J, Miadlikowska J, Miller A, Moncalvo JM, Mozley-Standridge S, Oberwinkler F, Parmasto E, Reeb V, Rogers JD, Roux C, Ryvarden L, Sampaio JP, Schüssler A, Sugiyama J, Thorn RG, Tibell L, Untereiner WA, Walker C, Wang Z, Weir A, Weiss M, White MM, Winka K, Yao YJ, Zhang N (2007) A higher-level phylogenetic classification of the Fungi. Mycol Res 111:509–547

Hoffmann C, Dollive S, Grunberg S, Chen J, Li H, Wu GD, Lewis JW, Bushman FD (2013) Archaea and fungi of the human gut microbiome: correlations with diet and bacterial residents. PLoS One 8(6):e66019

Hongoh Y, Ishikawa H (1997) Uric acid as a nitrogen resource for the brown planthopper, *Nilaparvata lugens*: studies with synthetic diets and aposymbiotic insects. Zool Sci 14:581–586

Hongoh Y, Ishikawa H (2000) Evolutionary studies on uricases of fungal endosymbionts of aphids and planthoppers. J Mol Evol 51:265–277

Hug LA, Baker BJ, Anantharaman K, Brown CT, Probst AJ, Castelle CJ, Butterfield CN, Hernsdorf AW, Amano Y, Ise K, Suzuki Y, Dudek N, Relman DA, Finstad KM, Amundson R, Thomas BC, Banfield JF (2016) A new view of the tree of life. Nat Microbiol 16048 doi:10.1038/nmicrobiol.2016.48

Hulcr J, Stelinsk LL (2017) The ambrosia symbiosis: from evolutionary ecology to practical management. Annu Rev Entomol 62:285–303

Hunt T, Bergsten J, Levkanicova Z, Papadopoulou A, John OS, Wild R, Hammond PM, Ahrens D, Balke M, Caterino MS, Gómez-Zurita J, Ribera I, Barraclough TG, Bocakova M, Bocak L, Vogler AP (2007) A comprehensive phylogeny of beetles reveals the evolutionary origins of a superradiation. Science 318:1913–1916

Ignatova EA, Nagornaia SS, Povazhnaia TN, Ianishevskaia GS (1996) The yeast flora of blood-sucking mosquitoes. Mikrobiol Z 58:12–15 (in Russian)

Imanishi Y, Jindamorakot S, Mikata K, Nakagiri A, Limtong S, Potacharoen W, Tanticharoen M, Nakase T (2008) Two new ascomycetous anamorphic yeast species related to *Candida friedrichii*—*Candida jaroonii* sp. nov., and *Candida songkhlaensis* sp. nov.—isolated in Thailand. A van Leeuwenhoek 94:267–276

Jeffries TW, Grigoriev IV, Grimwood J, Laplaza JM, Aerts A, Salamov A, Schmutz J, Lindquist E, Dehal P, Shapiro H, Jin Y-S, Passoth V, Richardson PM (2007) Genome sequence of the lignocellulose-bioconverting and xylose-fermenting yeast *Pichia stipitis*. Nat Biotechnol 25:319–326

Johnstone RA, Bshary R (2002) From parasitism to mutualism: partner control in asymmetric interactions. Ecol Lett 5:634–639

Jones KG, Blackwell M (1996) Ribosomal DNA sequence analysis places the yeast-like genus *Symbiotaphrina* within filamentous ascomycetes. Mycologia 88:212–218

Jordal BH, Cognato AI (2012) Molecular phylogeny of bark and ambrosia beetles reveals multiple origins of fungus farming during periods of global warming. BMC Evol Biol 12:133

Jurzitza G (1979) The fungi symbiotic with anobiid beetles. In: Batra LR (ed) Insect-fungus symbiosis. Allanheld, Osmun, Montclair, pp 65–76

Kepler RM, Sung G-H, Harada Y, Tanaka K, Tanaka E, Hosoya T, Bischoff JF, Spatafora JW (2012) Host jumping onto close relatives and across kingdoms by *Tyrannicordyceps* (Clavicipitaceae) gen. nov. and *Ustilaginoidea* (Clavicipitaceae). Am J Bot 99:552–561

Kerrigan J, Rogers JD (2003) Microfungi associated with the wood-boring beetles *Saperda calcarata* (poplar borer) and *Cryptorhynchus lapathi* (poplar and willow borer). Mycotaxon 86:1–18

Kerrigan J, Rogers JD (2013) Biology, ecology and ultrastructure of *Ascobotryozyma* and *Botryozyma*, unique commensal nematode-associated yeasts. Mycologia 105:34–51

Klassen R, Schaffrath R, Buzzini P, Ganter PF (2017) Antagonistic interactions and killer yeasts. In: Buzzini P, Lachance MA, Yurkov AM (eds) Yeasts in natural ecosystems: ecology. Springer, Heidelberg, pp 229–275

Klepzig KD, Adams AS, Handelsman J, Raffa KF (2009) Symbioses: a key driver of insect physiological processes, ecological interactions, evolutionary diversification, and impacts on humans. Environ Entomol 38:67–77

Kreger-van Rij NJW (ed) (1984) The yeasts a taxonomic study, 3rd edn. Amsterdam, Elsevier

Kurtzman CP (2001) Six new anamorphic ascomycetous yeasts near *Candida tanzawaensis*. FEMS Yeast Res 1:177–185

Kurtzman CP (2011) Discussion of teleomorphic and anamorphic ascomycetous yeasts and yeast-like taxa. In: Kurtzman CP, Fell JW, Boekhout T (eds) The yeasts, a taxonomic study, vol 2, 5th edn. Elsevier, Amsterdam, pp 293–307

Kurtzman CP, Fell JW (eds) (1998) The yeasts, a taxonomic study, 4th edn. Amsterdam, Elsevier

Kurtzman CP, Robnett CJ (1998) Identification and phylogeny of ascomycetous yeasts from analysis of nuclear large subunit (26S) ribosomal DNA partial sequences. A van Leeuwenhoek 73:331–371

Kurtzman CP, Fell JW, Boekhout T (eds) (2011) The yeasts, a taxonomic study, 5th edn, 3 vols. Elsevier, Amsterdam

Kurtzman CP, Robnett CJ, Blackwell M (2016) Description of *Teunomyces* gen. nov. for the *Candida kruisii* clade, *Suhomyces* gen. nov. for the *Candida tanzawaensis* clade and *Suhomyces kilbournensis* sp. nov. FEMS Yeast Res 16: fow041

Lachance M-A, Starmer WT (1998) Ecology and yeasts. In: Kurtzman CP, Fell JW (eds) The yeasts, a taxonomic study, 4th edn. Elsevier, Amsterdam, pp 23–30
Lawrence JF (1973) Host preference in ciid beetles (Coleoptera: Ciidae) inhabiting the fruiting bodies of basidiomycetes in North America. Bull Mus Comp Zool 145:163–212
Lawrence JF (1989) Mycophagy in the Coleoptera: feeding strategies and morphological adaptations. In: Wilding N, Collins NM, Hammond PM, Webber JF (eds) Insect-fungus interactions. Academic Press, San Diego, pp 1–23
Legendre F, Nel A, Svenson GJ, Robillard T, Pellens R, Grandcolas P (2015) Phylogeny of Dictyoptera: dating the origin of cockroaches, praying mantises and termites with molecular data and controlled fossil evidence. PLoS One 10(7):e0130127
Leschen RAB, Buckley TRB (2007) Multistate characters and diet shifts: evolution of Erotylidae (Coleoptera). Syst Biol 56:97–112
Lichtwardt RW, White MM, Cafaro MJ (1999) Fungi associated with passalid beetles and their mites. Mycologia 91:694–702
Liu XZ, Wang QM, Göker M, Groenewald M, Kachalkin AV, Lumbsch HT, Millanes AM, Wedin M, Yurkov AM, Boekhout T, Bai FY (2016) Towards an integrated phylogenetic classification of the Tremellomycetes. Stud Mycol 81:85–147
Locey KJ, Lennon JT (2016) Scaling laws predict global microbial diversity. Proc Natl Acad Sci U S A 113:5970–5975
Lodder J (ed) (1970) The yeasts: a taxonomic study, 2nd edn. Amsterdam, North-Holland
Lodder J, Kreger van Rij NJW (eds) (1952) The yeasts: a taxonomic study, 1st edn. Amsterdam, North-Holland
Lu M, Hulcr J, Sun J (2016) The role of symbiotic microbes in insect invasions. Annu Rev Ecol Evol Syst 47:487–505
Lücking R, Huhndorf S, Pfister D, Plata ER, Lumbsch H (2009) Fungi evolved right on track. Mycologia 101:810–822
Lutzoni F, Pagel M (1997) Accelerated evolution as a consequence of transition to mutualism. Proc Natl Acad Sci U S A 94:11422–11427
Malloch D, Blackwell M (1990) *Kathistes*, a new genus of pleomorphic ascomycetes. Can J Microbiol 68:1712–1721
Martin R, Gazis RO, Skaltsas D, Chaverri P, Hibbett DS (2015) Unexpected diversity of basidiomycetous endophytes in sapwood and leaves of *Hevea*. Mycologia 107:284–297
Mayers CG, McNew DL, Harrington TC, Roeper RA, Fraedrich SW, Biedermann PHW, Castrillo LA, Reed SE (2015) Three genera in the Ceratocystidaceae are the respective symbionts of three independent lineages of ambrosia beetles with large, complex mycangia. Fungal Biol 119:1075–1092
McHugh JV, Marshall CJ, Fawcett FL (1997) A study of adult morphology in *Megalodacne heros* (Say) (Coleoptera: Erotylidae). Trans Am Entomol Soc 123:167–223
McKenna DD, Wild AL, Kanda K, Bellamy CL, Beutel RG, Caterino MS, Farnum CW, Hawks DC, Ivie MA, Jameson ML, Leschen RAB, Marvaldi AE, McHugh JV, Newton AF, Robertson JA, Thayer MK, Whiting MF, Lawrence JF, Ślipiński A, Maddison DR, Farrell BD (2015) The beetle tree of life reveals that Coleoptera survived end-Permian mass extinction to diversify during the Cretaceous terrestrial revolution. Syst Entomol 40:835–880
Misof B, Liu S, Meusemann K, Peters RS, Donath A, Mayer C, Frandsen PB, Ware J, Flouri T, Beutel RG, Niehuis O, Petersen M, Izquierdo-Carrasco F, Wappler T, Rust J, Aberer AJ, Aspöck U, Aspöck H, Bartel D, Blanke A, Berger S, Böhm A, Buckley TR, Calcott B, Chen J, Friedrich F, Fukui M, Fujita M, Greve C, Grobe P, Gu S, Huang Y, Jermiin LS, Kawahara AY, Krogmann L, Kubiak M, Lanfear R, Letsch H, Li Y, Li Z, Li J, Lu H, Machida R, Mashimo Y, Kapli P, McKenna DD, Meng G, Nakagaki Y, Navarrete-Heredia JL, Ott M, Ou Y, Pass G, Podsiadlowski L, Pohl H, von Reumont BM, Schütte K, Sekiya K, Shimizu S, Slipinski A, Stamatakis A, Song W, Su X, Szucsich NU, Tan M, Tan X, Tang M, Tang J, Timelthaler G, Tomizuka S, Trautwein M, Tong X, Uchifune T, Walzl MG, Wiegmann BM, Wilbrandt J, Wipfler B, Wong TK, Wu Q, Wu G, Xie Y, Yang S, Yang Q, Yeates DK, Yoshizawa K,

Zhang Q, Zhang R, Zhang W, Zhang Y, Zhao J, Zhou C, Zhou L, Ziesmann T, Zou S, Li Y, Xu X, Zhang Y, Yang H, Wang J, Wang J, Kjer KM, Zhou X (2014) Phylogenomics resolves the timing and pattern of insect evolution. Science 346:763–767

Mittelbach M, Vannette RL (2017) Mutualism in yeasts. In: Buzzini P, Lachance MA, Yurkov AM (eds) Yeasts in natural ecosystems: ecology. Springer, Heidelberg, pp 155–178

Mittelbach M, Yurkov AM, Nocentini D, Nepi M, Weigend M, Begerow D (2015) Nectar sugars and bird visitation define a floral niche for basidiomycetous yeast on the Canary Islands. BMC Ecol 15:2

Molnar O, Schatzmayr G, Fuchs E, Prillinger H (2004) *Trichosporon mycotoxinivorans* sp. nov., a new yeast species useful in biological detoxification of various mycotoxins. Syst Appl Microbiol 27:661–671

Morais CG, Cadete RM, Uetanabaro APT, Rosa LH, Lachance M-A, Rosa CA (2013) D-xylose-fermenting and xylanase-producing yeast species from rotting wood of two Atlantic rainforest habitats in Brazil. Fungal Genet Biol 60:19–28

Moran NA (1996) Accelerated evolution and Muller's rachet in endosymbiotic bacteria. Proc Natl Acad Sci U S A 93:2873–2878

Moran NA, McCutcheon JP, Nakabachi A (2008) Genomics and evolution of heritable bacterial symbionts. Annu Rev Genet 42:165–190

Mueller UG, Gerardo N (2002) Fungus-farming insects: multiple origins and diverse evolutionary histories. Proc Natl Acad Sci U S A 99:15247–15249

Nagy LG, Ohm RA, Kovács GM, Floudas D, Riley R, Gácser A, Sipiczki M, Davis JM, Doty SL, de Hoog GS, Lange BF, Spatafora JW, Martin FM, Grigoriev IV, Hibbett DS (2014) Phylogenomics reveals latent homology behind the convergent evolution of yeast forms. Nat Commun 5:4471. doi:10.1038/ncomms5471

Nakase T, Itoh M, Takematsu A, Komagata K (1988) *Candida tanzawaensis*, a new species of yeast isolated from moss collected in Japan. Trans Mycol Soc Jpn 29:331–338

Nardi JB, Bee CM, Miller LA, Nguyen NH, Suh S-O, Blackwell M (2006) Communities of microbes that inhabit the changing hindgut landscape of a subsocial beetle. Arthropod Struct Dev 35:57–68

Nardon P, Grenier AM (1989) Endosymbiosis in Coleoptera: biological, biochemical, and genetic aspects. In: Schwemmler W, Gassner G (eds) Insect endocytobiosis: morphology, physiology, genetics, evolution. CRC Press, Boca Raton, pp 175–216

Nasir H, Noda H (2003) Yeast-like symbiotes as a sterol source in anobiid beetles (Coleoptera, Anobiidae): possible metabolic pathways from fungal sterols to 7-dehydrocholesterol. Arch Insect Biochem Physiol 52:175–182

Nguyen NH, Suh SO, Erbil CK, Blackwell M (2006a) *Metschnikowia noctiluminum* sp nov., *Metschnikowia corniflorae* sp nov., and *Candida chrysomelidarum* sp nov., isolated from green lacewings and beetles. Mycol Res 110:346–356

Nguyen NH, Suh SO, Marshall CJ, Blackwell M (2006b) Morphological and ecological similarities: wood-boring beetles associated with novel xylose-fermenting yeasts, *Spathaspora passalidarum* gen. sp. nov. and *Candida jeffriesii* sp. nov. Mycol Res 110:1232–1241

Nikoh N, Fukatsu T (2000) Interkingdom host jumping underground: phylogenetic analysis of entomoparasitic fungi of the genus *Cordyceps*. Mol Biol Evol 17:629–638

Nishino T, Tanahashi M, Lin C-P, Koga R, Fukatsu T (2016) Fungal and bacterial endosymbionts of eared leafhoppers of the subfamily Ledrinae (Hemiptera: Cicadellidae). Appl Entomol Zool 51:465–477

Nobre T, Koné NA, Konaté S, Linsenmair KE, Aanen DK (2011) Dating the fungus growing termites' mutualism shows a mixture between ancient codiversification and recent symbiont dispersal across divergent hosts. Mol Ecol 20:2619–2627

Noda H, Kodama K (1996) Phylogenetic position of yeast-like endosymbionts of anobiid beetles. Appl Environ Microbiol 62:162–167

Noda H, Koizumi Y (2003) Sterol biosynthesis by symbiotes: cytochrome P450 sterol C-22 desaturase genes from yeastlike symbiotes of rice planthoppers and anobiid beetles. Insect Biochem Mol Biol 33:649–658

Noda H, Omura T (1992) Purification of yeast-like symbiotes of planthoppers. J Invertebr Pathol 59:104–105

Noda H, Nakashima N, Koizumi M (1995) Phylogenetic position of yeast-like symbiotes of rice planthoppers based on partial 18S rDNA sequences. Insect Biochem Mol Biol 25:639–646

Pant NC, Fraenkel G (1950) The function of the symbiotic yeasts of two insect species, *Lasioderma serricorne* F. and *Stegobium* (*Sitodrepa*) *paniceum* L. Science 112:498–500

Pimentel D, Stachow U, Takacs DA, Brubaker HW, Dumas AR, Meaney JJ, O'Neil J, Onsi DE, Corzilius DB (1992) Conserving biological diversity in agricultural/forestry systems. Bioscience 42:354–362

Price PW (1988) An overview of organismal interactions in ecosystems in evolutionary and ecological time. Agric Ecosyst Environ 24:369–377

Price PW, Denno RF, Eubanks MD, Finke DL, Kaplan I (2011) Insect ecology: behavior, populations, and communities. Cambridge University Press, New York

Prillinger H (1987) Are there yeasts in Homobasidiomycetes? Stud Mycol 30:33–59

Prillinger H, Altenbuchner J, Laaser G, Dörfler C (1993) Yeasts isolated from Homobasidiomycetes (*Asterophora*, *Collybia*): new aspects for sexuality, taxonomy, and speciation. Exp Mycol 17:24–45

Prillinger H, Messner R, Koenig H, Bauer R, Lopandic K, Molnar O, Dangel P, Weigang F, Kirisits T, Nakase T, Sigler L (1996) Yeasts associated with termites: a phenotypic and genotypic characterization and use of coevolution for dating evolutionary radiations in asco- and basidiomycetes. Syst Appl Microbiol 19:265–283

Putignani L, Del Chierico F, Petrucca A, Vernocchi P, Dallapiccola B (2014) The human gut microbiota: a dynamic interplay with the host from birth to senescence settled during childhood. Pediatr Res 76:2–10

Ramsbottom, J (1953) Mushrooms and toadstools: a study of the activities of fungi. The New Naturalist's Series, Collins, London, United Kingdom

Ricci I, Damiani C, Scuppa P, Mosca M, Crotti E, Rossi P, Rizzi A, Capone A, Gonella E, Ballarini P, Chouaia B, Sagnon N, Esposito F, Alma A, Mandrioli M, Sacchi L, Bandi C, Daffonchio D, Favia G (2011) The yeast *Wickerhamomyces anomalus* (*Pichia anomala*) inhabits the midgut and reproductive system of the Asian malaria vector *Anopheles stephensi*. Environ Microbiol 13:911–921

Riley R, Haridas S, Wolfe KH, Lopes MR, Hittinger CT, Göker M, Salamov A, Wisecaver J, Long TM, Aerts AL, Barry K, Choi C, Clum A, Coughlan AY, Deshpande S, Douglass AP, Hanson SJ, Klenk H-P, LaButti K, Lapidus A, Lindquist E, Lipzen A, Meier-Kolthoff JP, Ohm RA, Otillar RP, Pangilinan J, Peng Y, Rokas A, Rosa CA, Scheuner C, Sibirny AA, Slot JC, Stielow JB, Sun H, Kurtzman CP, Blackwell M, Grigoriev IV, Jeffries TW (2016) Comparative genomics of biotechnologically important yeasts. Proc Natl Acad Sci U S A 113:9882–9887

Rodrigues Hoffmann A, Proctor LM, Surette MG, Suchodolski JS (2016) The microbiome: the trillions of microorganisms that maintain health and cause disease in humans and companion animals. Vet Pathol 53:10–21

Rosa CA, Péter G (eds) (2006) Biodiversity and ecophysiology of yeasts. Springer, Berlin, Heidelberg

Sabree ZL, Moran NA (2014) Host-specific assemblages typify gut microbial communities of related insect species. SpringerPlus 3:138

Sánchez Márquez S, Bills GF, Zabalgogeazcoa I (2008) Diversity and structure of the fungal endophytic assemblages from two sympatric coastal grasses. Fungal Divers 33:87–100

Sasaki K, Abid MR, Miyazaki M (1996) Deoxyhypusine synthase gene is essential for cell viability in the yeast *Saccharomyces cerevisiae*. FEBS Lett 384:151–154

Sasan RK, Bidochka MJ (2012) The insect-pathogenic fungus *Metarhizium robertsii* (Clavicipitaceae) is also an endophyte that stimulates plant root development. Am J Bot 99:101–107

Schoch CL, Sung G-H, López-Giráldez FL, Townsend JP, Miadlikowska J, Hofstetter V, Robbertse B, Matheny PB, Kauf F, Wang Z, Gueidan C, Andrie RM, Trippe K, Ciufetti LM,

Wynns A, Fraker E, Hodkinson BP, Bonito G, Groenewald JZ, Arsanlou M, de Hoog GS, Crous PW, Hewitt D, Pfister DH, Peterson K, Grysenhout M, Wingfield MJ, Aptroot A, Suh S-O, Blackwell M, Hillis DM, Griffith GW, Castlebury LA, Rossman AY, Lumbsch HT, Lücking RL, Büdel B, Rauhut A, Diederich P, Ertz D, Geiser DM, Hosaka K, Inderbitzin P, Kohlmeyer J, Volkmann-Kohlmeyer B, Mostert L, O'Donnell K, Sipman H, Rogers JD, Shoemaker RA, Sugiyama J, Summerbell RC, Untereiner W, Johnston PR, Stenroos S, Zuccaro A, Dyer P, Crittenden P, Cole MS, Hansen K, Trappe JM, Yahr R, Lutzoni F, Spatafora JW (2009) The Ascomycota tree of life: a phylum-wide phylogeny clarifies the origin and evolution of fundamental reproductive and ecological traits. Syst Biol 58:224–239

Schultz TR, Brady SG (2008) Major evolutionary transitions in ant agriculture. Proc Natl Acad Sci U S A 105:5435–5440

Sender R, Fuchs S, Milo R (2016) Revised estimates for the number of human and bacteria cells in the body. PLoS Biol 14(8):e1002533

Shen SK, Dowd PF (1991) Detoxification spectrum of the cigarette beetle symbiont *Symbiotaphrina kochii* in culture. Entomol Exp Appl 60:51–59

Shen XX, Zhou X, Kominek J, Kurtzman CP, Hittinger CT, Rokas A (2016) Reconstructing the backbone of the Saccharomycotina yeast phylogeny using genome scale data. G3 (Bethesda) 6:3927–3939

Six DL (2012) Ecological and evolutionary determinants of bark beetle—fungus symbioses. Insects 3:339–366

Spatafora JW, Blackwell M (1993) The polyphyletic origins of ophiostomatoid fungi. Mycol Res 98:1–9

Spatafora JW, Sung GH, Sung HM, Hywel-Jones L, White JF (2007) Phylogenetic evidence for an animal pathogen origin of ergot and the grass endophytes. Mol Ecol 16:1701–1711

Starmer WT (1981) A comparison of *Drosophila* habitats according to the physiological attributes of the associated yeast communities. Evolution 35:38–52

Starmer WT, Lachance MA (2011) Yeast ecology. In: Kurtzman CP, Fell JW, Boekhout T (eds) The yeasts: a taxonomic study, Elsevier, vol 1, 5th edn. New York, NY, pp 65–83

Stefanini I, Dapporto L, Legras J-L, Antonio Calabretta A, Di Paola M, De Filippo C, Viola R, Capretti P, Polsinelli M, Turillazzi S, Cavalieri D (2012) Role of social wasps in *Saccharomyces cerevisiae* ecology and evolution. Proc Natl Acad Sci U S A 109:13398–13403

Stefanini I, Dapporto L, Berná L, Polsinelli M, Turillazzi S, Cavalieri D (2016) Social wasps are a *Saccharomyces* mating nest. Proc Natl Acad Sci U S A 113:2247–2251

Streinzer M, Paulus HF, Spaethe J (2009) Floral colour signal increases short-range detectability of a sexually deceptive orchid to its bee pollinator. J Exp Biol 212:1365–1370

Suh S-O, Blackwell M (2005) The beetle gut as a habitat for new species of yeasts. In: Vega FE, Blackwell M (eds) Insect-fungal associations: ecology and evolution. Oxford University Press, New York, pp 244–256

Suh S-O, Noda H, Blackwell M (2001) Insect symbiosis: derivation of yeast-like endosymbionts within an entomopathogenic lineage. Mol Biol Evol 18:995–1000

Suh S-O, Marshall C, McHugh JV, Blackwell M (2003) Wood ingestion by passalid beetles in the presence of xylose fermenting gut yeasts. Mol Ecol 12:3137–3145

Suh S-O, McHugh JV, Blackwell M (2004a) Expansion of the *Candida tanzawaensis* yeast clade: 16 novel *Candida* species from basidiocarp-feeding beetles. Int J Syst Evol Microbiol 54:2409–2429

Suh S-O, White MM, Nguyen NH, Blackwell M (2004b) The identification of *Enteroramus dimorphus*: a xylose-fermenting yeast attached to the gut of beetles. Mycologia 96:756–760

Suh S-O, McHugh JV, Pollock D, Blackwell M (2005) The beetle gut: a hyperdiverse source of novel yeasts. Mycol Res 109:261–265

Suh S-O, Nguyen NH, Blackwell M (2006a) A yeast clade near *Candida kruisii* uncovered: nine novel *Candida* species associated with basidioma-feeding beetles. Mycol Res 110:1379–1394

Suh S-O, Blackwell M, Kurtzman CP, Lachance M-A (2006b) Phylogenetics of Saccharomycetales, the ascomycete yeasts. Mycologia 98:1008–1019

Suh S-O, Houseknecht JL, Gujjari P, Jianlong J, Zhou JJ (2013) *Scheffersomyces parashehatae* f. a., sp. nov., *Scheffersomyces xylosifermentans* f.a., sp. nov., *Candida broadrunensis* sp. nov. and *Candida manassasensis* sp. nov., novel yeasts associated with wood-ingesting insects, and their ecological and biofuel implications. Int J Syst Evol Microbiol 63:4330–4339

Sung G-H, Hywel-Jones NL, Sung J-M, Luangsa-ard JJ, Shrestha B, Spatafora JW (2007) Phylogenetic classification of *Cordyceps* and the clavicipitaceous fungi. Stud Mycol 57:5–59

Sylvester K, Wang Q-M, James B, Mendez R, Hulfachor AB, Hittinger CT (2015) Temperature and host preferences drive the diversification of *Saccharomyces* and other yeasts: a survey and the discovery of eight new yeast species. FEMS Yeast Res 15:fov002

Taylor TN, Hass H, Kerp H, Krings M, Hanlin RT (2005) Perithecial ascomycetes from the 400 million year old Rhynie chert: an example of ancestral polymorphism. Mycologia 97:269–285

Toussaint EFA, Seidel M, Arriaga-Varela E, Hájek J, Král D, Sekerka L, Short AEZ, Fikáček M (2017) The peril of dating beetles. Syst Entomol 42:1–10

Turnbaugh PJ, Ley RE, Hamady M, Fraser-Liggett CM, Knight R, Gordon JI (2007) The human microbiome project. Nature 449:804–810

Urbina H, Blackwell M (2012) Multilocus phylogenetic study of the *Scheffersomyces* yeast clade and characterization of the n-terminal region of xylose reductase gene. PLoS One 7(6):e39128

Urbina H, Frank R, Blackwell M (2013a) *Scheffersomyces cryptocercus*: a new xylose-fermenting yeast associated with the gut of wood roaches and new combinations in the *Sugiyamaella* yeast clade. Mycologia 105:650–660

Urbina H, Schuster J, Blackwell M (2013b) The gut of Guatemalan passalid beetles: a niche colonized by cellobiose- and xylose-fermenting yeasts. Fungal Ecol 6:339–355

Urubschurov V, Janczyk P (2011) Biodiversity of yeasts in the gastrointestinal ecosystem with emphasis on its importance for the host. In Grillo O, Venora G (eds). The dynamical processes of biodiversity—case studies of evolution and spatial distribution. InTech, pp 277–302

Valzano M, Cecarini V, Cappelli A, Capone A, Bozic J, Cuccioloni M, Epis S, Petrelli D, Angeletti M, Eleuteri A, Favia G, Ricci I (2016) A yeast strain associated to *Anopheles* mosquitoes produces a toxin able to kill malaria parasites. Malar J 15:21

van der Walt JP (1961) The mycetome symbiont of *Lasioderma serricorne*. A van Leeuwenhoek 27:362–366

van Tol S, Dimopoulos G (2016) Influences of the mosquito microbiota on vector competence. In Raikhel AS (ed) Progress in Mosquito Research, Adv Insect Physiol 51:243–291

Vega FE (2008) Insect pathology and fungal endophytes. J Invertebr Pathol 98:277–279

Vega FE, Dowd PF (2005) The role of yeasts as insect endosymbionts. In: Vega FE, Blackwell M (eds) Insect-fungal associations: ecology and evolution. Oxford University Press, New York, pp 211–243

Wan P-J, Tang Y-H, Yuan S-Y, Wang W-X, Lai F-X, Yu X-P, Fu Q (2016) ATP phosphoribosyltransferase from symbiont *Entomomyces delphacidicola* involved in histidine biosynthesis of *Nilaparvata lugens* (Stål). Amino Acids 48:2605–2617

Weeks AR, Turelli M, Harcombe WR, Reynolds KT, Hoffmann AA (2007) From parasite to mutualist: rapid evolution of *Wolbachia* in natural populations of *Drosophila*. PLoS Biol 5: e114

Weete JD, Abril M, Blackwell M (2010) Phylogenetic distribution of fungal sterols. PLoS One 5: e10899

Wiegmann BM, Trautwein MD, Winkler IS, Barr NB, Kim J-W, Lambkin C, Bertone MA, Cassel BK, Bayless KM, Heimberg AM, Wheeler BM, Peterson KJ, Pape T, Sinclair BJ, Skevington JH, Blagoderov V, Caravask J, Kutty SN, Schmidt-Ottm U, Kampmeier GE, Thompson FC, Grimaldi DA, Beckenbach AT, Courtney GW, Friedrich M, Meierl R, Yeates DK (2011) Episodic radiations in the fly tree of life. Proc Natl Acad Sci U S A 108:5690–5695

Witzgall P, Proffit M, Rozpedowska E, Becher PG, Andreadis S, Coracini M, Lindblom TU, Ream LJ, Hagman A, Bengtsson M, Kurtzman CP, Piskur J, Knight A (2012) "This is not an apple" – yeast mutualism in codling moth. J Chem Ecol 38:949–957

Xue J, Zhou X, Zhang C-X, Yu L-L, Fan H-W, Wang Z, Xu H-J, Xi Y, Zhu Z-R, Zhou W-W, Pan P-L, Li B-L, Colbourne JK, Noda H, Suetsugu Y, Kobayashi T, Zheng Y, Liu S, Zhang R, Yang L, Luo Y-D, Fang D-M, Chen Y, Zhan D-L, Lv X-D, Cai Y, Wang Z-B, Huang H-J, R-L ZX-C, Lou Y-H, Yu B, Zhuo J-C, Ye Y-X, Zhang W-Q, Shen Z-C, Yang H-M, Wang J, Wang J, Bao Y-Y, Cheng J-A (2014) Genomes of the rice pest brown planthopper and its endosymbionts reveal complex complementary contributions for host adaptation. Genome Biol 15:521

Chapter 14
Yeasts in Birds

Giancarlo Moschetti, Antonio Alfonzo, and Nicola Francesca

Abstract Wild animals, particularly birds, play a relevant role in disseminating microscopic organisms, including yeasts and filamentous fungi. The ecology of yeasts associated with birds has been studied mainly for the urban avifauna, and little is known about yeasts associated with wild birds including migratory species. Recent findings on the adaptation and persistence of yeasts in animals suggest that birds can play an important role in the ecology and distribution of yeasts.

It has been shown that birds can vector pathogenic fungi by disseminating fungal cells from the cloacae in the environment via their excretions. Wild birds, especially pigeons and their feces, represent a reservoir for pathogenic yeasts, and the genera *Cryptococcus* and *Candida* are the most frequently isolated yeast genera from birds. The contact with birds, in particular with their excretions, is considered one of the main causes of cryptococcosis and candidiasis.

Dispersal of nonclinical yeasts by birds is important for agriculture because microbial communities affect the quality of food products such as some fermented foods. Insect vectors have been recognized as source of extraordinary large yeast diversity in many studies. Migratory birds are able to travel over large distances and carry yeasts even between continents. Compared to insects, our knowledge of yeasts associated with birds is limited. The present chapter provides an overview on birds as a reservoir for pathogenic yeasts and reviews studies on migratory birds acting as dissemination vectors of yeasts.

Keywords Birds • Yeasts • Thermotolerance • Novel species • Ecological niche

Contents

G. Moschetti • A. Alfonzo • N. Francesca (✉)
Department of Agricultural and Forestry Sciences, University of Palermo, Viale delle Scienze 4, 90128 Palermo, Italy
e-mail: nicola.francesca@unipa.it

P. Buzzini et al. (eds.), *Yeasts in Natural Ecosystems: Diversity*,
DOI 10.1007/978-3-319-62683-3_14

14.1 Introduction

Birds are susceptible to be colonized by fungi, including yeasts. The ecology of yeasts associated with birds has been studied mainly for the avifauna of the urban areas (Kobayashi et al. 2005; Costa et al. 2010; Chae et al. 2012), whereas very few reports (e.g., Cafarchia et al. 2006a) have been published on yeasts associated with wild birds, including migratory species. The main focus of these studies has been directed to the evaluation of the health status of local birds and monitoring of bird diseases (Silvanose et al. 2001) but also to human infections transmitted by wild birds (Omenn 2010).

The number of migratory birds worldwide has been estimated to be 5 billions per year (Hahn et al. 2009). Thus, they should be considered important vectors of microfungi (including those pathogenic for humans) and are likely to play an important role in the environmental dissemination of yeasts. Birds can transmit microfungi from the cloacae in the environment via their excretions (Stevic 1962; Rosa et al. 2009; Basukriadi et al. 2010; Ricci et al. 2011; Stefanini et al. 2012; Chen et al. 2013; Hui et al. 2013). Accordingly, wild birds, especially pigeons, have been regarded as a reservoir of clinically relevant yeasts, especially *Candida* spp. and *Cryptococcus* spp., which are the most frequently reported genera from bird feces (Astorga et al. 1994; Dynowska and Kisicka 2005; Georgopoulou and Tsiouris 2008; Dynowska et al. 2013; Gogu-Bogdan et al. 2014). Yeasts associated with humans and pathogenic yeasts have been recently reviewed (Begerow et al. 2017; Inácio and Daniel 2017).

Yeast diversity vectored by birds results probably from the interaction with the environment birds live in and substrates they feed on. The few studies investigating birds feeding on flower nectar demonstrated that yeast communities found in bird-pollinated flowers differ substantially from those visited by insects but that phyllosphere-related yeasts have frequently been found in flowers visited by birds (Mittelbach et al. 2015). The study of nectar yeasts associated with plants on the Tenerife island (Canary Islands, Spain) showed that flowers visited by birds harbored on average also a more diverse yeast community than communities from insect-pollinated plants (Mittelbach et al. 2015). In another study performed on South African plants, abundance and frequency of observation of yeasts in nectar were positively correlated with bird pollination (de Vega et al. 2009). It has also been suggested that species vectored by migratory birds contribute to local yeast communities found in some agricultural ecosystems, like vineyards, where they can affect the quality of some fermented foods (Francesca et al. 2010). Some studies on yeasts vectored by animals have been focused on *Saccharomyces cerevisiae* (Goddard et al. 2010; Stefanini et al. 2012; Sampaio and Gonçalves 2017). These studies have shown that social wasps have a prominent role in the dissemination of this species (Stefanini et al. 2012). Recently, Dapporto et al. (2016) also demonstrated that social wasp intestines represent a reservoir of yeast diversity that should be preserved to safeguard mainly the natural variance of *S. cerevisiae* (see also Sampaio and Gonçalves 2017). Further details on yeasts associated with insects and flower nectar are provided in Mittelbach and Vannette (2017) and in Chap. 13 of this book. Despite their importance for yeast dispersal, social wasps, honey bees, and a few other prominent insect vectors do not travel over long distances, whereas migratory birds are able to travel between continents.

Bird migration is the regular seasonal movement between breeding and wintering grounds, which occur in autumn and spring, respectively. The phenomenon of migration is typical on some islands of the Mediterranean Basin where birds make a stopover to find food. These movements follow the seasonality of the availability of food resources. The main energy source for birds during flight is the body fat. Birds with low fat reserves necessarily have to stop in some resting sites (stopover) along the migration route to feed. In the Mediterranean Basin, Sicily, due to its geographical position, is the most important resting site for many bird species. Resting sites such as Sicily, where many different bird species coalesce, could represent a crossroads for dispersal routes of microbial propagules carried by the different bird species. The present chapter will provide an overview on yeast species vectored by birds and focus mainly on birds as reservoirs for pathogenic yeast, as well as on dissemination of yeasts by migratory birds. Furthermore, new insights on yeast ecology will be provided to show that migratory birds could represent a so far unknown source of thermotolerant yeasts.

14.2 Birds as Reservoir for Pathogenic Yeasts

14.2.1 Human Pathogens

Humans can be exposed to fungal pathogens through direct contact with infected animals (see also Inácio and Daniel 2017). Some emerging pathogens disseminated by animals, including birds, increasingly attract the attention of medical mycologists (Astorga et al. 1994; Dynowska and Kisicka 2005; Cafarchia et al. 2006b; Georgopoulou and Tsiouris 2008; Dynowska et al. 2013; Gogu-Bogdan et al. 2014). Birds are particularly susceptible to fungal colonization as they provide rather favorable conditions for fungal growth, i.e., poorly vascularized air sacs, very poor or even no antimycotic activity in the serum, and a thin skin with no sweat or sebaceous glands (Dynowska and Kisicka 2005; Dahlhausen 2006). However, despite these predisposing factors, the high internal temperature of bird body serves a barrier for many pathogenic fungi and yeasts. However, birds may still act as passive (mechanical) vectors of potential fungal pathogens (including some yeasts), which can be disseminated in the environment (Robert and Casadevall 2009). Many species of birds, namely pigeons and other wild birds, live or stay temporarily in highly anthropogenic and urban areas, and represent a large reservoir of potentially pathogenic fungi (Dynowska et al. 2013). In recent years, great attention has been given to the dissemination of *Cryptococcus* spp. by birds (e.g., Amirrajab et al. 2016). Irga et al. (2016) found a relationship between the diversity of fungal propagules occurring in air samples and sites with high urban bird frequency and suggested that the roosts of urban bird colonies could represent a potential health risk for humans. Both *Cryptococcus neoformans* and *Cryptococcus gattii* are considered the main causative agents of cryptococcosis, which is an important human and animal disease caused by inhalation of spores or encapsulated yeasts (Malik et al. 2003; Alves et al. 2016; Begerow et al. 2017; Inácio and Daniel 2017). Both species have been frequently isolated from warm-blooded animals worldwide (e.g., Blaschke-Hellmessen 2000). However, among pathogenic *Cryptococcus* species, *Cr. neoformans* is the most abundant species globally and has been frequently isolated from several species of birds (especially from feces), including pigeons, doves, passerines, and ducks. Birds can transmit *Cryptococcus* cells from their cloacae to the environment via their excretions (Filiú et al. 2002; Cafarchia et al. 2006b). Recently, Amirrajab et al. (2016) showed the presence of *Cr. neoformans* in cloaca, crop, and nasal secretion of migratory birds. The contact with birds' excretions, in particular those from pigeons, is considered one of the main causes of cryptococcosis in patients with immune deficiency syndrome (Fessel 1993; Kobayashi et al. 2005).

The ecology of fungal pathogens associated with the mallard (*Anas platyrhynchos*), one of the most abundant migratory water birds, has also been investigated (Meissner et al. 2015). The authors reported that mallards living in urban environments yielded more diverse fungal communities than those living in nonurban areas, probably because of the favorable nutritional and microclimatic conditions for fungal growth occurring in urban areas (Godlewska et al. 2009; Newbound et al. 2010). Meissner et al. (2015) isolated a total of 53 yeast species from the bill cavity and the cloacae of mallards. Among them, 19% were species potentially pathogenic to humans, namely, some *Cryptococcus* spp., *Candida albicans*, and *Candida tropicalis*, suggesting that mallards might act as vectors for the dissemination of pathogenic microfungi in the environment via their excretions.

14.2.2 Bird Pathogens

Yeast infections in birds are well documented for *Cryptococcus* and *Candida* species, which are also pathogenic for humans. Outbreaks of fungal diseases (sometimes due to pathogenic *Cryptococcus* species) have also been found in other wild birds (Burr 1981; Windingstad et al. 1989; Souza and Degernes 2005; Rippon et al. 2012), in birds held in captivity (Hubbard et al. 1985; Wolff et al. 1992), and in poultry (Akan et al. 2002; Khosravi et al. 2008; Lanteri et al. 2012). Fungal infections are frequently associated with high mortality of infected birds that occurs in the course of a few days (Pearson 1969; Khosravi et al. 2008). This high mortality in poultry also results in significant economic losses (Zafra et al.

2008; Arné et al. 2011). The asexual ascomycete *Macrorhabdus ornithogaster* has been isolated from a number of birds, where it infects the stomach growing in dense mats of parallel cells, giving the appearance of "logjams" (Phalen et al. 2011). This yeast however causes widespread infections and occasionally results in devastating disease outbreaks in some bird species, particularly *Melopsittacus undulatus* (the budgerigar) and *Forpus* spp. (the parrotlet) (reviewed in Phalen et al. 2011).

14.3 Migratory Birds as Vector of Yeasts: Sampling and Isolation Procedures

An annotated checklist of potential pathogenic microorganisms vectored by migratory birds has been provided by Hubalek (2004) and also included the yeasts *C. albicans* and *C. tropicalis*. The author concluded that migratory species could be involved in the dispersal of potential pathogens. Most often yeasts associated with birds have been isolated from their cloacae where they arrive after gastric transit (Casadevall and Pirofski 2000; Dynowska and Kisicka 2005; Cafarchia et al. 2006a; Francesca et al. 2010, 2012). Hence, the primary source of yeast contamination most likely originates from the food ingested by the bird. Fruits, seeds, and insects are commonly consumed by birds during their feeding activities (Snow et al. 1998), and they could represent the primary source of yeast colonization of bird's digestive tract. The presence of yeasts on surfaces of raw materials such as drupes of fruits, seeds, as well as in soil and water has been extensively reported (see also Chaps. 1, 2, 3, 4 and 8 of this book). Some of these yeasts are eliminated and digested, while others survive the passage through the gastrointestinal tract and are isolated from feces or bird cloaca (discussed below).

The migration of birds is a process involving billions of individuals on different continents, whereby the animals cover large distances and cross natural dispersal barriers such as deserts and oceans (Goymann et al. 2010). Bird migrations occur in both Northern and Southern Hemispheres in different seasons and are linked to the seasonality of food availability (Moreau 1961, 1972). In this framework, the islands in the Mediterranean Sea represent important resting sites (stopover), where birds replenish the fat lost during their flights between North Africa and Europe and vice versa (Goymann et al. 2010).

Isolation of yeasts from avian species should follow specific sampling protocol procedures (Francesca et al. 2010):

- Birds are captured using a mist net with four shelf-nets (100 m in length), which is placed in the proximity of woody areas in each site of ringing. To avoid injury and reduce stress, the nets are treated according to the guidelines of the "Istituto Superiore per la Protezione e la Ricerca Ambientale" (ISPRA: http://www.isprambiente.gov.it/it/ispra) (Fig. 14.1). Since there is a high risk of contamination by local material, ISPRA guidelines suggest to catch birds on the first stopover points visited on the migration route within the proximity of woody areas in each ringing site just at the moment of landing.

Fig. 14.1 Sampling procedures to catch migratory birds. Mist nets (**a**, **b**), *Saxicola rubetra* (whinchat) (**c**)

- Birds should be quickly ringed by expert ornithologists, and the cloacae are examined for yeast presence. Therefore, cloacae are plugged with sterile cotton swabs, and these are subsequently inoculated in a nutritive medium, which is suitable for cultivation of yeasts (Francesca et al. 2010).
- The procedure includes also the identification of birds at species level by phenotypic analysis (Mullarney et al. 1999), the classification of birds according to their migration strategy (trans-Saharan or partial; Svensson 1992), and visual biometric measurement of subcutaneous fat amount (SFA) of the abdominal region (Kaiser 1993). Generally, the values of SFA range between 0 and 5. Since the body fat is the first energy source dissipated by birds during migration, a stop is normally necessary when SFA values reach between 0 and 1 (Goymann et al. 2010). Accordingly, it is safe to assume that birds characterized by a very low level of SFA (0 or 1) make the first stop at the resting sites.
- Captured birds are quickly released after the analysis.

A comprehensive survey of yeast species associated with migratory birds has been conducted recently by Francesca et al. (2012, 2014). In selected experimental sites located in Sicily and surrounding smaller islands, hundreds of birds were sampled in spring and autumn during their migration across Mediterranean Basin. Yeasts were found in about one third of the sampled birds that is in agreement with the frequency of isolation reported in previous studies (Cafarchia et al. 2006a). The bird families showing the highest frequencies of yeast isolation were Sylviidae, Turdidae, and Muscicapidae (Table 14.1). With exception of species belonging to Fringillidae and Passeridae, which are insectivores in spring and granivorous in

Table 14.1 Birds captured in four Sicilian ringing sites to isolate yeasts

Bird family	Bird species	Migration type	No. of individuals sampled	No. of individuals carrying yeasts	SFA[a] of birds vectoring yeasts	
					0–1	2–7
Sylviidae	*Sylvia borin* (garden warbler)	Trans-Saharan migratory bird	127	61	34	27
	Sylvia curruca (lesser whitethroat)	Trans-Saharan migratory bird	1	0	0	0
	Hippolais polyglotta (melodious warbler)	Trans-Saharan migratory bird	1	0	0	0
	Hippolais icterina (icterine warbler)	Trans-Saharan migratory bird	32	8	6	2
	Acrocephalus scirpaceus (reed warbler)	Trans-Saharan migratory bird	12	5	1	4
	Acrocephalus arundinaceus (great reed warbler)	Trans-Saharan migratory bird	3	0	0	0
	Sylvia atricapilla (blackcap)	Trans-Saharan migratory bird	2	1	1	0
	Acrocephalus schoenobaenus (sedge warbler)	Trans-Saharan migratory bird	12	3	3	0
	Phylloscopus trochilus (willow warbler)	Trans-Saharan migratory bird	7	4	0	4
	Phylloscopus collybita (chiffchaff)	Non-trans-Saharan (short distance)	1	0	0	0
	Phylloscopus sibilatrix (wood warbler)	Trans-Saharan migratory bird	24	7	3	4
	Sylvia melanocephala (Sardinian warbler)	Trans-Saharan migratory bird	15	0	0	0
	Sylvia communis (whitethroat)	Trans-Saharan migratory bird	32	12	2	10
	Sylvia cantillans (subalpine warbler)	Trans-Saharan migratory bird	49	14	2	12
	Cettia cetti (cetti's warbler)	Trans-Saharan migratory bird	2	0	0	0
Muscicapidae	*Ficedula albicollis* (collared flycatcher)	Trans-Saharan migratory bird	8	0	0	0
	Ficedula hypoleuca (pied flycatcher)	Trans-Saharan migratory bird	30	5	2	3
	Muscicapa striata (spotted flycatcher)	Trans-Saharan migratory bird	56	13	10	3
Turdidae	*Phoenicurus phoenicurus* (redstart)	Trans-Saharan migratory bird	12	4	4	0
	Turdus philomelos (song thrush)	Trans-Saharan migratory bird	1	0	0	0
	Erithacus rubecula (robin)	Trans-Saharan migratory bird	1	1	1	0
	Luscinia megarhynchos (nightingale)	Trans-Saharan migratory bird	4	1	1	0
	Monticola solitarius (blue rock thrush)	Partial migratory bird	16	5	2	3
	Saxicola rubetra (winchat)	Trans-Saharan migratory bird	65	27	14	13

(continued)

Table 14.1 (continued)

Bird family	Bird species	Migration type	No. of individuals sampled	No. of individuals carrying yeasts	SFA[a] of birds vectoring yeasts	
					0–1	2–7
Hirundinidae	*Hirundo rustica* (swallow)	Trans-Saharan migratory bird	12	2	2	0
	Delichon urbicum (house martin)	Trans-Saharan migratory bird	7	4	2	2
Passeridae	*Passer montanus* (tree sparrow)	Trans-Saharan migratory bird	1	0	0	0
	Passer hispaniolensis (Spanish sparrow)	Partial migratory birds	8	4	3	1
Oriolidae	*Oriolus oriolus* (golden oriole)	Trans-Saharan migratory bird	3	0	0	0
Fringillidae	*Carduelis cannabina* (linnet)	Trans-Saharan migratory bird	2	0	0	0
Laniidae	*Lanius senator* (woodchat shrike)	Trans-Saharan migratory bird	2	0	0	0
Motacillidae	*Anthus trivialis* (tree pipit)	Trans-Saharan migratory bird	2	0	0	0
Paridae	*Paridae Parus major* (great tit)	Partial migratory birds	2	1	1	0
Total			**552**	**182 (32.97%)**	**94**	**88**

[a]SFA, subcutaneous fat amount. Data showed in the table refer to results reported by Francesca et al. (2012, 2013, 2014 and 2016)

autumn, all other birds vectoring yeasts were obligate insectivores (Snow et al. 1998). The majority of birds, which yielded yeasts, were characterized by very low SFA values (Francesca et al. 2012, 2014) indicating that the majority of yeasts isolated from birds originated from places other than those they were caught in. These results suggest that migratory birds can vector yeasts over long distances. Similarly, Alfonzo et al. (2013) analyzed a few hundreds of birds for the presence of filamentous fungi and found that about 62% and 33% of birds vectored filamentous fungi and yeasts, respectively (Table 14.1). Bird families showing the highest frequencies of isolation of filamentous fungi were, again, Sylviidae, Muscicapidae, and Turdidae (Alfonzo et al. 2013).

14.4 Yeast in Migratory Birds: Diversity and Recently Described Novel Species

Although several studies have characterized both yeasts and filamentous fungi vectored by migratory birds, only few works have investigated the microbial diversity of bird cloacae. The number of living microorganisms present in bird cloacae is estimated to be around 10 CFU cm^{-2} (Hubalek 2004; Cafarchia et al. 2006a, b). Several yeast species, mainly belonging to the former phenotypic genera *Candida* sp., *Cryptococcus* sp., *Rhodotorula* sp., and *Trichosporon* sp. (and taxonomically related new genera) were previously reported from cloacae of migratory birds (Cafarchia et al. 2006a, b). Yeast diversity in bird cloacae was also studied by Francesca et al. (2012, 2014, 2016) using restriction fragment length polymorphism (RFLP) analysis and sequencing of the rRNA internal transcribed spacer (including ITS1, 5.8S, and ITS2) and 26S (including D1/D2) regions. Results of the RFLP analysis of yeasts isolated from migratory birds are reported in Table 14.2. Considerable yeast diversity, consisting of 28 different restriction profiles, was found. Interestingly, a rather high number of yeasts exhibited yet unknown restriction profiles of 5.8S–ITS region, as compared to previous studies (Fernandez-Espinar et al. 2000; Solieri et al. 2007; Tofalo et al. 2009). The sequencing of D1/D2 domains revealed the presence of 19 fungal species belonging to 13 genera, *C. albicans*, *Candida glabrata*, *Candida inconspicua*, *Candida stellimalicola*, *Cryptococcus albidus* var. *kuetzingii* (currently *Naganishia albida*), *Cryptococcus aureus* (*Papiliotrema aurea*), *Cryptococcus carnescens* (*Vishniacozyma carnescens*), *Cryptococcus magnus* (*Filobasidium magnum*), *Cryptococcus victoriae* (*Vishniacozyma victoriae*), *Debaryomyces hansenii*, *Hanseniaspora guilliermondii*, *Metschnikowia pulcherrima*, *Pichia kudriavzevii*, *Pichia terricola*, *Pseudozyma aphidis* (*Moesziomyces aphidis*), *Rhodotorula mucilaginosa*, and *S. cerevisiae*, and the dimorphic fungi *Aureobasidium pullulans* and *Sporisorium penniseti* (Table 14.2; Francesca et al. 2012, 2014, 2016). The ascomycetous taxa comprised more than 80% of isolates found in bird cloacae, and the prevalence of *C. albicans*, *C. glabrata*, and *C. inconspicua* is consistent with

Table 14.2 Molecular identification of yeast isolates from birds with RFLP-PCR and distribution of isolates per each bird species

		Size of restriction fragments				Size of restriction fragments				
R.P.	5.8S–ITS PCR[a]	CfoI	HaeIII	HinfI	26S PCR	HinfI	MseI	ApaI	Species (% identity)[a]	Accession number
I	580	260 + 160 + 80	440 + 150	240 + 180 + 140	1100	470 + 390 + 180 + 55	585 + 210 + 160 + 95 + 60	n.c.	*Aureobasidium pullulans* (100)	KF880791
II	600	180 + 165 + 90	430 + 150	230 + 175 + 125	1100	490 + 410 + 190	610 + 360 + 95	n.c.	*Aureobasidium pullulans* (99)	HQ641272
III	530	285 + 55	438 + 91	277 + 261	1100	490 + 403 + 186	570 + 405	555 + 410 + 130	*Candida albicans* (99)	KF880792
IV	550	290 + 260	460 + 90	280 + 270	1100	490 + 400 + 190	610 + 425	n.c.	*Candida albicans* (99)	HQ641284
V	880	380 + 160 + 140	650 + 220	350 + 260 + 55	1100	490 + 215 + 195 + 55	675 + 375 + 75	725 + 430	*Candida glabrata* (99)	KF880793
VI	900	380 + 165 + 140	660 + 220	350 + 260 + 50	1050	500 + 220 + 200 + 180	700 + 370	700 + 420	*Candida glabrata* (100)	HQ641276
VII	480	105 + 90 + 75 + 56	480	265 + 220	1100	485 + 235 + 180 + 130	n.c.	710 + 385	*Candida inconspicua* (99)	KF880794
VIII	480	105 + 90 + 75 + 55	450	260 + 210	1150	490 + 240 + 180 + 130 + 110	1000 + 90	700 + 390	*Candida inconspicua* (99)	HQ641283
IX	460	200 + 180 + 80	450	220 + 190 + 50	1100	490 + 400 + 100 + 90	390 + 375 + 285 + 75	n.c.	*Candida* spp. (95)	HQ641271
X	525	200 + 155 + 130	515	290 + 220	1100	500 + 400 + 180	n.c.	n.c.	*Candida stellimalicola* (99)	HQ641277
XI	700	325 + 300	510 + 70	280 + 255 + 95	1100	335 + 275 + 215 + 155	400 + 360 + 245 + 65	n.c.	*Naganishia albida* (formerly *Cryptococcus albidus* var. *kuetzingii*) (99)	KF880795
XII	520	250 + 200 + 70	520	290 + 170	1160	470 + 290 + 210 + 190	400 + 370 + 270	n.c.	*Papiliotrema aurea* (*Cryptococcus aureus*) (99)	HQ641274
XIII	550	280 + 240	350 + 90 + 65	250 + 250	1100	440 + 280 + 220 + 205	425 + 380 + 260 + 75	n.c.	*Vishniacozyma* (*Cryptococcus*) *carnescens* (99)	HQ641265
XIV	650	350 + 300	520 + 90	280 + 235 + 140	1100	255 + 200 + 175 + 160 + 145 + 75 + 55	400 + 365 + 244	n.c.	*Filobasidium magnum* (*Cryptococcus magnus*) (99)	KF880796
XV	650	355 + 295	495 + 95 + 60	270 + 240 + 140	1100	280 + 200 + 180 + 150	400 + 370 + 270	n.c.	*Filobasidium magnum* (*Cryptococcus magnus*) (99)	HQ641280

XVI	530	290 + 245	360 + 125	269 + 180 + 85	1100	420 + 275 + 210 + 200	405 + 285 + 245	n.c.	*Vishniacozyma (Cryptococcus) victoriae* (99)	KF880797
XVII	650	300 + 300 + 50	420 + 150 + 90	325 + 325	1100	490 + 410 + 190	610 + 320 + 115 + 75	n.c.	*Debaryomyces hansenii* (99)	HQ641266
XVIII	750	320 + 310 + 105	750	350 + 200 + 180	1000	415 + 395 + 190 + 105	610 + 500 + 440 + 100 + 75	n.c.	*Hanseniaspora guilliermondii* (99)	HQ641270
XIX	400	205 + 100 + 95	280 + 100	200 + 190	1100	380 + 260 + 240	580 + 270 + 140	600 + 420	*Metschnikowia pulcherrima* (98)	HQ641286
XX	500	115 + 90 + 75 + 55	325 + 90 + 75	270 + 225	1100	500 + 250 + 180	n.c.	700 + 400	*Pichia kudriavzevii*	HQ641275
XXI	450	130 + 100 + 90 + 85 + 45	290 + 125	240 + 105 + 105	1100	500 + 350	800 + 200	n.c.	*Pichia terricola*	HQ641279
XXII	480	240 + 220	380 + 115	404	1100	335 + 310 + 210 + 185 + 75	n.c.	n.c.	*Phaffomyces* sp. (97)	KF719195
XXIII	780	220 + 170 + 150 + 130	420 + 320	440 + 340	1100	480 + 250 + 200	400 + 370 + 270	n.c.	*Moesziomyces (Pseudozyma) aphidis* (99)	HQ641278
XXIV	640	300 + 225	404 + 217	346 + 215	1100	495 + 410 + 205	355 + 270 + 235 + 140	n.c.	*Rhodotorula mucilaginosa* (99)	KF880798
XXV	640	320 + 240 + 80	425 + 215	340 + 225 + 75	1100	500 + 400 + 200	400 + 300 + 250	n.c.	*Rhodotorula mucilaginosa (99)*	HQ641269
XXVI	850	375 + 335 + 140	320 + 240 + 170 + 130	370 + 130 + 110	1100	500 + 220 + 180	n.c.	n.c.	*Saccharomyces cerevisiae* (99)	HQ641267
XXVII	825	370 + 250 + 150	300 + 240 + 160 + 60	400 + 200 + 130	1100	500 + 260 + 200	400 + 380 + 300	n.c.	*Anthracocystis (Sporisorium) penniseti* (97)	HQ641273
XXVIII	640	610	560 + 80	330 + 310	1080	500 + 240 + 180 + 160	n.c.	n.c.	*Wickerhamomyces* sp. (97)	KF880799

[a]According to BLASTN search of D1/D2 26S rRNA gene sequence in NCBI database

Data showed in the table refer to results reported by Francesca et al. (2012, 2013, 2014 and 2016). All values for the 5.8S–ITS PCR, 26S PCR, and restriction fragments are given in bp. *R.P.* restriction profile, *n.c.* not cut

previous studies (Cafarchia et al. 2006a; Lord et al. 2010). Interestingly, some of the abovementioned species are commonly reported to be associated with winemaking (Loureiro and Malfeito-Ferreira 2003; Gonzalez-Pombo et al. 2008; Moreira et al. 2008; Francesca et al. 2010; Ocón et al. 2010). These findings suggest birds ingest the isolated yeasts with foods (i.e., seeds, grapes, fruits, etc.) and transmit them over large distances, e.g., in vineyards and related environments. The novel species *Wickerhamomyces sylviae*, *Phaffomyces usticensis*, and *Jaminaea phylloscopi* were found among strains isolated from bird cloacae (Francesca et al. 2013, 2014, 2016), thus confirming that migratory birds represent an interesting source of yeasts, including yet unknown species.

14.5 Yeast in Migratory Birds: Unusual Phenotypic Traits and Thermotolerance

Phenotypic features of yeasts isolated from migratory birds have been studied recently (Francesca et al. 2013). All eight strains of *W. sylviae* isolated from birds exhibited the ability to grow at 37 °C, a behavior that is in contrast to the closely related species *Wickerhamomyces anomalus*, *Wickerhamomyces linferdii*, and *Wickerhamomyces subpellicolosus*. Since the internal body temperature of migratory birds is reported to be around 42 ± 1 °C (Gwinner 1990), the ability of some strains to grow at 40 and 42 °C has been also tested in laboratory experiments. Surprisingly, some strains showed intense growth at both temperatures and at low pH (Table 14.3), thus suggesting that *W. sylviae* is able to survive and grow at the bird internal body temperature (Francesca et al. 2014). Although some authors reported the presence of *Wickerhamomyces* strains in the gut of insects (Rosa et al. 2009; Ricci et al. 2011; Hui et al. 2013), no thermotolerant isolates have been found so far. Thermotolerance and low pH tolerance were mainly exhibited by ascomycetous yeasts isolated from birds. These features could be an adaptation to survival and proliferation after ingestion and transit through the gastrointestinal tract (Francesca et al. 2014, 2016) and would also explain the higher frequency of Ascomycota from the cloacae of both migratory and nonmigratory birds (Hubalek 2004; Cafarchia et al. 2006a; Lord et al. 2010; Francesca et al. 2012). On the other hand, among basidiomycetous yeasts, only *Cr. albidus* var. *kuetzingii* (*Na. albida*) grew at 37 °C and at low pH (Francesca et al. 2014, 2016).

14.6 The Long Travel of "Migratory Yeasts"

Some yeasts possess adaptations allowing them to safely pass the gastrointestinal tract of birds (see Sect. 14.4). This potentially allows them to disperse over large geographic distances, as migratory birds often fly long times without stopping.

Table 14.3 Phenotypic characterization of yeasts isolated from migratory birds

Characteristics	*Aureobasidium pullulans* (*n* = 5)			*Candida albicans* (*n* = 12)			*Candida glabrata* (*n* = 16)			*Candida incospiscua* (*n* = 17)		*Naganishia albida* (*Cryptococcus albidus* var. *kuetzingii*) (*n* = 2)				*Filobasidium magnum* (*Cryptococcus magnus*) (*n* = 2)			*Vishniacozyma victoriae* (*Cryptococcus victoriae*) (*n* = 3)			*Jaminaea phylloscopi* (*n* = 5)			*Phaffomyces usticensis* (*n* = 4)			*Rhodotorula mucilaginosa* (*n* = 13)			*Wickerhamomyces sylviae* (*n* = 13)		
	+	W	−	+	W	−	+	W	−	+	W	−	+	W	−	+	W	−	+	W	−	+	W	−	+	W	−	+	W	−	+	W	−
Growth on/at:																																	
37°C	5			12			16			17			2					2			3	4	1		4				11	2	13		
40°C	5			10	2		16					17			2			2			3		5		3		1			13	13		
42°C	4	1		6		6	11	5				17			2			2			3		5				4			13			13
pH 3.0–25°C	5			12			16			9	8		2					2			3		5		3	1		13			13		
pH 3.0–37°C	3	2		12			16			6	2	9	2					2			3		5				4		5	8	7	6	
pH 3.0–40°C	5			10	2		14	2				17			2			2			3		5				4			13		10	3
pH 3.0–42°C	5			6		6	11	5				17			2			2			3		4	1			4			13			13
pH 2.5–25°C	3	2		12			16				3	14			2			2			3		5			2	2	13			3	9	1
pH 2.5–37°C			5	12			16					17	2					2			3		5				4		5	8		7	6
pH 2.5–40°C			5	10	2		16					17			2			2			3		5				4			13		10	3
pH 2.5–42°C			5	6		6	13	3				17			2			2			3		5				4			13		1	12
1% acetic acid—25°C			5			12			16			17			2			2			3	n.d.	n.d.	n.d.		2	2			13		2	11
1% acetic acid—37°C			5			12			16			17			2			2			3	n.d.	n.d.	n.d.			4			13		2	11
1% acetic acid—40°C			5			12			16			17			2			2			3	n.d.	n.d.	n.d.			4			13			13
1% acetic acid—42°C			5			12			16			17			2			2			3	n.d.	n.d.	n.d.			4			13			13

+ = positive growth; − = negative growth; *W* weak. Data showed in the table refer to results reported by Francesca et al. (2012, 2013, 2014, 2016). *n.d.* not determined

Thereby, they can carry viable yeast inoculum between overwintering sites and breeding grounds. However, there is still little knowledge about transportation of particular yeast species and how far the journeys of these "migratory yeasts" actually are.

The first experimental evidence suggesting long-term survival of yeasts in the gastrointestinal tract of migratory birds tested the duration of the persistence of *S. cerevisiae* in both cloacae and mouth of migratory birds (Francesca et al. 2010). An average timing for yeast dissemination by birds was calculated as up to 12 h after ingestion. On the basis of these results, the spatial dissemination of yeasts by migratory birds was estimated to range approximately from 300 to 500 km (Francesca et al. 2014). This distance is sufficient for a travel from sub-Saharan areas to the first stopover sites in Italy (i.e., Sicily and surrounding islands; Moreau 1961, 1972). Among yeasts associated with migratory birds, the isolation of *Phaff. usticensis* from cloacae of migratory birds caught in Sicily is of special interest. This is the first report of *Phaffomyces* yeasts in Europe and also from a non-cactus environment (Francesca et al. 2014). This finding also supports the notion of such "migratory yeasts" being transported from the sub-Saharan area (where cacti have been introduced) to Sicily (Stocker 1976). The isolation of a supposedly cactophilic yeasts from bird cloacae could be due to the feeding habits of birds, which ingest fruits, seeds, and insects, including also drosophilids (Snow et al. 1998). The importance of the cactus-feeding drosophilids for yeast vectoring is well documented in the literature (e.g., Fellows and Heed 1972; Starmer et al. 1988; see also Chap. 8 of this book). Thus, *Drosophila* flies (and possibly other insects) can be viewed as important sources of yeast colonization of the gastrointestinal tract of birds.

14.7 Yeast and Nectar-Feeding Birds

Hummingbirds (Trochilidae) are among the smallest birds feeding on nectar. They are well known for their very peculiar ability to collect nectar while hovering near a flower in the air rapidly flapping their wings. These birds are spread throughout North America and South America and have diversified with the plants they evolved to feed on (REF). Recently, a few studies have investigated fungi and yeasts associated with hummingbirds (Belisle et al. 2012, 2014). This work has been largely inspired by the previous research on yeasts in different flower-insect systems (see also Chaps. 6 and 13 of this book). Yeasts have been isolated from nectar of flowers visited by hummingbirds (Belisle et al. 2012) and from hummingbird bills (Belisle et al. 2014). The two studies also analyzed distribution patterns of yeasts assuming their nonrandom transportation between flowers. Although the composition of nectar yeast communities correlated with the sampling site (Belisle et al. 2012), no significant correlation has been observed between yeast communities on hummingbird bills and spatial distance, habitat type (impact of deforestation), species of hummingbirds, and their primary habitat, i.e., forest dependency

(Belisle et al. 2014). Similar to insect pollinators, hummingbirds are able to modify microbial communities in flowers they feed on. The bacteria *Gluconobacter* sp. growing in nectar of the hummingbird-pollinated plant *Mimulus aurantiacus* distracted pollinators, while the nectar-borne yeast *Metschnikowia reukaufii* had no negative effect on pollination success, seed set, and nectar consumption by pollinators (Vannette et al. 2013). How does bird visitation affect late-arriving pollinators is largely unknown since our knowledge about yeasts associated with hummingbirds is very limited. A total of 38 species have been identified from 408 isolates obtained from 890 birds and bats by Belisle et al. (2012). The most abundant species isolated were the ascomycetous species *A. pullulans* (dimorphic), *Candida intermedia*, *Candida metapsilosis*, *Candida quercitrusa*, *Candida rancensis*, *Clavispora lusitaniae*, *Lodderomyces elongisporus*, and *Metschnikowia koreensis* and the basidiomycetous species *Papiliotrema flavescens*. The most abundant yeasts found in flowers of the hummingbird-pollinated plant *Mimulus aurantiacus* belonged to the ascomycetous species *M. reukaufii*, *C. rancensis*, *Candida parapsilosis*, *Hanseniaspora valbyensis*, and *Metschnikowia kunwiensis* (Belisle et al. 2012).

Nectar-feeding birds are not limited to hummingbirds, and the shift in plant pollination toward ornithophily has occurred independently in many lineages of flowering plants involving also bird families Nectariniidae (the sunbirds), Meliphagidae (the honeyeaters), Icteridae (the American orioles), Thraupidae (the honeycreepers), and Zosteropidae (the white-eyes), among others (Cronk and Ojeda 2008). It has been shown that nonspecialized birds such as *Phylloscopus canariensis* (Canarian chiffchaff), *Parus caeruleus* (blue tit), and *Sylvia melanocephala* (Canary Sardinian warbler) can also feed on nectar of plants of the Canary Islands, even in the absence of probably extinct nectarivore sunbirds (Cronk and Ojeda 2008). It is important to mention that plant adaptation to bird pollination includes the production of abundant dilute nectar containing a high proportion of sucrose, whereas insect-pollinated flowers typically produce more concentrated nectars dominated by fructose and glucose (Nicolson and Fleming 2003; Cronk and Ojeda 2008; Mittelbach et al. 2015). Lower sugar content could favor the growth of yeast species, which otherwise would not resist the high osmotic pressure in nectar (discussed in Mittelbach et al. 2015). Contamination of flower nectar with nonspecialized yeast species vectored by birds and reduced environmental stress in nectar both explain higher observation frequency and diversity of yeast in bird-pollinated flowers (de Vega et al. 2009; Mittelbach et al. 2015). Specifically, the low concentrated monosaccharide nectar preferred by opportunistic nectar-feeding Canary birds facilitated the growth of inoculated transient species compared to insect-pollinated flowers (Mittelbach et al. 2015). The most abundant species in flowers visited by birds were *Cystofilobasidium alribaticum* (cited as *Cystofilobasidium capitatum*), *Vishniacozyma* spp., and *M. reukaufii* (Mittelbach et al. 2015). Other transient yeasts were Basidiomycota in the genera *Bullera*, *Cryptococcus*, *Curvibasidium*, *Erythrobasidium*, *Fonsecazyma*, *Kwoniella*, and *Papiliotrema* (Mittelbach et al. 2015).

14.8 Concluding Remarks

The reports available to date suggest that birds (in particular their gastrointestinal tract) can represent an ecological niche suitable for both pathogenic and nonpathogenic microorganisms and show birds to be potential vectors for yeasts. Feces of pigeons and other wild birds suit a natural reservoir for pathogenic yeasts of the genera *Cryptococcus* and *Candida*, and bird excretions are considered one of the main causes of cryptococcosis and candidiasis. Environmental dissemination of nonpathogenic yeasts by birds is also supposed to contribute to the microbial diversity of agricultural ecosystems (e.g., vineyards). Although insect vector and hosts are commonly recognized as a large reservoir of yeast diversity, yeasts associated with migratory birds deserve additional studies in the future. Unlike insect vectors, migratory birds can transport yeasts between continents and over long distances. The peculiar phenotypes found among yeasts vectored by migratory birds confirmed yeast adaptation to conditions in the gastrointestinal tract of birds. In particular, it has been demonstrated that migratory birds may represent a good ecological niche for thermotolerant yeasts. Because seasonal bird migration is a common phenomenon, the dissemination of yeasts by migratory birds might contribute to biodiversity in many geographical locations and should be considered in biodiversity assessments. Further analyses of birds should be directed to repeated samplings in different years and sites to obtain a more clear picture of yeast diversity vectored by migratory birds. Additional experiments on the persistence of yeasts in gastrointestinal tract and cloacae of birds are important to better understand mechanisms of the yeast migration with birds.

Acknowledgments Pietro Buzzini, Martin Kemler, Moritz Mittelbach, and Andrey Yurkov are acknowledged for their assistance in preparation of this book chapter.

References

Akan M, Hazroğlu R, Ilhan Z, Sareyyüpoğlu B, Tunca R (2002) A case of aspergillosis in a broiler breeder flock. Avian Dis 46:497–501

Alfonzo A, Francesca N, Sannino C, Settanni L, Moschetti G (2013) Filamentous fungi transported by birds during migration across the Mediterranean Sea. Curr Microbiol 66:236–242

Alves I, Arenas M, Currat M, Hanulova AS, Sousa VC, Ray N, Excoffier L (2016) Long-distance dispersal shaped patterns of human genetic diversity in Eurasia. Mol Biol Evol 33:946–958

Amirrajab N, Haghani I, Rasuli M, Shokohi T (2016) Migratory birds as a potential reservoirs of *Cryptococcus neoformans*. Int J Environ Res 10:459–464

Arné P, Thierry S, Wang D, Deville M, Le Loc'h G, Desoutter A, Féménia F, Nieguitsila A, Huang W, Chermette R, Guillot J (2011) *Aspergillus fumigatus* in poultry. Int J Microbiol 2011:1–14

Astorga RJ, Cubero MJ, Leon L, Maldonado A, Arenas A, Tarradas MC, Perea A (1994) Serological survey of infections in waterfowl in the Guadalquivir marshes (Spain). Avian Dis 38:371–375

Basukriadi A, Sjamsuridzal W, Putra BB (2010) Molecular identification and diversity of yeasts associated with *Apis cerana* foraging on flowers of *Jatropha integerrima*. Microbiol Indones 4:44–48

Begerow D, Kemler M, Feige A, Yurkov AM (2017) Parasitism in yeasts. In: Buzzini P, Lachance MA, Yurkov AM (eds) Yeasts in natural ecosystems: ecology. Springer, Heidelberg, pp 179–210

Belisle M, Peay KG, Fukami T (2012) Flowers as islands: spatial distribution of nectar-inhabiting microfungi among plants of *Mimulus aurantiacus*, a hummingbird-pollinated shrub. Microb Ecol 63:711–718

Belisle M, Mendenhall CD, Oviedo Brenes F, Fukami T (2014) Temporal variation in fungal communities associated with tropical hummingbirds and nectarivorous bats. Fungal Ecol 12:1–7

Blaschke-Hellmessen R (2000) *Cryptococcus* species etiological agents of zoonoses or sapronosis? Mycoses 43:48–60

Burr EW (1981) Enzootic aspergillosis in wild red vented cockatoos in the Philippines. Mycopathologia 73:21–28

Cafarchia C, Camarda A, Romito D (2006a) Occurrence of yeasts in cloacae of migratory birds. Mycopathologia 161:229–234

Cafarchia C, Romito D, Iatta R (2006b) Role of birds of prey as carriers and spreaders of *Cryptococcus neoformans* and other zoonotic yeasts. Med Mycol 44:485–492

Casadevall A, Pirofski LA (2000) Host-pathogen interactions: basic concepts of microbial commensalism, colonization, infection, and disease. Infect Immunity 68:6511–6518

Chae HS, Jang GE, Kim NH, Son HR, Lee JH, Kim SH, Park GN, Jo HJ, Kim JT, Chang KS (2012) Classification of *Cryptococcus neoformans* and yeast-like fungus isolates from pigeon droppings by colony phenotyping and ITS genotyping and their seasonal variations in Korea. Avian Dis 56:58–64

Chen L, Zhang L, Li ZH, Hui FL (2013) *Sympodiomycopsis yantaiensis* sp. nov., a basidiomycetous yeast isolated from insect frass. Int J Syst Evol Microbiol 63:3501–3505

Costa AK, Sidrim JJ, Cordeiro RA, Brilhante RS, Monteiro AJ, Rocha MF (2010) Urban pigeons (*Columba livia*) as a potential source of pathogenic yeasts: a focus on antifungal susceptibility of *Cryptococcus* strains in Northeast Brazil. Mycopathologia 169:207–213

Cronk Q, Ojeda I (2008) Bird-pollinated flowers in an evolutionary and molecular context. J Exp Bot 59:715–727

Dahlhausen RD (2006) Implication of mycoses in clinical disorders. In: Harrison GJ, Lightfoot T (eds) Clinical avian medicine. Spix Publishing, Palm Beach, FL, pp 691–704

Dapporto L, Stefanini I, Rivero D, Polsinelli M, Capretti P, De Marchi P, Viola R, Turillazzi S, Cavalieri D (2016) Social wasp intestines host the local phenotypic variability of *Saccharomyces cerevisiae* strains. Yeast 33:277–287

de Vega HCMM, Herrera CM, Johnson SDD (2009) Yeasts in floral nectar of some South African plants: quantification and associations with pollinator type and sugar concentration. S Afr J Bot 75:798–806

Dynowska M, Kisicka I (2005) Participation of birds in the circulation of pathogenic fungi descend from water environment: a case study of two species of *Charadriiformes* birds. Ecohydrol Hydrobiol 5:173–178

Dynowska M, Wojczulanis-Jakubas K, Pacynska J, Jakubas D, Ejdys E (2013) Potentially pathogenic yeast isolated from the throat and cloacae of an Arctic colonial seabird: the little auk (*Alle alle*). Polar Biol 36:343–348

Fellows DP, Heed WB (1972) Factors effecting host plant selection in desert-adapted cactophilic *Drosophila*. Ecology 53:850–858

Fernandez-Espinar MT, Esteve-Zarzoso B, Querol A, Barrio E (2000) RFLP analysis of the ribosomal transcribed spacers and the 5.8S rRNA gene region of the genus *Saccharomyces*: a fast method for species identification and the differentiation of flor yeasts. Antonie Van Leeuwenhoek 78:87–97

Fessel WJ (1993) Cryptococcal meningitis after unusual exposures to birds. New Engl J Med 328:1354–1355
Filiú WF, Wanke B, Agüena SM, Vilela VO, Macedo RC, Lazera M (2002) Avian habitats as sources of *Cryptococcus neoformans* in the city of Campo Grande, Mato Grosso do Sul, Brazil. Rev Soc Bras Med Trop 35:591–595
Francesca N, Chiurazzi M, Romano R, Aponte M, Settanni L, Moschetti G (2010) Indigenous yeast communities in the environment of 'Rovello bianco' grape variety and their use in commercial white wine fermentation. World J Microbiol Biotechnol 26:337–351
Francesca N, Canale DE, Settanni L, Moschetti G (2012) Dissemination of wine-related yeasts by migratory birds. Environ Microbiol Rep 4:105–112
Francesca N, Carvalho C, Almeida PM, Sannino C, Settanni L, Sampaio JP, Moschetti G (2013) *Wickerhamomyces sylviae* f.a., sp. nov., an ascomycetous yeast species isolated from migratory birds. Int J Syst Evol Microbiol 63:4824–4830
Francesca N, Carvalho C, Sannino C, Guerreiro MA, Almeida PM, Settanni L, Massa B, Sampaio JP, Moschetti G (2014) Yeasts vectored by migratory birds collected in the Mediterranean Island of Ustica and description of *Phaffomyces usticensis* f.a. sp. nov., a new species related to the cactus ecoclade. FEMS Yeast Res 14:910–921
Francesca N, Guerreiro MA, Carvalho C, Coelho M, Alfonzo A, Randazzo W, Sampaio JP, Moschetti G (2016) *Jaminaea phylloscopi* sp. nov. (Microstromatales), a basidiomycetous yeast isolated from migratory birds in the Mediterranean Basin. Int J Syst Evol Microbiol 66:824–829
Georgopoulou I, Tsiouris V (2008) The potential role of migratory birds in the transmission of zoonosis. Vet Ital 44:671–677
Goddard MR, Anfang N, Tang R, Gardner RC, Jun C (2010) A distinct population of Saccharomyces cerevisiae in New Zealand: evidence for local dispersal by insects and human-aided global dispersal in oak barrels. Environ Microbiol 12:63–73
Godlewska R, Wiśniewska K, Pietras Z, Jagusztyn-Krynicka EK (2009) Peptidoglycan-associated lipoprotein (Pal) of Gram-negative bacteria: function, structure, role in pathogenesis and potential application in immunoprophylaxis. FEMS Microbiol Lett 298:1–11
Gogu-Bogdan M, Damoc I, Pall E, Niculae M, Spinu M (2014) Wild birds as potential vectors for pathogen dissemination on migration routes in the Danube Delta Wetlands. Int J Curr Microbiol App Sci 3:890–897
Gonzalez-Pombo P, Perez G, Carrau F, Guisan JM, Batista-Viera F, Brena BM (2008) One-step purification and characterization of an intracellular β-glucosidase from *Metschnikowia pulcherrima*. Biotechnol Lett 30:1469–1475
Goymann W, Spina F, Ferri A, Fusani L (2010) Body fat influences departure from stopover sites in migratory birds: evidence from whole Island telemetry. Biol Lett 6:478–481
Gwinner E (1990) Bird migration: physiology and ecophysiology. Springer, Berlin
Hahn S, Bauer S, Liechti F (2009) The natural link between Europe and Africa: 2.1 billion birds on migration. Oikos 118:624–626
Hubalek Z (2004) An annotated checklist of pathogenic microorganisms associated with migratory birds. J Wildl Dis 40:639–659
Hubbard GB, Schmidt RE, Eisenbrandt DL, Witt WM, Fletcher KC (1985) Fungal infections of ventriculi in captive birds. J Wildl Dis 21:25–28
Hui FL, Chen L, Chu XY, Niu QH, Ke T (2013) *Wickerhamomyces mori* sp. nov., an anamorphic yeast species found in the guts of wood-boring insect larvae. Int J Syst Evol Microbiol 63:1174–1178
Inácio J, Daniel HM (2017) Commensalism: the case of the human zymobiome. In: Buzzini P, Lachance MA, Yurkov AM (eds) Yeasts in natural ecosystems: ecology. Springer, Heidelberg, pp 211–228
Irga PJ, Armstrong B, King WL, Burchett M, Torpy FR (2016) Correspondence between urban bird roosts and the presence of aerosolised fungal pathogens. Mycopathologia 181:689–699

Kaiser A (1993) A new multi-category classification of subcutaneous fat deposits of songbirds. J Field Ornithol 64:246–255

Khosravi AR, Shokri H, Ziglari T, Naeini AR, Mousavi Z, Hashemi H (2008) Outbreak of severe disseminated aspergillosis in a flock of ostrich (*Struthio camelus*). Mycoses 51:557–559

Kobayashi CC, Souza LK, Fernandes Ode F, Brito SC, Silva AC, Sousa ED, Silva Mdo R (2005) Characterization of *Cryptococcus neoformans* isolated from urban environmental sources in Goiania, Goias State, Brazil. Rev Inst Med Trop São Paulo 47:203–207

Lanteri G, Macrì F, Rapisarda G, Basile F, Reale S, Marino F (2012) Systemic candidiasis in farm-reared red-legged partridges (*Alectoris rufa*) caused by *Leucosporidium* spp. BMC Vet Res 8:81

Lord ATK, Mohandas K, Somanath S, Ambu S (2010) Multidrug resistant yeasts in synanthropic wild birds. Ann Clin Microbiol Antimicrob 9:11

Loureiro V, Malfeito-Ferreira M (2003) Spoilage yeasts in the wine industry. Int J Food Microbiol 86:23–50

Malik R, Krockenberger MB, Cross G, Doneley R, Madill DN, Black DP, Mcwhirter A, Rozenwax K, Rose M, Alley D, Forshaw I, Russell-Brown AC, Johnstone P, Martin CR, O'Brien DN (2003) Avian cryptococcosis. Med Mycol 41:115–124

Meissner W, Dynowska M, Goralska K, Rzyska H (2015) Mallards (*Anas platyrhynchos*) staying in urban environments have higher levels of microfungi biota diversity than do birds from nonurban areas. Fungal Ecol 17:164–169

Mittelbach M, Vannette RL (2017) Mutualism in yeasts. In: Buzzini P, Lachance MA, Yurkov AM (eds) Yeasts in natural ecosystems: ecology. Springer, Heidelberg, pp 155–178

Mittelbach M, Yurkov AM, Nocentini D, Nepi M, Weigend M, Begerow D (2015) Nectar sugars and bird visitation define a floral niche for basidiomycetous yeast on the canary Islands. BMC Ecol 15:2

Moreau RE (1961) Problems of Mediterranean-Saharan migration. Ibis 103:373–427

Moreau RE (1972) The Palaearctic-African bird migration systems. Academic Press, London

Moreira N, Mendes F, Guedes de Pinho P, Hogg T, Vasconcelos I (2008) Heavy sulphur compounds, higher alcohols and esters production profile of *Hanseniaspora uvarum* and *Hanseniaspora guilliermondii* grown as pure and mixed cultures in grape must. Int J Food Microbiol 124:231–238

Mullarney K, Svensson L, Zetterstrom D, Grant P (1999) Bird guide. Collins, London

Newbound M, Mccarthy MA, Lebel T (2010) Fungi and the urban environment: a review. Landsc Urban Plan 96:138–145

Nicolson SW, Fleming PA (2003) Nectar as food for birds: the physiological consequences of drinking dilute sugar solutions. Plant Syst Evol 238:139–153

Ocón E, Gutiérrez AR, Garijo P, Tenorio C, López I, Santamaría P (2010) Quantitative and qualitative analysis of non-*Saccharomyces* yeasts in spontaneous alcoholic fermentations. Eur Food Res Technol 230:885–891

Omenn GS (2010) Evolution and public health. Proc Natl Acad Sci USA 107:1702–1709

Pearson GL (1969) Aspergillosis in wintering mallards. Bull Wildl Dis Assoc 5:404–405

Phalen DN, Hannafusa Y, Tomazewski EK (2011) *Macrorhabdus* Tomaszewski, Logan, Snowden, Kurtzman and Phalen (2003). In: Kurtzman CP, Fell JW, Boekhout T (eds) The yeasts. A taxonomic study. Elsevier, Amsterdam, pp 1299–1301

Ricci I, Damiani C, Scuppa P, Mosca M, Crotti E, Rossi P, Rizzi A, Capone A, Gonella E, Ballarini P, Chouaia B, Sagnon N, Esposito F, Alma A, Mandrioli M, Sacchi L, Bandi C, Daffonchio D, Favia G (2011) The yeast *Wickerhamomyces anomalus* (*Pichia anomala*) inhabits the midgut and reproductive system of the Asian malaria vector *Anopheles stephensi*. Environ Microbiol 13:911–921

Rippon RJ, Alley MR, Castro I (2012) *Candida albicans* infection in free-living populations of hihi (stitchbird; *Notiomystis cincta*). N Z Vet J 58:299–306

Robert VA, Casadevall A (2009) Vertebrate endothermy restricts most fungi as potential pathogens. J Infect Dis 200:1623–1626

Rosa CA, Morais PB, Lachance MA, Santos RO, Melo WGP, Viana RHO, Braganca MA, Pimenta RS (2009) *Wickerhamomyces queroliae* sp. nov. and *Candida jalapaonensis* sp. nov., two yeast species isolated from Cerrado ecosystem in North Brazil. Int J Syst Evol Microbiol 59:1232–1236

Sampaio JP, Gonçalves P (2017) Biogeography and ecology of the genus *Saccharomyces*. In: Buzzini P, Lachance MA, Yurkov AM (eds) Yeasts in natural ecosystems: ecology. Springer, Heidelberg, pp 131–153

Silvanose CD, Bailey TA, Naldo JL, Howlett JC (2001) Bacterial flora of the conjunctiva and nasal cavity in normal and diseased captive bustards. Avian Dis 42:447–451

Snow DW, Perrins CM, Gillmor R (1998) The birds of the Western Palearctic: passerines. Oxford University Press, Oxford

Solieri L, Cassanelli S, Giudici P (2007) A new putative *Zygosaccharomyces* yeast species isolated from traditional balsamic vinegar. Yeast 24:403–417

Souza MJ, Degernes LA (2005) Mortality due to aspergillosis in wild Swans in Northwest Washington state, 2000–2002. J Avian Med Surg 19:98–106

Starmer WT, Phaff HJ, Bowles JM, Lachance MA (1988) Yeasts vectored by insects feeding on decaying saguaro cactus. Southwest Nat 33:362–363

Stefanini I, Dapporto L, Legras JL, Calabretta A, Di Paola M, De Filippo C, Viola R, Capretti P, Polsinelli M, Turillazzi S, Cavalieri D (2012) Role of social wasps in *Saccharomyces cerevisiae* ecology and evolution. Proc Natl Acad Sci USA 109:13398–13403

Stevic S (1962) The significance of bees (*Apis* sp.) and wasps (*Vespa* sp.) as carriers of yeast for the microflora of grapes and the quality of wine. Arkh Poljjoprivredne Nauke 50:80–92

Stocker O (1976) The water-photosynthesis syndrome and plant life ecological studies and the geographical plant distribution in the Saharan deserts. Water and plant life. Ecol Stud 19:506–521

Svensson L (1992) Identification guide to European passerines. British Trust for Ornithology, Stockholm

Tofalo R, Chaves-Lopez C, Di Fabio F, Schirone M, Felis GE, Torriani S, Paparella A, Suzzi G (2009) Molecular identification and osmotolerant profile of wine yeasts that ferment a high sugar grape must. Int J Food Microbiol 130:179–187

Vannette RL, Gauthier M-PL, Fukami T (2013) Nectar bacteria, but not yeast, weaken a plant-pollinator mutualism. Proc Biol Sci 280:20122601

Windingstad RM, Cole RJ, Nelson PE, Roffe TJ, George RR, Dorner JW (1989) *Fusarium* mycotoxins from peanuts suspected as a cause of sandhill crane mortality. J Wildl Dis 25:38–46

Wolff PL, Petrini KR, Kolmstetter C (1992) An outbreak of aspergillosis in crested wood partridges (Rollulus rouloul). J Zoo Wildl Med 23:108–112

Zafra R, Pérez J, Pérez-Ecija RA, Borge C, Bustamante R, Carbonero A, Tarradas C (2008) Concurrent aspergillosis and ascites with high mortality in a farm of growing broiler chickens. Avian Dis 52:711–713

Chapter 15
Census of Yeasts Isolated from Natural Ecosystem and Conserved in Worldwide Collections

Marizeth Groenewald, Kyria Boundy-Mills, Neža Čadež, Rikiya Endoh, Sasitorn Jindamorakot, Carolina Pohl-Albertyn, Carlos A. Rosa, Benedetta Turchetti, and Andrey Yurkov

Abstract There are many well-known public yeast repositories as well as a large number of smaller, less-known collections worldwide; most of these are with the primary goal to preserve the yeast biodiversity in a specific region and the strains from a range of species that are important environmental strains, food spoilage organisms, or strains that play a role in food preparation and human or animal pathogens. In order to have an overview on how many yeast strains are isolated

---

M. Groenewald (✉)
Westerdijk Fungal Biodiversity Institute, Uppsalalaan 8, Utrecht 3584 CT, The Netherlands
e-mail: m.groenewald@westerdijkinstitute.nl

K. Boundy-Mills
Phaff Yeast Culture Collection, University of California, Davis, CA, USA

N. Čadež
Biotechnical Faculty, University of Ljubljana, Ljubljana, Slovenia

R. Endoh
Japan Collection of Microorganisms, RIKEN BioResource Center (RIKEN BRC-JCM), Tsukuba, Ibaraki, Japan

S. Jindamorakot
BIOTEC, National Science and Technology Development Agency, Pathumthani, Thailand

C. Pohl-Albertyn
UNESCO MIRCEN Yeats Culture Collection, Department of Microbial, Biochemical and Food Biotechnology, University of the Free State, Bloemfontein, South Africa

C.A. Rosa
Departamento de Microbiologia, ICB, Universidade Federal de Minas Gerais, Belo Horizonte, Brazil

B. Turchetti
Department of Agricultural, Food and Environmental Sciences, Industrial Yeasts Collection DBVPG, University of Perugia, Perugia, Italy

A. Yurkov
Leibniz Institute DSMZ - German Collection of Microorganisms and Cell Cultures, Braunschweig, Germany

P. Buzzini et al. (eds.), *Yeasts in Natural Ecosystems: Diversity*,
DOI 10.1007/978-3-319-62683-3_15

from natural ecosystems and are preserved in collections worldwide, curators of public and private fungal/yeast culture collections were contacted to participate in this survey. Curators of 41 collections from 27 countries supplied data representing a total of 58,095 strains. This includes information on the collection itself, type of environment the strains were isolated from, the countries of origin of the strains, and also the taxonomic information. The ecosystems that are well represented according to the data of preserved strains in the participating collections are plants, insects/invertebrates, aquatic habitats, soil, and extreme cold and extreme warm/dry habitats. The strains have been isolated from a large number of countries worldwide (countries of origin), but it is clear that many parts of the world's ecosystems are not yet well sampled for yeast diversity. A challenge during this survey was to list the genera and species due to the current and constant changes in taxonomic names. The outcome of this survey is discussed in this chapter.

Keywords Collections • Environment • Origin • Preserved specimens • Taxonomy • Worldwide

Contents

15.1 Introduction

There are many well-known public yeast repositories worldwide. Many of these collections are listed at the World Data Centre for Microorganisms (WDCM; http://www.wdcm.org/), which is compiled and validated by the World Federation for Culture Collections (WFCC; http://www.wfcc.info). Thirty-seven of the largest collections are listed in Boundy-Mills et al. (2016), most of these with the primary goal to preserve the yeast biodiversity in a specific region. In addition to these collections, there are also a large number of smaller, less-known collections that also preserve strains from a range of species that are important in different areas such as environmental strains, food spoilage or fermentation, and human or animal pathogens. Many of the public collections have their own searchable online catalog.

The Global Catalogue of Microorganisms (GCM; Wu et al. 2013) is a searchable catalog in which the strain catalogs of more than 100 fungal/yeast/bacterial collections are merged. Different fields can be searched, including the taxonomic name, country of origin, and the isolation source. Unfortunately, GCM is not updated frequently with the current data of the collections.

Natural ecosystems of yeasts vary from aquatic to terrestrial habitats, polluted environments, invertebrates and vertebrates, and their surroundings as well as extreme cold to extreme warm and dry living conditions. Since the 1790s strains have been isolated from natural habitats and described (Kurtzman et al. 2015). Some of those strains are preserved in world-renowned public collections, but it is clear that a large part of the earth's biomes has not yet been sampled for yeasts (Kurtzman et al. 2015), and, unfortunately, many of the thousands of collected strains have not been preserved in culture collections for future generations to use.

In order to have an overview on how many yeast strains, isolated from natural ecosystems, are preserved in collections worldwide, curators of public and private fungal/yeast culture collections were contacted to participate and supply the necessary information on yeast strains isolated from natural habitats in their collections. Curators of 41 collections from 27 countries supplied data representing a total of 58,095 strains. This includes information on the collection itself, type of environment the strains were isolated from (e.g., extreme cold and extreme warm environments, aquatic habitats, plants, soil, forests, grassland, etc.), the countries of origin of the strains, and also the taxonomic information linked to the strains. The outcome of this survey is discussed in this chapter.

15.2 Global Effort to Preserve Yeast Strains for Future Generations

Curators of more than 80 collections from different parts of the world were asked to supply data for this survey. While just over half of the curators responded, some of the collections contacted do not preserve yeast strains or only have a limited number of strains, whereas others mainly preserve industrial yeasts or human and animal pathogenic species, which were not of interest for this survey. Other curators did not provide data due to the lack of an appropriate database from which this type of data could be retrieved easily or the lack of manpower to retrieve the data.

Based on survey responses from 41 collections in 27 countries, the information of 58,095 strains obtained from natural ecosystems was made available (Table 15.1). It is important to take into account that one strain can be preserved in two or more collections, as it was logistically not possible to eliminate this factor. From the 41 collections, most collections provided data that could be used in all further analyses (Tables 15.2 and 15.4). Table 15.1 provides an overview of the participating collections with full collection names and what type of data was used in the analyses of the environmental data (Table 15.2), the data from the countries

Table 15.1 Details of collections that contributed data to this survey

Region	Country	Full name of collection	Acronym	# of strains	Data included		
					Table 15.2	Table 15.3	Table 15.4
Africa (882)	South Africa	UFS Yeast Culture Collection	UFS	882	X	X	X
Asia (7588)	China	China General Microbiological Culture Collection Center	CGMCC	1343	X	X	
	Japan	Japan Collection of Microorganisms	JCM	1572	X	X	X
	Thailand	BIOTEC Culture Collection	BCC	3951	X	X	X
	Thailand	Thailand Bioresource Research Center	TBRC	516	X	X	X
	Thailand	TISTR Culture Collection, Bangkok MIRCEN	TISTR	206	X	X	X
Australia and New Zealand (326)	Australia	Plant Pathology Herbarium	BRIP	99	X	X	
	New Zealand	International Collection of Microorganisms from Plants	ICMP	227	X	X	X
Europe (15,055)	Belgium	MUCL Environmental and Applied Mycology	BCCM/ MUCL	1077	X	X	
	France	Centre International de Ressources Microbiennes-Levures	CIRM	354	X	X	
	Germany	Leibniz-Institut DSMZ-Deutsche Sammlung von Mikroorganismen und Zellkulturen GmbH	DSMZ	278	X	X	
	Greece	University of Athens/Hellenic Collection of Pathogenic Fungi	UOA/HCPF	92	X	X	
	Hungary	National Collection of Agricultural and Industrial Microorganisms	NCAIM	336	X	X	X
	Italy	Industrial Yeasts Collection DBVPG	DBVPG	3010	X	X	X
	Netherlands	CBS Collection, Westerdijk Fungal Biodiversity Institute	CBS	4074	X	X	X
	Portugal	Portuguese Yeast Culture Collection	PYCC	2085	X	X	
	Slovakia	Culture Collection of Yeasts	CCY	265	X	X	X
	Slovenia	Microbial Culture Collection Ex	EX	3003	X	X	X
	Slovenia	ZIM Collection of Industrial Microorganisms	ZIM	139	X	X	X
	Spain	Coleccion Espanola de Cultivos Tipo	CECT	141	X	X	
	Switzerland	Culture Collection of Switzerland	CCOS	47	X	X	
	United Kingdom	National Collection of Yeast Cultures	NCYC	154	X	X	

North America (17,809)	Canada	Department of Plant Sciences, University of Western Ontario	UWOPS	4473	X	X	X
	USA	USDA Agricultural Research Service Culture Collection	NRRL	7463	X	X	X
	USA	Culture Collection of Meredith Blackwell	Blackwell	2040	X	X	X
	USA	Phaff Yeast Culture Collection	UCDFST	3833	X	X	X
South America (11,533)	Argentina	Culture Collection of Diego Libkind	Libkind	960		X	X
	Brazil	Coleção de Culturas Microbianas Carlos Rosa	CCR	2268	X	X	X
	Brazil	Culture Collection of Microorganisms and Cells	UFMG	4492	X	X	X
	Brazil	Laboratory of Molecular Diversity of Federal University of Alagoas	UFAL	257	X	X	X
	Brazil	UNESP Microbial Resources Center	CRM-UNESP	374	X	X	X
	Brazil	Universidade Federal de Pernambuco	URM	55	X	X	X
	Brazil	Universidade Federal do Rio Grande do Sul Yeast Culture Collection	UFRGS	103	X	X	X
	Colombia	Yeast Culture Collection of Universidad del Valle	UNIVALLE	15	X	X	X
	Ecuador	Colección de Levaduras Quito Católica	CLQCA	3009	X	X	X
Former USSR (4702)	Belarus	Belarus Collection of Microorganisms	BIM	240	X	X	
	Russia	Yeast collection of Lomonosov Moscow State University	KBP	1374	X	X	
	Russia	Collection of the Institute of Viticulture and Oenology	Magarach	886	X	X	
	Russia	All-Russian Collection of Microorganisms	VKM	1089	X	X	
	Russia	Russian National Collection of Industrial Microorganisms	VKPM	482	X	X	
	Ukraine	Ukrainian Collection of Microorganisms	UCM	831	X	X	

The total number of strains per region is indicated with the region name of origin (Table 15.3), and the taxonomic information on the strains from the different environments (Table 15.4).

The highest numbers of strains obtained from natural ecosystems are preserved in collections in North America (17,809), Europe (15,055), and South America (11,533). The collections that have preserved large numbers of natural isolates are NRRL (7463), UFMG (4492), UWOPS (4473), CBS (4074), BCC (3951), UCDFST (3833), DBVPG (3010), CLQCA (3009), and EX (3003). The majority of the curators provided a short description of their collections, including some history, interesting facts, and core activities.

15.2.1 African Collection

Most strains in the UFS collection (South Africa) were isolated by J.P. van der Walt and co-workers during 1953–1987. The collection contains almost 3000 strains of diverse taxonomic affinities, of which many were collected from fynbos (shrubland) ecosystems. It constitutes the largest collection of its kind in Africa. In 1996, the UNESCO awarded a MIRCEN status to the department of the collection to develop and support microbiology and microbial biotechnology in Southern African countries, thereby making industrial biotechnology the primary focus of UFS.

15.2.2 Asian Collections

CGMCC (China) was founded in 1979 as the central collection in the network of various collections in China. The main focus is to preserve, supply, and maintain living microbial resources and contribute to scientific communities. The collection maintains approximately 40,000 strains (as of late 2013), of archaea, bacteria, and fungi including yeasts. One of the aims is to improve and expand the collection continuously by exploiting new microbial resources from various natural environments.

JCM (Japan) is the biggest culture collection in Japan. It was established in 1981. Since 2004 this collection has participated in the RIKEN BioResource Center (BRC). Its mission is to contribute to scientific communities by maintaining and serving high-quality bacterial, archaeal, and fungal resources for various research fields, particularly in health and environmental sciences. It maintains more than 26,000 microbial strains of which 16,000 are publically available that include 2600 yeast strains.

BCC (Thailand) was founded in 1996. Its primary objective is to collect and maintain microorganisms and their relevant data for in-house research at the National Center for Genetic Engineering and Biotechnology (BIOTEC). Microorganisms are tested to find valuable products such as secondary metabolites, enzymes, and bioactive short peptides. The collection maintains more than

80,000 strains of which approximately 75% are taxonomically and ecologically diverse filamentous fungi/yeasts. The largest group is those isolated from insects.

TBRC (Thailand) was established in 2015 and operates under the jurisdiction of BIOTEC, the National Science and Technology Development Agency (NSTDA), and under the Ministry of Science and Technology (MOST). It preserves, provides, and facilitates the coordination of exchange of biological information and resources and develops mechanisms enabled by information technologies. The TBRC Network represents the commitment to generate economic value and enhance the competitiveness to the ASEAN Economic Community (AEC) from its rich biological resources. Approximately 2500 strains of bacteria, yeasts, and fungi are preserved in this collection.

TISTR (Thailand) was created with UNESCO/UNEP's support since 1976 to serve as an MRC (Microbial Resource Centre) for Southeast Asia with the aim of establishing a holding center for agriculturally and industrially useful microbial strains. It has been devoted to collection, preservation, and distribution of microorganisms and consists of more than 5000 strains of bacteria, yeasts, fungi, and microalgae, covering a wide range of known or potential applications for biotechnology development, agriculture, environment, and industry.

15.2.3 European Collections

DSMZ (Germany) originated in 1968 when the microbial service collection was launched at the Institute for Microbiology in Göttingen. The collection, under the name "DSM—Deutsche Sammlung von Mikroorganismen," became independent in 1976. In 1987 the collections of actinomycetes and Gram-positive bacteria were transferred to DSM from Darmstadt and Munich, respectively. The collection became an independent institution in the end of 1988. It is now one of the 89 institutes of the Leibniz Association. The collection holds presently 700 yeast strains (Boundy-Mills et al. 2016), which were accumulated from German universities and institutions. DSMZ also maintains patent and type strains.

The UOA/HCPF (Greece) was founded in 1995. It is the only registered national culture collection for yeast/fungal human and animal pathogens, their genetic material, and metabolites. The collection currently holds 3225 strains of yeasts and 1765 strains of filamentous fungi. Services include the distribution of strains; isolation and identification of strains from clinical specimens and from substrates related to public health, safety and welfare, consulting service, and screening; and characterization of fungal strains for useful properties.

The predecessor of NCAIM (Hungary), the Hungarian Microbiological Gene Bank (HMGB), was established in 1974. In 1985 NCAIM became a centralized culture collection. The mission of this collection includes (among others) participating in ex situ conservation of microorganisms and supplying microorganisms to interested parties. The main research activity of NCAIM is directed to the investigation of biodiversity and systematics of yeasts. Currently it maintains 1760 yeast, 1390 bacteria, and 330 filamentous fungal strains.

In 1912 DBVPG (Italy), or at that time "Collezione dei Lieviti Vinari dell'Istituto di Microbiologia Agraria e Tecnica" (acronym IMAT), was founded. The main objectives are the acquisition, classification, preservation, and distribution of strains. Research focuses on taxonomy, biodiversity, and biotechnology of yeasts. It maintains about 6000 yeast and yeastlike strains associated with the alcohol fermentation industry and strains isolated from different environmental and food-associated habitats including many psychrophilic and psychrotolerant strains.

CBS (the Netherlands) is a collection of the Westerdijk Fungal Biodiversity Institute, formerly known as the CBS-KNAW Fungal Biodiversity Centre, that maintains a collection of filamentous fungi, yeasts, and bacteria and is the biggest collection of its kind. The institute's principal focus is on the taxonomy and evolution of fungi as well as on functional aspects of fungal biology and ecology. The collection maintains approx. 100,000 strains, of which more than 10,000 are yeasts, representing a large percentage of the species cultured to date. It also maintains patent and type strains. In diversity of species, it is unchallenged as a reference center for mycological research. CBS has several active research groups and gives advice on mycological problems of a scientific, health-related, or industrial nature.

PYCC (Portugal) was founded 1952; it is a service and research collection that provides support to R&D activities in Portugal and abroad. The collection serves as a repository of yeast biodiversity and genetic resources, with emphasis on Mediterranean foods, beverages, and natural habitat. It holds approximately 4000 yeast strains that represent most of the known diversity of this group of microorganisms. About 1500 strains are unique in this collection and were obtained in ecological studies carried out by researchers of the laboratories that housed the collection.

The CCY (Slovakia) is the largest general yeast collection in Slovakia. It holds 3600 strains of yeasts and yeastlike organisms including biotechnologically important strains, strains isolated from different contaminated sources, from various ecological niches, strains with specific characteristics, mutant strains, and type and patent strains. The collection is expanding with industrially and scientifically interesting strains. The biotechnology, food, and pharmaceutical industries have had a long and profitable association with the yeasts in the collection.

EX (Slovenia) was established in 1998. It is one of the few collections in the world that are specialized in the isolation and preservation of extremophilic microorganisms, including hypersaline environments such as different salterns. Extremophilic fungi represent the majority of strains, although extremophilic bacteria and archaea are also preserved by EX.

The ZIM (Slovenia) was systematically organized in 1991, although activities started in 1947. It maintains 2470 yeast strains, 259 filamentous and wood-decay fungi, and 932 bacteria, all isolated mostly from various ecological niches in Slovenia. Objectives are preservation of strains as genetic potential for biotechnological and food applications, taxonomic studies, and application of identification techniques for the detection of prevalence, abundance, and origin of individual groups of microorganisms.

CECT (Spain) is the only public microbial Biological Resource Centre (mBRC) in Spain. It was started by J. R. Villanueva as a research collection in 1960. It serves as a repository and provider of bacteria, archaea, yeast, and filamentous fungi and has grown steadily and currently registers over 8490 strains. It aspires to act as an interface connecting Spain with worldwide efforts in this field and works together with other collections to help boost European competitiveness in biotechnology.

The CCOS (Switzerland) is the Swiss national public culture collection for bacteria, yeasts and fungi, cell lines, and plasmids. It contains over 1000 strains, and its mission is to preserve the microbial diversity and to enhance the value and quality of biological material for science and industry. CCOS was initiated in a Swiss CTI project and was founded as a corporation in 2010.

NCYC (United Kingdom) originated as a collection of British brewing yeasts with the aim to safeguard brewery microbial sources. It became a National Collection in 1948 and continued to broaden as numerous yeast strains associated with food production and food spoilage and from the natural environment became of great interest. The collection now contains more than 4000 strains including brewing yeasts, yeasts associated with food spoilage, yeasts of medical and industrial importance, as well as strains from natural environments.

15.2.4 New Zealand Collection

ICMP had its origin in 1952 as the personal collection of plant pathogenic bacteria and rhizobia of D. W. Dye. In 1992, the collection was transferred to Landcare Research, one of eight government-owned "Crown Research Institutes." The collection focuses on globally sourced plant pathogens and fungi of New Zealand and the South Pacific. The over 20,000 strains are roughly evenly split between bacteria and fungi/yeasts, including 900 type and pathotype strains.

15.2.5 North American Collections

UCDFST (USA) is the fourth largest collection of its kind in the world (Boundy-Mills et al. 2016), containing over 7500 strains. This collection contains 3695 strains of yeasts belonging to more than 550 species which were originally isolated from natural ecosystems including plants, insects, animals, air, soil, and water. This collection was compiled primarily by eminent yeast ecologist and taxonomist H. J. Phaff (1913–2001). The collection contains over 700 strains of *Saccharomyces cerevisiae* from environmental sources and over 400 strains of the cactus-associated yeast *Pichia cactophila*.

NRRL (USA) opened in 1940 as the Northern Regional Research Laboratory. The goal of the laboratory was to utilize agricultural commodities. One component was to develop fermentation processes for crop utilization, and because of this, the culture collection was established. The collection specializes in yeasts, filamentous fungi, bacteria, and actinomycetes. Current work includes genome sequencing of yeasts, filamentous fungi, and bacteria, especially those of agricultural and biotechnological importance.

The most important collection of yeasts isolated from insects is the Blackwell collection (USA). This collection is the outcome of the "beetle belly yeast project" under the supervision of S.-O. Suh during the years 1996–2007. Methods for dissecting the beetles and isolating the yeasts from their gut were developed, and during the first years, 650 strains of 290 different genotypes were isolated from insects or their habitats. Among these were 200 undescribed taxa. In 2014, approximately 2000 strains were sent to CBS for preservation.

A. Lachance began the UWOPS collection (Canada) in 1979 with *Kluyveromyces* type strains received from the Phaff collection and strains isolated in his yeast course. Additions to the collection were from local cherry black knots and oak exudates as well as strains from different collection trips to the Caribbean, Arizona and Florida (1985), Hawaii (1987), and Mexico (1991, 1994). An enormous collection from floricolous insects of Eastern Australia was added in 1995. In the late 1990s, yeasts from the South Pacific islands, Costa Rica, and Hawaii were added.

15.2.6 South American Collections

The Libkind (Argentina) yeast culture collection holds more than 3500 yeast strains from Patagonian natural environments, with special focus on extremophilic and polyextremophilic yeasts. The objectives of the collection are to preserve natural microbiota of Andean Patagonia pristine environments and the bioprospecting of biotechnological relevant traits. Although not publically available, this collection is still a very important collection of interesting yeasts isolated from natural environments in Argentina.

The CCR collection (Brazil) is a work collection linked to the Laboratório de Microbiologia Ambiental e Biotecnologia and Laboratório de Microbiologia Aplicada of the Universidade Federal do Tocantins. It was created in 2010 to provide a center for the deposit of strains from substrates in the wide area of Brazilian northern Cerrado and Amazonia. It has around 5000 yeast strains and 3000 filamentous fungal strains and also about 900 bacterial strains.

The CRM-UNESP collection (Brazil) started in 2013 to preserve microbial genetic resources isolated by researchers from UNESP. The majority of the microorganisms are filamentous fungi and yeasts from diverse sources such as social insects (mostly attine ants), terrestrial and marine Antarctic samples, and marine invertebrates from the Brazilian coast.

UFAL (Brazil) was founded in 2013. The first strains were obtained from water tanks of bromeliads of the Atlantic Rain Forest in Northeast Brazil. Other substrates include flowers, fruits, leaves of plants of the Atlantic Rain Forest and Caatinga (a type of dry scrubland vegetation), and marine invertebrates. The collection contains approximately 550 strains. The main objectives are to isolate the yeast communities of different Brazilian biomes, to describe possible new species, and to bioprospect these for biotechnological applications.

UFMG (Brazil) started in 2007 to preserve the yeasts isolated from natural habitats in Brazil. It has approximately 45,000 yeast strains, including the research collection, that make it the largest collection in Brazil and is the most representative collection of the existing tropical Brazilian biodiversity. The main goals are to know the yeast biodiversity in tropical habitats and Antarctica and to describe new yeast species together with their biotechnological exploitation. A total of 5985 yeast strains, with identities confirmed by sequencing, have been deposited in this collection.

UFRGS (Brazil) was founded in 2003, and its core purpose is to preserve yeast strains isolated from South Brazilian substrates. It has 103 strains in total, mostly isolated from the phylloplane, flowers and fruits of bromeliads, macrophytes associated with marshlands, and fermented products such as wine and cheese. Besides identification, all strains are tested for biotechnological potential (enzyme production, ester production, and oleaginous nature).

UNIVALLE (Colombia) began its activities in 2008, and its core purpose is to conserve yeast strains from different Colombian substrates, such as lakes, fruits, and flowers from plants and fermented beverages. Besides identification of novel strains and species, all strains were tested for biotechnological potential (enzyme production and fermentation).

Since 2006, CLQCA (Ecuador) is the repository of wild yeasts isolated on substrates located all over the 24 provinces in Ecuador, including the Galápagos Islands. It holds more than 3000 strains. It is not commercial but shares strains with other collections on the basis of research agreements. The curator, J. Carvajal Barriga, was the founder of the collection, and his main interests include biorefineries, microbial archaeology, and the production of beer and other fermented beverages.

15.2.7 Collections of the Former USSR

BIM (Belarus) is the central microbial service collection in Belarus. It holds type and reference material as well as strains potentially useful for industrial applications. The collection operates as a department of the Institute of Microbiology of the National Academy of Sciences of Belarus. The department was founded in 1975. The yeast collection holds more than 240 strains from diverse taxonomic groups, most of which are species of industrial interest.

KBP (Russia) was founded in the 1950s, and the majority of yeast strains were isolated from soil and other natural habitats by I. P. Babjeva, I. S. Reshetova, I. Y. Chernov, and co-workers. The collection is operated as the research collection that preserves the biodiversity of yeasts isolated during research projects of the Department of Soil Biology on the territory of mainly the USSR. KBP joined the project called "Noah's Ark" to consolidate available university research collections into a single depository.

Magarach (Russia) is focusing on yeasts for wine fermentation and was founded in 1893. Since 1907, yeast strains as well as the spoilage microorganisms were preserved and propagated in the laboratory for teaching and industrial applications in the Magarach wine school. This collection focused on industrial wine yeasts and spoilage agents. Yeast strains were mainly collected in the territory of the USSR.

The central service collection of the Soviet Union, VKM (Russia), was founded in 1958 (Golubev 2008). In 1964 they consolidated the material from the 30 culture collections then present in the former USSR. The yeast collection started from the strains obtained by V. I. Kudryavtsev in the 1930s across the USSR. The VKM was expanded mainly with the strains recovered by microbiologists from the republics of the Soviet Union and via exchange of the material with other collections from various countries. Presently, the VKM is the largest microbial collection in Russia.

VKPM (Russia) is a state service collection that specializes in deposition, identification, certification, storing, and distribution of strains, which are used for research purposes and serve the needs of biotechnological industry. The collection was founded in 1969 as a state repository of industrial microorganisms. Since 1991, the collection is named Russian National (also as All-Russian) Collection of Industrial Microorganisms. The collection received the BRC status in 2014.

UCM (Ukraine) started in 1929 as the collection of reference and research strains of the Institute of Microbiology and Epidemiology of the Academy of Sciences of the Ukrainian SSR (originally, All-Ukrainian Academy of Sciences). At present, the collection holds bacterial, mollicute, fungal, and yeast strains isolated from various sources. The Institute of Microbiology and Virology, which hosts the UCM, established the yeast collection, which mainly includes strains isolated in Ukraine.

15.3 Range of Natural Habitats of Preserved Strains

Table 15.2 lists the natural ecosystems from which 57,135 yeast strains were isolated. The ecosystems that are affiliated with yeasts, according to the data of preserved strains in the participating collections, are plants (31.7%), insects/invertebrates (20.1%), aquatic habitats (12.6%), and soil (9.5%). Extreme warm/dry

Table 15.2 Overview of the representation of different ecological niches in the contributing collections

Primary category	% of strains in category[a]	Subcategory	% of strains in subcategory[b]
[c]Vertebrates	2.2	Animals (no additional information)	89.3
		Bovine/equine/swine/poultry	4.1
		Reptiles/turtles	7.0
[c]Atmosphere	1.0	Outside/inside buildings	100.0
Aquatic habitats	12.6	Water (no additional information)	35.9
		Water from lakes, rivers, springs, and ponds	21.0
		Water and soil from sea	39.3
		Soil from rivers, swamps, and mangroves	2.8
		Soil from hot springs	1.2
Extreme cold habitats	5.4	Antarctic regions (no additional information)	13.6
		(Melted) ice, sediment, snow, or water	22.0
		Soil, mountains, rocks, or glaciers	61.7
		Seawater (Arctic)	2.5
Extreme warm and dry habitats	6.1	Desert plants, mostly cacti	87.7
		Desert soil or insects	11.8
		Soil from volcanos	0.1
Forests and woodlands	3.2	Forests (no additional information)	52.1
		Fruit or flowers from plants	22.4
		Forests near lakes, rivers or sea	8.3
		Soil from (rain) forests or woodlands	17.4
Fungi and lichens	2.0	Mushrooms/fungi/lichens	100
Insects and invertebrates	20.1	Insects (no additional information)	24.7
		Ants or termites	3.1
		Bees (including honey)	15.0
		Beetles	36.4
		Flies, moths, caterpillars	2.0
		Insects from plants in forests	16.5
		Insects living in the vicinity of water (lakes, rivers, ponds, meadows, and sea)	0.3
		Insects living in mountains	0.1
		Marine invertebrates	1.6
		Arthropods and isopods in vicinity of water	0.1
Plants	31.7	Plants (no additional information)	48.2
		Flowers (including wild and grass)	9.4
		Leaves or stems	2.7
		Fruit (mostly wild)	10.4
		Plant or tree exudates	1.8
		Rotting and decaying plants or wood	16.0
		Trees (including frass)	4.4
		Plant phytotelma	4.7
		Agricultural plants	2.5

(continued)

Table 15.2 (continued)

Primary category	% of strains in category[a]	Subcategory	% of strains in subcategory[b]
Soil	9.5	Soil (no additional information)	71.7
		Soil from hard ground such as caves, caverns, mountains, and rocks	15.1
		Soil from grasslands or meadows	2.1
		Sandy soil	0.8
		Agricultural soil	0.8
		Soil surrounding fynbos	0.7
		Acid boreal (podzolic) soil	0.3
		Uncultured soil	8.5
		Soil (no additional information)	71.7
[c]Traditional beverages	6.2	Fermented	100.0

[a]% of the 57,135 strains for which data was available for this table
[b]% the specific subcategory makes out from the primary category
[c]Categories not included in data of all collections

(6.1%) and extreme cold (5.4%) habitats were also included in the list. There are many important subcategories within the main ecological groups that are also well associated with yeasts: rotting and decaying plants or wood (16.0%); beetles (36.4%); bees (15.0%); soil in extreme cold habitats including glaciers, rocks, and mountains (61.7%); water and soil from the sea (39.3%); freshwater such as lakes and rivers (21%); and plants from extreme warm habitats, mostly cacti (87.7%). The percentages for the specific subcategories are made out from the primary category. Many of the strains from the EX collection (approx. 1500) were obtained from extreme cold environments and make up almost half of the strains from this environmental niche listed in Table 15.2. Almost 20% of the insect-related strains are from the Blackwell collection, and approximately half of the 3833 strains from the UCDFST collection were obtained from extreme warm and dry environments.

15.4 Countries of Origin

It is clear that many parts of the world's ecosystems are not yet well sampled for yeast diversity (Kurtzman et al. 2015; data from online collection catalog searches). Table 15.3 gives an overview of the countries of origin of 54,419 of the strains preserved in the participating collections. The remaining strains are from unknown origin. Strains from this survey were isolated from more than 200 countries or regions. Most strains were collected in the USA and Brazil (>7000 strains each), followed by Ecuador, Slovenia, and Thailand with 1000–2000 strains each and with

Table 15.3 Overview of the number of strains per country (left side of table) and number of strains per region (right side of table) present in the contributing collections

# of strains/ country listed	Countries	Regions or states	# of strains/ region listed
<30	Algeria; Azerbaijan; Bhutan; Bolivia ; Botswana; Brunei; Burkina Faso; Burma; Cambodia; Cameroon; Colombia; Congo; Dominica; Egypt; Estonia; Fiji; Guatemala; Guyana; Honduras; Ireland; Ivory coast; Kazakhstan; Kenya; Korea; Laos; Latvia; Liberia; Luxemburg; Malta; Mauritius; Monaco; Mongolia; Morocco; Mozambique; Nepal; Netherlands Antilles; Nigeria; Pakistan; Peru; Poland; Qatar; Romania; Saudi Arabia; Singapore; Somalia; South Africa; Marion island; South Korea; Suriname; Swaziland; Tajikistan; Trinidad and Tobago; Tunisia; Uganda; United Arab Emirates; Uruguay; Vanuatu; West Indies; Yugoslavia; Zimbabwe		
30–50	Antigua; Greenland; Indochina; Kyrgyzstan; Madagascar; Serbia; Sri Lanka; Turkey; Uzbekistan		
	Czech Republic	Bohemia	<30
	Georgia	Abkhazia	<30
	Scotland	Shetland Islands	<30
	Tanzania	Serengeti	<30
50–100	Bulgaria; Ethiopia; Haiti; India; Lesotho; Moldova; Montenegro; Turkmenistan; Venezuela; Belarus		
100–150	Armenia; Austria; Burundi; Croatia; Denmark; Dominican Republic; Greece; Iceland; Switzerland; Ukraine		
	Indonesia	Java	<30
150–200	Belgium; Belize; Iran; Israel; Jamaica; Malaysia; Taiwan		
200–300	Chile; Finland		
	France	Reunion	<30
300–400	Cuba; French Guiana; Hungary; Papua New Guinea; Slovakia; Sweden		
	France	Reunion	<30

(continued)

Table 15.3 (continued)

# of strains/ country listed	Countries	Regions or states	# of strains/ region listed
400–500	Bahamas; Costa Rica; Vietnam		
	New Zealand	Cook Islands	<30
	United Kingdom (UK)	Falkland islands; Virgin Islands, Wales	<30
		British Indian Ocean	50–100
		Cayman Islands	100–150
500–1000	Antarctica; Australia; Germany; Mexico; Norway; Spain; Ukraine		
	Netherlands	Saint Martin	<30
	Portugal	Madeira Islands	<30
1000–2000	Argentina; Canada; China; Japan; Panama		
	Italy	Sardinia	<30
	Russia	Crimea	100–150
	South Africa	Marion Island	<30
2000–3000	Ecuador; Slovenia; Thailand		
>7000	Brazil		
	United States of America (USA)	Wisconsin (WI); Illinois (IL); Maine (ME); Maryland (MD); Minnesota (MN); New Mexico (NM); New York (NY); Pennsylvania (PA); South Carolina (SC); Texas (TX); Utah (UT); Virgin Islands; Virginia (VA); Wyoming (WY)	<30
		Georgia (GA); Puerto Rico	30–50
		California (CA); Navassa Island; North Carolina (NC); Tennessee (TN)	50–100
		Florida (FL)	150–200
		Arizona (AZ)	400–500
		Louisiana (LA)	500–1000
		Hawaii (HI)	1000–2000

2000–3000 strains each for Argentina, Canada, China, Italy, Japan, Panama, Russia, and South Africa. The general trend is that many collections from a country/region preserve mostly strains from that region (Ecuador (CLQCA), Thailand (BCC, TBRC, TISTR), Slovakia (CCY), Slovenia (ZIM, EX), Ukraine (UCM), China (CGMCC) and most of the former USSR collections). The USA collections hold more than half of the strains obtained from the USA (Table 15.3). Many of the Argentinian strains are from the Libkind collection, and most of the

Brazilian strains are only preserved in the six Brazilian collections, with only a limited number kept in collections in other countries.

On the contrary, most European collections (e.g., CBS, BCCM/MUCL, CIRM, DSMZ, NCAIM, DBVPG, CECT) have a more evenly spread geographic distribution of strains; this is also true for the VKM (Russia) and JCM (Japan) collections. The NRRL (USA), UCDFST (USA), UWOPS (Canada), and UFS (South Africa) preserve many strains from their own country as well as from other countries.

15.5 Genera and Species Represented by Strains from Natural Environments

A challenge during this survey was to list the genera and species and estimate the number of strains linked to them in the participating collections. Many name changes occurred in the past few years, especially for the basidiomycetous yeasts (Wang et al. 2015a, b; Liu et al. 2015). Many collections have not yet updated the taxonomic affiliation of their strains, and it is possible that strains with only genus names associated were only based on preliminary identifications. Without a species epithet, it was not possible to know in which correct genus it belongs. This may have an effect on the number of strains in specific genera such as *Cryptococcus*, *Rhodotorula*, *Pichia*, and *Candida*. It is also good to keep in mind that different methods were used to identify the strains in the collections. Not all strains are yet identified using rDNA sequencing as some are only identified on physiological characteristics. This may also have an effect on the data presented here, especially at the species level.

According to the data obtained, most of the strains found in nature and preserved in the participating culture collections are affiliated with ascomycetous genera (Table 15.4). Well-defined genera that are abundantly found are *Clavispora*, *Cyberlindnera*, *Debaryomyces*, *Hanseniaspora*, *Lipomyces*, *Meyerozyma*, *Saccharomyces*, and *Sporopachydermia* with 500–1000 strains each and *Metschnikowia* within the 1000–2000 strain range. Most of the species are well distributed among the different collections except for *Clavispora*, *Metschnikowia*, and *Sporopachydermia* strains which are abundantly found in the UWOPS collection; the *Aureobasidium* strains are mostly from the EX collections, whereas *Lipomyces* strains are found in large numbers in the UFS collection. Almost all strains of the species *Saccharomyces eubayanus*, one of the most common species found of *Saccharomyces* during this survey, are preserved in the Libkind collection. It is possible that many of the strains currently identified as *Cryptococcus* sp., *Rhodotorula* sp., *Pichia* sp., and *Candida* sp. from the different collections may be changed to other or newly described genera when species names are updated in conformity with recent taxonomic revisions. Some but not all collections have made these updates, so it is unfortunately not possible to take that into account for this survey. Interestingly, many of the human pathogenic *Candida* species,

Table 15.4 Overview of the number of strains per genus and the most abundant species present in the contributing collections

# of strains per genus[a]	Genera	Most abundant species[b]
10–30	*Bannozyma*; *Bensingtonia*; *Botryozyma*; *Brettanomyces*; *Bulleribasidium*; *Bulleromyces*; *Carlosrosaea*; *Citeromyces*; *Colacogloea*; *Curvibasidium*; *Danielozyma*; *Derxomyces*; *Diddensiella*; *Dipodascopsis*; *Effuseotrichosporon*; [c]*Endomyces*; [c]*Farysia*; *Fellomyces*; *Fibulobasidium*; *Glaciozyma*; *Groenewaldozyma*; *Holtermanniella*; *Kockovaella*; [c]*Kondoa*; *Martiniozyma*; *Meira*; *Middelhovenomyces*; *Moniliella*; *Mrakia*; *Nadsonia*; *Nakaseomyces*; *Naumovozyma*; *Occultifur*; *Saccharomycodes*; *Sakaguchia*; [c]*Saprochaete*; *Schizoblastosporion*; *Sterigmatomyces*; *Symmetrospora*; *Tausonia*; *Teunomyces*; *Tilletiopsis*; *Trigonopsis*; [c]*Ustilago*; *Vanderwaltozyma*; *Zygoascus*	
30–50	*Cutaneotrichosporon*; *Cystobasidium*; *Dekkera*; *Diutina*; *Eremothecium*; *Kloeckera*; *Lodderomyces*; *Millerozyma*; *Moesziomyces*; *Peterozyma*; *Priceomyces*; *Rhodosporidium*; *Solicoccozyma*; *Spathaspora*; *Spencermartinsiella*; *Tetrapisispora*; *Trichomonascus*; *Zygotorulaspora*	
50–100	*Ambrosiozyma*; *Bullera*; *Cystobasidium*; *Cystofilobasidium*; *Dipodascus*; *Filobasidium*; *Issatchenkia*; *Leucosporidiella*; [c]*Magnusiomyces*; *Naematelia*; *Phaffomyces*; *Sporidiobolus*; *Suhomyces*; *Tortispora*; [c]*Tremella*; *Udeniomyces*	
	Goffeauzyma	*Goffeauzyma agrionensis*
	Kregervanrija	*Kregervanrija fluxuum*
	Phaffia	*Phaffia rhodozyma*
	Yarrowia	*Yarrowia lipolytica*
100–150	*Apiotrichum*; *Dioszegia*; *Geotrichum*; *Hannaella*; *Nakazawaea*; *Saccharomycopsis*; *Saturnispora*; *Sugiyamaella*	
	Hyphopichia	*Hyphopichia burtonii*
	Komagataella	*Komagataella pastoris*
	Kuraishia	*Kuraishia capsulata*; *Kuraishia molischiana*
	Kwoniella	*Kwoniella heveanensis*
	Leucosporidium	*Leucosporidium scottii*
	Naganishia	*Naganishia albida*
	Rhodosporidiobolus	*Rhodosporidiobolus ruineniae*
	Starmera	*Starmera caribaea*
	Vishniacozyma	*Vishniacozyma victoriae*

(continued)

Table 15.4 (continued)

# of strains per genus[a]	Genera	Most abundant species[b]
150–200	[c]*Aureobasidium*	*Aureobasidiummelanogenum; Aureobasidium pullulans*
	Blastobotrys	*Blastobotrys parvus*
	Galactomyces	*Galactomyces geotrichum*
	[c]*Hortaea*	*Hortaea werneckii*
	Kurtzmaniella	*Kurtzmaniella cleridarum*
	Myxozyma	*Myxozyma mucilagina*
	Starmerella	*Starmerella meliponinorum*
	Yamadazyma	*Yamadazyma mexicana*; *Yamadazyma scolyti*
200–300	*Barnettozyma*	*Barnettozyma californica*
	Pseudozyma	*Pseudozyma hubeiensis*
	Schwanniomyces	*Pseudozyma polymorphus* var.*africanus*; *Pseudozyma vanrijiae*
	Sporobolomyces	*Sporobolomyces ruberrimus*
	Wickerhamiella	*Wickerhamiella australiensis*; *Wickerhamiella occidentalis*
	Zygosaccharomyces	*Zygosaccharomyces bailii*; *Zygosaccharomyces rouxii*
300–400	*Kazachstania*	*Kazachstania exigua*; *Kazachstania pintolopesii*; *Kazachstania unispora*
	Ogataea	*Ogataea polymorpha*
	Saitozyma	*Saitozyma podzolica*
	Trichosporon	*Trichosporon asahii*
	Vanrija	*Vanrija humicola*
400–500	*Cryptococcus*	
	Kodamaea	*Kodamaea anthophila*; *Kodamaea ohmeri*
	Lachancea	*Lachancea fermentati*; *Lachancea kluyveri*; *Lachancea thermotolerans*
	Papiliotrema	*Papiliotrema flavescens*; *Papiliotrema laurentii*
	Scheffersomyces	*Scheffersomyces stipitis*
	Torulaspora	*Torulaspora delbrueckii*
	Wickerhamomyces	*Wickerhamomyces pijperi*
500–1000	*Clavispora*	*Clavispora lusitaniae*; *Clavispora opuntiae*
	Cyberlindnera	*Cyberlindnera fabianii*; *Cyberlindnera subsufficiens*
	Debaryomyces	*Debaryomyces hansenii*; *Debaryomyces nepalensis*

(continued)

Table 15.4 (continued)

# of strains per genus[a]	Genera	Most abundant species[b]
	Hanseniaspora	*Hanseniaspora guilliermondii*; *Hanseniaspora opuntiae*; *Hanseniaspora thailandica*; *Hanseniaspora uvarum*
	Kluyveromyces	*Kluyveromyces lactis*; *Kluyveromyces marxianus*
	Lipomyces	*Lipomyces starkeyi*; *Lipomyces tetrasporus*
	Meyerozyma	*Meyerozyma caribbica*; *Meyerozyma guilliermondii*
	Saccharomyces	*Saccharomyces cerevisiae*; *Saccharomyces eubayanus*; *Saccharomyces paradoxus*
	Sporopachydermia	*Sporopachydermia cereana* complex
1000–2000	*Metschnikowia*	*Metschnikowia borealis*; *Metschnikowia continentalis*; *Metschnikowia gruessii*; *Metschnikowia hawaiiensis*; *Metschnikowia ipomoeae*; *Metschnikowia koreensis*; *Metschnikowia lochheadii*; *Metschnikowia pulcherrima*; *Metschnikowia reukaufii*; *Metschnikowia santacecilia*
	Rhodotorula	*Rhodotorula mucilaginosa*; *Rhodotorula paludigena*
>3000	*Pichia*	*Pichia deserticola*; *Pichia cactophila*; *Pichia heedii*; *Pichia kluyveri*; *Pichia kudriavzevii*; *Pichia membranifaciens*; *Pichia norvegensis*
>6000	*Candida*	*Candida albicans*; *Candida apicola*; *Candida boidinii*; *Candida glabrata*; *Candida intermedia*; *Candida natalensis*; *Candida orthopsilosis*; *Candida parapsilosis*; *Candida parazyma*; *Candida sake*; *Candida sonorensis*; *Candida tropicalis*

[a]Different methods were used to identify strains

[b]Most abundant species were only indicated when the number of strains of a species was more than 50 and more than 100 for *Candida* and *Pichia* species

[c]Genera and species data not included in yeast data of all collections

namely, *Candida albicans*, *Candida glabrata, Candida parapsilosis*, and *Candida tropicalis*, are often isolated from natural ecosystems other than from humans. Two *Pichia* species that are commonly found in extreme warm environments are *P. cactophila* and *Pichia deserticola*, two of the most frequently found species of this genus during this survey.

There are some genera included in Table 15.4 that not all of the participating collections supplied data for, as many of the curators categorize them as dimorphic or filamentous fungi rather than yeast (e.g., *Aureobasidium, Endomyces, Farysia, Kondoa, Magnusiomyces, Saprochaete, Tremella, Ustilago*), so the actual numbers of strains preserved in the collections for these genera may be much higher than indicated in Table 15.4.

15.6 Concluding Remarks

Although the number of participating collections only represents a small portion of the total number of collections worldwide preserving yeast strains, the presented data give a good indication of what type of strains is preserved to allow innovation and discovery by future generations. It is clear that yeasts have a wide range of habitats, from the cold Antarctic regions to the hot and dry deserts. The close association of yeasts with insects and plants and survival of yeasts in soil, fresh, and salt waters that is documented through the scientific studies described in other chapters of this book is supported by this survey. Furthermore, some culture collection online catalogs as well as GCM list publications associated with specific yeast strains, which allows researchers to more easily find ecological information associated with specific yeast strains.

The strains included in this survey were obtained from countries all over the world. However, it is clear that many regions such as the tropical lowland rainforests in the Amazon, Indonesia, and Central Africa; Arabian, Chinese, and Australian deserts; the arctic forests and tundras in Eurasia; and montane and submontane ecosystems, such as those present in the Himalaya, the Rockies, and the Andes are still underrepresented (this survey, Kurtzman et al. 2015). These surveys show that the close collaboration between depositors of yeasts and the worldwide culture collections is making excellent progress in preserving yeast biodiversity for future generations. The numbers of species presented in this survey are unfortunately only a drop in the ocean when compared to the predicted number of yeast species globally (Lachance 2006; Kurtzman et al. 2015), either waiting to be isolated, or isolated by not yet properly named, or already isolated and properly named but not preserved for future studies. The lack of suitable databases to facilitate data queries for specific data was one of the biggest obstacles encountered in this survey. The need for effective and public databases, whether local or as part of international efforts such as GCM, will become more urgent with emerging globalization.

Acknowledgments We would like to thank the following curators of the participating collections for sharing the information on their yeast collections: J.C. Barriga (CLQCA), M. Blackwell (Blackwell), S. Casaregola (CRIM), L. Chatanon (TISTR), H.-M. Daniel (BCCM/MUCL), T.N. Golovach and O.D. Yaneva (UCM), V.I. Golubev (VKM), A.V. Kachalkin (KBP), C.P. Kurtzman (NRRL), A. Lachance (UWOPS), M. Landell (UFAL), S. Landert (CCOS), D. Libkind (Libkind), X. Liu (CGMCC), L. López-Ocaña (CECT), P.B. Morais (CCR), G. Novik (BIM), F.C. Pagnocca (CRM-UNESP), Y. Pei (BRIP), G. Péter (NCAIM), I. Roberts (NCYC), J.P. Sampaio and C. Carvalho (PYCC), T.N. Tanashcuk (Magarach), R. Vadkertiova (CCY), P. Valente (UFRGS), A. Velegraki (UOA/HCPF), M.M. Vustin (VKPM), B. Weir (ICMP), A. van Wyk (UFS), and P. Zalar and N. Cimerman-Gunde (EX).

References

Boundy-Mills KL, Glantschnig E, Roberts IN, Yurkov A, Casaregola S, Daniel H-M, Groenewald M, Turchetti B (2016) Yeast culture collections in the twenty-first century: new opportunities and challenges. Yeast 33:243–260

Golubev WI (2008) Taxonomical, ecological, and geographical diversity of yeast cultures in the all-Russia collection of microorganisms (VKM). Appl Biochem Microbiol 44:84–88

Kurtzman CP, Quintilla Mateo R, Kolecka A, Theelen B, Robert V, Boekhout T (2015) Advances in yeast systematics and phylogeny and their use as predictors of biotechnologically important metabolic pathways. FEMS Yeast Res 15:fov050

Lachance MA (2006) Yeast biodiversity: how many and how much? In: Rosa CA, Peter G (eds) Biodiversity and ecophysiology of yeasts. Springer, Berlin, pp 1–9

Liu X-Z, Wang Q-M, Göker M, Groenewald M, Kachalkin AV, Lumbsch HT, Millanes AM, Wedin M, Yurkov AM, Boekhout T, Bai F-Y (2015) Towards an integrated phylogenetic classification of the Tremellomycetes. Stud Mycol 81:85–147

Wang Q-M, Begerow D, Groenewald M, Liu X-Z, Theelen B, Bai F-Y, Boekhout T (2015a) Multigene phylogeny and taxonomic revision of yeasts and related fungi in the Ustilaginomycotina. Stud Mycol 81:55–83

Wang Q-M, Yurkov AM, Göker M, Lumbsch HT, Leavitt SD, Groenewald M, Theelen B, Liu X-Z, Boekhout T, Bai F-Y (2015b) Phylogenetic classification of yeasts and related taxa within Pucciniomycotina. Stud Mycol 81:149–189

Wu L, Sun Q, Sugawara H, Yang S, Zhou Y, McCluskey K, Vasilenko A, Suzuki K-I, Ohkuma M, Lee Y, Robert V, Ingsriswang S, Guissart F, Desmeth P, Ma J (2013) Global catalogue of microorganisms (gcm): a comprehensive database and information retrieval, analysis, and visualization system for microbial resources. BMC Genomics 14:933

Subject Index

P. Buzzini et al. (eds.), *Yeasts in Natural Ecosystems: Diversity*,
DOI 10.1007/978-3-319-62683-3

Microorganism Index

P. Buzzini et al. (eds.), *Yeasts in Natural Ecosystems: Diversity*,
DOI 10.1007/978-3-319-62683-3

R

T

U

V

MIX
Papier aus verantwortungsvollen Quellen
Paper from responsible sources
FSC® C105338

If you have any concerns about our products,
you can contact us on
ProductSafety@springernature.com

In case Publisher is established outside the EU,
the EU authorized representative is:
Springer Nature Customer Service Center GmbH
Europaplatz 3, 69115 Heidelberg, Germany

Printed by Libri Plureos GmbH
in Hamburg, Germany